Polymerization: Synthesis and Characterization

Polymerization: Synthesis and Characterization

Edited by **Mick Reece**

NY RESEARCH PRESS
New York

Published by NY Research Press,
23 West, 55th Street, Suite 816,
New York, NY 10019, USA
www.nyresearchpress.com

Polymerization: Synthesis and Characterization
Edited by Mick Reece

International Standard Book Number: 978-1-63238-364-8 (Hardback)

Printed in China.

Contents

Preface

In my initial years as a student, I used to run to the library at every possible instance to grab a book and learn something new. Books were my primary source of knowledge and I would not have come such a long way without all that I learnt from them. Thus, when I was approached to edit this book; I became understandably nostalgic. It was an absolute honor to be considered worthy of guiding the current generation as well as those to come. I put all my knowledge and hard work into making this book most beneficial for its readers.

Polymerization science has evolved to acquire a position of great significance in various fields including manufacturing sciences, biology, pharmaceuticals etc. This book contains contributions from number of authors engaged in the field of polymerization. The topics included are polymer characterization by atomic force microscopy of the polymer network structure generated in Ferroelectric Liquid Crystals Cells; polymerization of olefins; ring opening polymerization; polymerization by microwave irradiation technique of acrylamide/starch/acrylic acid; cationic polymerization of vinyl monomers; plant cell wall polymers; block and graft copolymerization by living/controlled polymerization; emulsion polymerization; rheology of biomaterials; polyADP-Ribosylation in post-fertilization and genome reprogramming; and fabrication of doped microstructures by two-photon polymerization. The aim of this book is to serve as a valuable source of reference and inspire readers to pursue research and in-depth study in this field.

I wish to thank my publisher for supporting me at every step. I would also like to thank all the authors who have contributed their researches in this book. I hope this book will be a valuable contribution to the progress of the field.

Editor

Section 1

Polymer Characterization

Atomic Force Microscopy Observations of the Polymer Network Structure Formed in Ferroelectric Liquid Crystals Cells

M. Petit

Additional information is available at the end of the chapter

1. Introduction

Polymer network stabilized liquid crystals (PSLCs) have attracted increasing interest over the past decade because of their potential applications mainly in electro-optic devices such as displays and light shutters [1-5]. The main motivation to incorporate a polymer network in liquid crystal cells was to bulk-stabilize a desired director configuration against any mechanical shock and distortions which can irreversibly alter the functionality of the cells. The PSLCs are composite materials in which a low density polymer network is dispersed within liquid crystal medium [6,7]. The polymer network is formed by chemical crosslinking of a small amount (few percent) of photo-reactive monomers dissolved in low molecular weight mesogenic material, through a polymerization reaction photochemically activated by a UV illumination. When the polymerization occurs in an aligned geometry, the resulting polymer network is roughly aligned parallel to the direction initially imposed by the liquid crystal medium in which the network has been formed [8]. Depending on the type of the reactive mesogen, the morphology of the polymer network may correspond to an open structure consisting of anisotropic fibrils [8-10]. The lateral size of fibrils is of the order of a few tenths of a micron [11-13]; their density increases with the initial reactive monomer concentration. The polymer fibrils, by creating a large internal boundary, provide a bulk anchoring mechanism which allows a control of the liquid crystal alignment in the bulk. Application of an electric field causes a distortion of the liquid crystal host, which corresponds to a field-induced director rotation, without any reorientation of the fibrils [8] considered rigid by the authors because the network is heavily cross-linked. The structure of the polymer network has been observed using different techniques. The scanning electron microscope (SEM) observations, observations between crossed polarizer after removal of the un-reacted species and observations at high temperature, all provide information on the network structure. The SEM and crossed polarizers observations give a detailed of the lateral size of fibrils [11-13] and the width of the fibers. Most of these observations have been reported in the nematic [14] or cholesteric [15] liquid crystals phases. However, a few information has been reported in the literature about the vertical distance between fibers especially when the polymer network was formed in a ordering helical ferroelectric liquid crystals (SmC*) and paraelectric SmA phases (PSFLC). The PSFLC composites have been investigated by different methods, electro optic technical [16-22] and dielectric method [23-26]. It is know that the dielectric properties of the ferroelectric liquid crystals in the ideal unbounded sample are nowadays well understood. The dielectric dispersion is essentially dominated by the dynamic of the soft mode, responsible for the phase transition from the paraelectric phase SmA to the ferroelectric phase SmC* phase, and the Goldstone mode connected with the helical structure in the ferroelectric SmC* phase [27]. The subject is already compiled in monographs [28,29].

Detailed dielectric spectroscopic study over a wide temperature and frequency ranges in the PSFLC systems reveal different molecular dynamics of this type of composites [23-25]. It was reported [23-25], that the dielectric strength of the Goldstone mode decreases with increasing polymer concentration, however, the relaxation frequency was found to be higher for the PSFLC composite film compared to that of the corresponding pure FLC. However, the behavior of the soft mode dielectric strength is not completely explained yet. Kundu et al. [25] are showed for polymer stabilized ferroelectric liquid crystal (PSFLC) systems that the soft mode dielectric strength remains unchanged when the FLC cells are stabilized by a polymer network formed from a nonmesogenic reactive monomer. The most of these studies have not given a quantitative interpretation of the dielectric response until our work published in [26].

In our previous study [26], although the range of the SmA phase is very narrow, the effect of the polymer network on the soft mode is clearly observed. By increasing the polymer concentration the dielectric strength is reduced and the relaxation frequency is increased [26].

It is knows that, in a thin enough sample when the sample thickness becomes comparable with the pitch of the helix, the entire structure in helix-free. The director twist-bend fixed by the surface polar anchoring can still exist as well as in the helicoidal samples. This spaced modulation is called a twisted structure [30]. However when the twist helical pitch is confining with the polymer network in a thin thickness, logically there is no reason that the twisted structure still remains in spite of the confinement effects. In this chapter we will illustrate that in spite of the confinement of the twist short helical structure in a polymer network the ferroelectric liquid crystal conserves the helical structure. This behavior was interpreted by the type of the morphology of the polymer network which is formed in the liquid crystals medium. After describing the dielectric responses, in sect.3-2-A we will describe in detail the effect of the polymer network density formed in a planar alignment of the ferroelectric liquid crystals cells [30] on the dielectric responses namely soft and Goldstone mode. The results of the dielectric responses were interpreted by the morphology of the polymer network using atomic force microscopy investigations. In sec.3-2-B the effect of the applied electric field during the photopolymerisation on the dielectric responses is reported, the observations of the structure of the polymer network by AFM agree well with all dielectric responses.

2. Experiments

The liquid crystal compound used in these studies is the ROLIC 8823 (Rolic research ltd) which exhibits the following phase sequence: Crystal (Cr) -27 °C SmC* 63.5 °C SmA 65°C Isotropic (Figure 1). In the SmC* phase (Figure 2), at low temperature the helical pitch is about 0.3μm and the spontaneous polarization (P_s is close to 100 nC/cm²). To prepare the polymer network we have used a photoreactive diacrylate mesogen as photocurable monomer which presents a nematic (N) phase between Cr and Isotropic phases Cr–88 °C–N–118 °C– Isotropic. The chemical structure of the monomer is presented in figure 3.

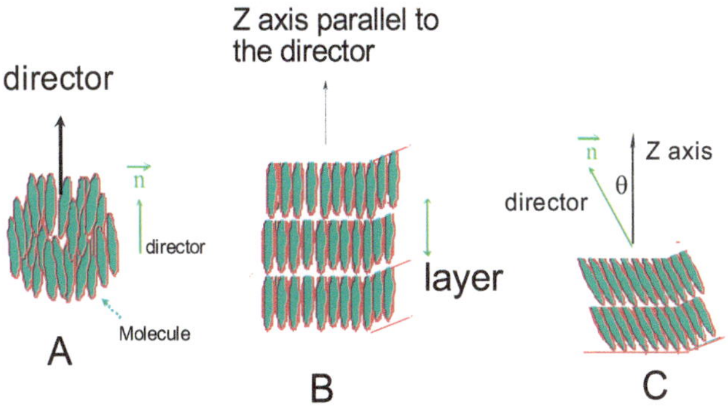

Figure 1. A schematic representation of the nematic phase (A), Paraelectric phase or smectic A (B) and Smectic C (C).

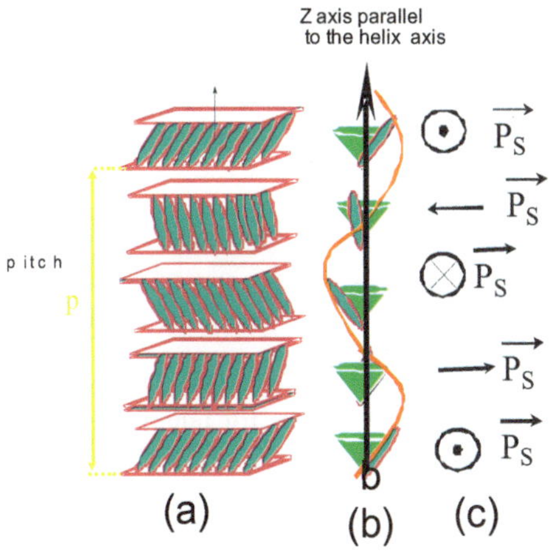

Figure 2. A schematic representation of a smectic C* phase : molecular order in the layer (a), helical structure (b) and the direction of the ferroelectric spontaneous polarization in the smectic layers planes (c).

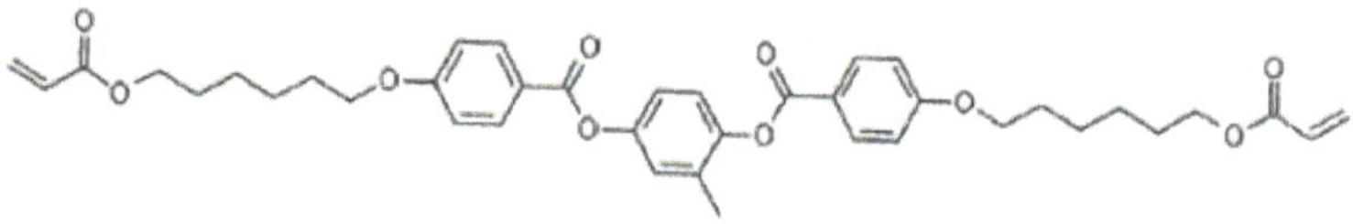

Figure 3. The molecular structure of the monomer.

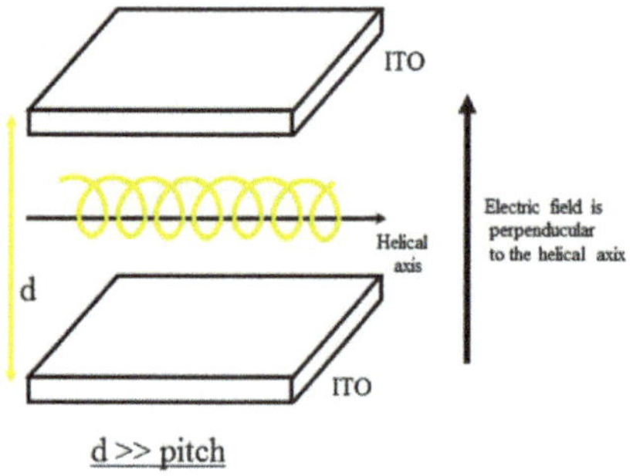

Figure 4. The Planar configuration of the helical structure in our system, the electric field is applied perpendicular to the helical axis.

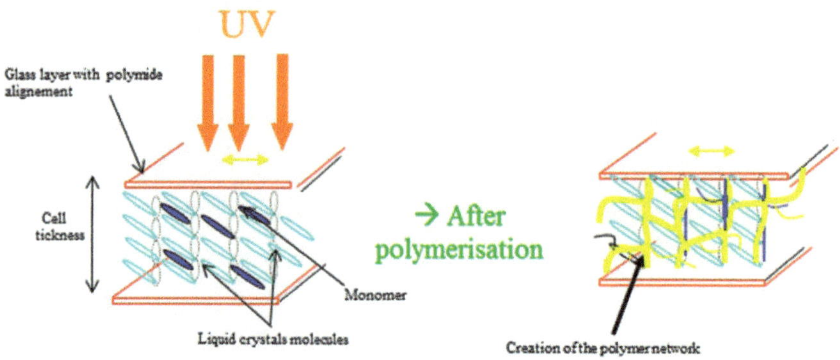

Figure 5. A schematic illustration the formation of the polymer network, before de photopolymerisation (on the left) and after the pohopolymerisation (on the right).

The PSFLC mixture was prepared by mixing the Diacrylate monomer with weight concentrations between 2 and 7%. The ferroelectric liquid crystals (FLC) compound and the diacrylate were dissolved in the isotropic phase to make a homogeneous mixture. A 5 μm thick EHC Inc, Japon-cell (two glass faces were treated with polyimide to favorite a planar alignment (Figure 4)) was filled by the mixture in its isotropic phase. In order to obtain a good alignment in the SmC* phase, the cell was slowly cooled (0.1 °C/minute) from the isotropic phase under an applied electric field (5V/μm) into the SmC* phase. For the section

3.2.A, the sample cells were then exposed to ultraviolet light (wavelength λ = 365nm) at 25 °C with an intensity of 5 (mW/cm^2) for 30 minutes without any applied electric field. Here, polymer phase separation and network formation take place (Figure 5). During these studies, the cell was placed on a hot stage (Linkam TMS 93) for temperature control. The texture observations of the cells were carried out by means of a polarized optical microscope (POM)(LEICA DMRXP).

Dielectric measurements were performed in the frequency range of 10 Hz–13 MHz HP 4192A. In a linear dielectric response, the time dependent polarization $P(t)$ of the sample, being induced by a weak measuring electric field $E(t)$, is proportional to the field :

$$P(t)= \varepsilon_0 (\varepsilon^*-1) E(t) \text{ where } E(t)=E_0 exp(-i\omega t)$$

Is a sinusoidal electric field applied to the dielectric under test and ε^* is the complex dielectric permittivity, ω is the angular frequency, and f is the frequency of measuring electric field displayed in the channel A of most impedance analyzers, and ε_0 is the dielectric permittivity of free space. In linear dielectric spectroscopy the amplitude of the measuring electric field should be chosen so that it does not suppress the helicoidal structure of the SmC* phase. Generally for ferroelectric liquid crystals two processes contribute to the dielectric spectra. We are interesting here only to the collective processes namely soft and Goldstone modes.

In order to obtain the characteristic dielectric strengths and relaxation frequencies of the ferroelectric relaxation modes, the dielectric spectra were fitted simultaneously by the Cole-Cole function :

$$\varepsilon^* = \varepsilon_\infty + (\Delta \varepsilon_G) / (1+jf/f_G)^{1-\alpha_G} + (\Delta \varepsilon_S) / (1+jf/f_S)^{1-\alpha_S} + (\sigma/j2\pi f\varepsilon_0) \quad (1)$$

Where f is the frequency, ε_∞ is the high frequency limit of the dielectric permittivity, $\Delta\varepsilon_G$ and $\Delta\varepsilon_S$ represent the dielectric strengths corresponding to Goldstone and soft modes, respectively ; f_G and f_S represent the relaxation frequencies of the two modes, α_G and α_S are the distribution parameters, and σ is the electric conductivity. The temperature dependencies of different dielectric processes are reported and discussed below. To image the topography of polymer networks a Veeco Multimode Atomic Force Microscopy (AFM) equipped with a Nanoscope IIIa controller was used. All AFM scans were taken in tapping mode with commercially available tips made of Phosphorus doped Silicon.

3. Results and discussion

3.1. Optical observations

The first objective of this study was to investigate the effect of the applied electric field and phase order (before the polymerization) on the alignment on the polymer network formed in the FLC host. To illustrates the effect of the applied electric field during the cooling from the isotropic phase to the SmC* phase on the alignment of the SmC* layers we present on Figure 6.

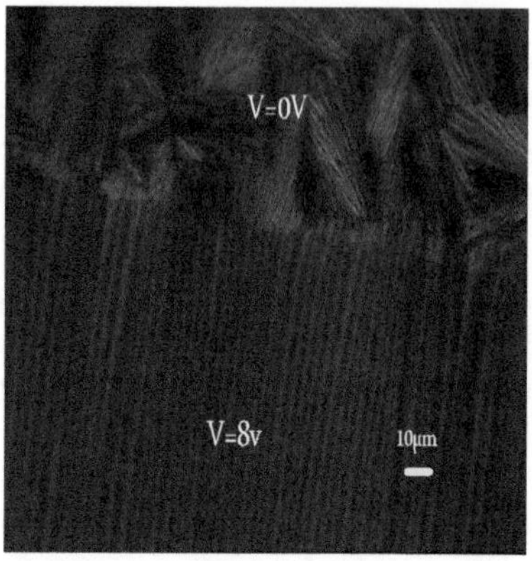

Figure 6. Optical micrographs of PSFLC samples obtained between crossed polarizers at T = 70 °C, for 7% polymer concentrations formed at 25°C. (V=0V) indicate the region where no field was applied and (V=8 v) indicate the region where an electric field was applied.

Figure 6 shows example of optical micrographs representing the observed textures of the PSFLC cells obtained for 7% initial monomer concentration. These micrographs were obtained at a temperature of 70 ° C (above the SmA- Isotrope transition temperature) so that only the birefringence associated to the polymer network would appear. The micrographs of Figure 6 clearly show the anisotropic structure of the polymer network. The optical observation of this anisotropic structure is due not only to the residual birefringence of polymer fibrils, but also to the remaining birefringence of the surrounding FLC molecules which are still aligned by the polymer structure. Two regions have been observed (Figure 6), the first region when no electric field has been applied (V=0V), the polymer network was randomly distributed. However, in the region where an electric field is applied (V=8V) the polymer network presents a good alignment. To illustrate the effect of the phase order on the formation of the polymer network, two cells of 7% polymer concentration were polymerized in two different temperatures. The first cell was polymerized at 25°C. The measured helical pitch at 25°C is about 0.25 μm (Figure 7). However, the second cell was polymerized at high temperature (T=58°C) where the helical pitch diverges (Figure 7). Figure 8 shows the optical micrographs representing the PSFLC cells obtained for 7% polymer concentration which are polymerized at two different temperatures. As seen in this figure, the dechiralisation lines [31] have been clearly observed on the structure of the polymer network (dechiralisation lines are perpendicular to the polymer fibers (Figure 8 (a))).

We can remember here, that the dechiralisation lines are well known lines defects can be observed in a planar samples filled by a highly twisted FLC [30-32]. In a smectic C* sample, competition between a strong surface anchoring and a helicoidal configuration in the bulk induces a double lattice of singular lines. These lines are located near both the boundary surfaces and are parallel to the plane of the layers [30-32]. Those lines are named the dechiralisation lines. Hence, these results illustrate that the polymer structure conserves locally a lines defect print. We can also indicate here that the fact that the helical structure of the FLC at T=25°C is very shorter no twisted structure on the polymer fiber has been observed by polarizing microscopy. However, when the polymer network was formed at high temperature, the helical structure of the polymer fibers is easy observed by polarizer microscopy (Figure 8 (b)). One can seen that, at high temperature, the dechiralisation lines were not printed on the polymer fibers. That illustrate that the order and structure of the liquid crystal phase are transferred onto the polymer network. It has been already shown by Archer et al [33] that the defect of the FLC twisted grain boundary were found printed on the polymer network structure. However, for short-pitch ferroelectric liquid crystals, the transfer of the lines defects has not been observed. Figure 8 shows also that the polymer network which was formed at high temperature, the dechiralisation lines were not printed on the polymer fibers. Because, when approaching the phase transition SmC*- SmA, the helical pitch diverges. The divergence of the helical pitch (unwinding of the helix) is seen as disappearance of dechiralisation lines [31].

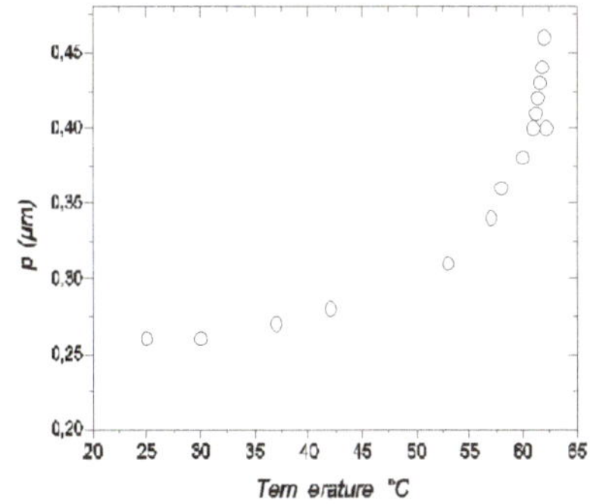

Figure 7. Temperature dependence of the helical pitch of the FLC measured by mean of Grandjean-Cano method [32].

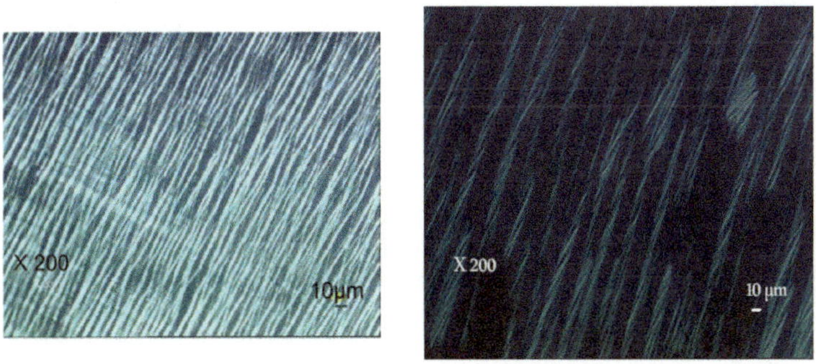

(a) (b)

Figure 8. Optical microscopy images between crossed polarizers at T = 70 °C, for 7% polymer concentration. (a) polymerization at 25°C (dechiralisation lines are clearly observed and correspond to the straight lines which are perpendicular to the polymer fibers). (b) polymerization at 58°C (the polymer fibers present a helical structure).

This implies that even after polymerization, the characteristic property of the host phase in which polymerization was carried out was effectively retained.

3.2. Dielectric studies

3.2.1. Effect of the polymer network density on the dielectric responses

*Goldstone mode of the SmC**

Figures 9 (a) and (b) show examples of the dispersion, $\varepsilon'(f)$ and absorption, $\varepsilon''(f)$ dielectric spectra obtained in the SmC* phase at low temperatures for different polymer concentrations. For all of the concentrations studied, two relaxation mechanisms were detected. The first, at low frequencies between (1 and 3 kHz) with a high amplitude, is due to the Goldstone mode; whereas the second observed at high frequencies (> 1 MHz) with a weak amplitude is an artifact due to the indium tin oxide (ITO) conducting layers. As shown in Fig. 9(a), at low frequencies, the dielectric response shows a very strong polymer concentration dependence; at 100 Hz, for example, $\varepsilon'(f)$ decreases from 100 to 30 when the polymer concentration increases from 0% to 7%.

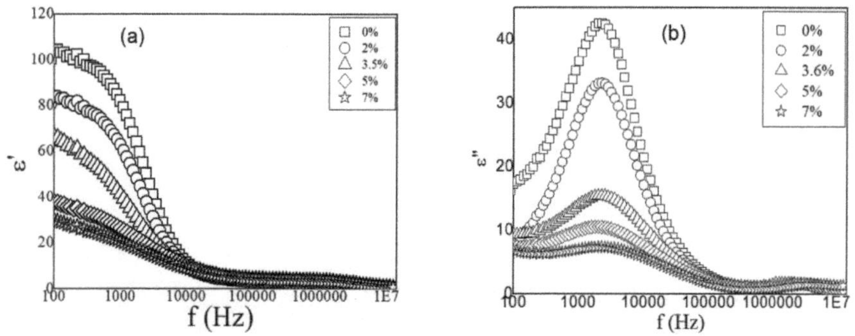

Figure 9. Frequency dependence of (a) the real and (b) the imaginary parts of complex permittivity in the smectic C* phase for different poclymer concentrations at T=25 °C.

This effect is also clearly illustrated in the behavior of the absorption peak observed in the $\varepsilon''(f)$ spectra [Fig. 9 (b)]; the absorption peak strongly decreases from 48 to 8 when the polymer concentration is varied from 0% to 7%. The parameters $\Delta\varepsilon_G$ and f_G obtained from the curve-fit procedure are displayed in Figs. 10 (a) and (b). The behavior of $\Delta\varepsilon_G$ versus temperature showed the same general features for all the samples Fig.10 (a) ; $\Delta\varepsilon_G$ slightly increases to reach a maximum at a temperature called T_{max} 3 °C below T_c then decreases abruptly above T_{max}. The behavior of $\Delta\varepsilon_G$ versus temperature is dependent on that of the helical pitch of the FLC Fig. 7. Usually the maximum observed in $\Delta\varepsilon_G(T)$ is related to that exhibited by the helical pitch Fig. 7 at temperatures close to T_c and indicates that the helical structure of the FLC is preserved in all our PSFLC systems.

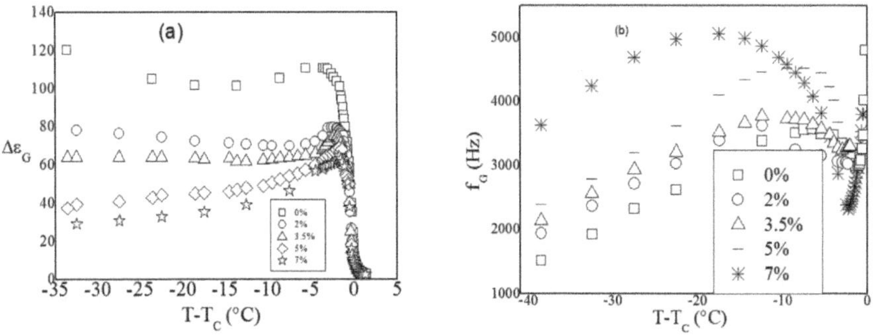

Figure 10. Temperature dependence (a) of $\Delta \varepsilon_G$ and (b) the relaxation frequency of the Goldstone mode for different polymer concentrations.

The temperature dependence of the Goldstone relaxation frequency Fig. 10 (b) shows that f_G slightly increases with temperature, reaches a maximum, and then rapidly decreases to a minimum value at a temperature corresponding to T_{max}, after T_{max}, an abrupt increase in f_G is observed for a temperature close to T_c. Qualitatively, the thermal behavior of the Goldstone mode is not affected by the polymer network. However, quantitative differences were observed for $\Delta\varepsilon_G$ and f_G when the polymer network density increases. To illustrate this effect, we present on Figs. 11 (a) and 11 (b) the evolution at room temperature of $\Delta \varepsilon_G$ and f_G as a function of the polymer concentration. It can be seen from these figures that the increase in the polymer concentration from 0% to 7% leads to a breakdown of $\Delta \varepsilon_G$ from 120 to 27 and to an increase in f_G from 1.5 to 3.5 kHz. Changes in the dynamic of the Goldstone mode have already been observed in PSFLC systems by Gasser et al. [35] and in other composite based FLC, as a random network formed from dispersions of aerosil particles within FLC media [36–38]. For aerosil/FLC composites, a decrease in $\Delta\varepsilon_G$ and a shift of f_G toward high frequencies [37,38] with increasing the density of aerosol particles were observed.

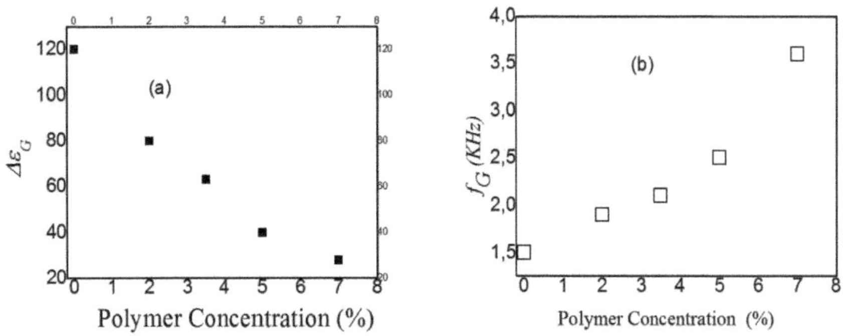

Figure 11. The dielectric strength ($\Delta \varepsilon_G$) (a) and relaxation frequency f_G (b) of the Goldstone mode at 25°C as function of the polymer concentration

The Goldstone mode even disappears in these systems for a sufficiently high aerosil density. The authors have interpreted the behavior of the dielectric response in these systems by size effects on smectic domains [36–38]; the reduction in the Goldstone mode strength and the increase in the relaxation frequency with increasing the concentration of aerosil particles are due, according to these authors, to the formation of smaller smectic domains where fluctuations are quenched by surface interactions, leading to a deformation of the helix. Additionally, the orientation of these smectic domains becomes randomly distributed so that fewer domains are therefore preferentially oriented in the direction of the applied electric field. We believe that the interpretation given above cannot explain the behavior of our PSFLC systems, despite similar changes in dielectric relaxation being observed. First, the polymer network in PSFLC cells are anisotropic and stabilizes the configuration of smectic domains oriented preferentially to the direction of the electric field. We think that the changes observed in dielectric response in our systems are essentially governed by elastic effects. We expect that the network-FLC interactions enhance the apparent elasticity of the PSFLC films, and accordingly, causes the increase in the relaxation frequency f_G and the reduction in dielectric strength $\Delta\varepsilon_G$. In fact, the dielectric strength and the relaxation frequency of the Goldstone mode in the case of a pure SmC* phase are expressed as [39] :

$$\Delta\varepsilon_G = (P_s/\theta)^2 / (2\varepsilon_0 K_\varphi q^2_0) \tag{2}$$

$$f_G = (K_{eff}\, q^2_0) / (2\pi\gamma_\varphi) \tag{3}$$

where γ_φ is the rotational viscosity, K_{eff} is the effective elastic constant. θ and P_s are the tilt angle and spontaneous polarization, respectively. $q_0 = 2\pi/p_0$ with p_0 is the helical pitch of the FLC. If we think that the Equations (2) and Equation (3) are applicable for the PSFLC composite, q_0 is considered here constant. The reduction in $\Delta\varepsilon_G$ as a function of polymer concentration is attributed to the variation in (P_s/θ) and/or K_{eff}. To clarify that, we have plotted in Fig. 12 the ratio (P_s/θ) as a function of polymer concentration. This figure shows that this ratio can be considered independent of the polymer concentration. As a consequence, the observed decrease in the $\Delta\varepsilon_G$ with polymer concentration can be explained rather by the increase in the effective elastic constant K_{eff}.

On the other hand, the relaxation frequency of the Goldstone mode Equation 3 is controlled both by the elastic ($K_{eff}\, q^2_0$) and viscous (γ_{eff}) forces. From the equation 3, the increase in f_G with polymer concentration can be explained by the decrease in the Goldstone rotational viscosity (γ_{eff}) and/or the increase in the effective elastic constant K_{eff}. From Eqs. 2 and 3, the effective rotational viscosity can be expressed as $(\gamma_{eff}) = (P_s\, \theta^{-1})^2 / 4\pi\varepsilon f_G\, \Delta\varepsilon_G$. According to this expression and using the experimental data of ($\Delta\varepsilon_G$ x f_G) (Fig 13(b)) and (P_s/θ) (Fig 12), γ_{eff} was evaluated as a function of temperature for all polymer concentrations studied. Generally, γ_{eff} of the PSFLC films increases with the polymer network density. At room temperature, for example, γ_{eff} increases from 0.2 to 0.6 *Pa. s* when the polymer concentration increases from 0% to 7% (Fig. 13(b)).

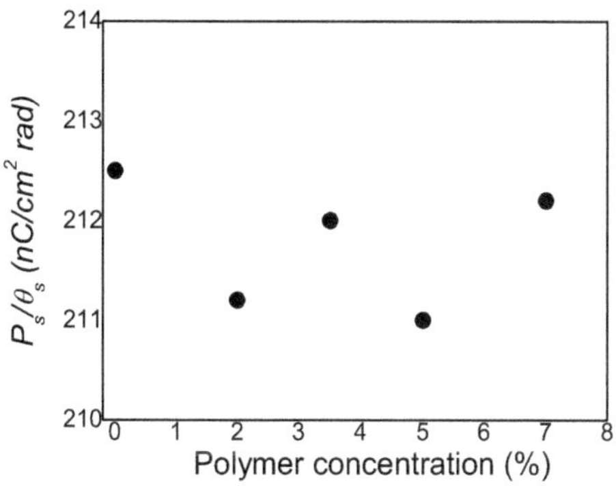

Figure 12. The (P_s/θ) ratio as function of the polymer network density

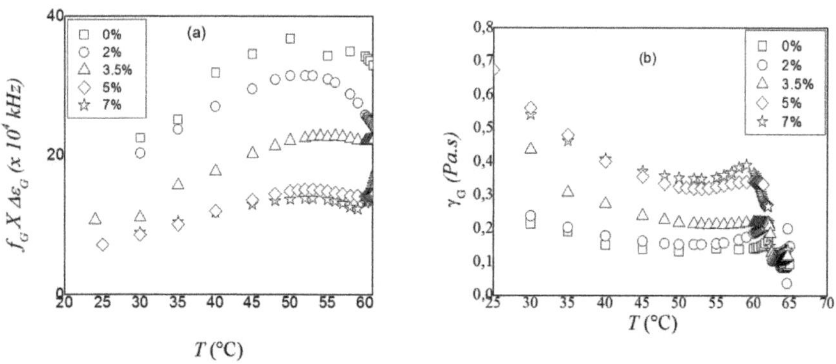

Figure 13. Temperature dependence of the Goldstone mode rotational viscosity (γ_{eff}) (b) and the product $\Delta\varepsilon_G$ x f_G (a) for different polymer concentrations.

Consequently, the increase in the relaxation frequency with the network density is certainly due to the increase in the effective elastic constant K_{eff}. To illustrate this, K_{eff} was evaluated at room temperature from Eqs. 2 and 3; the results are displayed in Fig. 14. This figure shows that K_{eff} linearly increases from $0.5\ 10^{-11}$ to 2.310^{-11} **N** when the polymer concentration increases from 0% to 7%. The values of K_{eff} found here compare well with those obtained for the same PSFLC systems from the electro-optic measurements [18,20].

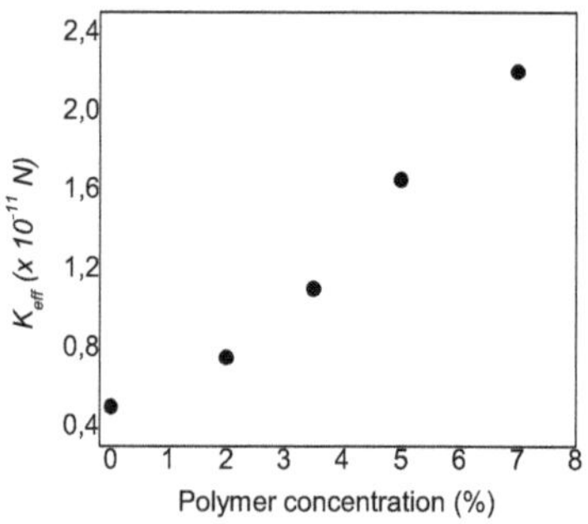

Figure 14. The effective twist elastic constant K_{eff} versus polymer concentration.

In conclusion, the increase in the relaxation frequency and the reduction in the dielectric strength of the Goldstone mode for the PSFLC films seem to be due to the increase in the twist elastic energy, resulting from the strong interaction between liquid crystal molecules and the polymer network liquid crystal molecules and the polymer network.

Polymer network morphology by AFM investigations:

The principal result surprising in this party (section 3-2-A) is the role played by a polymer network to stabilize of the ferroelectric order. Indeed Bayth et al [34] which are reported that when the helical pitch of the smectic C* was confined between two parallel glass with a homogeneous planar anchoring causes an unwound of the helical structure which is proportional to the cell thickness. This transition is due to the result of the competition between the energy cost of the lines of dechiralisation slightly dependent on the thickness and that to the unwound of helix which is highly depends to the cell thickness. A similar mechanism could be compared in the case of our results by schematizing the fibril like cylinders with a homogeneous planar anchoring along their axis. In this case, the work which is reported by Baytch et al should remain valid once subsisted the thickness of the cell by the distance between cylinders. We think that the fibril seem to have a twisted morphology. In this case, these morphology could stabilizes the smectic phase. The increase of the apparent elasticity coefficients can be explained from a simple energetic argument if we take into account the polymer network morphology. One must note here that the polymer network was formed within the SmC* phase in which the director field is highly twisted in a helix. This helical structure of the FLC is transferred on the polymer network as shown in Figure 15.

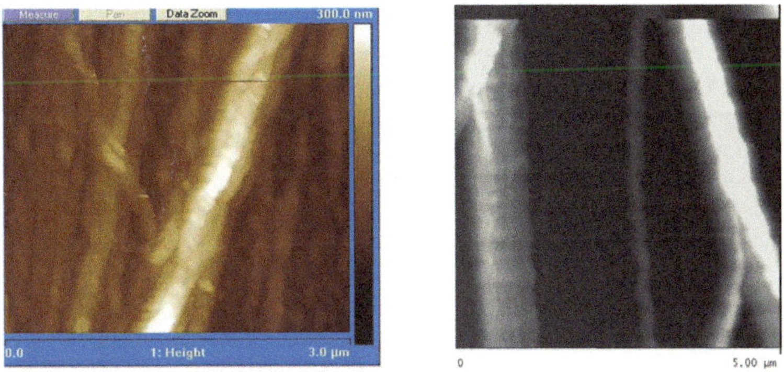

Figure 15. Tapping mode AFM height image of polymer fibers formed from a 5% polymer concentration into the short-pitch FLC (at T= 25°C). The twisted structure of fibrils is clearly observed indicating that the helical structure of the FLC from which the network was formed is printed on the polymer fibers.

For AFM experiments, the PSFLC cells were disassembled and flushed with solvent to remove the FLC. Figure 15 (on the left) presents a tapping mode AFM image (Veeco Instrument Inc.) of a 3×3 μm^2 region, and shows a fibrillar structure of the polymer network. The width of the fibers ranges from 150 to 300 *nm*. The AFM images clearly reveal a twisted structure of the fibers with a periodicity of about 0.24 μm which approximately corresponds to the pitch of the FLC helix used in this study. Hence, the polymer structure conserves locally a helical print which in turn stabilizes the FLC structure present during the polymerization. It has been already shown by I. Dierking et al. [12], and by G.A. Held et al.[40] that for long-pitch cholesteric liquid crystals, the helical superstaructure was transferred onto a polymer network. However, for short-pitch ferroelectric liquid crystals, this transfer has not been observed. Recently, K. Akagi et al [41] were confirmed the possibility of the printing of the short helical pitch on the polymer network. Archer et al [33] showed that the defect of the FLC twisted grain boundary were found printed on the polymer network structure. In [21] the chevron pattern has been observed on the polymer network. The printing of the helical structure on the polymer network is also reported by [42]. One could not observe any twisting of the polymer strands when polymerized in the Sm A* phase [43]. This implies that even after polymerization, the characteristic property of the host phase in which polymerisation was carried out was effectively retained. The helical aspect of the fiber structure certainly influences and explains the electro-optical: Deformed Helix ferroelectric liquid crystals (DHF) [18] and electroclinic effects [19,20].

Sof mode in the SmA phase

Figures 16 (a) and 16 (b) show examples of the dispersion, $\varepsilon'(f)$ and absorption, $\varepsilon''(f)$, dielectric spectra obtained in the SmA phase at T_c for each polymer concentration. Two relaxations mechanisms are detected. The first, at frequencies between 10 and 30 kHz with a weak strength, is attributed to the soft-mode relaxation mechanism; whereas the second observed at high frequencies 1 MHz is due to the ITO conducting layers. As shown in Fig.16, at 1 kHz frequency, the dielectric response shows a very strong polymer concentration dependence; ε' decreases from 23 to 12 when the polymer concentration increases from 0% to 7%. This effect is also clearly demonstrated from the behavior of the absorption peak observed in Fig. 16 (b); the absorption peak decreases from 12 to 3 when the polymer concentration is varied from 0% to7%. We present in Figs. 17 (a) and 17 (b) the temperature dependence of the dielectric strength, $\Delta\varepsilon_s$, and the relaxation frequency, f_s, of the soft mode. For all studied concentrations, the behavior of versus temperature shows the same general features fig. 17 (a). A rapid increase in $\Delta\varepsilon_s$ is observed close to T_c for all concentrations studied. The increase in $\Delta\varepsilon_s$ at and close to T_c is dependent on the polymer concentration. Note that $\Delta\varepsilon_s$ becomes relatively weakly affected by the network as temperature increases from T_c (Fig. 17 (a)). The relaxation frequency, f_s, exhibits a linear temperature dependence (Fig. 17 (b)). At T_c, f_s increases from 10 to 36 kHz when the polymer concentration increases from 0% to 7%. This effect seems to be less dependent on the polymer network density at relatively higher temperatures $\geq T_c + 0.5$ °C (Fig. 17 (b)). Similar behaviors have been observed in the case of dispersed silica particles on FLC matrix near the SmA-SmC phase transition [6–8]. However, Kundu et al. [25] showed for other PSFLC systems that the soft-mode dielectric strength remains unchanged when the FLC cells are stabilized by a polymer network formed from a nonmesogenic reactive monomer. These authors did not provide any indications of the network structure of their systems. However Beckel et al. [44] demonstrated that nonmesogenic monomers give rise to polymer chains which microseparate from FLC molecules in the smectic layers leading to a layer swelling. Obviously, this polymer network structure is completely different to that obtained in our systems Fig. 15. In order to examine how the polymer network influences the soft-mode dielectric strength of the PSFLC, we used the model previously developed [20] to explain the electroclinic behavior of PSFLC films.

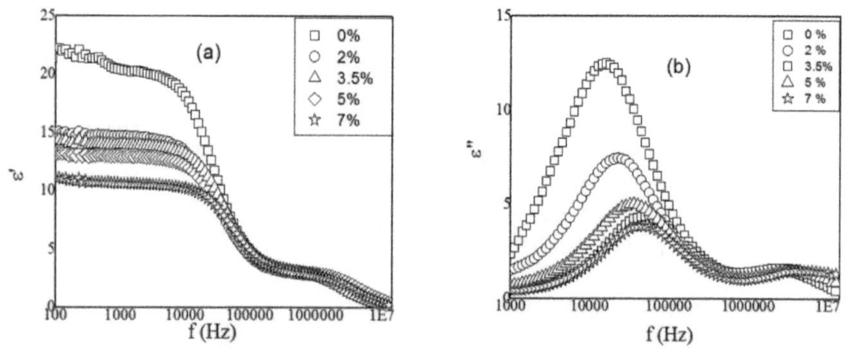

Figure 16. Frequency dependence of (a) the real and (b) the imaginary parts of the complex permittivity at the SmC-SmAphase-transition temperature, Tc, for different polymer concentrations.

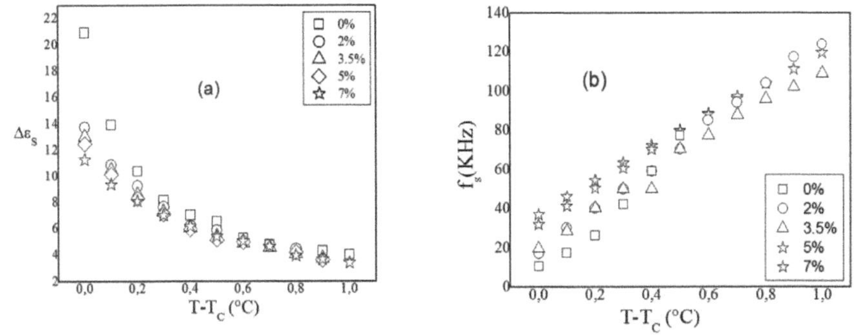

Figure 17. Temperature dependence of (a) the dielectric strength and (b) the relaxation frequency f_s of the soft mode in the smectic Aphase for different polymer concentrations.

Although polymer networks formed in liquid crystal media generally have a complex structure, they have been previously modeled as an assembly of parallel cylinders randomly distributed within the liquid crystal media [7]. The cylinders are interconnected via a chemical cross linking which ensure the network stability. In the framework of the rigid model of the network introduced by Li et al. [17], they considered that FLC molecules are affected by the bulk anchoring force from network. Li et al. interpreted this bulk anchoring force in terms of a field like effect, where the orientation of the FLC molecules was coupled to that of the anisotropic polymer network. This model gives a macroscopic description of the polymer network and was successfully applied to describe the "V-shaped" electro-optic properties of FLC gels [17]. We adopt in our theoretical approach this model structure of random polymer network of Li et al. We will consider below as a characteristic parameter of the polymer network structure the average intercylinder distance, which we call L_c, (Figure 18) and we try to analyze the effect of the applied electric field on smectic-A layers confined between two successive cylinders. The basic relation giving the free energy density of a chiral SmA phase near T_c in the presence of a small electric field E was expressed by Garoff et al. [45,46] as:

$$f_E = f_0 + (1/2)\,\alpha\,(T-T_c)\,\theta^2 - CP\theta + (P^2/\varepsilon_0\chi C) - PE \quad (4)$$

f_0 represents contributions to free energy density from the undistorted SmA phase. α is the mean-field coefficient and C is related to the piezoelectric coupling between the polarization P and the induced electroclinic tilt θ. ε_0 is the dielectric constant and χ is the electric susceptibility. Moreover, to the free energy density term expressed in Eq.4, two other contributions are required to describe the effect of electric field on SmA blocks confined between the polymer fibers. The first one consists of the free energy density f_{ps} arising from the bulk polymer stabilization, and may be written as [17]

$$f_{ps} = (1/2)\,W_p \sin^2\theta \approx (1/2)\,W_p\,\theta^2 \quad (5)$$

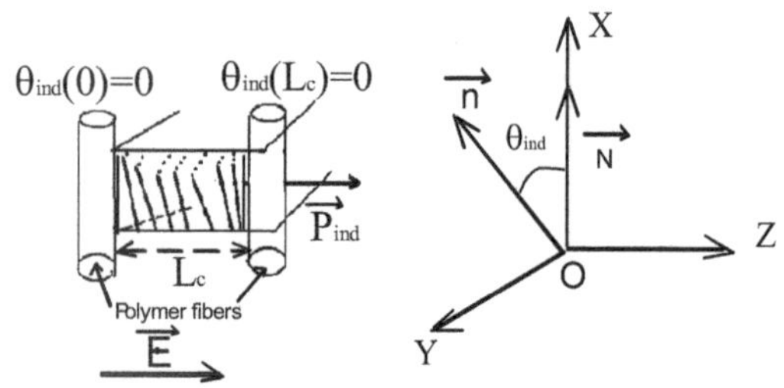

Figure 18. Tilt angle induced by electric field in the SmA* phase; distortion due to the anchorage at the polymer boundaries (on the left). L_c is the average distance between two successive polymer fibers parallel to the direction of the electric field. Cell normal and layer normal are parallel to the Z and X axis respectively.

W_p is the coupling coefficient for the interaction between the polymer network and the liquid crystal molecular director. Equation 5 is an approximative expression of f_{ps} because we are interested in low field regime where the angle is small. The second contribution comes from the elastic free energy density f_{el} arising from a director distortion upon application of E. In our system, the smectic layers are arranged perpendicular to the direction of the fibers. Between two successive groups of fibers separated by the average distance L_c (Figure18), due to anchoring forces between the liquid crystal molecules and the polymer network, the rotation of the director can be reasonably assumed not to be uniform: It is larger at or close to $L_c/2$ and weaker near the surface fibers. We must remark here that this theoretical approach of our PSFLC system is based on a one-dimensional model. We neglect then any splay deformation of the director, and we only consider the elastic energy arising from a twist deformation. f_{el} can then be given by the following expression:

$$f_{el} = (1/2)\, K_2\, (\partial\theta/\partial z)^2 \tag{6}$$

K_2 is the twist elastic constant, and Z denotes the coordinate along the axis parallel to the direction of the applied electric field. The total free energy density $f_t = f_E + f_{el} + f_{ps}$ is then expressed as :

$$f_t = f_0 + (1/2)\,\alpha\,(T-T_c)\theta^2 - CP\theta + (P^2/\varepsilon_0\chi C) - PE + (1/2)\, K_2\,(\partial\theta/\partial z)^2 + (1/2)\, W_p\, \theta^2 \tag{7}$$

The equilibrium values of P and θ are found by minimizing the free energy f_t with respect to P and θ, respectively. This leads to the following equations:

$$\alpha\,(T-T_c)\theta - CP + W_p\,\theta - K_2\,(\partial^2\theta/\partial z^2) = 0 \tag{8}$$

$$C\theta - (P/\varepsilon_0\chi) + E = 0 \tag{9}$$

Inserting Eq. 9 into Eq. 8 gives

$$K_2\,(\partial^2\theta/\partial z^2) - \alpha\,(T-T'_c)\theta + C\,\varepsilon_0\chi\, E = 0 \tag{10}$$

where $T'_c = T_c - W_p / \alpha$ is the SmC*–SmA transition temperature of the PSFLC system; $T_c = T_0 + C^2\chi\varepsilon_0 / \alpha$ is the SmC*–SmA transition temperature of pure FLC. Equation 10 governing the director distortion induced by an applied electric field in the SmA phase the anisotropic polymer fibers in PSFLC systems constrain the molecular orientation at their surfaces as the screw dislocations at the grain boundaries in the TGBA phase do. To solve Eq. 10, we assume the anchoring of the molecules at the fiber surfaces to be rigid. The boundary conditions at these surfaces are : $\theta(z=0)= \theta(z=L_c)=0$

Equation 10 has a solution :

$$\theta(z)=(\varepsilon\chi CE)(\alpha(T-T'_c))^{-1} [1-exp(z/a)((1-exp(L_c/a))^{-1} - exp(-z/a)(1+exp(-L_c/a))^{-1}] \quad (11)$$

Where

$$a= (K_2/(\alpha(T-T_c))^{1/2}$$

Averaging the θ (z) [15] and $P(z)$ values over the $0 \le z \le L_c$ domain (Figure 18) gives the expression of the mean induced tilt and polarization :

$$<\theta^{PSFLC}>=(\varepsilon_0\chi CE)/\alpha(T-T'_c)[1- tanh(L_c/2a) / (L_c/2a)] \quad (12)$$

$$<P^{PSFLC}> = (\varepsilon_0\chi E)+ (\varepsilon_0\chi^2 C^2 E) / \alpha(T-T'_c) [1- tanh(L_c/2a) / (L_c/2a)] \quad (13)$$

Where

$$a= K_2 (\alpha(T-T'_c))^{-1}$$

The average induced polarization can also be written as :

$$<P^{PSFLC}> = (\varepsilon_0\chi E)+ (\varepsilon_0 E\, \Delta\varepsilon^{PSFLC}) \quad (14)$$

The identification between equations (13) and (14) gives the expression of the dielectric strength of the soft mode, $\Delta\varepsilon^{PSFLC}$, as a function of L_c :

$$\Delta\varepsilon^{PSFLC} \approx (\varepsilon_0\chi^2 C^2 E) / \alpha(T-T'_c) [1-H] \quad (15)$$

Where $H = tanh(L_c/2a)/(L_c/2a)$ which we called the elastic parameter, depends on the network density via L_c. For a same given reduced temperature, $(T-T'_c)$, the equation (10) can be expressed as:

$$\Delta\varepsilon^{PSFLC} \approx \Delta\varepsilon^{FLC} [1-H] \quad (16)$$

$\Delta\varepsilon^{FLC}$ denotes the soft mode dielectric strength of the pure FLC. Equation (16) shows that, in the SmA* phase, the main parameter that governs the soft mode dielectric strength in the PSFLC films is L_c. This means that the stored elastic energy arising from the distortion of the director upon application of electric field is as much larger than the L_c becomes smaller, which causes a decreases of the soft mode dielectric strength (equation (16)). The effect of the polymer concentration on the dielectric strength is in accordance with the theoretical results (Equation 16). According to this equation, the $\Delta\varepsilon_s$ is reduced by the increase of the polymer network density or by decreasing the distance L_c. To explain quantitatively the behavior of the reduction of the soft mode dielectric strength we can use the expression given by equation (16). From this equation, we conclude that the main parameter which governs the soft mode dielectric strength is the vertical distance which separated between two successive groups of polymer fibers L_c. We used the values of $\alpha \approx 8.8\ 10^{+3}\ N\ m^{-2}K^{-1}$, $a = 0.12\mu m$ [20], and the values of $\Delta\varepsilon_s$ presented in the Figure 17(a) at $T\text{-}T_c = 0.1°\ C$, with a reasonable value of the twist elastic constant, typically, $K_2 \approx 10^{-11}$ N, the equation (16) was graphically resolved to evaluate L_c. We obtain a mean inter-fiber L_c. The results are displayed in Fig. 15. The calculated values were compared to those directly measured from the topography of the polymer networks obtained by means of AFM experiments (Fig. 19). Figure 19 shows example of AFM images and height profiles on the z direction in Fig. 18 of the polymer network. The height profile in the **z** direction indicates two successive groups of fibers. The mean distance L_c Fig. 18 between them was evaluated for each polymer concentration and displayed in Fig. 20(a). The measured values of L_c linearly decrease with the polymer density and agree well with those calculated from the model. It seems from these results that the fibrillar and anisotropic nature of the network stabilizes, at long-range scale, the SmA order and opposes the electric field effect on the deformation of the SmA director. These results are in accordance with previous works [19,20] of the electroclinic effect in the same PSFLC systems, which demonstrate that the electroclinic susceptibility of these systems is reduced with the increase in the polymer network density. It must be noted here that the average lateral separation distance (y direction in Fig.18) between polymer strands is estimated to be about 10 μm (Fig. 19), which is 1 order of magnitude higher than L_c.

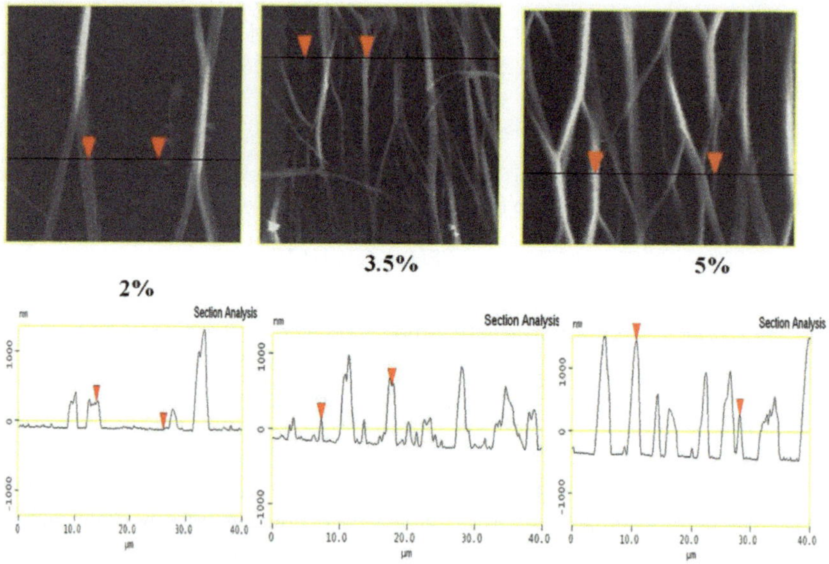

Figure 19. Tapping mode AFM images of polymer network structure of 40 x 40 μm ^{2} (upper) and the height profile of the network structure (lower) of the 2%, 3.5% and 5%, respectively polymer concentration formed at T=25°C.

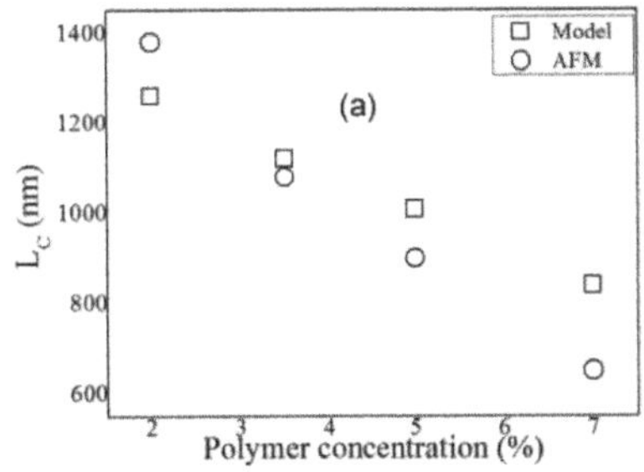

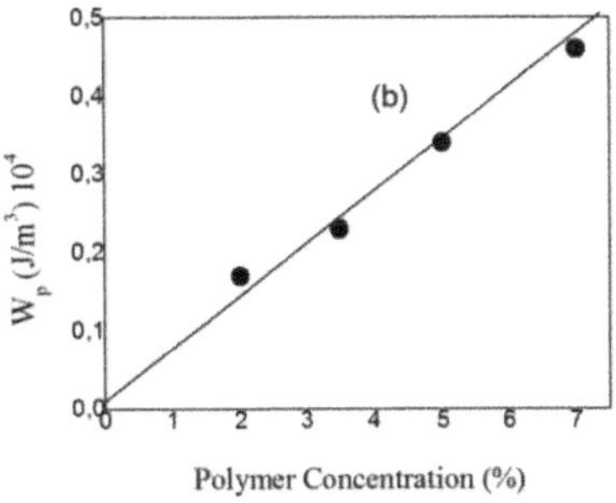

Figure 20. The average distance between two successive groups of fibers L_c as a function of polymer concentration (a) and the coupling interaction Polymer-Liquid crystals (W_P)(b).

This difference between interfiber distances in the two directions is an unexpected result. In fact, during the photopolymerization process, the mobility of the reactive monomers could be comparable within the smectic layers so that they present the same ability to come together and react to form the network. This ability could be significantly different across the smectic layers. Therefore, it would be reasonable to suspect that the average distance between polymer fibers could be of the same order of magnitude in lateral direction as well as in the z direction. The result found here is not yet clear, and the physical and chemical mechanisms governing the formation of the network could provide an explanation of our finding. This is not the aim of the work presented in this chapter. From the shift of the transition temperature, $\Delta T=T_c-T'_c = W_p/\alpha$, the coupling coefficient, W_p, characterizing the interaction energy between the FLC and the polymer network can be estimated. ΔTc values of 0.2 °C, 0.5 °C, 0.7 °C, an 0.9 °C were found for the polymer concentration of 2%, 3.5%, 5%, and 7%, respectively. The values of W_p are displayed in Fig. 20 (b). W_p linearly increases with the polymer concentration. The linear behavior of Wp in PSFLC system was already reported in earlier works [18-20]. The value of W_P found here are within 1 order of magnitude of those reported by Furue et al. [21] and Li et al. [17] on other PSFLC systems.

3.2.2. Effect of the polymer network morphology created under electric field in the short pitch ferroelectric liquid crystal on the dielectric responses

In this study we show that, both the dielectric intensity and the relaxation frequencies of the collective relaxations mechanisms namely soft and the Goldstone modes are strongly affected not only by the presence of the polymer network but also by the polymerization conditions.

Two samples cells with7% polymer concentration were then exposed to ultraviolet light (wavelength λ = 365nm) at 25 °C with an intensity of 18 (mW/cm^2) for 30 minutes. One of each is polymerized without the presence of electric field, however, the other cell is polymerized with the presence of AC electric field at $5\ V\ \mu m^{-1}$ and for $f= 1\ Hz$.

*Goldstone mode of the SmC**

In Figures 21 (a) and (b) the parameters $\Delta\varepsilon_G$ and f_G obtained from the curve-fit procedure are displayed. For all studied samples, the behavior of $\Delta\varepsilon_G$ versus temperature shows the same general features (Figure 21 (a)). As seen in this figure, $\Delta\varepsilon_G$ increases slightly to reach a maximum at T_{max} (3°C below T_c). Above T_{max}, $\Delta\varepsilon_G$ abruptly decreases.

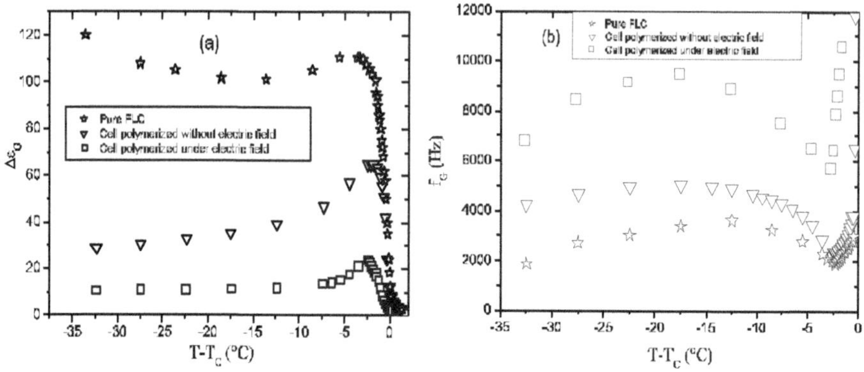

Figure 21. Temperature dependence of the Goldstone mode dielectric intensity (a) and the relaxation frequency (b).

Figure 21(b). This figure shows that the f_G slightly increases with temperature, reaches a maximum, then rapidly decreases to a minimum value at a temperature corresponding to T_{max}. It can be see from the figure 21 (b) that at room temperature, the f_G values are about 1.5 *kHz*, 4 *kHz* and 7 *kHz* for pure FLC, cell polymerized without and that polymerized with electric field, respectively. Significant differences were reported by Kaur et al [23] in another PSFLC systems. These authors were reported that the relaxation frequency f_G for the case when an bias field is applied during the polymerisation is very lower than in the case for sample when no bias was applied during the polymerisation. From the Equation (2) and (Equation 3), the values of K_{eff} are found about 2 10^{-11} N and 5.6 10^{-11} N for mixtures polymerized without and those polymerized with electric field, respectively. In conclusion, a lower dielectric strength (or larger relaxation frequency) could appear for a result of an increase of the elastic constant.

Soft mode of the SmA phase*

In Figures 22 (a) and (b) the temperature dependences of the dielectric strength, $\Delta\varepsilon_s$, and the relaxation frequencies, f_s, are shown for the soft mode detected in the SmA phase. One can see that at $(T\text{-}T_c)=0.1°\ C$, an increase in the polymer concentration leads to the strong reduction of the soft mode dielectric strength. At the same temperature, $\Delta\varepsilon_s$ for a mixture polymerized without electric field is three times higher than those polymerized under electric field.

It can be seen that, at $T\text{-}T_c=0.1°C$, an increase of the polymer concentration from 0 to 7% leads to an increase of f_s from 10 *KHz* to 40 *KHz*, respectively. It can be seen from this figure that, at $T\text{-}T_c=0.1°\ C$ the cell polymerized with electric field presents a high f_s value about 80 *KHz (Figure 22(b)*. To explain quantitatively the behavior of the reduction of the soft mode dielectric strength we can use the expression given by equation (16). We used the values of $\alpha\approx 8.8\ 10^{+3}$ N $m^{-2}K^{-1}$, a =0.12μm [19,20], and the values of $\Delta\varepsilon_s$ presented in the Figure 22(a) at $T\text{-}T_c=0.1°\ C$, with a reasonable value of the twist elastic constant, typically, $K_2 \approx 10^{-11}$ N, the equation (16) was graphically resolved to evaluate L_c. We obtain a mean inter-fiber distance $L_c \approx 500$ *nm* and 200 *nm*, for samples polymerized without and those polymerized with electric field respectively.

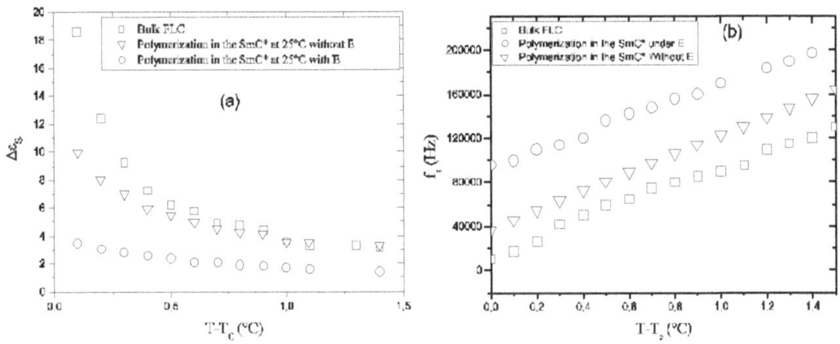

Figure 22. Temperature dependence of the dielectric intensity (a) and the relaxation frequency (b) of the soft mode in the smectic A* phase.

Atomic force microscopy observations

Figure 23 shows examples of AFM images and height profiles in the *z* direction of the Figure 18 of the polymer networks. These figures reveal that the separation distance between two polymer fibers (in the *Y* direction of the Figure 18) varies from 1 *μm* to 3*μm* for the cell polymerized without electric field and from 300 *nm* to 500 *nm* for sample polymerized under electric field. The diameter of the fibrils was found varied between 500 *nm* and 1*μm* for sample polymerized without electric field. However, it is found between 200*nm* and 700 *nm* for sample polymerized with electric field.

Figures 3 and 4 (lower) represent the height profiles (on the *z* direction in the figure2) of the polymer networks. The height profiles show two successive groups of fibers. The mean distance, L_c between them was evaluated for each sample and are about 480*nm* and 250 *nm* for samples when no field was applied during the polymerisation and those polymerized under electric field respectively. In conclusion, it must be noted from these observations that the polymerization under electric field is a principal factor affecting the polymer network structure. These calculated values are in accordance with those measured by the model. ΔT_c values of the sample polymerized without Electric and those polymerized with E are 0.9°C and 1.4°C, respectively. The coupling interaction W_p values are 0.46 10^4 (J/m^3) and 12,3 10^4 (J/m^3) for sample polymerized without electric field and those polymerized with electric field respectively.

In conclusion, the main parameter which governs the dielectric strength and the relaxation frequency in our PSFLC systems is the distance L_c which separated between two layers of the polymer network in the Z direction parallel to the direction of the electric field. The changes observed on the dielectric response are confirmed by AFM pictures. The polymerization under electric field is another factor affecting the structure of the polymer network formed in liquid crystals.

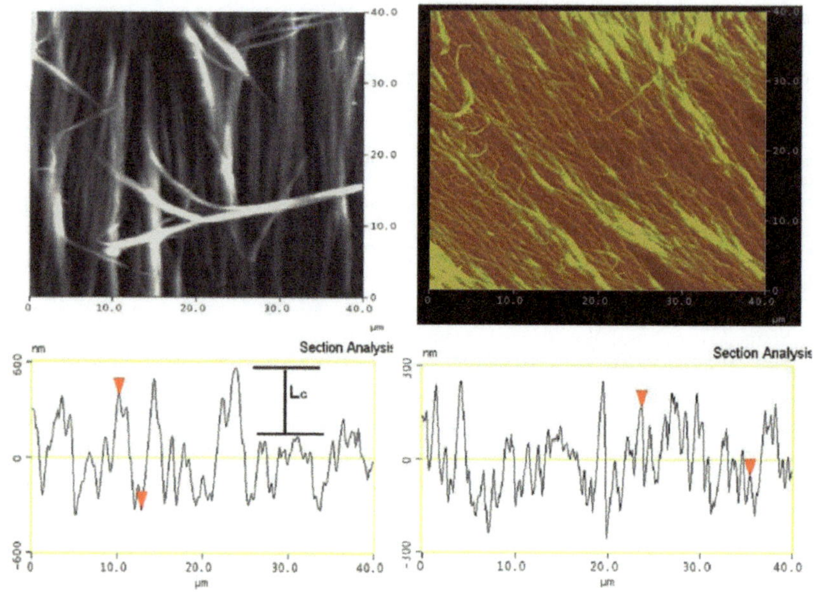

Figure 23. Tapping mode AFM images of polymer network structure of 40 x 40 μm ^{2} (upper) and the height profile of the network structure (lower) of the 7% polymer concentration (polymerization without electric field) on the left and that polymerized under electric field (on the right).

Author details

M. Petit
Université 20 Août 1955-Skikda, Algérie

4. References

[1] Drzaic PS (1986) Polymer dispersed nematic liquid crystal for large area displays and light valves. J. Appl. Phys., 60 : 2142.

[2] Fergason JL (1985) Polymer encapsulated nematic liquid crystals for display and light control applications. SID Dig. Tech. Pap, 16, 68.

[3] Drzaic PS (1995) Liquid Crystal dispersions, World Scientific, NJ

[4] Doane JW, Vaz NA, Wu BG, Zumer S (1986) Field controlled light scattering from nematic microdroplets. Appl. Phys. Lett., 48: 296.

[5] Doane JW, Golemme A, West JL, Whitehead JB, Wu BG (1988) Polymer Dispersed Liquid Crystals for Display Applications. Mol. Cryst. Liq. Cryst., 165: 511.

[6] Broer D, Gossink R, Hikmet R.A.M (1990) Oriented polymer networks obtained by photopolymerization of liquid-crystalline monomers. Angew. Makromol. Chem. 183: 45.

[7] Crawford G.P, Zumer S. (1996) Liquid Crystals in Complex Geometries Formed by Polymer and Porous Networks (Taylor & Francis, London).

[8] Hikmet RAM, Boots HMJ, (1995) Domain structure and switching behavior of anisotropic gels. Phys. Rev. E 51: 5824-5831.
[9] Chang CC, Chien LC, Meyer RB (1997) Electro-optical study of nematic elastomer gels. Phys. Rev. E 56: 595 599.
[10] Ma RQ, Yang DK (2000) Fréedericksz transition in polymer-stabilized nematic liquid crystals. Phys. Rev. E 61: 1567-1573.
[11] Fung YK, Yang DK, Sun Y, Chien LC, Zumer S, Doane JW (1995) Polymer networks formed in liquid crystals. Liq. Cryst. 19: 797-801.
[12] Dierking I, Kosbar LL, Afzali Adakani A, Lowe AC., Held GA (1997) Two-stage switching behavior of polymer stabilized cholesteric textures. J. Appl. Phys. 81: 3007.
[13] Rajaram CV, Hudson SD, Chien LC,
[14] Doane, Chien, Yang, Bos – see chapters 1, 4, 5, 11, 12, 13 of Liquid Crystals in Complex Geometries, edited by GP Crawford & S. Zumer (Taylor & Frances, London. 1996)
[15] Rajaram CV, Hudson SD, Chien LC (1996) Effect of Polymerization Temperature on the Morphology and Electrooptic Properties of Polymer-Stabilized Liquid Crystals. Chem. Mater. 8: 2451.
[16] Dierking I, Osipov MA, Lagerwall ST (2000) The effect of a polymer network on smectic phase structure as probed by polarization measurements on a ferroelectric liquid crystal. Eur. Phys. J. E 2: 303.
[17] Li J, Zhu X, Xuan L, Huang X (2002) "V-Shaped" Electro-Optic Characteristics in FLC Gels Ferroelectrics 277: 85-105.
[18] Petit M, Daoudi A, Ismaili M, Buisine JM, (2006) Distortion and unwinding of the helical structure in polymer-stabilized short-pitch ferroelectric liquid crystal. Eur. Phys.J. E 20: 327.
[19] Petit M, Daoudi A, Ismaili M, Buisine JM, Da Costa A (2008) Effect of the Polymer Network Density Formed in Short Pitch Ferroelectric Liquid Crystal on the Electroclinic Effect Mol. Cryst. Liq. Cryst. 487: 61-73.
[20] Petit M, Daoudi A, Ismaili M, Buisine JM (2006) Electroclinic effect in a chiral smectic- *A* liquid crystal stabilized by an anisotropic polymer network .Phys. Rev. E 74: 061707.
[21] Furue H, Takahashi T, Kobayashi S (1999) Monostabilization of Surface-Stabilized Ferroelectric Liquid Crystal Using Polymer Stabilization. Jpn. J. Appl. Phys., Part 1 38: 5660-5663.
[22] Takahashi T, Umeda T, Furue H, Kobayashi S (1999) Modeling and Computer Simulation of the Electrooptic Response of Polymer-Stabilized Ferroelectric Liquid Crystal Cells. Jpn. J.Appl. Phys., Part 1 38: 5991-5995.
[23] Kaur S, Dierking I, Gleeson HF (2009) Dielectric spectroscopy of Polymer Stabilised Ferroelectric Liquid Crystals. *Eur. Phys. J. E*, 30: 265-274.
[24] Mukherjee A, Bhattacharyya SS, Chaudhuri BK, Wu SL (2009)Effect of polymer relaxation on dielectric spectroscopic study of polymer–ferroelectric liquid crystal composites. *Journal of Molecular Liquids*, 148, 127,
[25] Kundu S, Ray T, Roy SK, Haase W, Dabrowski R (2003) Effect of UV Curable Polymer on The Dielectric & Electro-Optic Properties of Ferroelectric Liquid Crystal. *Ferroelectrics*. 282, 239-248
[26] Petit M, Hemine J, Daoudi A, Ismaili M, Buisine JM, Da Costa A (2009) Effect of the network density on dynamics of the soft and the Goldstone modes in short-

ferroelectric liquid crystals stabilized by an anisotropic polymer network .*Phys. Rev. E*, 79, 031705-9
[27] Blinc R, Zeks B (1978) Dynamics of helicoidal ferroelectric smectic-C* liquid crystals. Phys. Rev. A.18. 740-745.
[28] Goodby JW et al.∴Ferroelectric Liquid Crystals: Principales, Properties and Applications. (Gordon and Breach Science Publishers 1991)
[29] Musevic I, Blinc R, Zeks B (200)The Physics of Ferroelectric and Antiferroelectric Liquid Crystals.(World Scientific).
[30] Glogarova M, Pavel J (1984) The Behaviour of Thin Samples of Ferroelectric Liquid Crystals. Mol. Cryst. Liq. Cryst 114, 249-257.
[31] Brunet M, Parodi O (1982)Défauts dans les smectiques C chiraux. - II- Double paroi de déchiralisation, J. Phys. France 43, 515-522.
[32] Bouligand Y (1975) Defects and Textures in Cholesteric Analogues Given by Some Biological Systems *J. Phys.*, 36, C1-331.
[33] Archer P, Dierking I (2009) Polymer stabilisation of twisted smectic liquid crystal defect states *Soft Matter*, 5, 835.
[34] Baytch N, Selinger RLB, Selinger JV, Shashidhar R (2003) Simulations of helix unwinding in ferroelectric liquid crystals.Phys. Rev. E. 68, 041702.
[35] M. Gasser, A. Gembus, D. Ganzke, and I. Dierking (2000) Collective dynamics of polymer-network stabilized ferroelectric liquid crystals. Mol. Mater. 12, 347.
[36] Kutnjak Z, Kralj S, Žumer S (2002)Effect of dispersed silica particles on the smectic-*A*–smectic-*C** phase transition.Phys. Rev. E 66, 041702.
[37] Rozanski SA, Thoen J (2005) Influence of dispersed aerosil particles on the collective dynamic modes in a ferroelectric liquid crystal with polarization sign reversal Non-Cryst. Solids 351, 2802.
[38] Rozanski SA, Thoen J (2005) Collective dynamic modes in ferroelectric liquid crystal-aerosil dispersions. Liq. Cryst. 32, 331-340 2005.
[39] Carlsson T, Zeks B, Filipic C, Levstik A (1990) Theoretical model of the frequency and temperature dependence of the complex dielectric constant of ferroelectric liquid crystals near the smectic-*C**–smectic-*A* phase transition. Phys. Rev. A 42, 877-889.
[40] Held GA, Kosbar LL, Dierking I, Lowe AC, Grinstein G, Lee V, Miller RD (1997) Confocal Microscopy Study of Texture Transitions in a Polymer Stabilized Cholesteric Liquid Crystal. Phys. Rev. Lett. 79, 3443-3446.
[41] Goh M, Kyotani M, Akagi K (2007) Highly twisted helical polyacetylene with morphology free from the bundle of fibrils synthesized in chiral nematic liquid crystal reaction field. *J. Am. Chem. Soc.*, 129 (27), 8519 -8527.
[42] Hoischen A, Benning SA, Kitzerow HS(2009) Electroconvection of liquid crystals: Tool for fabricating modulated polymer surfaces. J. Appl. Phys. 105, 013540.
[43] Kaur S, Dierking I, Gleeson HF(2009). Eur. Phys. J. E, 2, 3, (2009).
[44] Beckel E, Cramer NB, Harant AW, Bowman CN (2003) Electro-optic properties of thiol-ene polymer stabilized ferroelectric liquid crystals. Liq. Cryst. 30, 1343.
[45] Garoff S, Meyer RB (1977) Electroclinic Effect at the *A-C* Phase Change in a Chiral Smectic Liquid Crystal.Phys. Rev. Lett. 38, 848-851.
[46] Garoff S, Meyer RB (1979) Electroclinic effect at the *A-C* phase change in a chiral smectic liquid crystal. Phys. Rev. A 19, 338-347.

Section 2

Complex Macromolecular Systems

Chapter 2

Polymer Biocompatibility

Isabel Cristina Celerino de Moraes Porto

Additional information is available at the end of the chapter

1. Introduction

Polymers are very versatile materials and are used in many applications including pharmaceutical applications. Natural polymers, modified natural polymers, and synthetic polymers are used as excipients in the manufacture of cosmetics and systems for conventional and modified delivery of drugs, by altering the composition and physical properties such as molecular weight, polydispersity, crystallinity, and thermal transitions. They can be prepared to provide a wide range of degradation rates and mechanical properties (Amaral, 2005; Villanova & Oréfice, 2010).

More recently, polymers have been developed that can modulate and deliver drugs to target areas. Biodegradable polymers, bioadhesives, biomimetic materials, and responsive hydrogels have been included in pharmaceutical formulations (Villanova & Oréfice, 2010). Naturally derived polymers offer several advantages compared with synthetic polymers, namely biocompatibility, biodegradability, and biological activity, as most of them are present in the structural tissues of living organisms. However, the low mechanical strength and high rates of degradation of natural polymers often result in their use in combination with ceramics, or in subsequent cross-linking reactions to reduce the degradation rates. Synthetic polymers can be tailored to meet an absorption time requirement, potentially facilitating reproducibility and scale-up, with no concerns about disease transmission, which often constitutes a problem with naturally occurring polymers (Amaral, 2005).

Synthetic polymers present an attractive avenue for biocompatible biomaterials because of their well-studied syntheses and modifiable properties (Ouchi & Ohya, 2004; Puskas & Chen, 2004). Biocompatible polymers can be made into devices directly or incorporated into devices by coating to reduce the chance of rejection when incorporated into the body. There have been significant developments in recent years in shape memory materials, tissue engineering, and coronary stents for use as biocompatible materials (Quansah, 2004).

Biodegradable polymers, especially those belonging to the family of polylactic acid and polyglycolic acid, play an increasingly important role in orthopedics. These polymers degrade by hydrolysis and enzymatic activity and have a range of mechanical and physical properties that can be engineered to suit a particular application. Their degradation characteristics depend on several parameters including their molecular structure, crystallinity, and copolymer ratio. These biomaterials are also rapidly gaining recognition in the fledging field of tissue engineering because they can be fashioned into porous scaffolds or carriers of cells, extracellular matrix components, and bioactive agents. Although their future appears to be bright, several questions regarding the biocompatibility of these materials linger and should be addressed before their wide-scale use (Athanasiou et al., 1996). The most important requirement for a biodegradable polymer to be used in medical applications is its compatibility not only in terms of physical and chemical properties but also the properties that define their behavior at the time they come into contact with the body (Silva et al., 2004).

2. Biocompatibility

In general, a biomaterial is defined as any substance, except food and medications, that can be used for a length of time as part of a system that aims to treat or to replace any tissue, organ, or body function. Few materials, if any, are totally inert from a physiological standpoint; most materials present a variety of components with potential toxic or irritating properties. In addition, chemical reactions that occur during setting of the material may also produce noxious effects (Anusavice, 2003).

Biomaterials need to satisfy a number of prerequisites before that can be used in applications, including biocompatibility. To verify this feature, its components should be subjected to different tests (Schmalz, 2002), performed as recommended by various organizations and federations. These tests consist of a sequence of research protocols, described and regulated in many countries, for correct use of experimental materials under evaluation, thereby determining their safety for clinical application in humans (Costa, 2001).

Biocompatibility may be defined as (Williams, 2008):

> *"ability of a biomaterial to perform its desired function with respect to a medical therapy, without eliciting any undesirable local or systemic effects in the recipient or beneficiary of that therapy, but generating the most appropriate beneficial cellular or tissue response to that specific situation, and optimizing the clinically relevant performance of that therapy."*

Biocompatibility can also be defined as the relationship between a material and the organism so that neither produces undesirable effects. Biocompatibility has been mentioned in many works with increasing interest in evaluating the characteristics of medical and dental materials and devices and responses caused by its components. An ideal pattern for determining these properties has not yet been determined, however, various methods have been suggested for this purpose.

Biocompatibility is a term that encompasses many aspects of the material, including its physical, mechanical, and chemical properties, as well as potential cytotoxic, mutagenic, and allergenic effects, so that no significant injuries or toxic effects on the biological function of cells and individuals arise (Costa, 2001; Lemmons & Natiella, 1986; Schmalz, 2002).

Until the biocompatibility of a material is proven, it must be subjected to various studies ranging from in vitro assays to clinical trials, involving distinct areas such as pharmaceutics, biology, chemistry, and toxicology.

Biomaterials or their degradation byproducts should not trigger an adverse inflammatory reaction or immune response once implanted. Surface properties such as chemistry, roughness, and surface energy play a major role in cell–material interactions, particularly when considering an absorbable material, which is always presenting a new surface. These surface characteristics determine not only how biological molecules adsorb to the surface, but also their spatial orientation on adsorption. This is of particular importance in host–implant interactions, as the adsorption of proteins present in physiological fluids, such as albumin, immunoglobulin, fibrinogen, and fibronectin, dictate the subsequent inflammatory response and the fate of the implant (Amaral, 2005).

2.1. Methods of biocompatibility testing

Biocompatibility testing has sought to standardize biological tests for biomaterials, to find an effective and safe testing protocol that is more reliable for comparing results from different studies.

These tests are divided into 3 groups, corresponding to primary (level I), secondary (level II), and preclinical (level III) tests, which include analysis of the cytotoxicity and irritant potential of systemic toxicity in animals through intramuscular and subcutaneous implants, and usage tests by observation of tissue reactions after insertion of the material, for example, in human teeth.

Level I tests can be done both in vitro and in vivo. In vitro tests assess the properties of the material directly in cultured cells that react to the effects of the experimental material. Many constituents judged initially as cytotoxic may be modified or have their use controlled by manufacturers to prevent cytotoxicity. In vivo tests are mainly based on the implantation of materials into subcutaneous or intramuscular areas in rats and rabbits to evaluate the tissue response to the implanted material after a period of observation (Anusavice, 2003).

To conduct these tests, it is necessary to involve health researchers for research methodology development and evaluation of the tissue and researchers for development of materials and their properties, such as engineers and chemists.

2.2. In vitro and in vivo tests

The ideal biological research methodology consists of in vivo experiments, despite the ethical aspects involved. Nevertheless, although in vitro studies provide responses limited by the absence of biological and physiological components that are impossible to reproduce entirely, they continue to be used and are suitable for determining whether a material contains significant quantities of extractable biologically harmful components (Porto et al., 2011).

Clinically, the results are divergent. Studies in vitro could report no damage (Büyükgüral & Cereli, 2008), moderate damage (Nayyar et al., 2007) or intense damage (Costa et al., 2002). This reinforces the effect of the physiology present in in vivo systems, but does not diminish the relevance of in vitro research. In vitro research remains important for pointing out pathways for studies of the adverse reactions recorded clinically.

In in vitro cell cultures, the complex physiology of an organism performing various functions simultaneously is not present. Thus, the buffer capacity of complex humoral and cellular systems in the intact organism is absent; a biomaterial may not work well in the in vitro test, but may be biocompatible in vivo (Kirkpatrick et al., 2005; Libonati et al., 2011). This highlights the necessity of integrated in vitro and in vivo tests for valuable predictive estimation of the toxicity of complex materials (Libonati et al., 2011).

2.2.1. *In vitro biocompatibility testing*

In the field of biomaterials, it is necessary to consider aspects of biosecurity, such as elimination of cytotoxicity and other harmful effects of the material to be used (Kirkpatrick et al., 2005). By definition, the cytotoxicity of a material or device refers to the toxicological risks caused by a material or its extract in a cell culture (Cao et al., 2005). To perform these tests, mammalian cells, usually of mouse or human origin, obtained from a commercial supplier, are cultured in the laboratory in flasks using nutrient culture media. These cultured mammalian cells reproduce by cellular division and can be subcultured to produce multiple flasks of cells for use in evaluating the cytotoxicity of materials. For cytotoxicity tests in vitro, permanent cell lines or primary cultures are recommended. Although there are difficulties with isolation and maintenance, the primary cells are very important for biological assays because of their similarity to the original tissue (Wallin, 1998).

Cytotoxicity tests are considered a rapid, sensitive and standardized method to determine the toxicity of a material or if it contains significant amounts of biologically harmful leachable compounds. The presence in cultures of isolated cells and the absence of important physiological effect present in vivo systems, which help to protect cells within the body, produces a test with high sensitivity. Culture medium of mammalian cells is the preferred method for the extraction of substances that can be released from a material, because it is a physiological solution capable of extracting a wide range of chemical structures, not only those soluble in water.

Several types of cells can be used for cytotoxicity tests such as cell lines - human and mouse fibroblasts, lymphocytes, queratinocytes, mouse odontoblast-like cells, mouse macrophages, rat submandibular salivary gland acinar cells - (Bakoupoulou et al., 2007; Franz et al., 2009; Kostoryz et al., 2003; Lin et al., 2007; Roll et al., 2004; Samuelsen et al., 2008), and primary cell types - human lymphocytes, polymorphonuclear leukocytes, and mixed leukocytes, mouse blastocysts, mouse macrophages, and mouse embryo cells - (Becher et al., 2006; Huang & Chang, 2002; Jonhson et al., 1985; Libonati et al., 2011; Porto et al., 2009, 2011; Prica et al., 2006).

Macrophages present throughout the body, including the oral tissues, are involved in inflammation and presentation of antigen during the reaction to infectious agents and foreign bodies. They also amplify the inflammatory response by signaling other cells, and play a central role in the pathogenesis of inflammatory response, therefore are relevant for testing the biocompatibility of materials in vitro (Becher et al., 2006). When stimulated, the macrophages are subjected to a process known as macrophagic activation in which they increase their metabolic, motility, and phagocytic activity. Many new proteins are synthesized under activation, including inducible nitric oxide synthase. Nitric oxide (NO), a product of nitric oxide synthase, plays an important role in the defensive function of macrophages (Parslow et al., 2001). Therefore, possible factors such as toxic levels of leachable compounds from materials or devices can be indicated by nitric oxide secretion, cell death, reduction in or loss of cellular function.

Interactions of materials and their components with cells at a molecular level are responsible for tissue reactions, such as inflammation, necrosis (Accorinte et al., 2005), immunological alterations, genotoxicity (Kleinsasser et al., 2004), and apoptosis (Paranjpe et al., 2005).

Among the 3 categories of tests for assessing cytotoxicity that are listed in ISO 10993-5 (2009) (extract test, direct contact test, indirect contact test), it is possible apply a wide variety of experimental protocols. The choice of one or more of these categories depends on the nature of the sample to be evaluated, the potential site of use, and the nature of the use. This choice then determines the details for the preparation of the samples to be tested, the preparation of the cultured cells, and the way in which the cells are exposed to the samples or their extracts. At the end of the exposure time, evaluation of the presence and extent of the cytotoxic effect is undertaken.

The various methods and parameters used in determining cytotoxicity can be grouped into the following categories of evaluation: assessment of cell damage by morphological means, measurement of cell damage; measurement of cell growth; measurement of specific aspects of cellular metabolism.

There are several ways to produce results in each of these 4 categories. The investigator should be aware of the test categories and into which category a particular technique fits, so that comparisons can be made with other results on similar devices or materials at both the intra- and interlaboratory levels. Quantitative evaluation of cytotoxicity can be done using cell death, inhibition of cell growth, cellular proliferation and colony formation, cell number, amount of protein, enzyme released, vital dye release, vital dye reduction, or any other measurable parameters that can be quantified by objective means (ISO 10.993-5, 2009).The biochemical methods (DNA synthesis, protein synthesis, and ATP activity) demonstrated good agreement in toxicity ranking of the materials, regardless of which cell culture was used, and the cell cultures responded similarly for each method. Methods that measured the functional characteristics of cells (adhesion and phagocytosis) were highly sensitive but had low toxicity ranking agreement and reproducibility. Assays (defined as method and cell culture combinations) using cell lines were more reproducible than assays using primary cell types. Significant differences in sensitivity were noted among the assay systems for particular material types (Johnson et al., 1985).

Relative sensitivity of in vitro biocompatibility test systems was explored by Johnson et al. (1983) and showed cellular responses of 12 standardized cell lines to 20 materials representing a range of toxicity. Results of the tissue culture assays were compared with those obtained for the same materials in vivo using a 5-day rabbit intramuscular implant assay. Methods involving measurement of cellular growth (colony counts or presence of confluence) in serum-fortified media extracts of test samples were generally more sensitive and discriminating than those in which test materials were placed directly in cell cultures (measurement of zone of growth inhibition). Based on sensitivity, reproducibility, ability to discriminate materials, and grader agreement, 4 of the 12 cell lines and 2 of the 4 test methods appeared most suitable for screening and evaluation of materials. Agreement of results using these 4 cell lines with intramuscular implantation tests for the 30 materials ranged from 60 to 90% (Johnson et al., 1983).

The pattern of apoptotic response, cellular glucose, oxygen consumption, and gene expression after exposure to the various compounds, single substances or biomaterials has been used to evaluate cytotoxicity in previous studies (Becher et al., 2006; Lin et al., 2007; Nooca et al., 2007; Sangsanon et al., 2007) and can be used to assess cytopathic effects of multicomponent extracts from cured polymers. The use of different methods to evaluate cytotoxicity may result provide data for a knowledge base to clarify how experimental materials affect cell behavior (Porto et al., 2009).

Using the MTT assay, cell viability can be assessed by the cytochemical demonstration of succinic dehydrogenase enzyme activity, which is a measure of the mitochondrial respiration of the cells. Polymers and its components may alter the enzyme activity of primary cells (Becher et al., 2006; Chen et al., 2003; Porto et al., 2009, 2011) or immortal cell lines (Demirci et al., 2008; Lin et al., 2007; Poskus et al., 2009).

One of the most studied alternative in vitro testing methods for identification of developmental toxicity is the embryonic stem cell test. A study conducted by Van Dartel and Piersma (2011) presents the progress that has been made with regard to the prediction of developmental toxicity using the embryonic stem cell test combined with transcriptomics. Although the embryonic stem cell test has been formally validated, the applicability domain as well as the predictability of the model needs further study to allow its successful implementation as an alternative testing method in regulatory toxicity testing.

Development of additional aspects required for further optimization of the embryonic stem cell test, including kinetics, the use of human embryonic stem cells and computational toxicology, and the current and future use of the embryonic stem cell test model for prediction of developmental toxicity in testing strategies and in regulatory toxicity evaluations should be also discussed. Genomics technologies have already provided proof of principle of their value in identification of toxicants such as carcinogenic compounds. Within the embryonic stem cell test, gene expression profiling has shown its value in the identification of developmental toxicity and in the evaluation of factors critical for risk assessment, such as dose and time responses. It is expected that the implementation of genomics in the embryonic stem cell test will provide a more detailed end point evaluation compared with the classic morphological scoring of differentiation cultures. Therefore, genomics may contribute to improvement of the embryonic stem cell test, both in terms of the definition of its applicability domain and its predictive capacity (Van Dartel & Piersma, 2011).

The embryonic stem cell test is a high-throughput in vitro screening assay for developmental toxicity free of animal use. The embryonic stem cell test uses the ability of murine embryonic stem cells to differentiate into the mesodermal cardiac lineage in combination with 2 cytotoxicity test systems. Validation of the embryonic stem cell test showed that the test system is very promising as an alternative to animal testing, however to optimize predictability and increase knowledge on the applicability domain of the embryonic stem cell test, improvements to the method were proposed and studied. The authors discuss the first definition of the embryonic stem cell test and the innovative approaches that have been proposed to increase the predictivity of the embryonic stem cell test, including implementation of molecular end points in the embryonic stem cell test, such as OMICS technologies and the addition of alternative differentiation models to the testing paradigm, such as neural and osteoblast differentiation and the use of human stem cells. These efforts to improve the embryonic stem cell test increase the value of embryonic stem cells used as in vitro systems to predict developmental toxicity (Theunissen & Piersma, 2012).

2.2.2. *In vivo biocompatibility testing*

Level II tests are based on tissue assessment of animals that received implants subcutaneously and intramuscular injection of a material with potential to cause systemic toxicity by inhalation, skin irritation, among other responses. Dermal toxicity tests are important because of the large number of chemicals with which we have daily contact. When a material, product, or toxic component is identified, it can be replaced, diluted, or neutralized to reduce the level of toxicity.

Despite their high cost, controversy, and ponderous bureaucratic challenges, animal tests are critical for assessing the biological responses to a new material before it is used in humans. Many aspects of clinical biological responses cannot be modeled by in vitro tests (Anderson, 2001). Animal tests offer evidence about these types of effects without putting humans at risk. Animal tests may be structured to mimic human clinical use to some degree, are commonly less expensive than human clinical trials, can be finished more quickly in many cases, and can be controlled to a greater degree. Animals may be exposed to materials or their degradation products with routes of administration or doses that would be unethical to consider in humans. Animal tests may be used to determine responses that are difficult or impossible to ethically test in humans and may be tested at many phases of life (for example, embryos or 'children') in a way that is not possible in humans (Wataha, 2012).

Functional properties and biocompatibility are basic issues in the development of biomaterials. There are several methodologies to analyze these properties, however, in all cases the human host has been the ultimate test (Natiella & Lemmons, 1986).

Biocompatible materials cannot be mutagenic or influence inflammatory mediators causing systemic responses, including toxicity, tissue injury, teratogenic or carcinogenic effects. Such materials must be free of agents that may cause allergic responses to individuals sensitive to these substances. After a material has successfully passed the tests for levels I and II, it should be tested in humans (level III test) to evaluate its performance and the favorable or unfavorable reactions that may present under normal clinical conditions (Anusavice, 2003).

3. Biocompatibility for specific polymeric systems

3.1. Chitosan

Chitosan is a biopolymer type polysaccharide and has a chemical structure similar to vegetal fiber cellulose. It is derived from chitin, a polysaccharide that is extremely abundant in nature and is the main component of the exoskeletons of insects and crustaceans, and is found in the cell walls of some species of fungi. Chitosan has several important uses due to its antifungal and antimicrobial activity, its ability to inhibit tumor cells and act as a controlled releaser of drugs in the body.

VandeVord et al. (2002) examined the biocompatibility of chitosan scaffolds using a mammalian implantation model. Early migration of neutrophils into the implantation area, which resolved with increasing implantation time, was reported. Besides this early accumulation of neutrophils, cells that are usually associated with acute inflammation, no evidence of other signs associated with an inflammatory response, such as erythema and edema, were found.

Endotoxins were not detected and a chronic inflammatory response did not develop. In addition, a very low incidence of specific immune reactions was observed, as determined by lymphocyte proliferation assays and antibody binding responses measured using ELISA techniques. Formation of normal granulation tissue associated with accelerated angiogenesis, the typical healing response, was observed. The results from this study indicated that chitosan has a high degree of biocompatibility. These results are in accordance with previous studies (Rao & Sharma, 1997; Tomihata & Ikada, 1997), in which the biocompatibility of chitosan films with different degrees of acetylation is reported, even for highly deacetylated chitin derivatives.

A wide variety of cells have been successfully cultured in/on chitosan matrices, among them keratinocytes (Chatelet et al., 2001), chondrocytes (Denuziere et al., 1998), osteoblasts (Lahiji et al., 2000), endothelial cells, hepatocytes (Elçin et al., 1998), Schwann cells (Yuan et al., 2004), and NH3T3 cells (Li et al., 2012), indicating the cytocompatibility of chitosan toward these cells. In addition, no cytotoxic products are released from chitosan matrices, as shown by the maintenance of cell metabolic activity (Amaral, 2005).

3.2. Dental polymers

In vitro and in vivo studies have demonstrated that the cytotoxic effects of polymeric dental materials, such as light-cured methacrylate polymers, depend on the quantity of unconverted resinous monomers that remain after polymerization (Annunziata et al., 2006; Falconi et al., 2007).

Among the components capable of being lixiviated from dental polymers are bisphenol A-glycidyl methacrylate (bis-GMA), triethylene glycol dimethacrylate (TEGDMA), and 2-hydroxyethyl methacrylate (HEMA) (Lin et al., 2007). These monomers can alter cellular metabolism at concentrations well below the toxic level. The changes induced can be considered a potential mechanism for clinical and subclinical effects (Nooca et al., 2007).

Bis-GMA is the most toxic among them. A mechanism suggested for the cytotoxicity of this monomer is alteration of the lipid layer of the cell membrane, which affects membrane permeability (Lefebvre et al., 1996). In addition, bis-GMA undergoes hydrolysis, producing methacrylic acid as a metabolite. Methacrylic acid is soluble in water and capable of inducing cytotoxicity via stimulation of tumor necrosis factor alpha release (Kostoryz et al., 2003).

HEMA is an amphoteric monomer that displaces water in dentin and is capable of diffusing rapidly through the dentin (Bouillaguet et al., 1996). It is also miscible with most of the monomers used in composites (Bouillaguet, 2004). Therefore, the possibility of other components of the adhesive system acting synergistically with HEMA to increase cytotoxicity needs to be investigated, especially because other relatively hydrophobic resinous composites are soluble in HEMA and may be carried through the dentin (Bouillaguet et al., 1996).

HEMA has been shown to be a potent mediator of cell death by apoptosis at concentrations ranging from millimolar to micromolar (Cetinguç et al, 2007). If released at low concentrations for a prolonged period of time, HEMA could reduce the rate of cellular proliferation and result in apoptosis, possibly as a response to DNA damage (Samuelsen et al., 2008).

Dental adhesive and composites components may alter the succinate dehydrogenase enzyme activity of primary cells (Becher et al., 2006; Chen et al., 2003; Porto et al., 2009, 2011) or immortal cell lines (Demirci et al., 2008; Lin et al., 2007; Poskus et al., 2009).

Despite have adverse effects measured in vitro tests, dental polymers may or not may trigger noxious reactions in vivo, and it depends on the amount released compounds, concentration and other factors. Thus after many in vitro tests and animal tests be completed is possible that the hazards and risks are acceptable for human usage.

3.3. Polylactide acid and polyglycolide acid

Synthetic polymers, such as polylactic acid and polyglycolic acid and their copolymer are becoming more and more attractive for tissue engineering applications and have already been approved by the US Food and Drug Administration. Since the 1960s, polylactic acid and polyglycolic acid have been applied in medicine as controlled drug delivery systems, mostly injectable as microspheres, orthopedic fixation devices such as pins and screws, and as scaffold to mimic the extracellular matrix for cells (Ashammaki & Rokkanen, 1997; Gunatillake & Adhikari, 2003; Jain, 2000; Richardson et al., 2001; Thomson et al., 1995).

Depending on the lactic and glycolic portion of the polylactide-co-glycolide copolymer, various physical and mechanical properties can be produced and the degradation rate can be specifically regulated (Wintermantel & Ha, 2002). Polylactic acid shows a high crystallinity and is attractive as a biodegradable polymer because degradation products naturally occur in the body and are resorbed through the metabolic pathway. Both polymers demonstrate a similar degradation rate caused by random hydrolysis of their ester linkage producing either lactic acid from polylactic acid or glycolic acid from polyglycolic acid, which can be excreted in urine. The degradation rate of the material depends on various criteria besides the copolymer ratio, such as configurational structure, crystallinity, morphology, stress, the amount of residual monomer, porosity, and the site of the implantation (Jain, 2000). The desired mechanical properties are also based on the polymer ratio and on the molecular weight and crystallinity of the scaffold (Mauth et al., 2007).

Various in vitro and in vivo studies have approved the biocompatibility and the biodegradability of these polymers. Mild inflammatory reactions have been observed after a massive release of acidic degradation in vivo depending on the amount and degradation rate of the material (Bostman et al., 1990; Grayson et al., 2004; Holy et al., 1999; Hutmacher, 2000; Thomson et al., 1999; Young et al., 2002). However, the use of synthetic biocompatible material has the enormous advantage of reproducible synthesis and the mechanical and chemical properties including the structure, size, viscosity, and porosity, as well as degradation rate of the desired scaffold can be controlled. Furthermore, bioactive molecules can be incorporated and locally applied by controlled release of the biodegradable polylactic acid, polyglycolic acid, or polylactide-co-glycolide copolymer system and so influence the cell phenotype expression (Mauth et al., 2007).

In their study Athanasiou et al. (1996) undertook an extensive literature review on the toxicity and biocompatibility of polylactic acid-polyglycolic acid biomaterials. In general, polylactic acid/polyglycolic acid biomaterials showed satisfactory biocompatibility, although some reduction in cell proliferation has been noted. Many studies have successfully demonstrated biocompatibility of polylactic acid/polyglycolic acid biomaterials in vivo (Athanasiou et al., 1996).

4. Conclusion

For the biocompatibility of a material to be proved, it must be subjected to various studies ranging from in vitro assays to clinical trials and involving distinct areas such as pharmaceutics, biology, chemistry, and toxicology. The use of standardized tests allows better comparison between the results of different studies to clarify the behavior of the materials and their safety in relation to cells and tissues.

For biodegradable polymers, long-term tests should be prioritized to enable the evaluation of the effects of continuous release of metabolites resulting from their degradation and to observe the type and extent of local and systemic changes.

Only a combination of various in vitro and in vivo tests can provide an overview of the interaction of biomaterials with the host.

Author details

Isabel Cristina Celerino de Moraes Porto
Federal University of Alagoas, CESMAC University Center, Brazil

5. References

Accorinte, M.L.R., Loguercio, A.D., Reis, A., Muench, A., & Araújo, V.C. (2005). Adverse effects of human pulps after direct pulp capping with the different components from a total-etch three-step adhesive system. *Dental Materials,* Vol. 21, No. 7, (July 2005), pp. 599–607, ISSN 0109-5641.

Amaral, I.M.S.R.F. (2005). Chitosan matrices for cell-based bone regenerative therapies. PhD thesis, Faculdade de Engenharia, Universidade do Porto, Portugal.

Anderson, J.M. (2001). Biological responses to materials. *Annual Review of Materials Research,* Vol. 31, pp. 81-110.

Annunziata, M., Aversa, R., Apicella, A., Annunziata, A., Apicella, D., Buonaiuto, C., & Guida, L. (2006). In vitro biological response to a light-cured composite when used for cementation of composite inlays. *Dental Materials,* Vol. 22, No. 12, (December 2006), pp. 1081–1085, ISSN 0109-5641.

Anusavice, K.J. (2003). *Phillip's Science of Dental Materials* (11th edition), Saunders, ISBN 9780721693873, Philadelphia, PA.

Ashammakhi, N. & Rokkanen, P. (1997). Absorbable polyglycolide devices in trauma and bone surgery. *Biomaterials.* Vol. 18, No. 1, pp. 3–9.

Athanasiou, K.A., Agrawal, C.M., Barber, F.A., & Burkhart, S.S. (1998). Orthopaedic applications for PLA-PGA biodegradable polymers. *Arthroscopy*. Vol. 14, No. 7, (October 1998), pp. 726–737, ISSN 0749-8063.

Athanasiou, K.A., Niederauer, G.G., Agrawal C.M. (1996). Sterilization, toxicity, biocompatibility and clinical applications of polylactic acid/ polyglycolic acid copolymers. *Biomaterials*. Vol. 17, No. 2, (January 1996), pp. 93–102, ISSN 0142-9612.

Bakopoulou, A., Tsiftsoglou, A., Galaktidou, G., Markala, D., Triviai, I., Garefis, P. (2007). Patterns of cell death and cell cycle profiles of cultured WEHI 13 var fibroblasts exposed to eluates of composite resins used for direct and indirect restorations. *European Journal of Oral Sciences*, Vol. 115, No. 5,(October 2007), pp. 397–407.

Becher, R., Kopperud, H.M., Al, R.H., Samuelsen, J.T., Morisbak, E., Dahlman, H.J., et al. (2006). Pattern of cell death after in vitro exposure to GDMA, TEGDMA, HEMA and two compomer extracts. *Dental Materials*, Vol. 22, No. 7, (July 2006), pp. 630–640, ISSN 0109-5641.

Bostman, O., Hirvensalo, E., Mäkinen, J., & Rokkanen, P. (1990). Foreign-body reactions to fracture fixation implants of biodegradable synthetic polymers. *Journal of Bone and Joint Surgery British Volume*, Vol. 72, No. 4, pp. 592–596.

Bouillaguet, S. (2004). Biological risks of resin-based materials to the dentin-pulp complex. *Critical Reviews in Oral Biology and Medicine*, Vol. 15, No. 1, (January 2004), pp. 47–60, ISSN 1045-4411.

Bouillaguet, S., Wataha, J., Hanks, C.T., Ciucchi, B., & Holz, J. (1996). In vitro cytotoxicity and dentin permeability of HEMA. *Journal of Endodontics*, Vol. 22, No. 5, (May 1996), pp. 244–248, ISSN 0099-2399.

Büyükgüral, B. & Cehreli, Z. (2008). Effect of different adhesive protocols vs calcium hydroxide on primary tooth pulp with different remaining dentin thickness: 24-month results. *Clinical Oral Investigations*, Vol. 12, No. 1, pp. 91–96, ISSN 1432-6981.

Cao, T., Saw, T.Y., Heng, B.C., Liu, H., Yap, A.U., & Ng, M.L. (2005). Comparison of different test models for assessment of cytotoxicity of composites resins. *Journal of Applied Toxicology*, Vol. 25, No. 2, pp. 101–108.

Çetinguç, A., Olmez, S., & Vural, N. (2007). HEMA diffusion from dentin bonding agents in young and old primary molars in vitro. *Dental Materials*, Vol. 23, No. 3, pp. 302–307.

Chatelet, C., Damour, O., & Domard, A. (2001). Influence of the degree of acetylation on some biological properties of chitosan films. *Biomaterials*, Vol. 22, No. 3, pp. 261–268.

Chen, R.-S., Liu, C.-C., Tseng, W.-Y., Jeng, J.-H., & Lin, C.-P. (2003). Cytotoxicity of three dentin bonding agents on human dental pulp cells. *Journal of Dentistry*, Vol. 31, No. 3, pp. 223–229.

Costa, C.A.S. (2001). Testes de citotoxicidade em culturas de células. In: *Metodologia Científica. Ensino e Pesquisa em Odontologia*, Estrela, C. (Ed.), pp. 145–160. São Paulo: Artes Médicas.

Costa, C.A.S., Nascimento, A.B.L., & Teixeira, H.M. (2002). Response of human pulps following acid conditioning and application of a bonding agent in deep cavities. *Dental Materials*, Vol. 18, pp. 543–551.

Demirci, M., Hiller, K.-A., Bosl, C., Galler, K., Schmalz, G., & Schweikl, H. (2008). The induction of oxidative stress, cytotoxicity, and genotoxicity by dental adhesives. *Dental Materials*, Vol. 24, No. 3, pp. 362–371.

Denuziere, A., Ferrier, D., Damour, O., & Domard, A. (1998). Chitosan-chondroitin sulfate and chitosan-hyaluronate polyelectrolyte complexes: biological properties. *Biomaterials*, Vol. 19, No. 14, pp. 1275–1285.

Elçin, Y.M., Dixit, V., & Gitnick, G. (1998). Hepatocyte attachment on biodegradable modified chitosan membranes: In vitro evaluation for the development of liver organoids. *Artificial Organs*, Vol. 22, No. 10, pp. 837–846.

Falconi, M., Teti, G., Zago, M., Pelotti, S., Breschi, L., & Mazzotti, G. (2007). Effects of HEMA on type I collagen protein in human gingival fibroblasts. *Cell Biology and Toxicology*, Vol. 23, No. 5, (September–October 2007), pp. 313–322, ISSN 0742-2091.

Franz, A., König, F., Lucas, T., Watts, D.C., Schedle, A. (2009). Cytotoxic effects of dental bonding substances as a function of degree of conversion. *Dental Materials*, Vol. 25, No. 2, pp. 232–239.

Grayson, A.C., Voskerician, G., Lynn, A., Anderson, J.M., Cima, M.J., & Langer R. (2004). Differential degradation rates in vivo and in vitro of biocompatible poly(lactic acid) and poly(glycolic acid) homo- and co-polymers for a polymeric drug-delivery microchip. *Journal of Biomaterials Science Polymer Edition* Vol. 15, No. 10, pp. 1281-1304.

Gunatillake, P.A. & Adhikari, R. (2003). Biodegradable synthetic polymers for tissue engineering. *European Cells & Materials*, Vol. 5, pp. 1–16.

Holy, C.E., Dang, S.M., Davies, J.E., & Shoichet, MS. (1999). In vitro degradation of a novel poly(lactide-co-glycolide) 75/25 foam. *Biomaterials*, Vol. 20, No. 13, pp. 1177–1185.

Huang, F.M., Chang, Y.C. (2002). Cytotoxicity of dentine-bonding agents on human pulp cells in vitro. *International Endodontic Journal*, Vol. 35, No. 11, pp. 905-909.

Hutmacher, D.W. (2000). Scaffolds in tissue engineering bone and cartilage. *Biomaterials*, Vol. 21, No. 24, pp. 2529–2543.

ISO: No. 10993-1 (2009). Biological evaluation of medical devices

Jain, R.A. (2000). The manufacturing techniques of various drug loaded biodegradable poly(lactide-co-glycolide) (PLGA) devices. *Biomaterials*, Vol. 21, No. 23, pp. 2475–2490.

Johnson, H.J., Northup, S.J., Seagraves, P.A., Garvin P.J., & Wallin, R.F. (1983). Biocompatibility test procedures for materials evaluation in vitro. I. Comparative test system sensitivity. *Journal of Biomedical Materials Research*, Vol. 17, No. 4 (July 1983), pp. 571–586.

Johnson, H.J., Northup, S.J., Seagraves, P.A., Atallah, M., Garvin, P.J., Lin, L. et al. (1985). Biocompatibility test procedures for materials evaluation in vitro. II. Objective methods of toxicity assessment. *Journal of Biomedical Materials Research*, Vol. 19, No. 5, pp. 489–508, doi: 10.1002/jbm.820190503.

Kirkpatrick, C.J., Peters, K., Hermanns, M.I., Bittinger, F., Krump-Konvalinkova, V., Fuchs, S. et al.. (2005). In vitro methodologies to evaluate biocompatibility: status quo and perspective. *ITBM*, Vol. 26, No. 3, pp. 192–199

Kleinsasser, N.H., Wallner, B.C., Harréus, U.A., Kleinjung, T., Folwaczny, M., Hickel, R. et al. (2004). Genotoxicity and cytotoxicity of dental materials in human lymphocytes as assessed by the single cell microgel electrophoresis (comet) assay. *Journal of Dentistry*, Vol. 32, No. 3, pp. 229–234.

Kostoryz, E.L., Eick, J., Chappelow, C.C., Glaros, A.G., Wetmore, L., & Yourtee, D.M. (2003). In vitro effect of light-cure dental adhesive on Il-6 release from LPS-stimulated and unstimulated macrophages. *Journal of Biomedical Materials Research Part A*, Vol. 65, No. 1, pp. 89–94.

Lahiji, A., Sohrabi, A., Hungerford, D.S., & Frondoza, C.G. (2000). Chitosan supports the expression of extracellular matrix proteins in human osteoblasts and chondrocytes. *Journal of Biomedical Materials Research*, Vol. 51, No. 4, pp. 586–595.

Lefebvre, C.A., Schuster, G., Rueggeberg, F.A., Tamareselvy, K., & Knoernschild, K.L. (1996). Responses of oral epithelial cells to dental resin components. *Journal of Biomaterials Science Polymer Edition*, Vol. 7, No. 11, pp. 965–976.

Lemmons, J. & Natiella, J. (1986) Biomaterials, biocompatibility and peri-implant considerations. *Dental Clinics of North America*, Vol. 30, No. 1 (January 1986), pp. 3–23.

Li, X., Kong, X., Zhang, Z., Nan, K., Li, L., Wang, X. et al. Cytotoxicity and biocompatibility evaluation of N,O-carboxymethyl chitosan/oxidized alginate hydrogel for drug delivery application. *International Journal of Biological Macromolecules*, 2012, doi:10.1016/j.ijbiomac.2012.03.008.

Libonati, A., Marzo, G., Klinger, F.G., Farini, D., Gallusi, G., Tecco, S. et al. (2011). Embryotoxicity assays for leached components from dental restorative materials. *Reproductive Biology and Endocrinology*, 2011, pp. 9:136. doi:10.1186/1477-7827-9-136.

Lin, N.J., Bailey, L.O., Becker, M.L., Washburn, N.R., & Henderson, L.A. (2007). Macrophage response to methacrylate conversion using a gradient approach. *Acta Biomaterialia*, Vol. Vol. 104, No. 2, pp. e45–e52.

Mauth, C., Huwig, A., Graf-Hausner, U., & Roulet, J.-F. (2007). Restorative applications for dental pulp therapy. In: *Topics in Tissue Engineering*, Ashammakhi, N., Reis, R., & Chiellini, E. (Eds.), Vol. 3, Chapter 3.

Nayyar, S., Teari, S., & Arora, B. (2007). Comparison of human pulp response to total-etch and self-etch bonding agents. *Oral Surgery, Oral Medicine, Oral Pathology, Oral Radiology, and Endodontology*, Vol. 104, No. 2, pp. e45–e52.

Nocca, G., De Palma, F., Minucci, A., De Sole, P., Martorana, G.E., Callà, C. et al. (2007). Alterations of energy metabolism and glutathione levels of HL-60 cells induced by methacrylates present in composite resins. *Journal of Dentistry*, Vol. 35, No. 3, pp. 187–194.

Ouchi, T. & Ohya, Y. (2004). Design of lactide copolymers as biomaterials. *Journal of Polymer Science. Part A Polymer Chemistry*, Vol. 42, No. 3, (February 2004). pp. 453–462, ISSN 1099-0518.

Paranjpe, A., Bordador, L.C., Wang, M.Y., Hume, W.R., Jewett, A. (2005). Resin monomer 2-hydroxyethyl methacrylate (HEMA) is a potent inducer of apoptotic cell death in human and mouse cells. *Journal of Dental Research*, Vol. 84, No. 2, pp. 172–177.

Parslow, T.G., Stites, D., Terr, A.I., & Imboden, I.B. (2001). *Medical Immunology*. New York: Lange Medical Books/McGraw-Hill Medical Publishing Division.

Porto, I.C., Andrade, A.K., Guênes, G.M., Ribeiro, A.I., Braz, R., Castro, C.M. (2009) In vitro potencial cytotoxicity of an adhesive system to alveolar macrophages. *Brazilian Dental Journal,* Vol. 20, No. 3, pp. 195–200.

Porto, I.C., Oliveira D.C., Raele, R.A., Ribas, K.H., Montes, M.A., & De Castro, C.M. (2011). Cytotoxicity of current adhesive systems: In vitro testing on cell cultures of primary murine macrophages. *Dental Materials,* Vol. 27, No. 3, (March 2011), pp. 221–228.

Poskus, L.T., Lima, R.S.M.S., Lima, I.R., Guimarães, J.G.A., Silva, E.M., & Granjeiro, J.M.G. (2009). Cytotoxicity of current adhesive systems: *in vitro* testing on cell culture of L929 and balb/c 3T3 fibroblasts. *Revista Odonto Ciências,* Vol. 24, No. 2, pp. 129–134.

Prica, D., Galič N, Želježič D, Prica A. (2006). Genotoxicity evaluation of five different dentin bonding agents by chromosomal aberration analysis. *Journal of Oral Rehabilitation,* Vol. 33, No. 6, pp. 462–471.

Puskas, J.E. & Chen, Y. (2004). Biomedical application of commercial polymers and novel polyisobutylene-based thermoplastic elastomers for soft tissue replacement. *Biomacromolecules,* Vol .5, No. 4, (July–August 2004), pp. 1141–1154, ISSN 1525-7797.

Quansah, J.K. Synthetic polymers for biocompatible biomaterials. *Materials literature seminar.* Sep. 23 2004. Available in: www.chemistry.illinois.edu/research/materials/seminar_abstract_/2004_2005/Quansah. Abstract.LitSeminar.pdf. April 13 2012.

Rao, S.B. & Sharma, C.P. (1997). Use of chitosan as a biomaterial: studies on its safety and hemostatic potential. *Journal Biomedical Materials Research,* Vol. 34, No. 1, pp. 21–28.

Richardson, T.P., Murphy, W.L., & Mooney, D.J. (2001). Polymeric delivery of proteins and plasmid DNA for tissue engineering and gene therapy. *Critical Reviews in Eukaryotic Gene Expression,* Vol. 11, Nos. 1–3, pp. 47–58.

Roll EB, Dall J, Runningen G, Morisbak E. (2004). In vitro cell death induced by irradiation and chemicals relevant for dental applications; dose–response and potentiation effects. *European Journal of Oral Sciences.* Vol. 112, No. 3, pp. 273–279.

Sangsanog,P.,Waleetorncheepsawat, S., Suwantong, O., Wutticharoenmongkol, P., Weeranantanapan, O., Chuenjitbuntaworn, B., et al. (2007). In Vitro Biocompatibility of Schwann Cells on Surfaces of Biocompatible Polymeric Electrospun Fibrous and Solution-Cast Film Scaffolds. *Biomacromolecules,* Vol. 8, pp. No. 5, 1587-1594.

Samuelsen, J.T., Holme, J.A., Becher, R., Karlsson, S., Morisbak, E., & Dahl, J.E. (2008). HEMA reduces cell proliferation and induces apoptosis in vitro. *Dentals Materials,* Vol. 24, No. 1, pp. 134–140.

Schmalz, G. (2002). Material science: biological aspects. Journal of Dental Research, Vol. 81, No. 10, pp. 660–663.

Silva, G.A., Marques, A.P., Gomes, M.E., Coutinho, O.P., & Reis, R.L. (2004). Cytotoxicity screening of biodegradable polymeric systems. In *Biodegradable Systems in Tissue Engineering and Regenerative Medicine,* Reis, R.L. & San Róman, J. (Eds.), Chapter 19. Boca Raton, FL: CRC Press.

Theunissen, P.T. & Piersma, A.H. (2012). Innovative approaches in the embryonic stem cell test (EST). Frontiers in Bioscience, Vol. 17, (January 2012), pp. 1965–1975. DOI:http://dx.doi.org/10.2741/4032.

Thomson, R.C., Wake, M.C., Yaszemki, M.J., & Mikos, A.G. (1995), Biodegradable polymer scaffolds to engineer trabecular bone. *Journal of Biomaterials Science Polymer Edition,* Vol. 7, No. 1, pp. 23–38.

Thomson, R.C., Mikos, A.G., Beahm, E., Lemon, J.C., Satterfield, W.C., Aufdemorte, T.B. et al. (1999). Guided tissue fabrication from periosteum using preformed biodegradable polymer scaffolds. *Biomaterials,* Vol. 20, No. 21, pp. 2007–2018.

Tomihata, K. & Ikada, Y. (1997). In vitro and in vivo degradation of films of chitin and its deacetylated derivatives. *Biomaterials,* Vol. 18, No. 7, pp. 567–575.

Van Dartel, D.A. & Piersma, A.H. (2011). The embryonic stem cell test combined with toxicogenomics as an alternative testing model for the assessment of developmental toxicity. *Reproductive Toxicology,* Vol. 32, No. 2, (September 2011), pp. 235–244.

VandeVord, P.J., Matthew, H.W.T., DeSilva, S.P., Mayton, L., Wu, B., & Wooley, P.H. (2002). Evaluation of the biocompatibility of a chitosan scaffold in mice. *Journal of Biomedical Materials Research,* Vol. 59, No. 3, pp. 585–590.

Villanova, J.C.O. & Oréfice, R.L. (2010). Aplicações farmacêuticas de polímeros. *Polímeros: Ciência e Tecnologia,* Vol. 20, No. 1, pp. 51–64.

Wallin, R.F. A pratical guide to ISO 10993-5: cytotoxicity. *Medical device & diagnostic industry magazine.* (1998). Available in: www.mddionline.com/article/practical-guide-iso-10993-5-cytotoxicity. March 10 2012.

Wataha, J.C. (2012). Predicting clinical biological responses to dental materials. *Dental Materials*. Vol. 28, No. 1, pp. 23-40.

Williams, D.F. (2008). On the mechanisms of biocompatibility. *Biomaterials.* Vol. 29, No. 20, pp. 2941-2953.

Wintermantel, E. & Ha, S.W. (2002). Medizintechnik mit biokompatiblen Werkstoffen und Verfahren, Vol. 3, pp. 192–232. Berlin: Springer.

Young, C.S., Terada, S., Vacanti, J.P., Honda, M., Bartlett, J.D., Yelick, P.C. (2002). Tissue engineering of complex tooth structures on biodegradable polymer scaffolds. *Journal of Dental Research,* Vol. 81, No. 10, pp. 695–700.

Yuan, Y., Zhang, P.Y., Yang, Y.M., Wang, X., Gu, X.S. (2004). The interaction of Schwann cells with chitosan membranes and fibers in vitro. *Biomaterials,* Vol. 25, No. 18, pp. 4273–4278.

Chapter 3

Analysis of the Miscibility of Polymer Blends Through Molecular Dynamics Simulations

Inger Martínez de Arenaza, Emiliano Meaurio and Jose-Ramon Sarasua

Additional information is available at the end of the chapter

1. Introduction

Molecular Dynamics Simulations are important tools in the prediction of the properties of polymer materials and have therefore become invaluable aids in the design of new materials suited to particular applications. The combination of Molecular Simulations and the Flory-Huggins theory allows the study of the compatibility of polymer blends. MD Simulations constitute the simplest type of simulation for complex systems containing different types of interactions between bonded and non bonded atoms. These interactions are summarized in the forcefield files. The systems may be very sensitive to some details of the model; particularly the choice of the forcefield. Among different properties, the interaction parameters can be calculated from the numerical trajectories of polymer blends, which constitute statistical samples.

Polylactides (PLAs) have been studied extensively for a number of applications due to their potential utility in a number of growing technologies. PLAs are biocompatible and biodegradable aliphatic polyesters. They have been used in a variety of applications such as biomedical materials for tissue regeneration, orthopedics, drug delivery matrices, sutures and scaffolds (Gupta A.et al., 2007). In addition, PLAs synthesized from renewable resources are not harmful for the environment so they are desirable materials for packaging applications to reduce the impact of plastic packaging residues on the environment (Tsuji H. et al., 2005). PLAs include two optically active and crystallizable isomers, namely poly(L-lactide) (PLLA) and poly(D-lactide) (PDLA), and the optically inactive and amorphous mixture of isomers termed poly(DL-lactide), (PDLA). Semicrystalline PLA is a brittle material with good mechanical properties such as high tensile strength and modulus. The compatibility of PLAs in polymer mixtures is of great scientific and technological interest (Uras R., et al. 2010). However, only a few miscible counterparts have been reported for PLAs: poly(methylmethacrylate) (PMMA) (Eguiburu J.L. et al, 1998; Zhang G. et al., 2003) poly(methylacrylate) (PMA) (Eguiburu J.L. et al, 1998), poly(vinyl acetate) (PVAc), (Gajria A. M. et al., 1997; Yoon J.S. et al., 1990; Ogata N. et al, 1997, Park J.W., 2003) and poly(vinyl phenol) (PVPh) (Meaurio E. et al, 2005a, 2005b). In addition, the investigation of the phase behaviour of PLA/PVPh blends was controversial, since partial miscibility was initially reported (Zhang et al, 1998a, 1998b), but complete miscibility was later demonstrated for the PLA/PVPh blends (Meaurio E. et al, 2005a, 2005b). In addition, PLAs are also miscible with styrene-vinyl phenol copolymers (STVPh) (Zuza et al, 2008) within a wide range of compositions. In the PLA/STVPh blends, immiscibility was observed with the styrene rich copolymers and with pure Poly(styrene) (PS), (Zuza et al, 2008). The latter result was indeed expected considering the difference of solubility parameters (δ = 9.5 and 10.1 $(cal/cm^3)^{1/2}$ for PS and PLA respectively; Coleman M.M. et al, 1990), and considering also the lack of specific interactions in the PLA/PS system.

The present work reports the MD Simulation results obtained with the Discover and Amorphous Cell modules of the Materials Studio software suite for blends of Poly(DL-Lactide) (PDLLA) with styrene-co-vinyl phenol copolymers (STVPh). The introduction of the energetic results obtained using Molecular Dynamics Simulation techniques in a modified Flory-Hugging Theory makes possible foretelling the phase behaviour of PDLLA/STVPh blends of different composition.

The immiscibility of the PDLLA/PS blends was confirmed by our research group in a recent work. At the same time PDLLA is completely miscible with PVPh due to the presence of attractive hydrogen bonding interactions between the hydroxyl groups (-OH) of PVPh and the carbonyls groups (-C=O) of PDLLA (Meaurio E. et al, 2005a, 2005b). The present work deals with the effect of the insertion of hydroxyl groups in the styrene repeat units (rendering vinyl phenol (VPh) repeat units) of Poly(styrene) (PS) on the miscibility of the resulting styrene-vinyl phenol copolymers (STVPh) with PDLLA. The results provide a detailed description of the factors governing the miscibility of the system and allow assessing the number of –OH groups necessary in the copolymer to attain miscibility with PDLLA as a consequence of the establishment of intermolecular –OH···O=C hydrogen bonds. Our research work in this system covers both the experimental and the modelling approaches, allowing the comparison of the results obtained from both methods. Certainly, MD Simulation results are in good agreement with the experimental measurements.

2. Computational details

MD Simulations were performed with the commercial software suite Materials Studio (v 4.1) supplied by Accelrys (San Diego, CA, USA). The Discover (molecular mechanics and dynamics simulation) and Amorphous Cell program modules were used to perform the computational work. Discover module is a molecular simulation program that provides a broad range of simulation methods, giving the ability to study molecular systems and a variety of material types (http://accelrys.com). It also enables to perform structural characterization and property prediction for the modelled systems. Amorphous Cell module is a suite of computational tools that allows to construct representative models of complex amorphous systems and to predict key properties. Amorphous Cell module employs the combined use of the arc algorithm developed by Theodorou and Suter and the scanning method of Meirovitch. This algorithm is suitable for longer chains and builds molecules in two separate steps, involving the creation of an initial guess structure, followed by relaxation of the structure to a state of minimum potential energy (Theodorou et al, 1985; Meirovitch, 1983).

Choosing the appropriate force field is one of the most important factors in obtaining real and reproducible results. For this study, the COMPASS force field has been selected because it enables accurate prediction of structural, conformational, vibrational, and thermo physical properties for a broad range of molecules in the isolated and condensed phases, and under a wide range of conditions of temperature and pressure.
(http://www.scripps.edu/rc/softwaredocs/msi/cerius45/compass/COMPASSTOC.doc.html).

COMPASS (Condensed-phase Optimized Molecular Potentials for Atomistic Simulation Studies) is based on PCFF (Polymer Consistent Force Field) (Rigby D. et al., 1999). COMPASS is the first ab initio force field used for modelling interatomic interactions. The potential energy of a system can be expressed as a sum of valence (or bond), crossterm, and non-bond interaction energies given as:

$$E_{TOTAL} = E_{valence} + E_{crossterm} + E_{nonbond} \quad (1)$$

The energy of valence interactions ($E_{valence}$), corresponds to energies associated with bond stretching (E_{bond}), valence angle bending (E_{angle}), dihedral angle torsion ($E_{torsion}$), and inversion ($E_{inversion}$) or also called out-of plane interaction energy (E_{oop}). Modern forcefields like COMPASS include a new term named Urey–Bradley (E_{UB}), that considers interactions between atom pairs in 1–3 configurations (i.e., atoms bound to a common atom):

$$E_{valence} = E_{bond} + E_{angle} + E_{torsion} + E_{oop} + E_{UB} \quad (2)$$

Cross terms included in $E_{crossterm}$ increase the accuracy of the force-field by introducing correction factors to the valence energy to account for the interdependence existing between different valence terms. For example, the term named $E_{bond\text{-}bond}$ considers stretch-stretch interactions between two adjacent bonds. Similarly, the COMPASS force-field includes stretch–bend, bend–bend, stretch–torsion, bend–torsion and bend–bend–torsion terms.

The non-bond interaction term ($E_{non\text{-}bond}$), represents the secondary interactions between non-bonded atoms. It is defined as the sum of the van der Waals energy (E_{vdW}), the Coulomb electrostatic energy ($E_{Coulomb}$), and the hydrogen bond energy (E_{Hbond}), given as:

$$E_{non\text{-}bond} = E_{vdW} + E_{Coulomb} + E_{Hbond} \quad (3)$$

In COMPASS, E_{vdW} is described by the Lennard-Jones 6-12 potential and the electrostatic energy is obtained from the partial charges of atoms in the system as estimated by the charge-equilibration method (Rappe A.K. and Goddard W.A., 1991). Electrostatic interactions were calculated by Ewald summation method, highly accurate in the calculation of long range interactions (Jawalkar S.S. et al, 2008). An accuracy of 0.0001kcal/mol with an update width of 5 Ǻ was applied to evaluate electrostatic interactions (Meirovitch, 1983). Moreover, cell multi-pole method was used to calculate the non-bonded interactions due to its efficient capacity to simulate big systems.

2.1. Determination of parameters for the prediction of miscibility

The Hildebrand solubility parameter (δ), describes the attractive strength between the molecules of the material, exerting a huge influence on blend miscibility. The last few hundred picoseconds of the trajectory file were used to calculate the solubility parameter, described as the root of the cohesive energy density (CED), equation (4):

$$\delta = \sqrt{CED} \tag{4}$$

The molecular weight of the polymer is an important factor for the MD Simulation. In many cases they can not be performed using voluminous systems due to data storage space limitations of the computers. Therefore, calculating the minimum chain length representing the high molecular weight polymer is crucial. This minimum length can be determined from the solubility parameter values calculated for chains of the pure polymers with different lengths. The molecular size at which the solubility parameter becomes nearly independent from the length and reaches a nearly constant value can be assumed to represent the real polymer chain (Zhang M. at al, 2003).

The cohesive energy densities (CED) are known to vary considerably with the molecular weight of the polymer. If all intramolecular forces are eliminated, CED is defined as the increase in energy per mole of a material (Gestoso P. and Brisson J., 2001), given as:

$$CED = \left({}^{E_{coh}}\!\big/_{V} \right) \tag{5}$$

The computational results of CED allow the calculation of the energy of mixing, ΔE_{mix}, according to equation 6:

$$\Delta E_{mix} = \Phi_A \left(\frac{E_{coh}}{V} \right)_A + \Phi_B \left(\frac{E_{coh}}{V} \right)_B - \left(\frac{E_{coh}}{V} \right)_{mix} \tag{6}$$

Where the terms in parenthesis represent the cohesive energies of the pure polymers (A and B) and the blend (mix), and Φ_A and Φ_B represent the volume fractions of the polymers in the blend.

Finally, the Flory-Huggins interaction parameter, χ, can be calculated from ΔE_{mix} according to (Case F.H. et al, 1994):

$$\chi = \left(\frac{\Delta E_{mix}}{RT\phi_A \phi_B} \right) V_m \tag{7}$$

where V_m is the molar volume of the repeat unit chosen as reference (PLA). R is the molar gas constant and T is the temperature of the simulation in Kelvin. In general, a positive value of the Flory-Huggins interaction parameter is considered indicative of the immiscibility of high molecular weight polymer blends, but actually the critical value of χ obeys equation 8:

$$\left(\chi_{AB}\right)_{critical} = \frac{1}{2}\left(\frac{1}{\sqrt{m_A}} + \frac{1}{\sqrt{m_B}}\right)^2 \tag{8}$$

where m_A and m_B represent the degree of polymerization of the pure polymers. If the interaction parameter of the blend is smaller than $\chi_{critical}$ the system is miscible in the whole composition range. If χ is slightly larger than the critical value, the blend exhibits partial miscibility: two phases coexist containing both components. For larger values of χ, the components are completely immiscible. By comparison of the values of χ calculated by atomistic simulation with the critical value (equation 8) the miscibility behaviour of the system can be predicted (Jawalkar S.S. et al., 2005).

3. Results and discussion

3.1. Construction of the models

Initial models were generated using the Flory´s rotational isomeric state theory (RIS), (Flory P.J., 1989). This method can be used as a basis for generating the conformations of unperturbed chains. The RIS approach is a powerful tool to predict conformational properties of polymer chains when statistical weight matrices are known (Blomqvist J., et al, 2001). The major advantage of this method is that it allows the quick evaluation of the miscibility while simultaneously includes the specific interactions between dissimilar molecules in a mixture (Zeng F.L. et al, 2009). Minimization of the system was carried out with 5000 steps using the conjugate gradient method (CGM). CGM method improves the line search direction by storing information from the previous iteration, utilizing the Polak-Ribiere algorithm. The minimized structures were examined to attain a suitable distribution of the chains in the blend and a correct number of contacts between the molecules. Furthermore, to avoid excessive overlaps between the chains, modified conditional probabilities are used to account for the nonbonded interactions between the atoms to be placed and the rest of the system.

Three-dimensional cubic unit cells with periodic boundary conditions were constructed using the Amorphous Cell module. Then, cubic cells were minimized and refined by molecular dynamics calculations. MD simulation runs were equilibrated for 200ps at 298K with time step of 1fs in the NVT ensemble (constant number of particles, N, volume, V, and temperature,T). This protocol was followed for all simulated structures.

3.2. Analysis of amorphous models

As indicated before, the miscibility of poly(lactide)/poly(styrene) (PLA/PS) and poly(lactide)/poly(vinylphenol) (PLA/PVPh) blends was investigated by our research group in a recent paper using Molecular Modelling (Martínez de Arenaza et al., 2010). First, the solubility parameters of the pure polymers (PLA, PS and PVPh) were calculated. Polymer chains of different length were generated using the Amorphous cell module with the aim of determining the minimum representative length for the polymer chains. The minimum length at which the solubility parameter adopts a constant value can be considered the representative length for modelization purposes. As can be seen in Figure 1, the solubility parameter decreases as chain length increases. In case of PDLLA, the solubility parameter becomes nearly constant above 20 repeat units (M_w=1440g/mol). However, only 10 repeat units of PS (M_w=1040g/mol) and 10 units of PVPh (M_w=1200g/mol) are necessary.

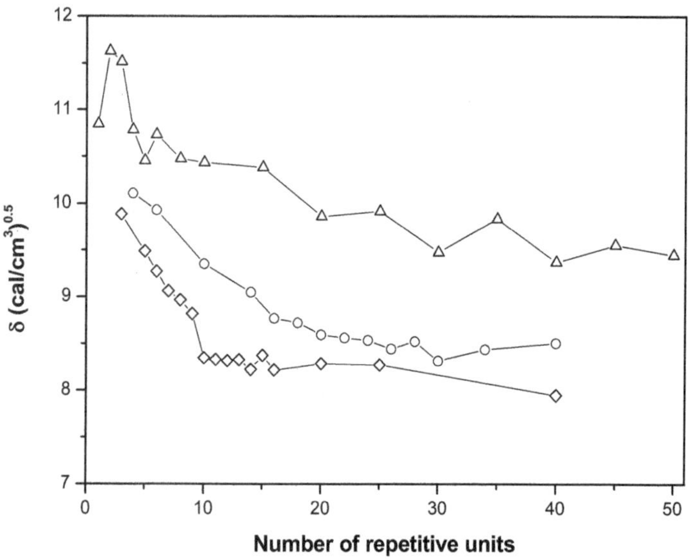

Figure 1. Calculated solubility parameter versus number of repeat units for (Δ) PVPh, (◊) PS and (○) PDLLA.

Table 1 summarizes naming conventions, molecular weights (Mw), and solubility parameter values obtained from the MD Simulation (δ_{MD}) and from experimental measurements (δ_{exp}), corresponding to the pure polymers and to the copolymers. In the STVPh-10 and STVPh-20 copolymers 10 and 20% of ST units have been replaced by VPh units respectively. As can be observed, the calculated solubility parameters for al the polymers investigated in this paper show negative deviations, about 10-20%, relative to the experimental values. This behaviour has been reported by other authors (Jawalkar S.S. and Aminabhavi T.M., 2006; Gestoso P., 2001; Mu D. et al., 2008).

The MD Simulations performed using polymer chains of representative length provide us the cohesive energy density (equation 5), necessary to calculate the energy of mixing (equation 6), from which the Flory-Huggins interaction parameter can be estimated (equation 7). In case of the PDLLA/PS system, calculated χ values were above the $\chi_{critical}$ line (see Figure 2), indicating immiscibility for these blends. On the contrary, negative interaction parameter values were calculated in the whole range of compositions for the PDLLA/PVPh system, indicating the miscibility of these blends (Figure 2). Figure 3 shows the snapshots corresponding to amorphous unit cells of 1:1 (mol:mol) compositions for PDLLA/PS and PDLLA/PVPh blends. In the snapshots carbon atoms are coloured grey, hydrogen atoms white and oxygen atoms are red coloured. The presence of -OH groups in PVPh allows the formation of hydrogen bonds with the carbonyl groups (-C=O) of the PDLLA as can be observed in Figure 4. These new strong interactions are the responsible for the compatibility of the PDLLA/PVPh blends, favouring the miscibility of the system. Numerous authors have studied the effect of hydrogen bonding in PVPh (Gestoso, 2003a and 2003b).

System	Acronyms	Repeat units	Mw (g/mol)	δ_{MD} $(cal/cm^3)^{0.5}$	δ_{exp} $(cal/cm^3)^{0.5}$
Poly(DL-lactide)	PDLLA	20	1440	8.6	10.6
Poly(styrene)	PS	10	1040	8.4	9.4
Poly(vinylphenol)	PVPh	10	1200	10.4	12.0
Poly(styrene-co-vinyl phenol) (10 mol% VPh)	STVPh-10	10	1056	8.8	-
Poly(styrene-co-vinyl phenol) (20 mol% VPh)	STVPh-20	10	1072	8.21	-

Table 1. Calculated and experimental solubility parameters for the polymers studied in this paper.

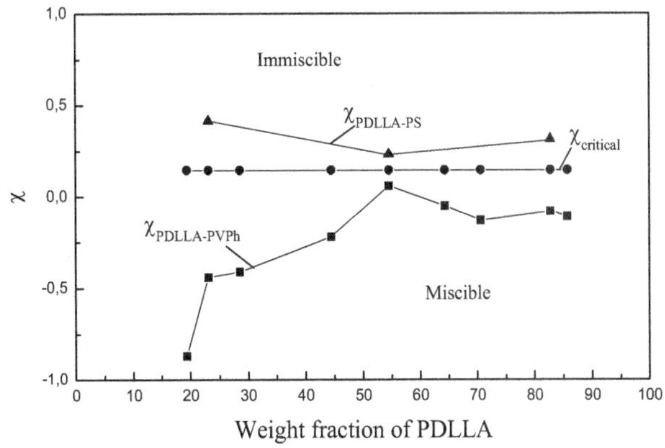

Figure 2. Difference between Flory-Huggins interaction parameters of PDLLA versus weight fraction of PDLLA

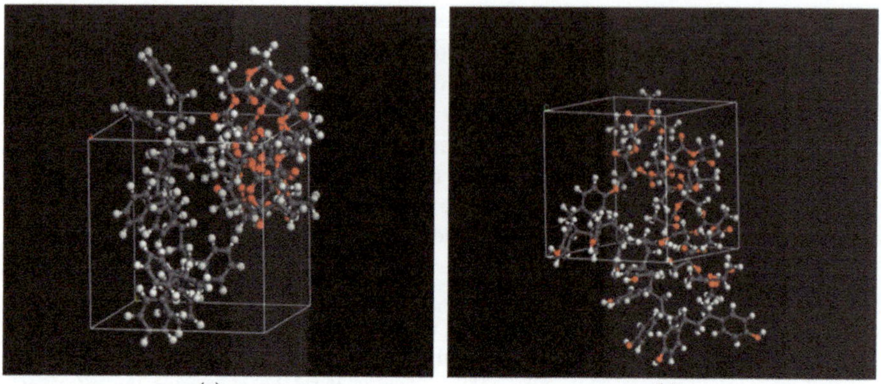

(a) (b)

Figure 3. Snapshots of amorphous unit cells of 1:1 (mol:mol) composition for (a) PDLLA/PS and (b) PDLLA/PVPh blends.

Figure 4. Snapshot showing the formation of hydrogen bonds between the -OH groups of PDLLA and the -C=O groups of PVPh (O-H···O=C).

In this work the analysis of the miscibility between PDLLA and PS or PVPh is extended to blends of Poly(DL-lactide) with Poly(styrene-co-vinyl phenol) copolymers (STVPh) in order to establish the number of vinylphenol units (VPh) necessary in the copolymer to achieve complete miscibility. Moreover, a comparison between the experimental and the modelling analyses is also provided here. The properties of the pure isolated polymers have been investigated from the modelization of single chains of the pure homopolymers (PDLLA, PS and PVPh) or copolymers (STVPh-10 and STVPh-20). Table 2 presents the experimental densities (Sarazin et al, 2003), molar volumes ($V_M = M_0/\rho$) and CED obtained from the MD simulations of the single-chain cells. The densities and molar volumes of the copolymers have been calculated assuming molar volume additivities. The model for the STVPh-10 copolymer consists on a single chain with 10 repeat units containing 9 styrene repeat units and 1 VPh repeat unit (10% styrene units replaced). Likewise, STVPh-20 symbolizes a chain of 10 units containing 8 styrene repeat units and 2 VPh repeat units (20% styrene units replaced; see Figure 5). Cell sizes depend on the system, being 26.51 Å for PDLLA, 12.12 Å for PS, 11.69 Å for PVPh, 14.82 Å for STVPh-10 and 14.77 Å for STVPh-20.

System Number	Polymer	Density (g/cm^3)	Molar Volume (cm^3/mol)	CED ($10^{-7} cal/m^3$)
1	PDLLA	1.247	57.7	7.388
2	PS	1.05	99.0	6.967
3	PVPh	1.25	96.0	10.89
4	STVPh-10	1.07	98.7	7.801
5	STVPh-20	1.09	98.4	5.425

Table 2. Experimental densities and calculated CED for the pure polymer and copolymers.

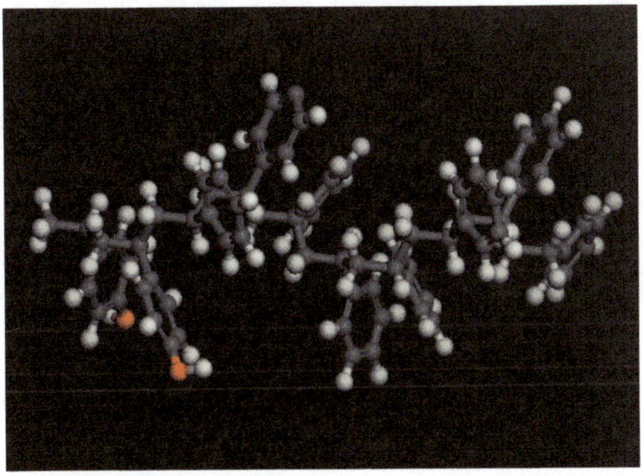

Figure 5. Snapshot of the STVPh-20 copolymer chain.

The next step in the miscibility investigation was the simulation of one 1:1 amorphous cell between PDLLA and two different STVPh copolymers: STVPh-10 and STVPh-20. The system was simulated using the specific parameters calculated previously from the individual polymers as the composition, molar volume and density. Table 3 summarizes the MD Simulation results for the two blends proposed. PDLLA/STVPh blends behave quite different when the percentage of ST units replaced by VPh units increases from 10 to 20%. As can be seen in table 3, according to MD simulations the PDLLA/STVPh blend obtained after replacing one styrene unit in PS by one VPh unit shows a positive interaction parameter of moderate magnitude, χ = 0.60. This value is above the critical data line at

0.1457, so the PDLLA/STVPh-10 blend is immiscible. On the contrary, the existence of two VPh units in the PS chain (20% VPh units) results in a negative value of χ, indicating the miscibility of the PDLLA/STVPh-20 blends. Figure 6 displays 1:1 (mol:mol) amorphous unit cells of PDLLA/STVPh-10 and PDLLA/STVPh-20 blends.

1:1	Cubic Cell	Density (g/cm³)	$\chi_{PDLLA/STVPh}$	COMPATIBILITY
	PDLLA/STVPh-10	1,158	0.60	IMMISCIBLE
	PDLLA/STVPh-20	1.148	-0.29	MISCIBLE

Table 3. Simulation details of two 1:1 (mol:mol) cubic unit cells of PDLLA/STVPh.

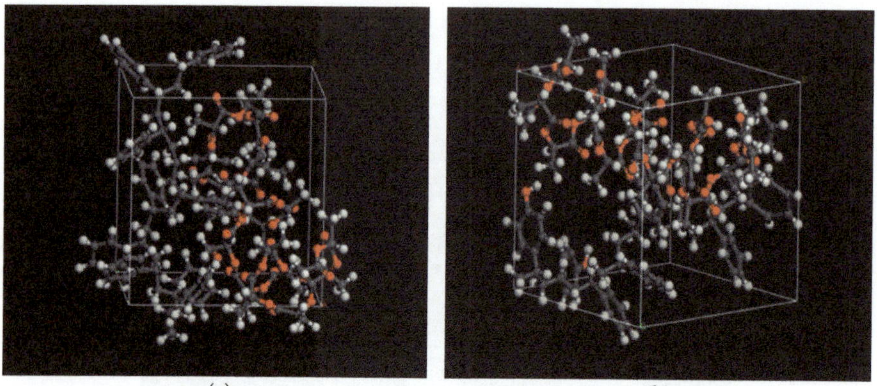

(a) (b)

Figure 6. Snapshots of 1:1 amorphous unit cell of (a) PDLLA/STVPh-10 blend (immiscible) and (b) PDLLA/STVPh-20 blend (miscible).

The accurate calculation of the minimum content of VPh units in the copolymer necessary to achieve miscibility requires the construction of additional models with intermediate compositions. With this aim, the polymer and copolymers chains lengths were doubled in order to increase the percentage of styrene units susceptible of substitution. Consequently, blends composed by PDLLA chains containing 40 repeat units and copolymer chains containing 20 repeat units were built to perform MD Simulations. Thus, the miscibility study was repeated with the new polymers and copolymers. As described before, the STVPh-5, STVPh-10, STVPh-15 and STVPh-20 acronyms represent the percentage of VPh units (5, 10, 15 and 20% respectively) inserted in the copolymer chain. Table 4 shows the CED obtained by MD Simulations for the new polymer chains, along with other selected properties.

Number	Systems	Repeat units	Mw (g/mol)	Density (g/cm^3)	CED (10^{-7} cal/m^3)
1	PDLLA	40	2880	1.247	6.766
2	STVPh-5	20	2400	1.06	6.932
3	STVPh-10	20	2720	1.07	6.970
4	STVPh-15	20	3040	1.08	7.544

Table 4. CED and other selected properties for the double-length polymers and copolymers.

Then, the mixtures were investigated showing again completely different behaviours with the degree of VPh in the copolymer chain. Table 5 displays the results of 1:1 cubic unit cells of blends. Recalling that, as discussed before (equation 8), $\chi_{critical}$ depends on the degrees of polymerization of the pure polymers (m_A and m_B), the $\chi_{critical}$ corresponding to the double-length chains decreases to 0.0728 (m_{PDLLA} and m_{PVPh-S} are 40 and 20 respectively). Molecular Modelling results for the PDLLA/STVPh-10 blends provide a χ value of 0.22 suggesting the immiscibility of the system according to the Flory-Huggins theory. However, χ for PDLLA/STVPh-15 blends was negative (-1.18), indicating the miscibility of these blends.

Cubic Unit Cell 1/1	Density (g/cm^3)	$\chi_{PDLLA/STVPh}$	COMPATIBILITY
PDLLA/STVPh-5	1.153	0.39	IMMISCIBLE
PDLLA/STVPh-10	1.158	0.22	IMMISCIBLE
PDLLA/STVPh-15	1.163	-1.18	MISCIBLE
PDLLA/STVPh-20	1.163	-1.33	MISCIBLE

Table 5. MD Simulation results for 1:1 cubit unit cell

Table 5 actually proves the miscibility of the 1:1 blends but does not tell about the rest of compositions. Therefore, we have carried out MD simulations for PDLLA/STVPh-15 blends of different composition, and Figure 7 shows the snapshots of two cubic amorphous unit cells for the PDLLA/STVPh-15 blends of 2:3 and 3:2 molar compositions. The range of compositions studied for this system actually includes the whole range of compositions, and the results are summarized in table 6.

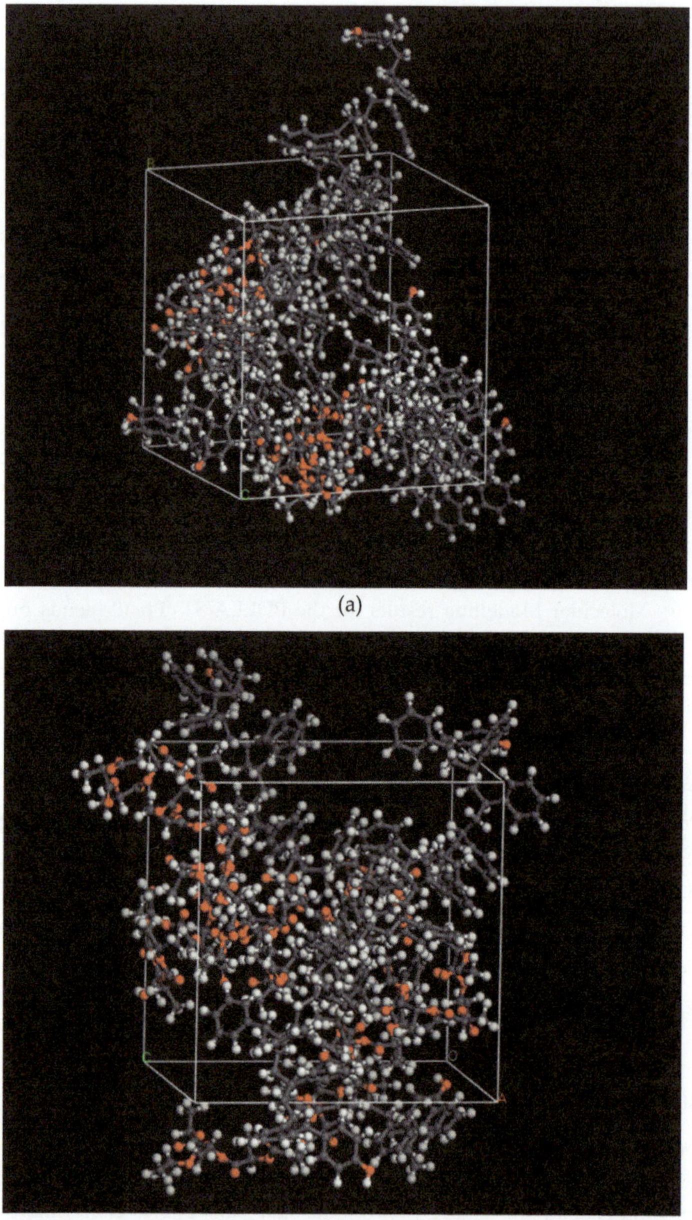

(a)

(b)

Figure 7. Snapshot of the cubic amorphous unit cells for PDLLA/STVPh-15 miscible blends of different molar composition: a) 2:3 and b) 3:2.

System number	PLA molar ratio (%)	Number of chains in the cubic cell PLA/STVPh-15	Density (g/cm³)	$\chi_{PDLLA/STVPh}$
1	0	0/1	1.07	-
2	14.88	1/5	1.099	-0.87
3	19.15	1/4	1.105	-1.06
4	24.00	1/3	1.114	-1.39
5	38.71	2/3	1.141	-1.25
6	44.83	1/1	1.158	-1.43
7	65.45	3/2	1.176	-0.73
8	65.29	2/1	1.888	-0.55
9	79.12	4/1	1.212	-1.62
10	82.56	5/1	1.217	-1.49
11	100	1/0	1.247	-

Table 6. Simulation details for the PDLLA/STVPh-15 blends.

As can be seen, in table 6, negative interaction parameters have been obtained across the whole range of compositions. Therefore, MD simulations predict complete miscibility for the PDLLA/STVPh-15 system. Figure 8 displays the dependence on composition for the interaction parameter. Considering the typical errors associated to these calculations, χ does not show a clear dependence with composition, it rather fluctuates about the value χ = -1.0.

These results can be compared with the experimental investigation of the PDLLA/STVPh system carried out recently by our research group (Zuza E. et al., 2008). High molecular weight STVPh copolymers of different composition were blended with high molecular weight PDLLA, and the analysis of the blends by Differential Scanning Calorimetry (DSC) showed that at least 16% of VPh units were necessary in the copolymer to achieve miscibility. Polystyrene and the STVPh copolymers containing less than 16% of VPh units were immiscible with PLA-s (Zuza E. et al., 2008). As can be seen, MD simulations show good agreement with the experimental results, and are a valuable tool in the prediction of the miscibility and phase behaviour of polymer blends.

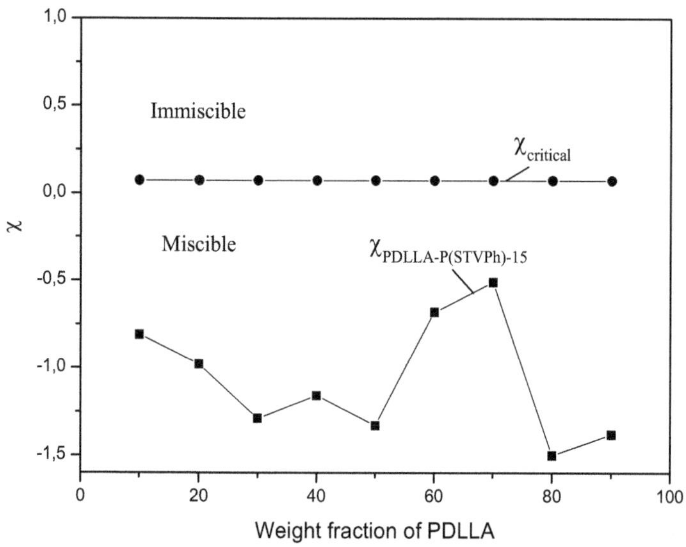

Figure 8. Dependence of the interaction parameter with composition for the PDLLA/STVPh-15 blends according to MD simulations.

4. Conclusions

The miscibility of Poly(DL-Lactide) (PDLLA) with styrene-co-vinyl phenol copolymers (STVPh) has been investigated using MD simulations and the results have been compared with the experimental information available. The MD simulations indicate the formation of hydrogen bonds between the –OH groups in the VPh repeat units and the C=O groups in PDLLA, in agreement with the results reported recently for these systems using FTIR spectroscopy (Meaurio E. et al, 2005a, 2005b; Zuza E. et al., 2008). According to the MD simulations, 10-15% VPh units must be introduced in the PS chain to achieve miscibility with PDLLA. This result is in very good agreement with the experimental results obtained in high molecular weight polymers, in which miscibility was only achieved for blends of PDLLA with STVPh copolymers containing at least 16 mol% of VPh repeat units. The good agreement between the modeling and the experimental results indicates that MD simulations are a valuable tool in the prediction of the miscibility and phase behaviour of polymer blends; particularly in cases where the polymers under consideration are not available.

Author details

Inger Martínez de Arenaza, Emiliano Meaurio and Jose-Ramon Sarasua
University of the Basque Country (UPV-EHU), Department of Mining-Metallurgy Engineering & Materials Science, School of Engineering, Bilbao, Spain

5. References

Ahmadi A., Freire J. J. (2008). Molecular dynamics simulation study of compatibility for the polyvinylmethylether/polystyrene mixture. *Molecular Simulation*, Vol. 34, No. 10-15, (December 2001), pp.1253-1258

Blomqvist J., Pielita L-O., Mannfors B. (2001). RIS Metropolis Monte Carlo studies of some aliphatic main chain and side group polyesters. *Polymer*, Vol. 42, No. 42, (January 2001), pp.109-116

Blümm E., Owen A.J. (1995). Miscibility, crystallization and melting of poly(3-hydroxybutyrate)/ poly(L-lactide) blends. *Polymer*, Vol. 36, No. 21, (1995), pp. 4077-4081

Case F.H., Honeycutt J.D. (1994). Will my Polymers Mix? - Applications of Modelling to Study Miscibility, Compatibility and Formulation. *Trends in Polymer Science*, Vol. 2, No. 8, (August 1994), pp. 259-266, (http://accelrys.com/resource-center/case-studies/archive/misc/misc.html; Case, F.)

Coleman M.M., Serman C.J., Bagwagar D.E., Painter P.C. (1990). A practical guide to polymer miscibility, *Polymer*, Vol. 31, No. 7, (July 1990), pp. 1187-1203

Eguiburu J.L., Iruin J.J., Fernandez-Berridi M.J., San Roman J. (1998). Blends of amorphous and crystalline polylactides with poly(methyl methacrylate) and poly(methylacrylate): A miscibility study. *Polymer*, Vol. 39, No. 26, (December 1998), pp. 6891-6897

Focarete M. L., Scandola M., Dobrzynski P., Kowalczuk M. (2002). Miscibility and Mechanical Properties of Blends of L-Lactide Copolymers with Atatic Poly(3-hydroxybutyrate). *Macromolecules*, Vol. 35, No. 22, (September 2005), pp. 8472-8477

Flory P.J. (1989). *Statistical Mechanics of Chain Molecules*, Hanser Gardner Pubns, ISBN 1569900191, Munich, Germany

Gestoso P., Brisson J. (2001). Effect of hydrogen bonds on the amorphous phase of a polymer as determined by atomistic molecular modelling. *Computational and Theoretical Polymer Science*, Vol. 11, No. 9, (September 2001), pp. 263-271

Gestoso P., Brisson J. (2001). Orientation of uniaxially stretched poly(vinyl phenol)/poly(vinyl methyl ether) blends. *Polymer*, Vol. 42, No. 20, (September 2001), pp. 8415-8424

Gestoso P., Brisson J. (2003). Investigation of the effect of chain rigidity on orientation of polymer blends: the case of poly(vinyl phenol). *Polymer*, Vol. 44, No. 25, (December 2003), pp. 7765-7776

Gestoso P., Brisson J. (2003). Towards the simulation of poly(vinyl phenol)/poly(vinyl methyl ether) blends by atomistic molecular modelling. *Polymer*, Vol. 44, No. 8, (April 2003), pp. 2321-2329

Gupta A.P., Kumar V. (2007). New emerging trends in synthetic biodegradable polymers-Polylactide: A critique, *European Polymer Journal*, Vol. 43, No. 10, (October 2007), pp. 4053-4074

Jawalkar S.S., Adoor S.G., Sairam M., Nadagouda M.N., Aminabhavi T.M. (2005). Molecular Modelling on the Binary Blend Compatibility of Poly(vinylalcohol) and Poly(methyl methacrylate): An Atomistic Simulation and Thermodynamic Approach. *Journal Physics Chemistry B.*, Vol. 109, No. 32, (16 June 2005), pp. 15611-15620

Jawalkar S.S., Aminabhavi T.M. (2006). Molecular modelling simulations and thermodynamic approaches to investigate compatibility/incompatibility of poly(L-lactide) and poly(vinyl alcohol) blends. *Polymer*, Vol. 47, No. 23, (September 2006), pp.8061-8071

Jawalkar S.S., Nataraj S.K., Raghu A.V., Aminabhavi T.M. (2008). Molecular Dynamics simulations on the blends of poly(vinyl pyrrolidone) and Poly(bisphenols-A-ether sulfone). *Journal of Applied Polymer Science*, Vol. 108, No. 6, (June 2008), pp. 3572-3576

Martínez de Arenaza I., Meaurio E., Coto B., Sarasua J. R. (2010). Molecular dynamics modelling for the analysis and prediction of miscibility in polylactide/polivinilphenol blends. *Polymer*, Vol. 51, No. 19, (July 2010), pp. 4431-4438

Meaurio E., Zuza E., Sarasua J. R. (2005). Direct Measurement of the Enthalpy of Mixing in Miscible Blends of Poly(DL-lactide) with Poly(vinylphenol). *Macromolecules*, Vol. 38, No. 22, (November 2005), pp. 9221-9228

Meaurio E., Zuza E., Sarasua J. R. (2005). Miscibility and Specific Interactions in Blends of Poly(L-Lactide) with Poly(vinylphenol). *Macromolecules*, Vol. 38, No. 4, (January 2005), pp. 1207-1215

Meirovitch, H. (1983). Computer simulation of self-avoiding walks: testing the scanning method. *Journal of Chemical Physics*, Vol. 79, No. 502, pp. 502-508.

Mu D., Huang X.R., Lu Z.Y., Sun C.C. (2008). Computer simulation study on the compatibility of poly(ethyleneoxide)/poly(methyl methacrylate) blends. *Chemical Physics*, Vol. 348, No. 1-3, (June 2008), pp.122-129

Nakafuku C., Sakoda M. (1993). Melting and crystallization of poly(L-lactic acid) and poly(ethylene oxide) binary mixture. *Polymer Journal*, Vol. 25, No. 9, pp. 909-917

Nakafuku C. (1994). High pressure crystallization of poly(L-lactic acid) in a binary mixture with poly(ethylene oxide). *Polymer Journal*, Vol. 26, No. 6, pp. 680-687

Nakafuku C. (1996). Effects of molecular weight on the melting and crystallization of poly(L-lactic acid) in a mixture with poly(ethylene oxide). *Polymer Journal*, Vol. 28, No. 7, pp. 568-576

Nijenhuis A. J., Colstee E., Grijpma D. W., Pennings A. J.(1996). High molecular weight poly(L-lactide) and poly(ethylene oxide) blends: thermal characterization and physical properties. *Polymer*, Vol. 37, No. 26, (1996), pp. 5849-5857

Ogata N., Jimenez T. (1997). Structure and thermal/mechanical properties of poly(L-lactide)-clay blend. *Journal of Polymer Science Part B: Polymer Physics*, Vol. 35, No. 2, (30 January 1997), pp. 389-396

Park J. W., Im S. S. (2003). Miscibility and morphology in blends of poly(L-lactic acid) and poly(vinyl acetate-*co*-vinyl alcohol). *Polymer*, Vol. 44, No. 15, (July 2003), pp. 4341-4354

Rappe A.K., Goddard W.A. (1991). Generalized Mulliken–Pauling Electronegativities. *Journal Physics Chemistry*, Vol. 95, No. 8340, pp. 3358-3363

Rigby D., Sun H., Eichinger B.E. (1999). Computer simulations of poly(ethylene oxide): force field, PVT diagram and cyclization behaviour. *Polymer International*, Vol. 44, No. 3, (November 1997), pp.311-330

Sarazin P, Favis BD. (2003). Morphology Control in Co-continuous Poly(L-lactide)/Polystyrene Blends: A Route towards Highly Structured and Interconnected Porosity in Poly(L-lactide) Materials. *Biomacromolecules*, Vol. 4 No 6, (November-December 2003) , pp. 1669-1679

Theodorou, D.N., Suter, U.W. (1985). Detailed Molecular Structure of a Vinyl Polymer Glass. *Macromolecules*, Vol. 18, No. 7, (July 1985), pp. 1467-1478.

Tsuji H., Smith R., Ikada Y. (2000). Porous biodegradable polyesters. I. Preparation of porous poly(L-lactide) films by extraction of poly(ethylene oxide) from their blends. *Journal Applied of Polymer Science*, Vol. 75, No.5,(January 2000), pp. 629-637

Uras R., Lim L-T., Selke S. E.M., Tsuji H. (2010). Poly(Lactic Acid): Synthesis, structures, properties, processing and applications. (1st edition), John Wiley & Sons, Inc., Hoboken, ISBN 978-0-470-29366-9, New Jersey and simultaneously in Canada

Yoon J. S., Oh S. H., Kim M., Chin I. J., Kim Y. H. (1999). Thermal and mechanical properties of poly(L-lactic acid)-poly(ethylene-co-vinyl acetate) blends. *Polymer*, Vol. 40, No. 9, (April 1999), pp. 2303-2313

Zeng F-L., Sun Y., Zhou Y., Li Q-K. (2009). Molecular simulations of the miscibility in binary mixtures of PVDF and POSS compounds. *Modelling and Simulation in Materials Science and Engineering*, Vol. 17, No. 075002, pp.1-13

Zhang M., Choi P., Sundararaj U. (2003). Molecular dynamics and thermal analysis study of anomalous thermodynamic behaviour of poly(ether imide)/polycarbonate blends. *Polymer*, Vol. 44, No. 6, (March 2006), pp.1979-1986

Zhang G., Zhang J., Wang S., Shen D. (2003). Miscibility and phase structure of binary blends of polylactide and poly(methyl methacrylate). *Journal Polymer Science Part B: Polymer Physics*, Vol. 41, No. 1, (January 2003), pp. 23-30

Zhang, L.; Goh, S. H.; Lee, S. Y. (1998). Miscibility and crystallization behavior of poly(L-lactide)/poly(p-vinylphenol) blends. *Polymer*, Vol. 39 No. 20 (September 1998) pp. 4841-4847

Zhang, L.; Goh, S. H.; Lee, S. Y. (1998). Miscibility and phase behavior of poly(D,L-lactide)/poly(p-vinylphenol) blends. *Journal of Applied Polymer Science,* Vol. 70 No. 4 (October 1998) pp. 811-816

Zuza E., Lejardi A., Ugartemendia J.M., Monasterio N., Meaurio E., Sarasua J.R. (2008). Compatibilization through Specific Interactions and Dynamic fragility in Poly(D,L-lactide)/Polystyrene blends. *Macromolecular Chemistry and Physics,* Vol. 209, No. 23, (1 December), pp. 2423-2433

http://accelrys.com/products/datasheets/compass.pdf

http://www.scripps.edu/rc/softwaredocs/msi/cerius45/compass/COMPASSTOC.doc.html

Chapter 4

Plant Cell Wall Polymers: Function, Structure and Biological Activity of Their Derivatives

Marisol Ochoa-Villarreal, Emmanuel Aispuro-Hernández,
Irasema Vargas-Arispuro and Miguel Ángel Martínez-Téllez

Additional information is available at the end of the chapter

1. Introduction

Plant cell walls represent the most abundant renewable resource on this planet. They are rich in mixed complex and simple biopolymers, which has opened the door to the development of wide applications in different technologic fields. In this regard the polymerization processes that allow the synthesis of the cell wall and their components in living models are relevant, as well as the properties of the polymers and their derivatives. Therefore this chapter outlines the basis of polymerization with a biological approach in the plant cell wall, highlighting the biological effects of plant cell wall derivatives and their current applications.

Plant cell wall is a dynamic network highly organized which changes throughout the life of the cell. The new primary cell wall is born in the cell during cell division and rapidly increases in surface area during cell expansion. The middle lamella forms the interface between the primary walls of neighboring cells. Finally, at differentiation, many cells elaborate with the primary wall a secondary cell wall, building a complex structure uniquely suited to the function of the cell. The functions of the plant cell wall may be grouped by its contribution to the structural integrity supporting the cell membrane, sense extracellular information and mediate signaling processes [1]. The main components of the plant cell wall involve different polymers including polysaccharides, proteins, aromatic substances, and also water and ions. Particularly, the different biomechanical properties of the plant cell wall are mainly defined by the content of the polymers cellulose, hemicelulloses and pectins and their interactions [2].

The rapid progress on plant cell wall research has allowed the comprehension of the different structures, their biosynthesis and functions. Nevertheless, there is a new prominent and worth line of research, the biological activity of some molecules derived from the primary cell wall polysaccharides. These active molecules named "oligosaccharins" by Albersheim in the mid 70s, include the biologically active oligosaccharides that are produced by partial hydrolysis of polymers of the cell wall. The main biologically active components of the cell wall are the pectin-derived oligosaccharides, and the hemicellulosic-derived oligosaccharides. The biological responses of plants to oligosaccharins can be divided into two broad categories: as modulators of plant defense, and plant growth and development.

This information has permitted the use of oligosaccharins as an alternative to improve different aspects such as yield and fruit quality, and may reach a higher impact in the study of the resistance of vegetable crops.

2. Plant cell wall polymers

Plant cell wall is a complex matrix of polysaccharides that provides support and strength essential for plant cell survival. Properties conferred by the cell wall are crucial to the form and function of plants. The main functions of the cell wall comprise the confer of resistance, rigidity and protection to the cell against different biotic or abiotic stresses, but still allowing nutrients, gases and various intercellular signals to reach the plasma membrane. The wall provides enough rigidity to support the heavy weight of high trees as large as 100 m height, but also is flexible and elastic allowing growth during expansion and differentiation. During growth, cell turgor pressure provides high tensile stress to the wall, enabling its enlargement due to the accumulation of polymers during a combination of stress relaxation cycles. The primary cell wall surrounds and protects the inner cell; it lies down the middle lamella during growth and expansion [2]. The primary wall is thought to contribute to the wall structural integrity, cell adhesion, and signal transduction. In this chapter we focus on the primary cell walls because it has been noted that most of their derivatives exert a biological function.

Plant cell wall is a dynamic and highly specialized network formed by a heterogeneous mixture of cellulose, hemicelluloses and pectins, and in some extent proteins and phenolic compounds. Wall composition in vascular plants is approximately 30% cellulose, 30% hemicellulose and 35% of pectin, with certain 1-5% structural proteins on dry weight basis. Cellulose and hemicelluloses polymers bring rigidity to the wall and pectin provides fluidity throw the gelatinous polysaccharides matrix. Cellulose and hemicelluloses are embedded in the amorphous pectin polymers and stabilized by proteins and phenolic compounds. Hemicelluloses bind to the surface of cellulose network preventing direct contact among microfibrils, and pectin are linked to hemicelluloses forming a gel phase.

3. Components and function of the primary constituents of plant cell wall

3.1. Cellulose

Cellulose is the main cell wall polymer that brings support to the plant. Cellulose is a linear insoluble unbranched polymer of β-(1,4)-D-Glucose residues associated with other cellulose chains by hydrogen bonding and Van der Waals forces. Cellulose chains aggregate together to form microfibrils, which are highly crystalline and insoluble structures, each one about 3 nm in diameter, chemically stable and resistant to enzymatic attack. Cellulose microfibrils comprise the core of the plant cell wall; one third of the total mass of wall is cellulose. The variation of dry weight of cellulose in a dicot such as *Arabidopsis thaliana* ranges from 15% of leaf to 33% of stem walls. The walls of monocot grass species have approximately 6–10% cellulose in leaves and 20–40% in stems [3].

Microfibrils comprise two types of cellulose called cellulose Iα and Iβ. The Iα has a single-chain triclinic unit cell, whereas cellulose Iβ has two chain monoclinic unit cell. In both forms cellulose in parallel and the terminal glucose residues rotated 180° forming a flat ribbon in which cellobiose (two glucose molecules linked by a β-(1,4) bond) is the repeating unit [4]. Cellulose chains may align in parallel (Type I) or antiparallel (Type II) orientation to each other. Only the Type I conformation is known to naturally occur in plants; however, concentrated alkaline treatments may cause Type II cellulose to form during harsh extraction procedures. The cellulose chains may form the Type Iα or Type Iβ conformation depending on the extent of staggering of the chains in relation to each other. Probably the interaction of cellulose microfibrils with hemicelluloses may affect the ratio of Type Iα to Type Iβ cellulose [5]. The microfibrilar disposition allows the existence of micro spaces between the microfibrils that are fulfilled by matricial polysaccharides according to the age and tissue type.

3.2. Hemicelluloses

Hemicelluloses are low molecular weight polysaccharides associated in plant cell walls with lignin and cellulose. These heterogenous group of polysaccharides that have β-(1,4)-linked backbones with an equatorial configuration at C1 and C4 and hence the backbones have structural similarity [6]. Hemicelluloses in dicotyledonous plants comprise xyloglucans, xilans, mannans and glucomannans, while the β-(1,3;1,4)-glucans are restricted to Poales and a few other groups. In addition, arabinoxylans are the main hemicellulosic polisaccharides in graminaceous species such as wheat and barley, and in grasses [7].

3.2.1. Xyloglucan

Xyloglucan (XyG) is the most abundant hemicellulose in primary cell walls found in every land plant species that has been analyzed. XyG are branched with α-D-xylose linked to C-6 of the backbone. The most frequently xyloglucan structure in dicotyledonous flowering plants is the repeating heptamer integrated by four glucans residues with α-D-xylose substituents in three constitutive glucans of the backbone, followed by a single unsubstituted glucan residue (Figure 1). The presence of this repeating heptamer block is an indicator of the presence of XyG polysaccharides in dicots species [8]. Beside the XyG residues, it may contain β-D-galactose and in less proportion L-fucose-α-(1,2)-D-galactose; in all cases the galactose residues are acetylated. The fact that all the substituents of xyloglucans are conserved denotes a highly biosynthesis control. On the other hand, in graminaceous monocots, XyG consist of 1 or 2 adjacent α-(1-6)-linked xylose residues with approximately 3 unsubstituted β-(1-4)-linked glucose backbone [9]. Despite the structural variability found in the species, the functions of the XyG in plants growth and development are hypothesized to be conserved among all species of flowering plants [10].

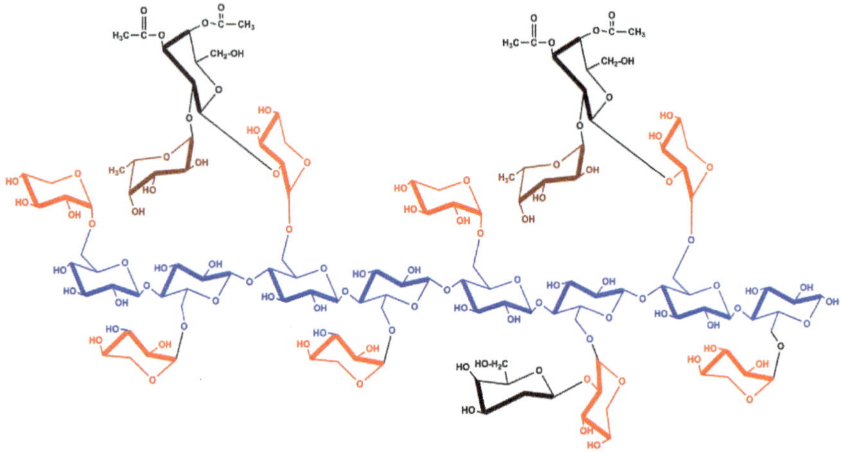

Figure 1. Structure of xyloglucan; principal component of the hemicelluloses. The heptamer block is shown (glucan$_4$-xylose$_3$). In blue backbone β-D-glucans; in red α-D-xylose; in black α-D-galactose and in brown α-L-fucose residues.

In dicotyledonous plants except for graminaceous, the cellulose and xyloglucan are in equal proportions. Some XyG chains are linked to the cellulose microfibrils supporting the important role of rigidity and maintenance of the cell, the rest XyG chains are cross-linked to cellulose microfibrils and pectic polymers, and altogether integrate the complex cell wall matrix. In addition XyG is thought to control cell wall enlargement potentially through the action of α-expansin, XyG endotransglucosylase or β-(1,4)-endoglucanases [11]. Several authors have revealed the XyG function by means of XyG-deficient mutants of *Arabidopsis thaliana,* whereas trying to elucidate the rol of xyloglucan in cell wall biomechanics and cell enlargement. More recently, mutations in two xylosyltransferase genes (*xx1/xx2)* involved in XyG synthesis in *Arabidopsis thaliana* resulted in a XyG-deficient mutant apparently normal but reduced in size. Hypocotyl walls were 20-50% weaker in *xx1/xx2* seedlings, suggesting that XyG plays a strengthening role in the cell wall [12]. It was also confirmed that when XyG is missing, pectins and xylans replace its role in cell wall biomechanics. The growth reduction in *xx1/xx2* plants may stem from the reduced effectiveness of α-expansin in the absence of XyG [13]. These results represent the complexity of the study of individual components in a plant cell wall matrix, it is necessary to point out the advantage of using multiple assays for a better comprehension in the wall extensibility function.

3.2.2. *Xylans*

Xylans are a diverse group of polysaccharides with the common backbone of β-(1,4)- linked xylose residues, with side chains of α-(1,2) linked glucuronic acid and 4-*O*-methyl glucuronic acid residues. Composition and distribution of the substitutions is wide variable according to the plant cell species. Xylans usually contain many arabinose residues attached to the backbone which are known as arabinoxylans and glucuronoarabinoxylans; high amounts of arabinoxylans are present in the endosperm of cereals [14]. In graminaceous species xylans may be linked to the cellulose microfibrils as the xyloglucan does in dicotyledonous plants, but the side chain branches are not attached; besides, the content of lateral substituents decrease gradually during cell growth.

3.2.3. *Mannans and glucomannans*

The β-(1,4)-linked polysaccharides rich in mannose or with mannose and glucose in a non-repeating pattern are the glucomannans and galactoglucomannans. Even though their presence in primary cell wall is low, mannans have been studied in their role as seed storage compounds, as evidenced by the embryo lethal phenotype in an *Arabidopsis* mutant that is lacking the major (gluco) mannan synthase in seeds [15].

3.3. Pectins

Pectins represent an outstanding family of cell wall polysaccharides with extraordinary versatile, but not yet fully known structures and functions. In plants the functions of pectins fulfills important biological functions such as: growth, development, morphogenesis, defense, cell–cell adhesion, wall structure, signaling, cell expansion, wall porosity, binding of ions, growth regulators and enzymes modulation, pollen tube growth, seed hydration, leaf abscission, and fruit development [16]. The extracted pectins of citrus peel and apples are used as a gelling and stabilizing agent in food and cosmetic industries. Pectins within the fruits and vegetables are part of the daily dietary fiber and have multiple positive effects on human health including lowering cholesterol, serum glucose levels, decrease occurrence of diabetes and cancer [17-19]. This points the relevance of pectins in diverse emerging fields of study, even in human health.

Pectins are the most wide complex family of polysaccharides in nature. They are present in primary walls of dicots and non-graminaceous monocots with approximately 35%; in grass and other commelinoid primary walls 2-10% and up to 5% in walls of woody tissues [20]. Pectins are formed with α-(1,4)-D-galacturonic acid residues. Galacturonic acid (GalA) comprises approximately 70% of pectin linked at the *O*-1 and the *O*-4 positions [16]. The structural classes of the pectic polysaccharides include homogalacturonan (HG), rhamnogalacturonan I (RG-I), and rhamnogalacturonan II (RG-II). Also xylogalacturonan (XGA) and apiogalacturonan (AGA) have been determined. AGA is found in the walls of aquatic plants such duckweeds (*Lemnaceae*) and marine seagrases (*Zosteraceae*) with apiose residues 2,3-linked to homogalacturonan. The XGA is more abundant, has HG substituted by D-xylose residues which has been determined in multiple species such as marine seagrasses, pea, apple, *Arabidopsis*, soybean with *O*-3 xylose and xylose branches at *O*-2 by another xylose residue [10].

3.3.1. Homogalacturonan

Homogalacturonan (HG) is the most abundant polymer of the pectins, it comprises nearly the 60% of pectins in plant cell wall [20]. HG is formed by long chains of linear 1,4-linked α-D-galacturonic acid, some of the carboxyl groups are partially methyl-esterified at C-6 and acetyl-esterified at positions *O*-2 and/or *O*-3 (Figure 2), depending on plant species. The linear units of HG in which more than 50% of the GalA are esterified with methyl (or methoxy) groups at the C-6 position are conventionally called high methyl-esterified HGs; otherwise they are referred as to low methyl-esterified HGs. The unmethylated HG is negatively charged and may ionically interact with Ca^{2+} to form a stable gel with other pectin molecules if 10> consecutive unmethyl-esterified GalA residues are coordinated; this is called the egg box (Figure 3). The egg box model can occur upon Ca^{2+} inducing gelling approximately in 70% of the pectin in plant cell walls [21].

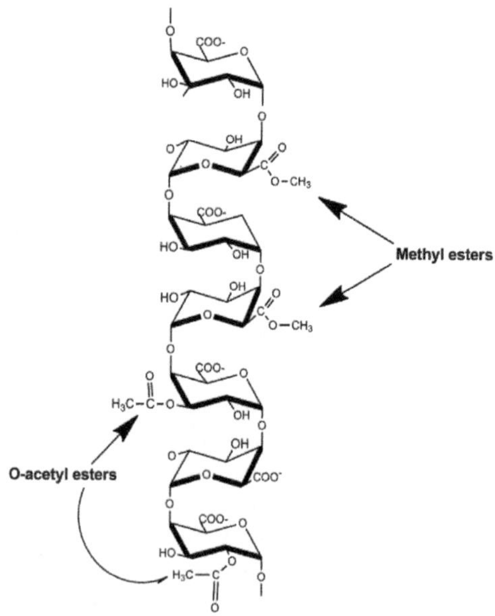

Figure 2. Homogalacturonan structure. Homogalacturonan is a linear polymer of α-(1,4)-D galacturonic acid with methyl-esterified at C-6 and acetyl-esterified at positions O-2 and/or O-3.

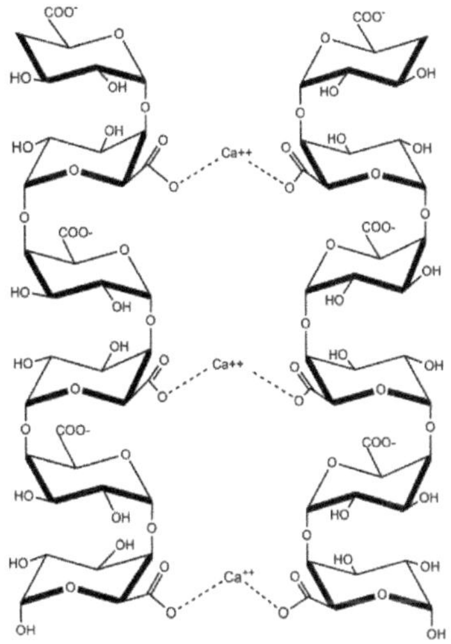

Figure 3. The egg-box model of calcium crosslinking in homogalacuronan polysaccharides.These properties favor the gellification process, therefore have been used by food technologists for the preparation of jams, candies and processed fruits products. The gels obtained using highly methyl-esterified HGs have a short and compact structure, are transparent and achieve a good preservation of original flavors; such gels are thermo-reversible. On the opposite, the low methyl-esterified HGs are thermally irreversible gels and only gell in the presence of multivalent ions (Ca^{2+}, Mg^{2+}).

3.3.2. Pectic branched polymers: Rhamnogalacturonans-I and Rhamnogalacturonans-II

Rhamnogalacturonan-I (RG-I) is a family of pectic polysaccharides that represent 20-35% of pectin. It contain a backbone of the repeating disaccharide galacturonic acid and rhamnose: [α-(1,2)-D-GalA-α-(1,4)-L-Rha]$_n$ partially substituted *O*-4 and/or *O*-3 positions of α-L rhamnose residues with single neutral glucosyl residues and with polymeric side c hains predominantly of α-(1,5)-L arabinans and β-(1,4)-D galactans, arabinogalactans-I (AG-I), arabinogalactans-II (AG-II) and possibly galacto-arabinans [22]. The backbone may be *O*-acetylated on C-2 and/or C-3 by α-L-rhamnose residues. In contrast, there is no compelling evidence that the residues are methyl-esterified, however, an enriched RG-I like wall fraction from flax has been reported to contain methyl esters [23]. The predominant side chains contain linear and branched α-L-arabinose and/or β-D-galactose residues [20] these two are linked to approximately half of the rhamnose residues of the RG-I backbone (Figure 4). The side chains showed a great heterogeneity according to the plant sources; the α-L-fucose, β-D-glucuronic acid and 4-*O*-methyl β-D-glucuronic acid, as ferulic and coumaric acid may be present [24]. The core of the polymeric side chains does not generally exceed 50 GalA residues although there are some exceptions like the galactan isolated from tobacco cell walls, with 370 units [25]. RG-I side chains are developmental and tissue-specific differentially regulated in the type of terminal sugars and oligosaccharides attached to the backbone, it is not well understood but suggests diverse functional specialization. Besides, the rhamnose residues may range from 20 to 80% depending on the pectin, plant source and extraction method [24]. It was suggested the RG-I functions as a linkage support to other pectic polysaccharides such as HG and Rhamnogalacturonans-II (RG-II), that are covalently attached as side chains [10].

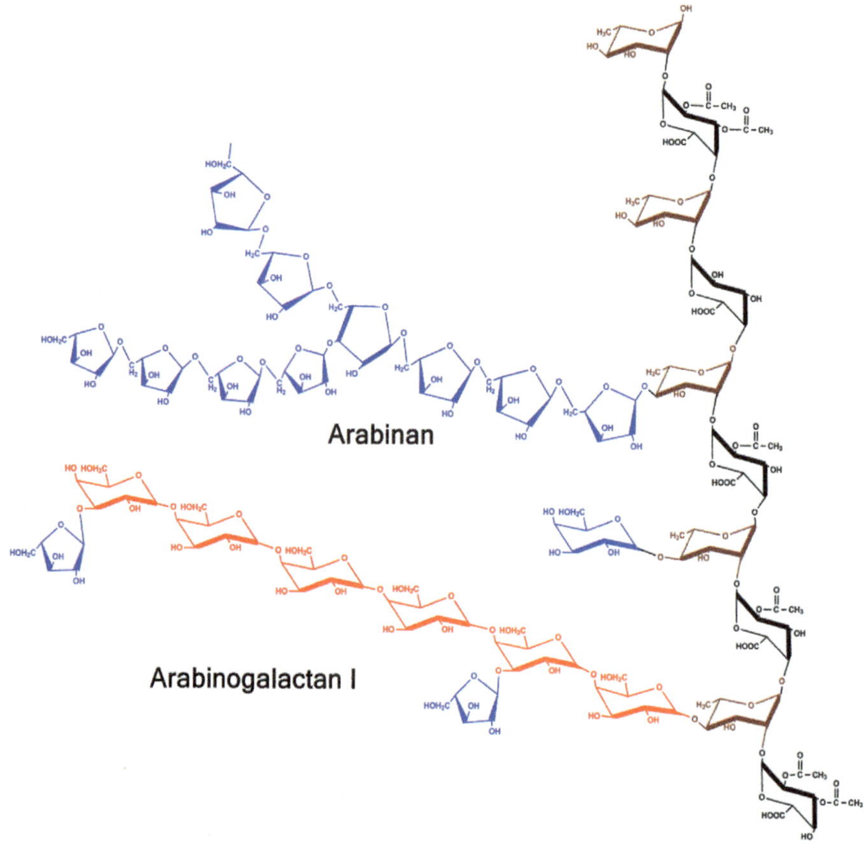

Figure 4. Major structural features of Rhamnogalacturonan I. The backbone is composed of the disaccharide repeating unit of α-(1,2)-D-Galacturonic Acid-α-(1,4)-L-Rhamnose. Branched oligosaccharides composed predominantly of α-L arabinose, in blue; and/or β-D-galactose residues, in red.

Rhamnogalacturonans-II are the most complex and branched polysaccharides of pectin. RG-II is a minor pectic component of plant cell walls with between 0.5 to 8% in dicots, non-graminaceous, monocots, and gymnosperms, and less than 0.1% in primary walls of commelinoid monocots [26]. The RG-II has a characteristic structure of seven to nine residues of α-D-galacturonic acid backbone with four branches clearly differentiated designated A, B, C and D (Figure 5).

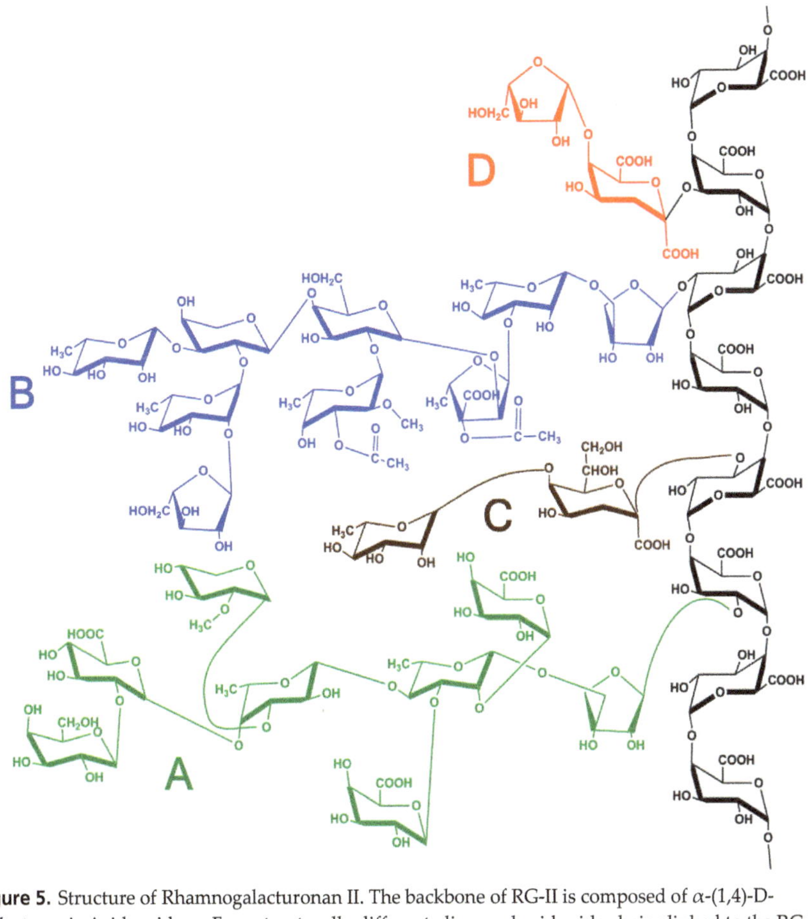

Figure 5. Structure of Rhamnogalacturonan II. The backbone of RG-II is composed of α-(1,4)-D-Galacturonic Acid residues. Four structurally different oligosaccharide side chains linked to the RG-II are represented in different colors (A-D).

RG-II has several kinds of substituents including 11 to 12 different glycosyl residues, some of them rare sugars in nature, like 2-*O*-methyl xylose, 2-*O*-methyl fucose, aceric acid, 2-keto-3-deoxy-D-lyxo heptulosaric acid (Dha) and 2-keto-3-deoxy-D-manno octulosonic acid (Kdo) [27-30]. About 28-36 individual sugars, interconnected by more than 20 different glycosidic linkages that makes a highly complex polymer with an α-D-galacturonic acid backbone partially methyl esterified at C-6 with galactosyl residues and branched with oligosaccharide chains [31]. Despite the complex structure and low amount of the RG-II, it has been related with important and specific roles in cell walls. RG-II polymers are self-associated with borate forming borate-cross-linked RG-II dimers firstly demonstrated in complexes derived from sugar-beet [32], which contributes to the mechanical properties of the primary wall with the three-dimensional pectic network *in muro* [33]. Nearly 95% of the RG-II polymers are in the dimmer complex form (dRG-II). The combination of the covalent crosslinking among the pectic polymers HG, RG-I and RG-II sets the network of the plant cell wall, bringing together strength, flexibility and functionality to the cells.

4. Biosynthesis of the plant cell wall polymers

Cell wall biosynthesis begins during cell division in the cytokinesis phase through the formation of the cell plate in the middle of the cell. Eventually, the primary cell wall is assembled by the deposition of polymers of cellulose, hemicelluloses and pectin. The biosynthesis of wall polymers starts in the nucleus with the transcription of genes coding for wall-related proteins and enzymes. Individual elements are channeled into the endomembrane system of endoplasmic reticulum and Golgi apparatus where they are polymerized and modified. The former polymers are then transported through vesicles and secreted outside plasma membrane for subsequent assembling and linkage to the wall. This mechanism is highly regulated along the process and depends on the physiological state of the cell and the interplay of signals going in and out of the cell [34].

Specifically, the long and rigid cellulose microfibrils of plant walls are synthesized from the inner face of the plasma membrane by cellulose synthase (CESA) complexes, which comprise multiple subunits forming a rosette structure of six globular CESA-containing complexes each of which synthesizes growing cellulose chains of 6–10 cellulose molecules (For review, see [4]). Newly synthesized microfibril is propelled by the action of the CESA, which polymerize the glucan chains in the specific positions driven by cortical microtubules [35]. Afterwards, microfibril is linked with xyloglucans (XyG) and pectic polysaccharides to form the cell wall complex network. Matrix polysaccharides not only cross-link microfibrils but also prevent the self association of new microfibrils into larger aggregates. XyG interacts with the formed microfibrils in the surface and also may be trapped inside them [36]. It has been observed that in primary walls, microfibrils linked to xyloglucan are smaller in diameter (less chain per fiber) than those in secondary walls. Besides, the binding between XyG and cellulose is known to weaken cellulose networks but increase their expansibility. The XyG is bound differently to three cellulose microfibrils domains. The first is available to endoglucanases, the second has to be solubilized by concentrated alkali and a third XyG is neither enzyme accessible nor chemical [36]. According to this, the type of hydrolysis used to obtain the fractions of polysaccharides (oligosaccharides) in some assays results in products of different degree of polymerization, which is related to some specific biological functions in the cell.

In contrast to cellulose, pectic polysaccharides are synthesized in the Golgi apparatus of the plant cell, and then are secreted to the apoplastic space through vesicular compartment.

Polysaccharides are transported from cis-face to trans-face of the Golgi where they are sorted and packaged into vesicles of the trans-Golgi network for transport to the plasma membrane. The movement of the vesicles containing the polymers is presumably along actin filaments that have myosin motors. It is no clear, how the synthesis of the pectin polysaccharides is initiated or whether lipid or protein donors are involved. The possible modification of the pectic glycosyl residues may be esterification, *O*-mehtylation, acetylation and feruloylation by feruloyltransferases in some *Chemopodiaceae* species [16].

To complete the biosynthesis of polysaccharides, it is necessary the assembly of the transported elements to form the functional matrix. This event involves both enzymatic and non-enzymatic mechanisms in the apoplast [37]. The physicochemical properties existing in the wall are dependent on the hydrophobic and hydrophilic domains given by the water and solutes. The hydrophobic domain is formed by the link of cellulose microfibrils to the hydrogen bonds that lead the exclusion of water from the interacting chains. The hydrophobic interactions may also be controlled by enzymes that diminish the branching of the xyloglucan linked to cellulose microfibrils such as xylosidases and glucanases [38]. Meanwhile, the hydrophilic domain of the wall is given by pectin polymers. Together, both domains contribute to the protoplast matrix medium leading the rearrangement of some polymers as the homogalacturonan. Linear homogalacturonans are synthesized in a highly methyl-esterified form in the Golgi and transported to the wall in membrane vesicles to be desesterified by wall localized pectin methylesterases. The conversion of the HG from the methylesterified form to the negatively charged form has been associated with the decrease of growth [39].

The glycosiltransferases and hydrolases are enzymes localized in the Golgi apparatus and work together to produce the xyloglucan precursors. Some changes take place after the synthesis of hemicelulloses in the Golgi. It has been shown that a specific apoplastic glycosidases are responsible for the trimming of new xyloglycan chains and this determines the heterogeneity of the polymer in the cell wall [6]. Hydrolases are likely to play an important role in determining hemicelluloses structures in the cell wall, and are coexpressed with polysaccharide biosynthetic enzymes. For detailed information about the cell wall related enzymes see [4,6,10,16,40].

In the last decades many biochemical approaches have enabled the identification and characterization of the structure of cell wall polymers and the enzymes involved in their biosynthesis. Beside classical molecular analyses, the development of *Arabidopsis thaliana* mutants has been a breakthrough to reveal specific functions of certain components of the wall. The advances in the determination of the structures of polymers through microscopy provide one of the best views of the organization and structure allowing the development of different plant cell wall models. Immunolocalization studies also have shown the location of some polysaccharides within the cell wall and in the apoplastic region. For instance, low methyl-esterified HGs are located in the middle lamella at the cell corners and around air spaces, whereas high methyl-esterified HGs are present throughout the cell wall [20]. Therefore, non-destructive methodologies such as NMR have been key techniques for elucidating the topology, dynamic and tridimensional arrangement of some cell structures, contributing to the cell wall knowledge.

5. Biological activity of plant cell wall derivatives

Exhaustive studies on the structure and function of the plant cell wall have led to the discovery of biologically active molecules derived from its polymeric carbohydrate components. These molecules are found in nature and can be released by acid, basic or enzymatic hydrolysis of the primary cell wall polysaccharides. Due to the complex combination of carbohydrate polymers in the cell wall of plants, there are variations among the physicochemical structure of hydrolyzed fragments, which exerts striking differences on activity and specificity to regulate some physiological processes in plants. These plant oligosaccharides with regulatory properties were called oligosaccharins and have been extensively studied by the workgroup of Albersheim since the mid-70s (reviewed in [41]). Among the oligosaccharins derived from plants, the most active and therefore the most studied are those derived from pectins and hemicelluloses, whose main regulatory functions depend on the degree of polymerization, chemical composition and structure, and can be divided in two broad categories: activation of plant defense mechanisms and plant growth and development.

In order to exert their regulatory properties, oigosaccharins must be first recognized by specific plant cell receptors which may be lectin-type proteins capable of transmitting the signal into the cell [42]. Even when the complete recognition mechanisms and signaling pathway for plant-derived oligosaccharins is far from being fully understood, protein receptors that recognize these molecules have been characterized in the model plant *Arabidopsis thaliana* [43-44] and its believed the downstream processes may occur via MAP kinases activity [45] (For review about the detailed perception mechanisms of oligosaccharines by plants, see [46]).

5.1. Pectin-derived oligosaccharins

Even when the most abundant component of pectins is the galacturonic acid, partial depolymerization of the pectic polysaccharides generates fragments that may (or not) contain other residues such as rhamnose, galactose, arabinose, xylose, glucose and mannose [20]. This combination confers variability to the structure, and thereby to the biological activity of the oligosaccharins. The oligosaccharins derived from homogalacturonan are called oligogalacturonides (OGAs), which are linear oligomers of galacturonic acid, where some residues may be methyl-esterified or acetylated. OGAs are elicitors of defense responses in plants, triggering the synthesis and accumulation of phytoalexins (antimicrobial compounds) and other molecular indicators of the activation of defensive patterns, such as the induction of pathogenesis related proteins and genes related to the hypersensitive reaction [47]. OGA-induced defense response patterns are summarized in Table 1.

OGAs trigger the rapid accumulation of reactive oxygen species (ROS) in plants, which is necessary for the deposition of callose, polysaccharide produced in response to wounding and pathogen infection. Furthermore, ROS are signaling molecules of several intracellular events. Therefore it was proposed ROS were involved in the OGA-induced resistance against fungal pathogens in three different ways: (1) directly exerting a cytotoxic effect to the invading pathogen, (2) inducing callose deposition for reinforcing the plant cell wall, and (3) mediating the signals leading to the expression of defense related genes and defensive metabolites [48]. Nevertheless, recent findings in *Arabidopsis thaliana* showed that the defensive gene activation was not directly correlated to the accumulation of hydrogen peroxide, and that OGA-induced resistance against the fungal pathogen *Botrytis cinerea* was independent of both the oxidative burst and callose deposition [49].

BIOLOGICAL ACTIVITY	PD	ORGANISM	REFERENCE
Defense Responses			
Phytoalexin synthesis	8-13	*Glycine max*	[77]
	3-12	*Glycine max*	[78]
	≥ 3	*Petroselinum sativum*	[79]
	9-15	*Phaseolus vulgaris*	[80,81]
Induction of phenylalanine ammonia-lyase	> 9	*Daucus carota*	[82]
	9-15	*Phaseolus vulgaris*	[80,81]
Induction of chalcone synthase	9-15	*Phaseolus vulgaris*	[81]
Induction of β-(1,3)-glucanase	≥ 3	*Petroselinum sativum*	[79]
Lignin synthesis	8-11	*Cucumis sativus*	[83]
	9-15	*Phaseolus vulgaris*	[81]
Protease inhibitors synthesis	2-3	*Lycopersicum esculentum*	[84]
Growth and Development			
Induction of ethylene production	5-19	*Lycopersicum esculentum*	[85]
	5-19	*Pyrus communis*	[86]
Steem growth inhibition	> 8	*Pisum sativum*	[87]
Protease inhibitors synthesis	10-14	*Nicotiana tabacum*	[88,89]
Quality Parameters			
Increase of the color and anthocyanin content	3-20	*Vitis vinifera*	[91]

Table 1. Biological activity exerted by oligogalacturonides with respect to the degree of polymerization

Plants treated with OGAs exhibit an enhanced resistance to pathogen infections. The induction of the defensive genes, peroxidase and β-(1,3)-glucanase has been related to the enhanced resistance of OGA-treated *Arabidopsis thaliana* against *Botrytis cinerea* [50-51]. Peroxidases are associated to the plant cell wall reinforcement by the synthesis of lignin, while β-(1,3)-glucanase could affect mycelium growth by hydrolyzing glucan chains from the wall of fungi [50]. In grapevine (*Vitis vinifera* L.), OGAs highly stimulated the enzymatic activity of chitinase and β-(1,3)-glucanase and induced the expression of defense related genes in different extent, which also led to a protection against *Botrytis cinerea* [47]. Some genes related to the formation of phytoalexins from the phenylpropanoid pathway were expressed rapidly and transient, various chitinase isoforms were expressed rapidly but their induction was more sustained, and some inhibitors of fungal hydrolytic enzymes were up-regulated later.

The degree of acetylation and methylation of OGAs has been less addressed but emerging research showed the influence of these functional group substituents on plant defense responses. The effect of the degree of acetylation of OGAs on the elicitation of defenses in wheat (*Triticum aestivum* L.) was studied by Randoux and coworkers [52]. It was found both acetylated and unacetylated OGAs induced accumulation of hydrogen peroxide at the site of fungal penetration, through activation of oxalate oxidase, which is also related to the enhanced peroxidase activity. Besides, the induction of lipoxigenase activity demonstrated the stimulation of the octadecanoid pathaway. Moreover, transgenic strawberries (*Fragaria vesca* L.) producing partially demethylated OGAs displayed an enhanced resistance against *Botrytis cinerea* [53].

Table 1 shows that OGAs modulate diverse growth and developmental processes in plants. Early responses related to the signaling transduction pathways comprise membrane depolarization, cytosolic acidification, apoplast alkalinization and calcium mobilization at the plasma membrane level, due to the activity of Ca^{2+} channels. Calcium ions are very important second messengers in plants and its level in intracellular compartments is determinant for the kind of physiological response. In tobacco cells OGAs induced different patterns of Ca^{2+} influx into cytosol, mitochondria and chloroplasts [54]. The increase in cytosolic free Ca^{2+} has been proposed to mediate the regulation of stomatal aperture and production of hydrogen peroxide in the guard cells of tomato (*Licopersicon esculentum* L.) and *Commelina communis* L. [55]. Calcium may also directly interact with long-sized OGAs promoting the formation of structures with an "egg-box" conformation, which may potentiate their biological activity [56]. Nevertheless the promotion of vegetative shoot formation in *Nicotiana tabacum* explants has been observed to be independent of exogenous Ca^{2+} [57].

OGAs regulate morphogenesis in plant tissues in a process associated with the action of auxins, which are growth-regulating phytohormones. Particularly, OGAs and auxins appear to play an antagonist role; since OGAs inhibit the expression of some auxin-inducible genes steps downstream of the auxin perception [58]. In this sense, root differentiation induced by OGAs was studied in *Arabidopsis thaliana* seedlings, where treatments decreased trichoblasts length but increased the number and length of root hairs [59]. Similar results were found in maize (*Zea mays* L.) seedlings where OGAs inhibited coleoptile growth and modified root architecture by inducing lateral root formation [59]. The growth inhibitory activity exerted by OGAs seems to be caused by (1) inhibition of cell elongation; since cell division in the primary root meristem is not altered [59] and (2) inactivation of a kinase enzyme implicated in the TOR signaling pathway, which integrates nutrient and growth factor signals in eukaryotic cells [60]. On the contrary OGAs induced primary root length growth in alfalfa (*Medicago sativa* L.) seedlings [61], which confirms that OGA-inducing activities are dependent on the plant species perceiving the signal.

Interestingly, the structure and stimulating activity of a rhamnogalacturonan I-derived oligosaccharide (RG-IO) isolated from flowers of *Nerium indicum* Mill. was investigated [62]. The structural features of the oligomer consisted in a rhamnogalacturonan backbone with several branches *O*-4 linked to L-rhamnose residues. To determine the structure of the branches the oligomer was partialy hidrolized and analyzed by mass spectrometry (ESI-MS). Branches were found to be mainly composed by β-(1,4)-D-galactan, a highly branched arabino β-(1,3;1,6)-D-galactan, and α-(1,5)-L-arabinan. Furthermore, RG-IO stimulated *in vitro* the production of nitric oxide in macrophage cells. Removal of some side chains from RG-IO reduced nitric oxide production, pointing out the relevance of the branches for its biological activity.

5.2. Hemicellulose-derived oligosaccharins

Xyloglucan is the main hemicellulosic component of the plant cell wall. Biological effects of xyloglucan derivatives are related to the intrinsic physiological function of polymeric xyloglucan in plant cells, comprising the control of extensibility and mechanics of the cell wall and cell expansion. Most research in this field highlights their regulatory activity on cell growth and elongation, which relies in the molecular size, distribution, and levels of substituted xylosyl units with galactosyl and fucosyl residues [63]. It has been observed active xyloglucan oligomers (XGOs) accelerate cell elongation in peeled stem segments of *Pisum sativum* [64], and in suspension-cultured cells of *Nicotiana tabacum*, expansion led to cell division [65]. On the contrary, treatments with polymeric xyloglucan suppressed cell elongation [64-65], indicating that molecular size is a determinant factor of response specificity.

Xyloglucan-derived octasaccharides promoted growth of coleoptiles in wheat seedlings and induced a rapid increase of α-L-fucosidase activity in *Rubus fruticosus* protoplasts [63]. In wheat immature embryos a xyloglucan-derived pentasaccharide induced rhizogenesis and stimulated the formation of callus and meristematic zones [66]. A fucose-galactose-xylose trisaccharide fragment of xyloglucan inhibited ethylene biosynthesis, stimulated embryogenesis in cell cultures of cotton (*Gossypium hirsutum* L.) and formation of callus [67]. In contrast, a mixture of XGOs induced ethylene production in whole fruits of persimmon (*Diospyros kaki* L.). Interestingly, non-fucosylated XGOs augmented ethylene levels in a greater extent than fucosylated XGOs [68]. Also, XGOs lacking the fucosyl residue were inactive to modulate potato resistance to disease [69], and to inhibit gibberellic acid-induced elongation of pea (*Pisum sativum* L.) epicotyls [70], when compared to fucosylated XGOs. These observations altogether may provide a clue to a better understanding of the relationship between the presence of certain substituents and the sensitivity of the response.

On the other hand, galactoglucomannan is composed by a backbone of glucose and mannose residues with side chains of galactose. More recent research about cell wall oligosaccharides derivatives has demonstrated a growth-regulating activity of galactoglucomannan-derived oligosaccharins at very low concentrations. Galactoglucomannan oligosaccharides (GGMOs) modulate root morphology in mung bean (*Vigna radiata* L.) [71] and *Karwinskia humboldtiana* [72] by a respective induction or inhibition of adventitious root formation. More recent research found GGMOs inhibited lateral roots formation but stimulated their elongation, without any effect on adventitious root elongation in mung bean seedlings [73].

Furthermore, GGMOs inhibited the elongation induced by exogenous phytohormones of pea stem segments [74], root and hypocotyl growth of mung bean [71, 73] and *K. humboldtiana* roots [72], indicating an antagonist activity of GGMOs against growth regulators, such as 2,4-dichlorophenoxyacetic acid, indole-3-acetic acid, indole-3-butyric acid, 1-naphthaleneacetic acid and gibberellic acid, at different extent. In addition, an increase in the activity of cell wall associated peroxidases has been reported during the GGMO-mediated inhibition of the elongation of hypocotyls in mung bean plants [73] and epicotyls in peas [75]. Which suggests the growth inhibition caused by GGMOs may be the result of processes catalyzed by plant cell wall peroxidases.

5.3. Cellulose-derived oligosaccharins

During many years it was thought that only non-cellulosic oligosaccharides derived from plant cell wall were biologically active. Surprisingly, it has been recently demonstrated that fragments of oligosaccharides released during cellulose degradation, called cellodextrins (CD), induce a variety of defense responses in grapevine cells. CD are oligomers of linear β-(1,4)-linked glucose residues. The induction of oxidative burst, transient elevation of cytosolic Ca^{2+}, expression of defense-related genes, and stimulation of chitinase and β-1,3-glucanase activities were triggered by CD in grapevine cells. Also, CD oligomers with a degree of polymerization ≥7 enhanced protection in detached leaves of grapevine against *Botrytis cinerea,* suggesting CD are important elicitors of defense reactions [76]. These results have opened the door to the study of many other biological processes where CD may be involved and to elucidate their action mechanisms in plants.

6. Current applications of plant cell wall-derived oligosaccharins

Increasing knowledge of the factors that modulate the biological activity of cell wall-derived oligosaccharides has naturally led to the development of technologies aimed to exploit the potential of these molecules in different fields. The first successful applications occurred in agriculture, where different crops can now be treated with commercially available preparations of pectin-derived-oligosaccharins in order to enhance the basal resistance of plants, and decrease the possible losses related to phytopatogenic infections. Another alternative is the use of oligosaccharins to improve the yield, as seen in tomato (*Lycopersicum esculentum* Mill.), where foliar treatments increased fruit yield by up to 40% with respect to the non-treated controls, and improved parameters of quality such as soluble solids content (SSC), acidity and firmness [90]. Encouraging results were found in *Vitis vinifera* L., since preharvest treatments of clusters with pectin-derived oligosaccharides enhanced the red color of table grapes cv. 'Flame Seedless' without affecting berry firmness neither SSC. Berry color enhancement was achieved due to a higher anthocyanin content in berry skin, given by the stimulation of the phenylpropanoid pathway [91]. Recent findings showed an increase in the antioxidant capacity of OGA-treated table grapes cv. 'Red Globe' and 'Flame Seedless', as a consequence of the induction of anthocyanins, flavonoids and phenolic compounds (non-published data). As a result of this research it was generated a register method for controlling coloration in table grapes based on oligogalacturonide (92). On the other hand, OGAs are currently being used in the industry of cosmetics. This emerging trend is based on the ability of OGAs to stimulate adhesion of keratinocytes to proteins of the dermoepidermal junction. This biological effect is apparently exerted by OGAs with a degree of polymerization ≤ 5 as indicated in the US patent [93]. However, further research is necessary to continue with the development of reliable applications.

Author details

Marisol Ochoa-Villarreal, Emmanuel Aispuro-Hernández and Miguel A. Martínez-Téllez*
Laboratorio de Fisiología Vegetal Molecular. Tecnología de Alimentos de Origen Vegetal, Centro de Investigación en Alimentación y Desarrollo, A.C., Hermosillo, Sonora, México

Irasema Vargas-Arispuro
Laboratorio de Ecología Química. Ciencia de los Alimentos, Centro de Investigación en Centro de Investigación en Alimentación y Desarrollo, A.C., Hermosillo, Sonora, México

Acknowledgement

The authors appreciate the support of Juan Reyes Calderón with the edition of the figures shown in this chapter.

7. References

[1] Hématy K, Cherk C, Somerville S (2009) Host-pathogen warfare at the plant cell wall. Curr Opin in Plant Biol. 12:406-413.
[2] McCann MC, Carpita NC. (2008) Designing the deconstruction of plant cell walls. Curr Opin in Plant Biol. 11:314-320.
[3] Smole MS, Kreze T, Strnad S, Kleinschek KS, Hribernik S. (2005) Characterization of grass fibres. J Mater Sci. 40:5349-5353.
[4] Somerville C. (2006) Cellulose synthesis in higher plants. Annu Rev Cell Dev Biol. 22:53-78.
[5] Jarvis MC. (2000) Interconversion of the Iα and Iβ crystalline forms of cellulose by bending. Charbohyd Res. 325:150-154.
[6] Scheller HV, Ulvkov P. (2010) Hemicelluloses. Annu Rev Plant Biol. 61:263-289.

* Corresponding Author

[7] Chistensen U, Alonso-Simon A, Scheller HV, Willats WGT, Harholt J. (2010) Characterization of the primary cell walls of seedlings of Brachypodium distachyon –A potential model plant for temperate grasses. Phytochemistry 71:62-69.

[8] Wilder BM, Albersheim P. (1973) The Structure of plant cell Walls. Plant Physiol. 51:889-893.

[9] Kato Y, Iki K, Matsuda K. (1981) Occurrence of a soluble and low molecular weight Xyloglucan and Its origin in etiolated mung Bean hypocotyls. Agric Biol Chem. 45:2745-2753.

[10] Caffall KH, Mohnen D. (2009) The structure, function, and biosynthesis of plant cell wall pectic polysaccharides. Carbohyd Res. 344:1879-1900.

[11] Cosgrove DJ. (2005) Growth of the plant cell wall. Nat Rev Mol Cell Biol. 6:850-861.

[12] Cavalier DM, Lerouxel O, Neumetzler L, Yamauchi K, Reinecke A, Freshour G, Zabotina OA, Hahn MG, Burgert I, Pauly M, Raikhel NV, Keegstra K. (2010) Disrupting two *Arabidopsis thaliana* Xylosyltransferase genes results in plants deficient in xyloglucan, a major primary cell wall component. Plant Cell 20:1519-1537.

[13] Park YB, Cosgrove DJ. (2012) Changes in cell wall biomechanical properties in the xyloglucan-deficient xxt1/xxt2 mutant of *Arabidopsis*. Plant Physiol. 158(1):465-75.

[14] Bochicchio R, Reicher F. (2003) Are hermicelluloses from Podocarpus lambertii of gymnosperms? Carbohyd Res. 53:127-136.

[15] Goubert F, Misrahi A, Park SK, Zhang ZN, Twell D, Dupree P. (2003) AtCSLA7, a cellulose synthase-like putative glycosyltranferase, is important for pollen tube growth and embryogenesis in *Arabidopsis*. Plant Physiol. 131:547-557.

[16] Mohnen D. (2008) Pectin structure and biosynthesis. Curr Opin in Plant Biol. 11:266-277.

[17] Jackson CL, Dreaden TM, Theobald LK, Tran NM, Beal TL, Eid M, Gao MY, Shirley RB, Stoffel M.T, Krumar MV, Mohnen D. (2007) Pectin induces apoptosis in human prostate cancer cells: correlation of apoptotic function with pectin structure. Glycobiology 17:805-819.

[18] Yapo BM, Koffi KL. (2008) Dietary fiber components in yellow passion fruit rinds-A potential fiber source. J Agric Food Chem. 54(7):2738-2744.

[19] Glinsky VV, Raz A. (2009) Modified citrus pectin antimetastatic properties: One bullet, multiple targets. Carbohyd Res. 344:1788-1791.

[20] Ridley BL, O'Neill MA, Mohnen D. (2001) Pectins: structure, biosynthesis, and oligogalacturonide-released signaling. Phytochemistry 57:929-967.

[21] Jarvis MC, Apperley DC. (1995) Chain conformation in concentrated pectic gels: evidence from 13C NMR. Carbohyd Res. 275:131-145.

[22] Obro J, Harholt J, Sheller HV, Orfila C. (2004) Rhamnogalaturonan-I in *Solanum tuberosum* tubes contains complex arabinogalactan structures. Phytochemistry 65:1429-1438.

[23] Rihouey C, Morvan C, Borissova L, Jauneau A, Demarty M, Jarvis M. (1995) Structural features of CDTA-soluble pectins from flax hypocotyls. Carbohyd Polym. 28:159-166.
[24] O'Neill M, Albersheim P, Darvill A. (1990) The pectic polysaccharides of primary cell walls. In: Dey. D.M. (Ed), Methods in Plant Biochemistry, Vol. 2. Academic Press, London, pp 415-441.
[25] Eda S, Katö K. (1980) Pectin isolated from the midrib of leaves of Nicotiana tabacum. Agric Biol Chem. 44(12):2793-2801.
[26] Matsunaga T, Ishii T, Matsunamoto S, Higuchi M, Darvill A, Albersheim P, O'Neill M. (2004). Occurrence of the primary cell wall polysaccharide rhamnogalacturonan-II in peridophytes, lycophytes, and bryophytes. Implications for the evolution of vascular plants. Plant Physiol. 134:339-351.
[27] Darvill AG, McNeil M, Albersheim P. (1978) Structure of Plant Cell Walls. VIII. A New Pectic Polysaccharide. Plant Physiol. 62(3):418-422.
[28] Spellman MW, McNeil M, Darvill AG, Albersheim P. (1983) Characterization of a structural complex heptasaccharide isolated from the pectic polysaccharide rhamnogalacturonan II. Carbohyd Res. 122:115-129.
[29] Stevenson TT, Darvill AG, Albersheim P. (1988) Structural features of the plant cell-wall polysaccharide rhamnogalacturonan II. Carbohyd Res. 179:269-288.
[30] York WS, Darvill AG, McNeil M, Albersheim P. (1985) 3-Deoxy-D-manno-2-octulosonic acid (KDO) is a component of rhamnogalacturonan II, a pectin polysaccharide in the primary cell walls of plants. Carbohyd Res. 138:109-126.
[31] Yapo BM. (2011) Pectic substances: From simple pectic polysaccharides to complex pectins- A new hypothetical model. Carbohyd Polym. 86:373-385.
[32] Ishii T, Matsunaga T. (2001) Pectic polysaccharide rhamnogalacturonan II is covalently linked to homogalacturonan. Phytochemistry 57(6):969-974.
[33] Fleischer A, O'Neill MA, Ehwald R. (1999). The pore size of non-graminaceous plant cell wall is rapidly decreased by borate ester cross-linking of the pectic polysaccharide rhamnogalacturonan-II. Plant Physiol. 121:829-838.
[34] Wojtaszek P. (2000) Genes and plant cell walls: a difficult relationship. Biol Rev. 75:437-475.
[35] Gutierrez R, Lindeboom JJ, Paredez AR, Emons AMC, Ehrhadt DW. (2009) Arabidopsis cortical microtubules position cellulose synthase delivery to the plasma membrane and interact with cellulose synthase trafficking compartments. Nat Cell Biol. 11:797-806.
[36] Pauly M, Albersheim P, Darvill AG, York WS. (1999) Molecular domains of the cellulose-xyloglucan network in the cell walls of higher plants. Plant J. 20:629-639.
[37] Cosgrove DJ. (1997) Assembly and enlargement of the primary cell wall in plants. Annu Rev Cell Develop Biol. 13:171-201.

[38] Nicol F, His I, jauneau A, Verhettes S, Canut H, Höfte H. (1998) A plasma membrane-bound putative endo-1,4-β-D-glucanase is required for normal wall assembly and cell elongation in *Arabidopsis*. EMBO J. 17:5563-5576.

[39] Derbyshire P, McCann MC, Roberts K. (2007) Restricted cell elongation in Arabidopsis hypocotyl is associated with a reduced average pectin esterification level. BMC Plant Biol. 7:31.

[40] Dwivany FM, Yulia D, Burton RA, Shirley NJ, Wilson SM, Fincher GB, Bacic A, Newbigin E, Doblin M. (2009) The CELLULOSE-SYNTHASE LIKE C (CSLC) family of barley includes members that are integral membrane proteins targeted to the plasma membrane. Mol Plant. 2(5):1025-1039.

[41] Albersheim P, Darvill A, Augur C, Cheong JJ, Eberhard S, Hahn MG, Marfa V, Mohnen D, O'Neill MA (1992) Oligosaccharins: oligosaccharide regulatory molecules. Chem Res. 25(2):77-83.

[42] Etzler ME. (1998) Oligosaccharide Signaling of Plant Cells. J Cell Biochem Supplem. 30:123-128.

[43] Brutus, A, Sicilia F, Macone A, Cervone F, De Lorenzo G. (2010) A domain swap approach reveals a role of the plant wall-associated kinase 1 (WAK1) as a receptor of oligogalacturonides. Proc Natl Acad Sci USA. 107:9452-9457.

[44] Gouget A, Senchou V, Govers F, Sanson A, Barre A, Rougé P, Pont-Lezica R, Canut H. (2006) Lectin receptor kinases participate in protein-protein interactions to mediate plasma membrane-cell wall adhesions in *Arabidopsis*. Plant Physiol. 140:81-90.

[45] Kohorn BD, Johansen S, Shishido A, Todorova T, Martinez R, Defeo E, Obregon P. (2009) Pectin activation of MAP kinase and gene expression is WAK2 dependent. Plant J. 60(6):974-82

[46] Enriquez-Guevara EA, Aispuro-Hernandez E, Vargas-Arispuro I, Martınez-Tellez MA. (2010) Cell wall oligosaccharine derivatives: biological activity and participation in the response of plant defense. Rev Mex Fitopatol. 28(2):75-86.

[47] Aziz A, Heyraud A, Lambert B. (2004) Oligogalacturonide signal transduction, induction of defense-related responses and protection of grapevine against *Botrytis cinerea*. Planta 218:767-774.

[48] Levine A, Tenhaken R, Dixon R, Lamb C. (1994) H2O2 from the oxidative burst orchestrates the plant hypersensitive disease resistance response. Cell 79(4):583-593.

[49] Galletti R, Denoux C, Gambetta S, Dewdney J, Ausubel F, De Lorenzo G, Ferrari S. (2008) The AtrbohD-mediated oxidative burst elicited by oligogalacturonides in *Arabidopsis* is dispensable for the activation of defense responses effective against *Botrytis cinerea*. Plant Physiol. 148(3):1695-1706.

[50] Rasul S, Dubreuil-Maurizi C, Lamotte O, Koen E, Poinssot B, Alcaraz G, Wendehenne D, Jeandroz S. (2012) Nitric oxide production mediates oligogalacturonide-triggered immunity and resistance to *Botrytis cinerea* in *Arabidopsis thaliana*. Plant Cell Environ. doi: 10.1111/j.1365-3040.2012.02505.x

[51] Ferrari S, Galletti R, Denoux C, De Lorenzo G, Ausubel F, Dewdney J. (2007) Resistance to *Botrytis cinerea* induced in arabidopsis by elicitors is independent of salicylic acid, ethylene, or jasmonate signaling but requires phytoalexin deficient. Plant Physiol. 144(1):367-379.

[52] Randoux B, Renard-Merlier D, Mulard G, Rossard S, Duyme F, Sanssené J, Courtois J, Durand R, Reignault P. (2010) Distinct defenses induced in wheat against powdery mildew by acetylated and nonacetylated oligogalacturonides. Phytopathology 100(12):1352-63.

[53] Osorio S, Castillejo C, Quesada MA, Medina-Escobar N, Brownsey GJ, Suau R, Heredia A, Botella MA, Valpuesta V. (2008) Partial demethylation of oligogalacturonides by pectin methyl esterase 1 is required for eliciting defence responses in wild strawberry (*Fragaria vesca*). The Plant J. 54:43-55.

[54] Manzoor H, Chiltz A, Madani S, Vatsa P, Schoefs B, Pugin A, Garcia-Brugger A. (2012) Calcium signatures and signaling in cytosol and organelles of tobacco cells induced by plant defense elicitors. Cell Calcium http://dx.doi.org/10.1016/j.ceca.2012.02.006.

[55] Lee S, Choi H, Suh S, Doo IS, Oh KY, Choi EJ, Schroeder Taylor AT, Low PS, Lee Y. (1999) Oligogalacturonic acid and chitosan reduce stomatal aperture by inducing the evolution of reactive oxygen species from guard cells of tomato and *Commelina communis*. Plant Physiol. 121(1):147-52.

[56] Cabrera JC, Boland A, Messiaen J, Cambier P, Van Cutsem P. (2008) Egg box conformation of oligogalacturonides: The time-dependent stabilization of the elicitor-active conformation increases its biological activity. Glycobiology 18(6):473-482.

[57] Falasca G, Capitani F, Della Rovere F, Zaghi D, Franchin C, Biondi S, Altamura MM. (2008) Oligogalacturonides enhance cytokinin-induced vegetative shoot formation in tobacco explants, inhibit polyamine biosynthetic gene expression, and promote long-term remobilisation of cell calcium. Planta 227(4):835-52.

[58] Mauro ML, De Lorenzo G, Costantino P, Bellincampi D. (2002) Oligogalacturonides inhibit the induction of late but not of early auxin-responsive genes in tobacco. Planta 215(3):494-501.

[59] Hernández-Mata G, Mellado-Rojas ME, Richards-Lewis A, López-Bucio J, Beltrán-Peña E, Soriano-Bello EL. (2010) Plant immunity induced by oligogalacturonides alters root growth in a process involving flavonoid accumulation in *Arabidopsis thaliana*. J Plant Growth Regul. 29:441-454.

[60] Peña-Uribe CA, García-Pineda E, Beltrán-Peña E, Reyes de la Cruz H. (2012) Oligogalacturonides inhibit growth and induce changes in S6K phosphorylation in maize (*Zea mays* L. var. Chalqueño). Plant Growth Regul. 67(2):151-159.

[61] Camejo D, Martí MC, Jiménez A, Cabrera JC, Olmos E, Sevilla F. (2011) Effect of oligogalacturonides on root length, extracellular alkalinization and O^{2-} accumulation in alfalfa. J Plant Physiol. 168:566-575.

[62] Dong Q, Liu X, Yao J, Dong X, Ma C, Xu Y, Fang J, Ding K. (2010) Structural characterization of pectic polysaccharide from *Nerium indicum* flowers. Phytochem. 71:1430-1437.

[63] Vargas-Rechia C, Reicher F, Sierakowski MR, Heyraud A, Driguez H, Liénart Y. (1998) Xyloglucan octasaccharide XXLGol derived from the seeds of *Hymenaea courbaril* acts as a signaling molecule. Plant Physiol. 116:1013-1021.

[64] Takeda T, Furuta Y, Awano T, Mizuno K, Mitsuishi Y, Hayashi T. (2002) Suppression and acceleration of cell elongation by integration of xyloglucans in pea stem segments. Proc Natl Acad Sci USA. 99(13):9055-9060.

[65] Kaida R, Sugawara S, Negoro S, Maki H, Hayashi T, Kaneko TS. (2010) Acceleration of cell growth by xyloglucan oligosaccharides in suspension-cultured tobacco cells. Mol Plant. 3(3):549-554.

[66] Pavlova ZN, Ash AO, Vnuchkova VA, Babakov AV, Torgov VI, Nechaev OA,Usov AI, Shibaev VN. (1992) Biological activity of a synthetic pentasaccharide fragment of xyloglucan. Plant Sci. 85:131–134.

[67] Rakitin VY, Dolgikh YI, Shaikina EY, Kuznetsov VV. Oligosaccharide inhibits ethylene synthesis and stimulates somatic embryogenesis in a cotton cell culture. Russ J Plant Physl+. 48(5):628-632.

[68] Cutillas-Iturralde A, Fulton DC, Fry SC, Lorences EP. (1998) Xyloglucan-derived oligosaccharides induce ethylene synthesis in persimmon (*Diospyros kaki* L.) fruit. J Exp Bot. 49(321):701-706.

[69] Perekhod EA, Chalenko GI, Il'inskaya LI, Vasyukova NI, Gerasimova N G, Babakov AV, Usov AI, Mel'nikova TM, Ozeretskovskaya OL. (1998) Modulation of potato resistance by means of xyloglucan fragments. Appl Biochem Micro+. 34(1):91-96.

[70] Warneck H, Seitz HU. (1993) Inhibition of gibberellic acid-induced elongation-growth of pea epicotyls by xyloglucan oligosaccharides. J Exp Bot. 44(7):1105-1109.

[71] Kollárová K, Henselová M, Lišková D. (2005) Effect of auxins and plant oligosaccharides on root formation and elongation growth of mung bean hypocotyls. Plant Growth Regul. 46:1-9

[72] Kollárová K, Lišková D, Lux A. (2007) Influence of galactoglucomannan oligosaccharides on root culture of *Karwinskia humboldtiana*. Plant Cell Tiss Organ Cult. 91:1-19.

[73] Kollárová K, Richterová D, Slováková L, Henselová M, Capek P, Lišková D. (2009) Impact of galactoglucomannan oligosaccharides on elongation growth in intact mung bean plants. Plant Sci. 177:324-330.

[74] Kollárová K, Lišková D, Capek P. (2006) Further biological characteristics of galactoglucomannan oligosaccharides. Biol Plant. 50(2):132-238.

[75] Kollárová K, Slováková L, Kollerová E, Lišková E. (2009) Galactoglucomannan oligosaccharides inhibition of elongation growth is in pea epicotyls coupled with peroxidase activity. Plant Sci. 64(5):919-922.

[76] Aziz A, Gauthier A, Bézier A, Joubert JM, Pugin A, Heyraud A, Baillieul F. (2007) Elicitor and resistance-inducing activities of β-1,4 cellodextrins in grapevine,

comparison with β-1,3 glucans and α-1,4 oligogalacturonides. J Exp Bot. 58(6):1463-1472.

[77] Davis KR, Darvill AG, Albersheim P, Dell A. (1986). Host-pathogen interactions: XXIX. Oligogalacturonides released from sodium polypectate by endopolygalacturonic acid lyase are elicitors of phytoalexins in soybean. Plant Physiol. 80:568-577.

[78] Komae K, Komae A, Misaki A. (1990) A 4,5-unsaturated low molecular oligogalacturonide as a potent phytoalexin-elicitor isolated from polygalacturonide of *Ficus awkeotsang*. Agric Biol Chem. 54(6):1477-1484.

[79] Davis KR, Hahlbrock K. (1987) Induction of defense responses in cultured parsley cells by plant cell wall fragments. Plant Physiol. 85:1286-1290.

[80] Dixon RA. (1986) The phytoalexin response: elicitation, signalling and control of host gene expression. Biolog Rev. 61:239-291.

[81] Tepper S., Anderson AJ. (1990) Interactions between pectic fragments and extracellular components from the fungal pathogen *Colletotrichum lindemuthianum*. Physiol Mol Plant P. 36:147-158.

[82] Messiaen J, Read ND, Van Cutsem P, Trewavas AJ. (1993) Cell wall oligogalacturonides increase cytosolic free calcium in carrot protoplasts. J Cell Sci. 104:365-371.

[83] Robertsen B. (1986) Elicitors of the production of lignin-like compounds in cucumber hypocotyls. Physiol Mol Plant P. 28:137-148.

[84] Moloshok T, Pearce G, Ryan CA. (1992) Oligouronide signaling of proteinase inhibitor genes in plants: structure activity relationships of di- and trigalacturonic acids and their derivatives. Arch Biochem Biophys. 294:731-734.

[85] Brecht JK, Huber DJ. (1988) Products released from enzymically active cell wall stimulate ethylene production and ripening in preclimacteric tomato (*Lycopersicon esculentum* Mill.) fruit. Plant Physiol. 88:1037-1041.

[86] Campbell AD, Labavitch JM. (1991) Induction and regulation of ethylene biosynthesis by pectic oligomers in cultured pear cells. Plant Physiol. 97:699-705.

[87] Branca C, De Lorenzo G, Cervone F. (1988) Competitive inhibition of the auxin-induced elongation by α-D-oligogalacturonides in pea stem segments. Physiol Plant. 72:499-504.

[88] Eberhard S, Doubrava N, Marf V, Mohnen D, Southwick A, Darvill A, Albersheim P. (1989) Pectic cell wall fragments regulate tobacco thin-cell-layer explant morphogenesis. Plant Cell 1:747-755.

[89] Marfa V, Gollin DJ, Eberhard S, Mohnen D, Darvill A, Albersheim, P. (1991) Oligogalacturonides are able to induce flowers to form on tobacco explants. Plant J. 1(2):217-225.

[90] García-Sahagún ML, Martínez-Juárez V, Avendaño-López AN, Padilla-Sahagún MC, Izquierdo-Oviedo H. (2009) Acción de oligosacáridos en el rendimiento y calidad de tomate. Rev Fitotec Mexicana 32:295-301.

[91] Ochoa-Villarreal M, Vargas-Arispuro I, Islas-Osuna MA, González-Aguilar G, Martínez-Téllez MA. (2011) Pectin-derived oligosaccharides increase color and anthocyanin content in Flame Seedless grapes. J Sci Food Agric. 91:1928-1930.

[92] Martínez-Téllez MA, Vargas-Arispuro I. (2010) Method for controlling colouration in table grapes base don oligogalacturonides. International patent WO/2010/044649.

[93] Lubrano C, Flavet L, Saintigny G, Robin J. (2007) Methods of treating aging of skin with oligosaccharides in cosmetic or dermatological compositions that stimulate adhesion of keratinocytes to major proteins of the dermoepidermal junction and restore epidermal cohesion. United Stated Patent No. US 2007/0293433A1.

Chapter 5

Comparison of Traditional Methods and Microwave Irradiation Method About Amylum/Acrylic Acid /Acrylamide Polymerization

Hao Ren, Zhen Niu, Juan Wang and Jing Ning

Additional information is available at the end of the chapter

1. Introduction

Based on previous work, the research in amylum grafted acrylic/acrylamide superabsorbent synthesis was made under traditional condition in this section. We synthetize binary grafted amylum superabsorbent and study its water absorption.

1.1. Experiment

1.1.1. Reagents and instruments

1.1.1.1. Reagents

Soluble Amylum (Corn)	A.R. Tianjin Kemiou Chemical Reagents Development Center
Wheat Amylum	F.G. Jinan Sanjia Sugar Co. LTD
Cassava Amylum	F.G. Jinan Sanjia Sugar Co. LTD
Potato Amylum	F.G. Qingzhou Zhengyi Seasoning Food Co. LTD
Pachyrhizus Amylum	F.G. Qingzhou Zhengyi Seasoning Food Co. LTD
Crylic Acid	C.P. Tianjin Kemiou Chemical Reagents Development Center
Acrylamide	A.R. Tianjin Kemiou Chemical Reagents Development Center
Potassium Hydroxide	A.R. Tianjin Kemiou Chemical Reagents Development Center
Potassium Peroxydisulfate	A.R. Tianjin Sitong Chemical Plant
N, N,-methylene biasacrylamide	C.P. Chemical Reagent of Traditional Chinese Medicine Co. LTD

1.1.1.2. Instruments

101-1 Constant Temperature Electrothermal Blowing Drying Oven	Shanghai Luda Experiment Instrument Co.LTD
TG328B Lightning Analytical Balance	Shanghai Balance Instrument Plant
JJ-1 Timing Electric Mixer	Jiangsu Zhongda Instrument Plant
TENSOR27 FTIR	Germany
Hitachi S-2500 SEM	Japanese Hitachi Co. LTD

1.1.2. Experiment principle and preparation method

1.1.2.1. Synthetic process flow of binary amylum grafted superabsorbent

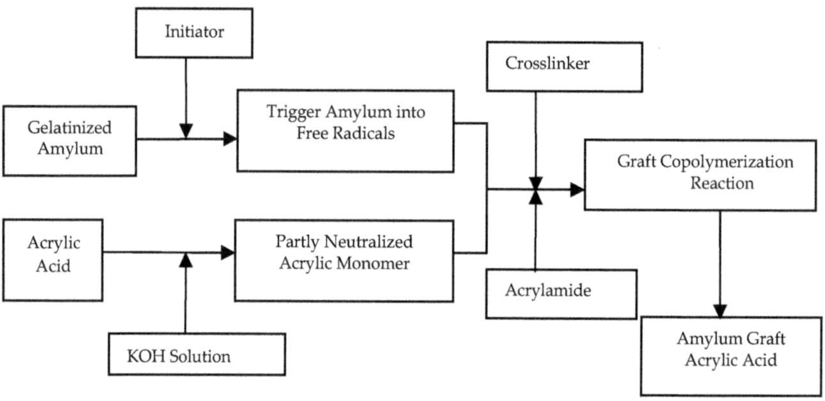

Scheme 1.

1.1.2.2. Major chemical reactions of binary grafted amylum superabsorbent synthesis

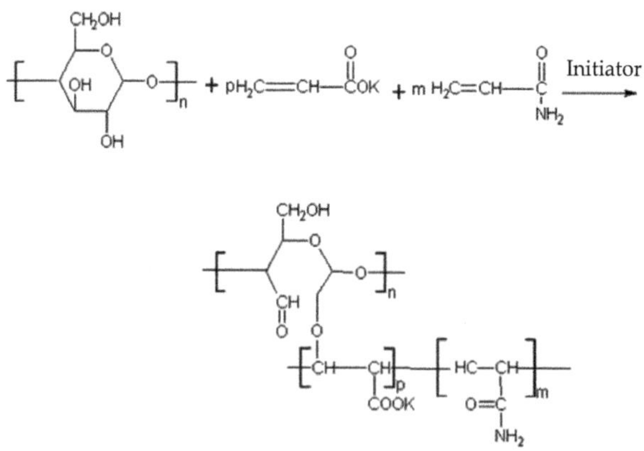

Scheme 2.

1.1.2.3. Process

1. Taking 2.0 g soluble amylum into four-mouth bottle, joined with deionized water.
2. Pasting 20 minute in 50 °C water.
3. Changing dosage of acrylamide in basic Experiment conditions.
4. Designing orthogonal experiment and selecting orthogonal table L_{25} (56) of 6 factors and 5 levels according to the single factor experiment results, such as table 1:
5. Adding right amount monomer, initiator and crosslinker according to orthogonal experiment requirements.
6. Beginning to react in experiment condition under the protection of nitrogen.
7. After the reaction, keeping it in 80°C drying oven to constant heavy.

	Monomer/g	T/℃	t/h	Neutralization degree/%	Initiator/g	Crosslinker/g
1	10	30	1.0	86	0.01	0.0009
2	11	40	1.5	88	0.02	0.0011
3	12	50	2.0	90	0.03	0.0013
4	13	60	2.5	92	0.04	0.0015
5	14	70	3.0	94	0.05	0.0017

Table 1. Orthogonal Experiment level Factor

1.1.3. Test method

1.1.3.1. Test of bibulous rate

There is a variety of methods testing which can test the water absorption of superabsorbent, such as low method, paper bag method, natural filtering method, sieve net method, and so on. There are many differences between data from various test methods and it is hard to compare different data of bibulous rate from various documents.

The sieve net method in this research has been widely used by many researchers. The procedures as follows: taking certain quality (m_1) product into the beaker, adding excessive deionized water and physiological saline, Standing before swelling completely, then screening and filtering the excessive water by 100 mesh screen cloth and adsorbing surface water of resin by paper absorbent. Then weighing (m_2) it, calculating bibulous rate according to equation 1(Hongke Tang& Qi Chen, 2007):

$$Q = (m_2 - m_1)/ m_1 \times 100\% \text{...........} \quad (1)$$

1.1.3.2. Structure characterization of bibulous resin

1. Infrared Spectrometry

Making corn amylum and absorbent resin dried fully and detecting it using infrared spectrophotometric method.

2. Scanning Electron Microscopy

The exterior characteristics were observed by the scanning electron microscopy (SEM) on the 1cm^2 resin.

1.2. Results and discussions

1.2.1. Ratio of acrylamide and acrylic acid

The result of Figure 1 shows that concentration of acrylamide in a certain range can improve the bibulous rate of resin. Different ionization degree affected water absorption of resin in the composition of monomer. In acrylamide molecule, $-CONH_2$ group is nonionic group and its dissociation degree in the water is small. So the effect of ion on it is small. The existence of $-CONH_2$ can improve water absorption because of synergy of groups. But if the dosage was too much, water absorption reduced because the $-CONH_2$ is less hydrophilic than $-COONa$.

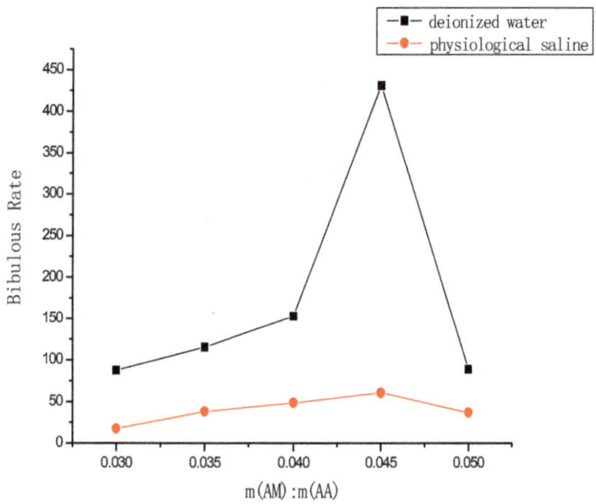

Figure 1. Effect of m(AM):m(AA)

1.2.2. Orthogonal Analysis

1.2.2.1. Analysis of the Orthogonal Experiment (Table 2)

	Monomer	T	t	Neutralization Degree	Initiator	Crosslinker	Water Absorption	Q
1	1	1	1	1	1	1	82.67	24.6
2	1	2	2	2	2	2	402.66	31.02
3	1	3	3	3	3	3	462.92	62.24
4	1	4	4	4	4	4	246.49	34.89
5	1	5	5	5	5	5	36.66	2.65
6	2	1	2	3	4	5	388.02	65.66
7	2	2	3	4	5	1	232.99	36.2
8	2	3	4	5	1	2	220.43	27.46
9	2	4	5	1	2	3	465.22	23.57
10	2	5	1	2	3	4	416.33	43.11
11	3	1	3	5	2	4	699.99	70.12
12	3	2	4	1	3	5	457.62	65.71
13	3	3	5	2	4	1	308.27	27.02
14	3	4	1	3	5	2	296.39	62.61
15	3	5	2	4	1	3	410.54	21.43
16	4	1	4	2	5	3	328.72	30.42
17	4	2	5	3	1	4	491.14	32.84
18	4	3	1	4	2	5	476.81	48.12
19	4	4	2	5	3	1	41.27	24.31
20	4	5	3	1	4	2	197.49	29.88
21	5	1	5	4	3	2	285.61	31.44
22	5	2	1	5	4	3	313.19	47.79
23	5	3	2	1	5	4	355.47	55.76
24	5	4	3	2	1	5	335.48	54.91
25	5	5	4	3	2	1	205.26	30.86
K1	266.28	357	317.08	311.69	308.05	174.09		
K2	344.6	379.52	319.96	358.29	449.94	280.52		
K3	434.56	364.78	385.77	368.75	332.75	396.12		
K4	307.09	296.97	311.7	330.49	290.69	441.88		
K5	299	253.26	317.38	262.31	250.05	329.92		
R	168.28	126.26	74.07	106.44	199.89	267.79		
K1'	31.08	44.45	45.25	39.9	32.25	28.6		
K2'	39.2	42.71	39.63	37.3	40.74	36.48		
K3'	49.38	44.12	50.65	50.84	45.36	37.09		
K4'	33.11	40.06	37.87	34.42	41.05	45.54		
K5'	44.15	25.59	23.5	34.46	37.53	47.41		
R'	16.2	18.86	27.15	16.42	13.11	18.81		

Table 2. Analysis of the orthogonal experiment

1.2.2.2 Analysis of range

Through the orthogonal experiment analysis, the influence of six factors of water absorption from big to small in order is: crosslinker concentration > initiator concentration > monomer concentration > temperature > neutralization degree > reaction time. The best combination is $A_3B_2C_3D_3E_2F_4$.

1.2.2.3. Analysis of monomer concentration

Bibulous rate was highest when dosage of acrylic acid and acrylamide was 12.0 g, bibulous rate reduced with the increase of monomer dosage after 12.0 g (Figure 2). Monomer in the certain range increased and the generation of copolymer increased. The better crosslinking, the better the water absorption will be.

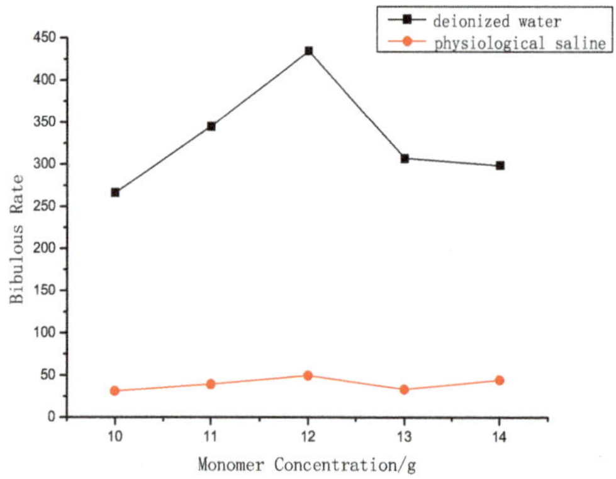

Figure 2. Effect of monomer concentration

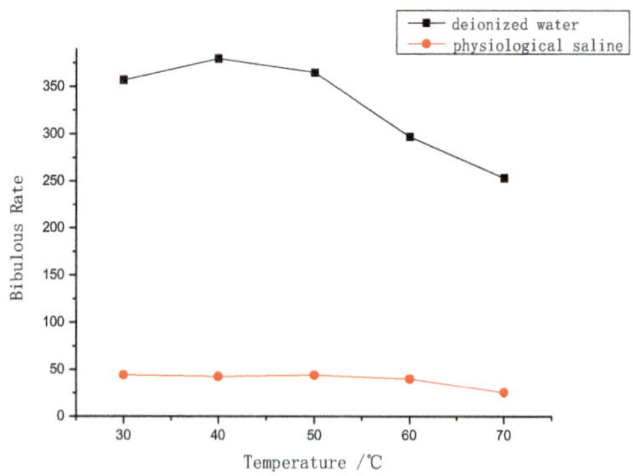

Figure 3. Effect of reaction temperature

1.2.2.4. Analysis of reaction temperature

We could see the best reaction temperature was 40°C from Figure 3. The temperature which is less than it was too low to facilitate the decomposition of initiator, as a result less free activity radicals were produced for the increase of the grafted rate and longer polymer chain reduced the bibulous rate of product. But when reaction temperature was more than 40°C, as chain transfer and chain termination speeded up, the grafted rate and bibulous rate (Yan Huang & Jiarui Shen, 1995) reduced.

1.2.2.5. Analysis of reaction time

Seeing from the Figure 4, the advisable polymerization time is 2 h. If the reaction time was too long, bibulous rate of resin was slightly lower, perhaps due to the reason that crosslinking density increased. When polymerization time was brief, the polymerization was incomplete and molecular mass of the product was low (Ge Y X & Zhang B Z, 2006), Network space became small after swelling, so the bibulous rate of resin reduced.

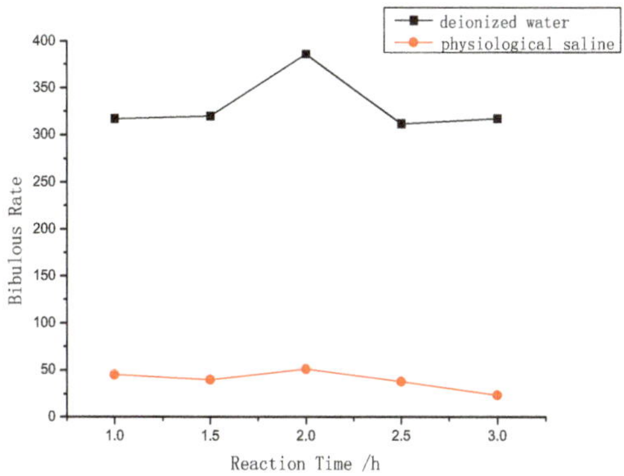

Figure 4. Effect of reaction time

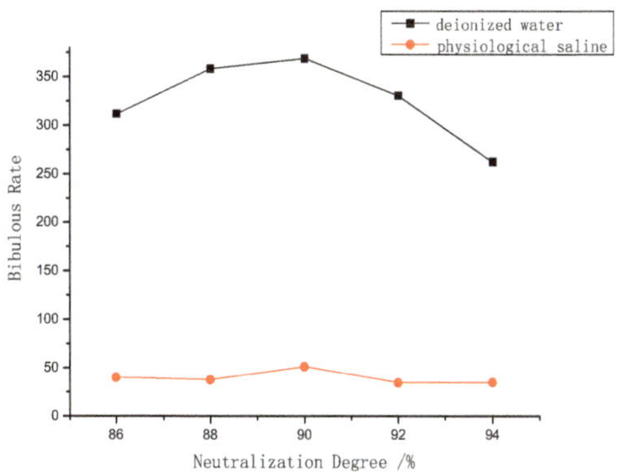

Figure 5. Effect of neutralization degree

1.2.2.6. Analysis of neutralization degree

Figure 5 indicates that the neutralization degree has great influence on the water absorption, and the bibulous rate is higher when the neutralization degree is 90%. Increasing the neutralization degree is increasing the concentration of the -COOK ion in the reaction system in essence. The dissociation ability of -COOK is far greater than -COOH in water, so the more -COOK led to more ionized -COO-. The increase of carboxylate ions in the lattice chain makes chain stretched and the repulsion of chains becomes strong, so the expansion of network polymerization increases, so does bibulous rate. However, with the increase of neutralization degree, the shielding effect of K^+ on carboxylate ion is gradually obvious in the system. And this kind of shielding effect will weaken repulsion between the chain and chain or adjacent carboxylate ions in the same chain, making the network expansion force reduced. Though neutralization degree is oversize, bibulous rate will reduce (Zhou M & Lin J M, 2000).

1.2.2.7. Analysis of initiator concentration

We could see that bibulous rate increased, along with the increase of the initiator concentration when concentration was less than 0.02 g, bibulous rate decreased instead when concentration was more than 0.02 g in Figure 6.

1.2.2.8. Analysis of crosslinker concentration

Figure 7 shows that bibulous rate was the highest when crosslinker concentration was 0.0015 g. That is because the crosslinking density is low, water molecules that easily seep into the resin make resin inflated, further closed to water to become gel and then appear the state with high bibulous. The longer the chain and the bigger crosslinking structure are, the larger swelling volume and the more water absorption are. But crosslinking density can't be too low, or you will make the product soluble in water. It is better to be in the lowest crosslinking degree under the condition of insolubility in water.

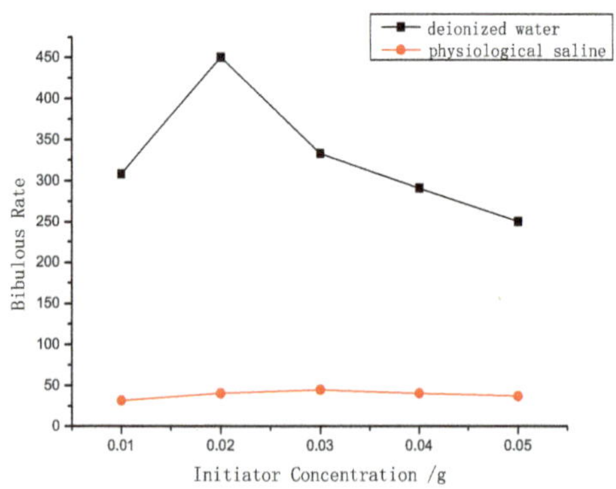

Figure 6. Effect of initiator concentration

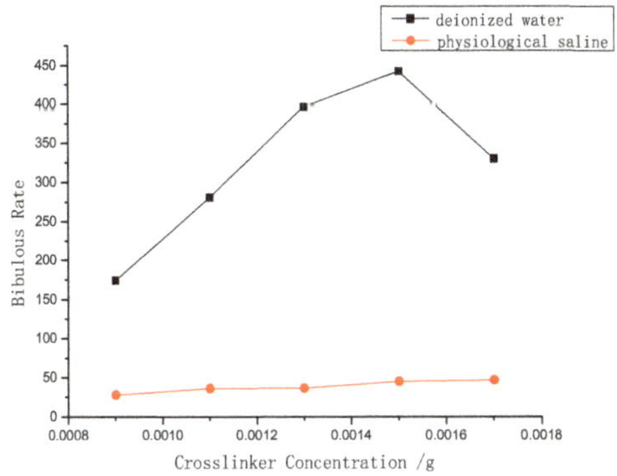

Figure 7. Effect of crosslinker concentration

1.2.3. Amylum Choice

It was clear that the water absorption of resin which was synthetized by corn amylum was much better than other kinds of amylum (Figure 8).

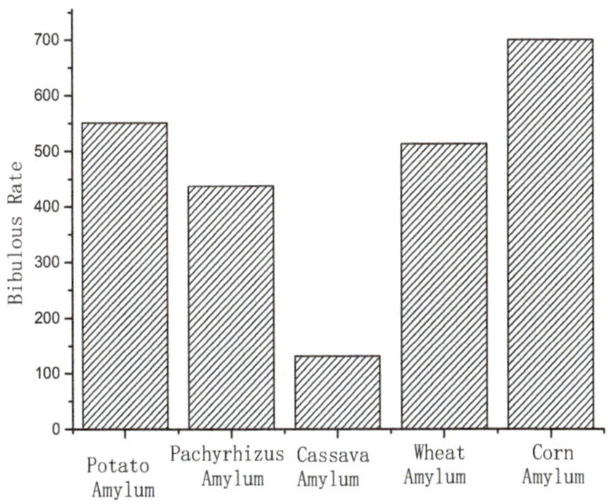

Figure 8. Effect of amylum

1.2.4. *FTIR and SEM*

1.2.4.1. *Spectrum of FTIR*

The infrared spectra (Figure 9) analysis shows that amylum and its grafted products appear -OH characteristic peak in 3490 cm^{-1}.Synthetic resin retained the amylum characteristic peak, and telescopic vibration absorption spectrum of C=O of carboxylic acid and acylamino appeared in 1571 cm^{-1} and 1714 cm^{-1} which are the characteristic absorption peaks of grafted acrylic acid and acrylamide. The telescopic vibration absorption spectrum of C-N of amide in 1407cm^{-1} is the characteristic absorption peak of grafted acrylamide. All of these are the proof of successful grafting.

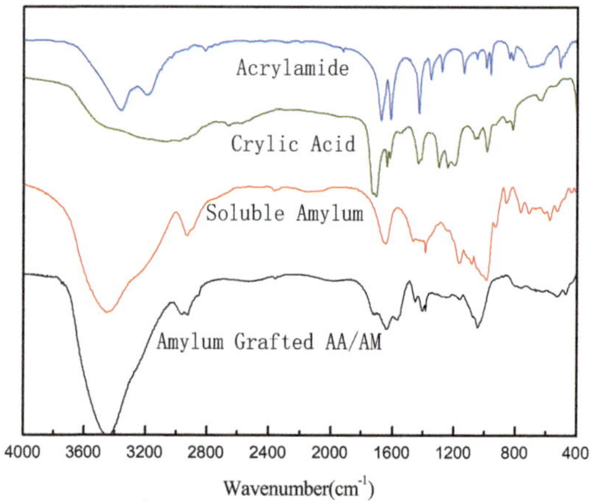

Figure 9. Spectrum of FTIR

1.2.4.2. *SEM micrograph of superabsorbent*

Grafting resin is network macromolecule formed by the skeleton polymers and branch polymers. Amylum, which belongs to hydrophilic semi-rigid bonds, makes polymer net into big space volume. This will form a developed hydrophilic network structure with a hole when amylum branch and grafting acrylic acid potassium, acrylamide branch chain interlace with each other.

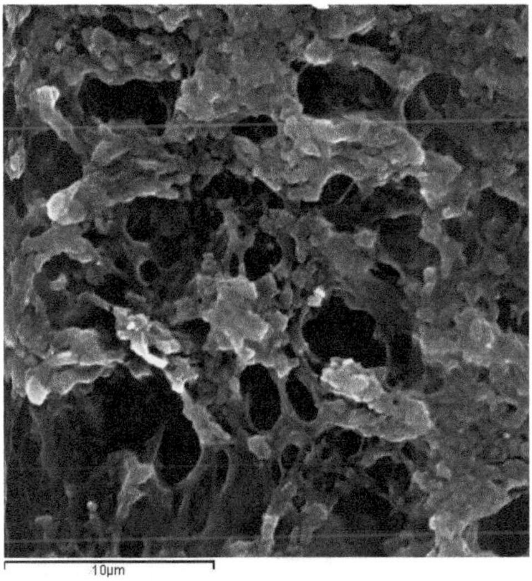

Figure 10. SEM micrograph of superabsorbent (×2000)

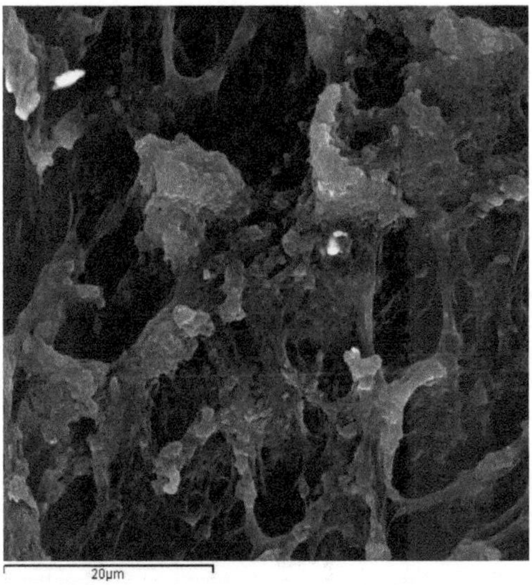

Figure 11. SEM micrograph of superabsorbent (×4000)

Figure 10 and Figure 11 are respectively the resin SEM micrograph pictures magnified 2000 times and 4000 times. We can see that the products form a three-dimensional network structure. The network structure is obviously dense and even. That is why binary grafted amylum superabsorbent has better water absorption than monobasic grafted amylum superabsorbent.

1.3. Summary

This chapter uses aqueous solution method for the study of amylum grafted acrylic/acrylamide superabsorbent resin. Through the exploration experiment, we make sure that choosing corn amylum as experiment material, potassium peroxydisulfate as the initiator and nitrogen gas as reaction protection gas. Known from the experiment result, amylum category, monomer concentration, initiator concentration, neutralization degree, polymerization temperature, polymerization time and crosslinker concentration and so on all have certain effect on water absorption.

Under the condition of determining gelatinization type and certain amylum, the chapter works out the best reaction condition of the preparation of corn amylum grafted acrylic/acrylamide superabsorbent resin .when monomer concentration is 12.0 g, reaction temperature is 30 °C, reaction time is 2 h, acrylic acid neutralization degree is 94.0%, initiator concentration is 0.02 g and crosslinker concentration is 0.0015 g, a better result can be obtained. The absorbent ratio of water is up to 699 times and the absorbent ratio of physiological saline is up to 70 times.

2. Synthesis of super-absorbent resin with the microwave method

2.1. Introduction

2.1.1. Introduction to Microwave Chemistry Mechanism

Microwave chemistry is a new cross-disciplinary study (Qinhan Jin, 1999) of chemical application. It is developed on the base of research on object property and interaction characteristics in the microwave field. It can be said that the microwave chemistry is a science, according to the electromagnetic field and the electromagnetic wave theory, the dielectric physical theory, condensed matter physics theory, plasma physics theory, material structure theory and chemical theory, using the modern microwave technology to research the physical and chemical behavior of material under microwave field.

Microwave (Fan Shouyong, 2003) is electromagnetic wave whose frequency is about 300 MHz-300 GHz, namely the wavelength within the scope of the 100 cm^{-1} mm. It is located between infrared radiation (light wave) of the electromagnetic spectrum and radio waves. Among it, the microwave band of 1-25 cm wavelength is specially used in radar, and the rest are used in telecommunications transmission. In order to prevent the interference of microwave power on radio communications, broadcast, television and radar, the microwave power frequency band of industry, scientific research, medicine, the household and civil has been ruled internationally. At present, the microwave technology has already been widely used in industrial and agricultural production, scientific research, medicine and so on.

Different from traditional Heating Method which heats conduction from the surface to the inside by thermal radiation, microwave heating is body heat caused by material dielectric loss in the electromagnetic field and at the same time electromagnetic energy transfer into heat energy. Due to the high efficiency and uniform heating characteristic of microwave heating, microwave heating can be applied to chemical reaction, influencing the reaction rate. Microwave, mainly includes microwave chemistry effect on condensed matter material, effect various chemical reactions by directly reacting with different kinds of chemical systems that are normally within the scope of microwave chemistry. The microwave chemistry also includes microwave effect on gaseous materials and microwave plasma chemistry.

2.1.2. Microwave

Gedye has found that microwave can promote organic reactions since 1986, microwave technology has been widely used in many organic reactions. A large number of Experiment results show that compared with traditional method, organic reactions under the action of microwave can speed up several times, several dozens of times even hundreds of times, and the biggest can be up to 1240 times. Now there are two different views about the reason why microwave can accelerate organic reactions in academia (Qinhan Jin, 1999).

Some academics thought that microwave heating is internal heating. It is fast, uniform and has no temperature gradient, hysteresis effect and many other characteristics. But microwave heating applied in chemical reactions is just a heating method, same as the traditional heating method. In a particular reaction, the kinetic will not change under the condition of invariable reactants, catalyst and product, without relation of heating method. They think that the microwave of 2450 MHz frequency used for chemical reaction belonged to the non-ionized radiation. When resonance with molecular chemical bond, it could not cause chemical bond fractured and make molecules promoted to higher rotational level and vibrative level. Therefore, the reason why microwave can accelerate organic reactions was attributed to the selective heating of the polarity organic, namely microwave radiation thermal effect.

Others believe that the effect of microwave on chemical reaction was extremely complex. On the one hand, the reactant molecule absorbed the energy of microwave, improved the movement speed of molecular. So the molecular motion became desultorily, which led to the increase in entropy. On the other hand the effect of microwave on polarity molecules force the molecular to move by the way which are under the function of the electromagnetic field and changing 2.45×10^3 times per second. Due to the decrease of the entropy, the effect mechanism of microwave on chemical reaction was not just described by microwave radiation thermal effect. Besides heating effect, microwave also had a kind of non-thermal effects that were not caused by the temperature. Organic reactions under the action of microwave changed the reaction kinetics and reduced the reaction activation energy.

Microwave application in organic synthesis has been more than 10 years, however, research on microwave accelerating mechanism is a new field, still in the initial stage now. Some reaction results are lack of experiment argumentations and many experiment phenomenon need to be explained more comprehensively, meticulously and systematically.

At present, the microwave research mainly focus on the following three aspects on organic synthetic chemistry:

1. Further improvement and establishment of new technology on microwave synthesis reaction technology.
2. Application and law of microwave organic synthesis reaction.
3. Systematic research of microwave chemistry theory.

2.1.3. *Application of microwave chemistry*

As is known to all, the traditional heating gradually heat conduction from the surface to the inside through the thermal radiation. In order to avoid excessive temperature gradient, heating speed often can't be too fast. What's more, it can't heat the different components of mixed materials in the same response device selectively.

Compared with traditional heating method, microwave heating has following characteristics [17]:

1. It can heat material from outer to inner gradually, and heating rate is quick.
2. High thermal efficiency and low energy consumption.
3. It has no hysteresis effect and it is easy to realize heating equably, also with low temperature gradient.
4. Heat each component of mixed materials selectively and realize automation control easily.
5. It can also promote endothermic reaction and exothermic reaction and has catalytic effect on some chemical reactions.
6. Reduce some chemical reaction temperature.
7. Make some high polymer materials to have ideal performance (Du T S & Kang S Z, 2000).

Microwave technology with such characteristics was widely used in organic synthesis, chemical analysis, food processing, medical science and many other fields.

The application of microwave organic synthesis technology has begun in the 1980s.After Gedye and mates found that applying microwave heating technology in organic synthesis of small molecules could significantly improve the reaction rate, using microwave technology to promote organic reactions became the focus of attentions. Nowadays the microwave heating organic synthesis reactions are consist of the alkylation reaction, esterification reaction, substitution and elimination reaction, sulphonation reaction, olefins addition reaction, condensation reaction, rearrangement reaction, pericyclic reaction, organic metal reaction and so on. The research shows that, microwave technology applied in organic synthesis not only can improve the reaction speed and shorten the reaction time, but also can help to have easy operation, high yield, pure product and other advantages (Wangxi Zhang, 2004).

In addition, microwave heating technology plays a support role on the synthesis and processing of polymer compound with high viscosity and low thermal conductivity, also the modification and curing of natural polymer.

2.2. Experiment

2.2.1. Reagents and instruments

2.2.1.1. Reagents

Soluble Amylum (Corn)	A.R. Tianjin Kemiou Chemical Reagents Development Center
Crylic Acid	C.P. Tianjin Kemiou Chemical Reagents Development Center
Potassium Hydroxide	A.R. Tianjin Kemiou Chemical Reagents Development Center
Potassium Peroxydisulfate	A.R. Tianjin Sitong Chemical Plant
N, N,-methylene biasacrylamide	C.P. Chemical Reagent of Traditional Chinese Medicine Co. LTD

2.2.1.2. Instruments

MAS-I Microwave Reactor	New Instrument Microwave Chemical Technology Co. LTD
101-1 Constant Temperature Electrothermal Blowing Drying Oven	Shanghai Luda Experiment Instrument Co.LTD
TG328B Lightning Analytical Balance	Shanghai Balance Instrument Plant
JJ-1 Timing Electric Mixer	Jiangsu Zhongda Instrument Plant
TENSOR27 FTIR	Germany
Hitachi S-2500 SEM	Japanese Hitachi Co. LTD

2.2.2. Experiment procedure

1. Take 2 g corn amylum into four-mouth bottle and add certain deionized water. Put the bottle into the microwave synthesis device after intensive mixing and connect it with condenser pipe and thermometer, stirring 20 minute under 50 °C, gelatinization with the microwave power in 800 W.
2. After the viscosity, cool it naturally to room temperature. The acrylic acid and acrylamide with a certain neutralization degree (which were neutralized by KOH), initiator and crosslinker were added.
3. Continue radiating in certain microwave power and temperature for some time.
4. When reaction was over, dismantle device, move the product to the small dish and dry it to constant weight.
5. Crush the product and do the performance measurement and structure characterization.

2.2.3. Test method

2.2.3.1. Test of bibulous rate

There are variety of methods to test the water absorption of superabsorbent, such as low method, paper bag method, natural filtering method, sieve net method and etc. The data from various test methods have big differences and it is hard to compare different data of bibulous rate from various documents.

The sieve net method is used widely by many researchers, which is known that take certain quality (m_1) product into the beaker and add excessive deionized water and physiological saline. Standing before swelling completely, then screen and filter the excessive water by 100 mesh screen cloth and adsorb surface water of resin using paper absorbent. Then weigh (m_2) it and calculate bibulous rate.

2.2.3.2. Structure Characterization of Bibulous Resin

1. Infrared Spectrometry

Making corn amylum and absorbent resin dried fully, potassium bromide was used for tabletting.

2. Scanning Electron Microscope Method

The exterior characteristics was selected and observed on the 1cm² resin by the scanning electron microscopy (SEM).

2.3. Results and discussions

2.3.1. Polymerization temperature

Reaction temperature has a certain effect on superabsorbent resin. When the temperature is lower, polymerization speed is slower, some resin crosslink incompletely and bibulous rate declines. When the temperature is too high, on the one hand initiator is beneficial to initiate polymerization. On the other hand chain transfer and chain termination speed up so that soluble part of the superabsorbent resin increases and bibulous rate declines. In this paper, we made a research on the effect of reaction temperature on water absorption under the fixed other conditions.

The study found that bibulous rate reached maximum when the reaction temperature was 35°C (Figure 12).

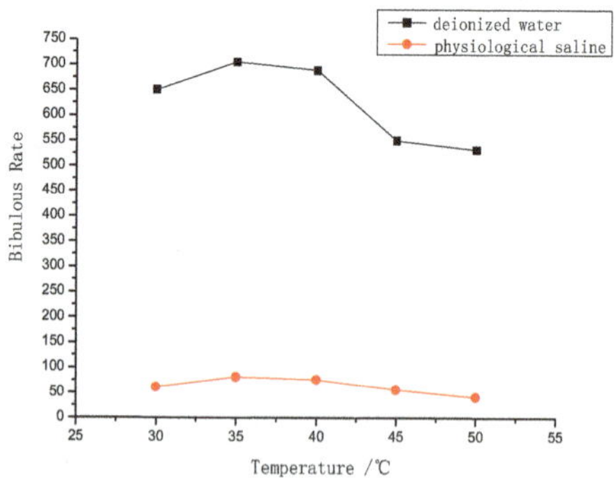

Figure 12. Effect of reaction temperature

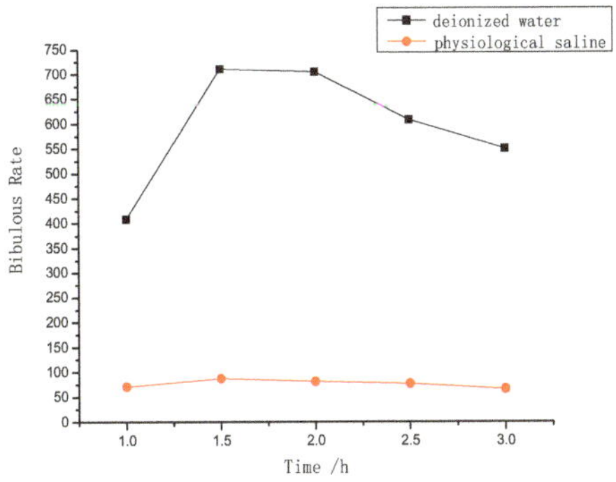

Figure 13. Effect of reaction time

2.3.2. *Polymerization time*

The reaction speed could be accelerated in the microwave condition, and reaction free radicals mechanism did not change. Microwave radiation made polymerization reaction speed increased, microwave inducing made reaction period shortened. Figure 13 is curve of the bibulous rate of synthetic resin with the change of microwave radiation time. The Experiment result showed that the microwave radiation could make the grafting copolymerization reaction speed greatly improved, reaction cycle shortened and water absorption performance was also improved on the whole. This may be caused by two reasons. One was the even synchronous heating characteristic of microwave, makinge the initiator and monomer of the system synchronously triggered, simultaneously reacted. And microwave heating rate was so quick that it can reach the needed temperature of system quickly. The other one was by means of microwave radiation response, may be "non-thermal effect" which made the water absorption performance improved. But with the increase of microwave reaction time, the bibulous rate of superabsorbent resin decreased. The reason may be that the radiation time was too long and resin crosslinking degree increased, so bibulous rate of synthetic resin was down.

2.3.3. *FTIR and SEM*

2.3.3.1. *Spectrum of FTIR*

The infrared spectra (Figure 14) analysis shows that amylum and its grafted products appear -OH characteristic peak in 3490 cm^{-1}. Synthetic resin retained the amylum characteristic peak, and telescopic vibration absorption spectrum of C=O of carboxylic acid and acylamino appeared in 1574 cm^{-1} and 1722 cm^{-1},which are the characteristic absorption peaks of grafted acrylic acid and acrylamide. The telescopic vibration absorption spectrum of C-N of amide in 1405 cm^{-1}, is the characteristic absorption peak of grafted acrylamide. All of these prove the success of this experiment.

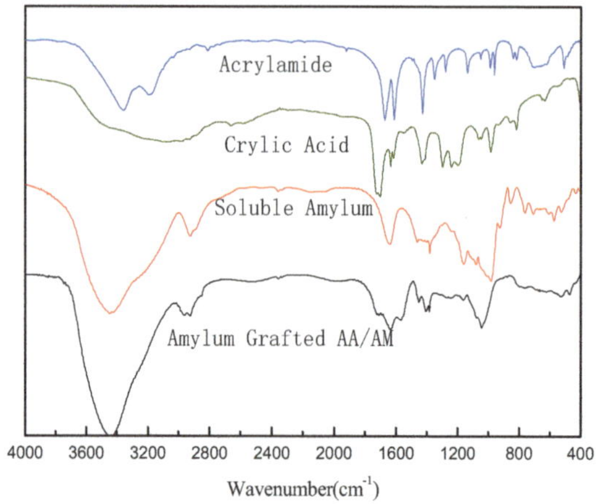

Figure 14. Spectrum of FTIR

2.3.3.2. Scanning picture

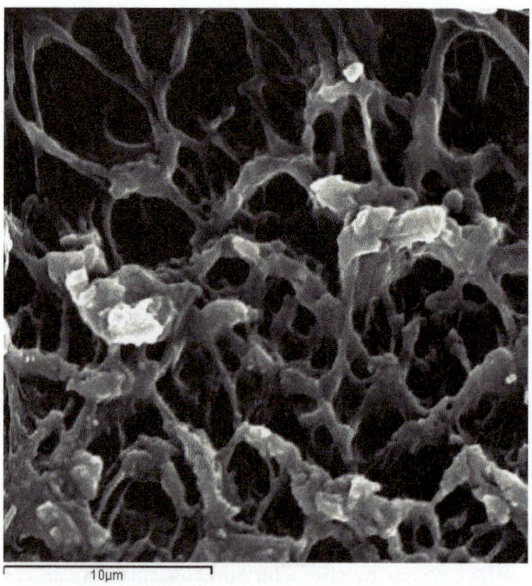

Figure 15. SEM micrograph of superabsorbent (×4000)

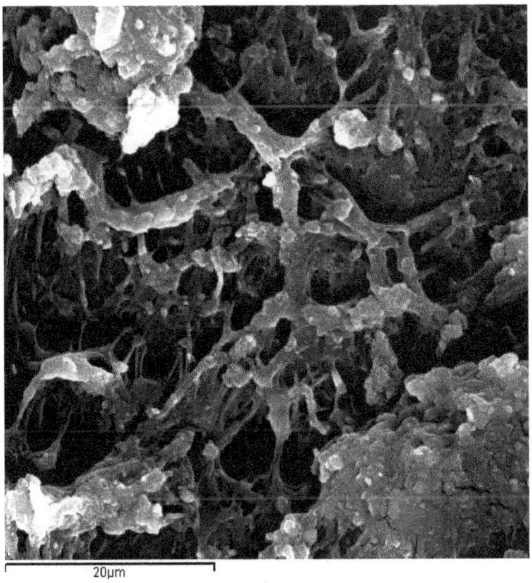

Figure 16. SEM micrograph of superabsorbent (×2000)

Figure 15 and Figure 16 are respectively the resin scanning photos magnified 4000 times and 2000 times. We can see that the product has already formed the uniform three-dimensional network structure. Compared with Figure 10 and Figure 11, the network structure is obviously dense and even. Microwave heating is used by dielectric loss principle, different from gradient transfer of the traditional heating method. Its internal and external synchronous heating causes product loose and porous. This is the reason why binary grafted amylum superabsorbent in microwave condition has better water absorption than that in traditional condition.

2.4. Summary

This chapter uses aqueous solution method for amylum grafted acrylic/acrylamide superabsorbent resin by microwave. Like traditional condition, polymerization temperature and polymerization time are fixed respectively. The chapter works out the best reaction condition of the preparation of corn amylum grafted acrylic/acrylamide superabsorbent resin is that reaction temperature is 35 °C and reaction time is 1.5 h. The absorbent ratio of water is up to 711 times and the absorbent ratio of physiological saline is up to 86 times.

The reaction time and temperature, the time of getting the product at the same bibulous rate by microwave is obviously slower than that using traditional method and the temperature is obviously lower. Under similar condition, the property of the product by microwave method is improved. What's more, reaction process doesn't need nitrogen protection under microwave condition. These phenomena explain that the synthesis of superabsorbent resin by means of microwave is of great value.

3. Exploration and research applications

3.1. Introduction

Superabsorbent resin is commonly used in soil superabsorbent resins and soil amelioraters in agriculture and forestry gardening areas. A large amount researches at home and abroad have shown that the use of superabsorbent resin can significantly improve water content of soil, volume expansion rate, total porosity and capillary porosity and soil structure. It can also ease change of ground temperature, microbial activities and soil fertility. It always increase crop output, improve survival rate of trees and promote plant growth. In water-shortage areas, the use of superabsorbent resin has significant water-saving effect.

China is listed as one of 13 short water countries and one of the world's most serious water-loss countries. The soil erosion areas are 3.67 million km^2, which occupy 38.2% of the total land areas. The desertification lands are throughout 13 provinces, cities and autonomous regions of the country and inferior soil exists generally. The composition balance of organic and inorganic substances in soils was destroyed by the unrestrained using of chemical fertilizers. At the same time, our country is a populous country in the world and the implementation of the soil and water conservation is one of our basic policies as well as a systematic engineering under such serious circumstances. With the characteristic of big water absorbability and high water retention, applying superabsorbent resin to soil erosion control, desertification, agricultural water-saving is a very significant work.

Now this technology is not large-scale used for water-saving agriculture and the main reason as follows. Firstly, the price of superabsorbent resin is expensive and using in agriculture will increase the product cost. It is difficult to promote. Secondly, a lot of superabsorbent resin which theoretically has high bibulous rate and salt tolerance are in fact unsatisfactory when used in complex soil environment. Thirdly, farmers in some drought areas who are often poor and backward have no perceptual knowledge on superabsorbent resin and can't accept it. These points have decided that promoting superabsorbent resins application on agriculture really needs quite a long time. It requires researchers not only to development a new superabsorbent resin with good comprehensive performance, but also to find a multi-function superabsorbent resin which is drought resistant, pest resistant, fit for plant growth and not destroying the soil structure under the soil environment and considering the nutrition demands of plant growth.

This chapter has made a research on amylum superabsorbent grafted AA/AM under traditional condition and microwave condition. There own water retention and water absorption ability, water steam suppression, expansion effect and water retaining capacity in the soil was studied in this experiment, hoping can offer some help on the future application experiment in large scale.

3.2. Experiment

3.2.1. Material and equipment

- Amylum Superabsorbent Grafted AA/AM by Traditional Method (50 mesh)

- Amylum Superabsorbent Grafted AA/AM by Microwave Method (50 mesh)
- Local Soil
- Commercial Grass Seed
- DZF-6021 Vacuum Drying Oven Shanghai Jinghong Experiment Instrument Co.LTD

3.2.2. *Experiment methods*

3.2.2.1. *Water retention of superabsorbent*

Taking bibulous saturated resin 50.0 g into Buchner funnel and putting a non-absorbent small board on it. Then putting weight on board, resting about 20 minute. The water mass was weight in different pressure.

3.2.2.2. *Pretreatment of experiment soil*

This study has selected the local loess as experiment object by air dry indoor and use drysievetest for pretreatment. Pat the soil slightly into small patch and dispose impurities. Then put the sample in 50 °C oven and drought it out.

Grinding dried soil sample through a 50 mesh sieve and setting it in dry place for standby application.

3.2.2.3. *Bibulous rate of superabsorbent resin in soil*

Add superabsorbent resin of 0.00%, 0.10%, 0.15%, 0.20%, 0.25%, 0.30% into 15.0 g soil respectively. Mix them equally and pour them into tea bag. Put them into beakers and take them out after soaking, hanging them up until no water drop and weight them. Calculate bibulous rate of superabsorbent resin in soil according to equation 2.

$$Q = (W_1 - W_0 - W_2)/W_2 \ldots \qquad (2)$$

Among:

- Q - bibulous rate of superabsorbent resin in soil
- W_0 - mass of water absorption of mixed soil /g
- W_1- mass of water absorption of blank soil /g
- W_2 - dosage of superabsorbent resin /g

3.2.2.4. *Experiment of soil spraying*

Taking five soils of 15.0 g, adding superabsorbent resin of 0.00 g, 0.03 g, 0.06 g, 0.09 g, 0.12 g respectively, mixing equally and pouring them into glass tube, making a 1.0 cm-diameter hole at the bottom of the glass tube and plugging the hole with cotton to prevent soil's leaking out. Spraying soil sample with 20.0 mL distilled water with the velocity of 5.0 mL/min by separating funnel on the top of glass tube and the flow of water from the glass tube is caught by measuring cylinder of 20.0 mL. The time and the water volume in the measuring cylinder was recorded when the water begins to flow out of the glass tube as soon as the flow stops.

3.2.2.5. Water steam suppression of soil

Taking five Experiment soils of 20.0 g into five same beakers respectively, adding superabsorbent resin of 0.00%, 0.15%, 0.30%, 0.45%, 0.60% of soil dosage, mixing well, joining with de-ionized water of 10.0 g and weighing it until the soil fully wet. Put soil sample into a 35 °C oven to dry it and weigh it every 30 min. Drawing curve by evaporation time and soil moisture and investigating effect of water steam suppression of soil. Calculating water content in soil according to equation 3:

$$W\% = (M_1 - M_2)/M_1 \times 100\% \ldots \quad (3)$$

Among:

- M_1 - initial mass /g
- M_2 - present mass /g

3.2.2.6. Expansion effect of superabsorbent resin

Taking five soils of 20.0 g, adding superabsorbent resin of 0.00 g, 0.03 g, 0.06 g, 0.09 g, 0.12 g respectively, mixing equally, measuring 15.0 mL mixed soil using measuring cylinder of 25.0 mL, adding enough distilled water, removing extra water by burette and measuring the volume of moisture soil samples. Calculating expansion rate according to equation 4:

$$V\,\% = (V-15)/15 \times 100\% \quad (4)$$

Among:

- V - Volume of wet soil sample /mL

3.2.2.7. Experiment of grass planting

Three experiment soils of 200.0 g were putted into three breathable flowerpots, adding 1.0 g superabsorbent resin prepared by traditional method in the first one, adding 1.0 g superabsorbent resin prepared by microwave method in the second one and the last one acts as blank soil. Choosing 30 plump grass seeds into three flowerpots averagely, adding water of 150.0 mL respectively, watering 10.0 mL every day and ensuring the sun exposure time, watering frequency and water dosage of three flowerpots are same. Observing them for 20 days and taking notes of grass growing condition. Planting test room temperature is kept 15-25°C. Planting irrigation water is local tap water.

3.3. Results and discussion

3.3.1. Resin self-protection water

Two resins both have good performance of water retention and products by microwave method are slightly better than that by traditional method from the Figure 17.

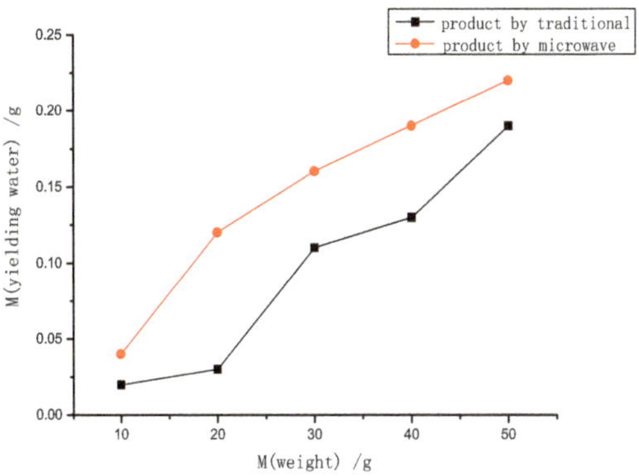

Figure 17. Test of water retention

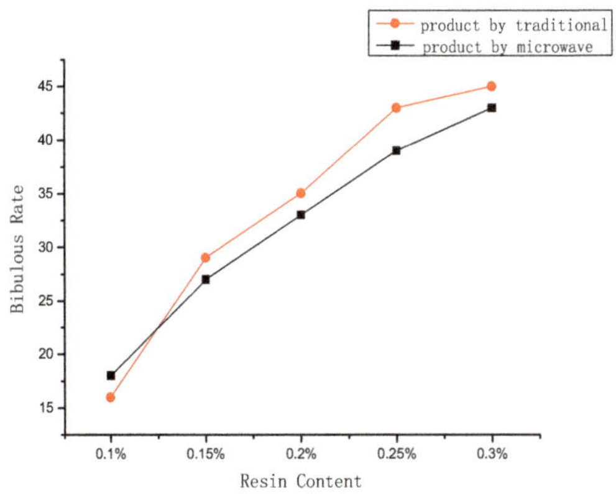

Figure 18. Bibulous rate of superabsorbent resin in soil

3.3.2. *Superabsorbent resin in soil*

We have found that with the increase of the dosage of resin, bibulous rate increases. This is because that with the increase of resin content in soil, the water absorption of resin is more obvious than water absorption of soil itself. Water absorption capacity of soil with superabsorbent resin is less than that without superabsorbent resin. This is because superabsorbent resin mixed in soil is influenced by soil oppression and free metal ions in the process of water swelling. (Figure 18)

The performance of products by microwave method is obviously better than that by traditional method. So products by microwave method are chosen for the following experiments.

3.3.3. *Experiment result of soil spraying*

The main purpose of soil spraying experiment is to test whether superabsorbent resin can form a water-resisting layer in soil of certain depth to prevent soil erosion in rainfall or irrigation cases and whether it can absorb a lot of water storing for use in drought.

From Table 3, we find that after adding superabsorbent resin, the water outflow decreases, so water retention ability of soil strengthens. We also see that after joining with superabsorbent resin, speed of water leakage has decelerated. It is because that the resin has crosslinked with its surrounding soil particles and formed a water-resisting layer, so it can prevent moisture infiltration. Therefore, water retention ability is stronger as less water flow.

It shows that soil mixed with superabsorbent resin can clearly improve the performance of water retention.

	0.00g	0.03g	0.06g	0.09g	0.12g
$V_{\text{Water Spraying}}$	20mL	20mL	20mL	20mL	20mL
V_{Spraying}	5mL/min	5mL/min	5mL/min	5mL/min	5mL/min
t_0	17min	21min	26min	35min	38min
$V_{\text{Yielding Water}}$	9.5mL	8.4mL	6.9mL	6.1mL	5.8mL

Table 3. Experiment result of soil spraying

3.3.4. *Water steam suppression of soil*

Superabsorbent resin has strong performance of water retention, even in the condition of pressure and heating. This experiment shows that different dosages of superabsorbent resin have inhibiting effects on water evaporation under 35 °C. Figure 19 show that soil without resin loses water faster when is heated. Water content is already 0.5% at 180 minute and the water in soil sample has nearly evaporated out. The water content of soil samples being added with resin of 0.15%, 0.3%, 0.45%, 0.6% are 2.3%, 3.3%, 3.5% and 4.1%.Obviously speed of water evaporation in soil has been inhibited by resin. With the increase of the resin content, inhibition of evaporation also increases.

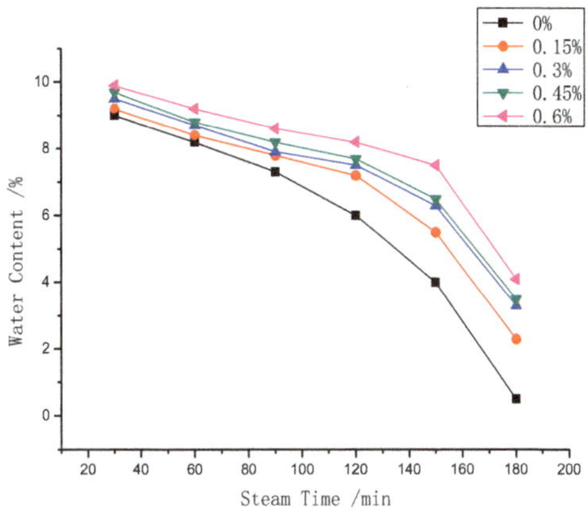

Figure 19. Effect of water steam suppression of soil

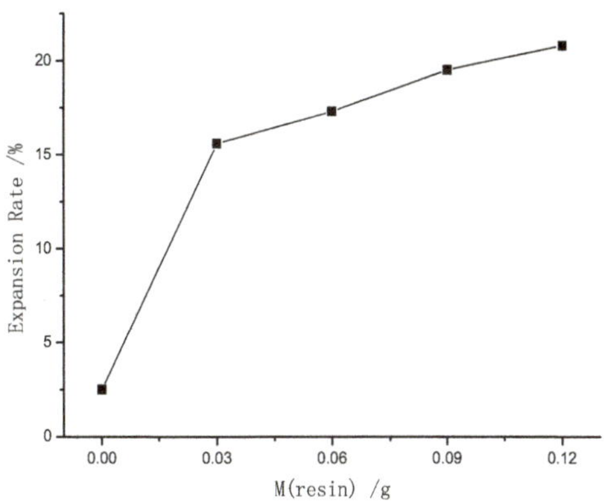

Figure 20. Effect of expansion effect of resin

3.3.5. Expansion effect of superabsorbent resin

The volume of soil has significantly increased after adding resin. The volume inflation rate has been amount to 20.8% when we add 0.12g resin in 20.0g soil. The research has found that the volume inflation rate increases with the increasing use of resin. This expansion effect is very good for improving soil compactness, regulating gas permeability of soil and making the soil structure favorable (Figure 20).

3.3.6. Experiment result of grass planting

In this experiment, we have not found that superabsorbent resin can fight for water with plants so that influence plant growth. The suitable dosage of superabsorbent resin has certainly facilitated plant growth and grass without resin is significantly weaker growth than grass with resin. They all have significantly effects with resin added either from germination rate or from average height (Table 4).

Type	Days	Seed Number	Germinative Number	Average Height/cm	Plant Growth
1	20	10	8	51.12	Bloom,Thick Stem
2	20	10	7	50.07	Bloom,Thin Stem
3	20	10	3	29.87	Poor, Thin Stem

Table 4. Experiment result of grass planting

3.4. Summary

Through testing water retention of resin, bibulous rate in soil, experiment of soil spraying, water steam suppression of soil, expansion effect of superabsorbent resin and experiment of grass planting, a simple research on application of superabsorbent resin has been discussed. The results show that: soils dashed with superabsorbent resin have improved the water retention of soil and soil compactness, regulate gas permeability of soil and make the soil structure favorable. The suitable dosage of superabsorbent resin has certainly facilitated plant growth.

Author details

Hao Ren, Zhen Niu, Juan Wang and Jing Ning
Shandong Provincial Key Laboratory of Fluorine Chemistry and Chemical Materials,
School of Chemistry and Chemical Engineering, University of Jinan, Shandong, China

Acknowledgement

This work was financially supported by the Publishing Fundation Project of University of Jinan. Special thanks are given to the research team for providing support in this study.

4. References

Du T S & Kang S Z. (2000). Situation and expectation of application of water-saving agent on agriculture. *Research on Agricultural Modernization,* 2000, 21(5):317-320.

Fushan Chen & Qian Gong. (2007). Influential Factors and Optimization of Synthesis of Cassava Amylum Superabsorbent Grafted AA/AM. *Leather Chemical Industry,* 2007, 24(5):38-42.

Ge Y X & Zhang B Z. (2006). Preparation and Characterization of Superabsorbent Resin. *Science and Technology Journal of Hebei University,* 2006, 27(4):285-287.

Hongke Tang & Qi Chen. (2007). Synthesis of Amylum Superabsorbent Grafted AA/AM. *Synthesis Chemistry,* 2007, 15(5):643-646.

Kabiri K & Zohuriaan-Mehr M J. (2003). Superabsorbent Hydrogel Composites. *Polymers for Advanced Technologies,* 2003, 14 (6):438-444.

Kabiri K, Omidian H & Zohuriaan-Mehr M J. (2003). Novel Approach to Highly Porous Superabsorbent Hydrogels: Synergistic Effect of Porogens on Porosity and Swelling Rate. *Polymer International,* 2003, 52(7):1158-1164.

Lokhande H T & Lyer V. (1992). Water-Superabsorbent Polymers through Gamma Radiation-Induced Graft-Copolymerization of Acrylontirile on Guar gum. *Journal of Applied Polymer Science,* 1992, 45:2031-2036.

Liu Z S & Rempel G L. (1997). Preparation of Superabsorbent Polymers by Crosslinking Acrylic Acid and Acrylamide Copolymers. *Journal of Applied Polymer Science,* 1997, 64 (7):1345-1353.

Lu S J, Duan M L & Lin S B. (2003). Synthesis of Superabsorbent Starch-Graft-Poly (potassium acrylate-co-acrylamide) and Its Properties. *Journal of Applied Polymer Science,* 2003, 88 (6):1536-1542.

Li D L & Yang P L. (2004). Experimental Studies on Water-saving Agent in Grape Cultivation. *Agricultural engineering journal,* 2004, 20(3): 51-54.

Pourjavadi A, Mahdavinia G R & Zohuriaan-Mehr M J. (2003). Modified Chitosan II H-chitoPAN, A Novel pH-responsive Superabsorbent Hydrogel. *Journal of Applied Polymer Science,* 2003, 90 (11):3115-3121.

Qinhan Jin. (1999). Microwave Chemistry. Beijing. *Science Press,* 1999.

Raju K M & Raju M P. (2001). Synthesis and Swelling Properties of Superabsorbent Copolymers. *Advances in Polymer Technology.*2001, 20(2):146-154.

Shouyong Fan. (2003). Microwave Technology and Microwave Circuit. Beijing. *Mechanical Industry Press,* 2003.

Suda K, Mongkolsawat K & Sonsuk M. (2002). Synthesis and property characterization of cassava starch grafted poly [acrylamide-co-(maleic acid)] superabsorbent via g-irradiation. *Polymer,* 2002, 43(14): 3915-3924.

Taden A, Tait A H & Kraft A. (2002). Synthesis and Polymerization of 5-(Methacrylamido) tatrazole, *a Water-soluble Acidic Monomer. Journal of Polymer Science,* 2002, 40 (23):4333-4343.

Wangxi Zhang. (2004). Application of Microwave in Chemical Reaction. *Synthesis technology and application,* 2004, 19(3):25-27.

Yan Huang & Jiarui Shen. (1995). Research on Thickening Ability of Amylum Grafted Acrylamide. *Polymer Materials Science and Engineering*, 1995, 11(3):98-102.

Yan Q Z & Su X T. (2005). Wave Synthesis of Cassava Amylum Superabsorbent Grafted AA/AM. *Chem J Chin Univ*, 2005, 26(7):1363-1365.

Zhou M & Lin J M. (2000). Research on Amylum Grafted Acrylamide. *Journal of Overseas University (Natural Science)*, 2000, 20 (4):362-365.

Section 3

Polymer Synthesis

Chapter 6

FI Catalyst for Polymerization of Olefin

S. Damavandi, S. Ahmadjo, R. Sandaroos and G.H. Zohuri

Additional information is available at the end of the chapter

1. Introduction

Since the pioneering by Karl Ziegler and Giulio Natta in the early 1950's on the polymerization of simple olefins, there has been intense interest in the application of early transition metal catalysts for the selective polymerization of inexpensive olefins. Following to Ziegler-Natta catalysts [1], metallocene catalysts were discovered in the late 1980's and resulted in numerous industrial processes for improving the properties of polyolefinic materials along with performance parameters. This field has been remarkably renewed with the use of catalysts based on early transition metals metallocene [2]. Development of new catalysts with transition metals has played a substantial role in the fast-growing polyolefins industry. Improvement of new, better performing, less costly polyolefins has often been a result of catalyst development. However, polymerization processes of olefins, beside the requirement of higher activity catalyst to control the particle size, particle size distribution, and morphology of the resultant polyolefin are quite important. In the other words, success in these developments requires an appropriate integration of catalyst selections with reactor type and process parameters [3-5].

As is well-known, most polyolefinic materials are produced using transition metal-catalyzed olefin polymerization technology. While the multisited heterogeneous Ziegler–Natta catalysts represented by $MgCl_2$-supported $TiCl_4$ catalysts currently dominate the market, molecular catalysts (single-site catalysts) represented by group 4 metallocene catalysts and constrained geometry catalysts (CGCs) are gaining an increasing presence in the market (Figure 1) [4-9].

As DFT (Density Functional Theory) calculations performed on a model metallocene catalyst $H_2SiCp_2ZrMe^+$ for ethylene polymerization (Figure 2) suggested that ethylene polymerization is a process that involves intense electron exchange between a ligand and a metal [10]. Accordingly and following the great success of the metallocene catalysts, significant efforts have been directed toward the discovery and application of new, highly active, single-site catalysts (post-metallocene catalysts) [10,11].

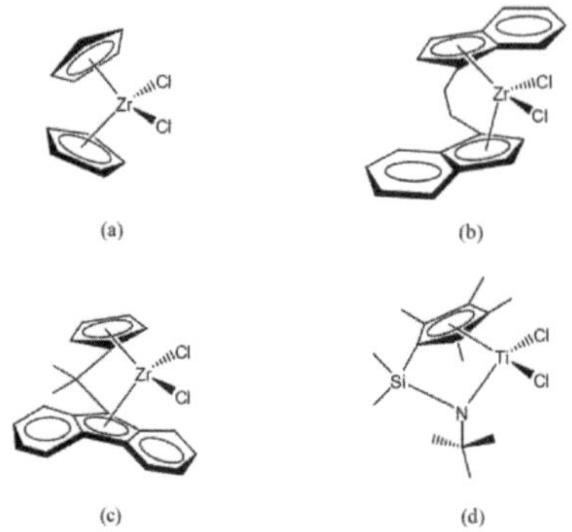

Figure 1. Metallocene catalysts for olefin polymerization: (a) Cp_2ZrCl_2; (b) *rac*-Et(Ind)$_2$ZrCl$_2$; (c) iPr(Flu)(Cp)ZrCl$_2$; (d) Constrained geometry catalyst (CGC) [10].

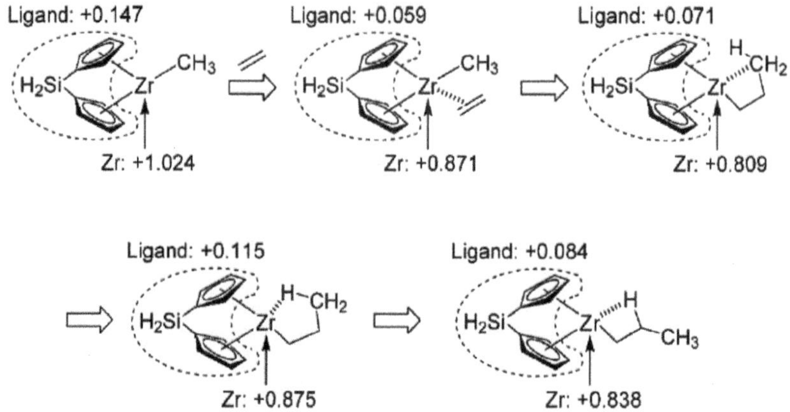

Figure 2. Calculated Charges of the ligand and Zr metal of a model metallocene catalyst for ethylene polymerization (DFT calculations) [10].

These research efforts have led to the introduction of quite a few high-activity single-site catalysts based on both early and late transition metal complexes with various ligand environments [12-18]. In association with appropriate cocatalysts, many of these catalysts show ethylene polymerization activities that are superior or comparable to those seen with early group 4 transition metals metallocene catalysts. These post-metallocene catalysts can produce a wide array of distinctive polymers (e.g., hyper-branched PEs, ethylene–methyl acrylate copolymers, monodisperse poly(1-hexene)s, and block copolymers based on α-olefins), many of which were inaccessible using metallocene catalysts [3,16-24].

Since the ligand structure has a central role in determining the activity as well as the stereospecifity of these types of catalysts, and as shown in Figure 2, flexible electronic nature of a ligand is a key requirement for achieving high activity, the research based on the ligand oriented catalyst design concept has resulted in the discovery of a number of highly active catalysts for the polymerization of ethylene, which include: phenoxy-imine ligand early transition metal complexes (FI catalysts), pyrrolide-imine ligand group 4 transition metal complexes (PI catalysts), indolide-imine ligand Ti complexes (II catalysts), phenoxy-imine ligand group 4 transition metal complexes (IF catalysts), phenoxy-ether ligand Ti complexes (FE catalysts), imine-pyridine ligand late transition metal complexes (IP catalysts), and tris(pyrazolyl) borate ligand Ta complexes (PB catalysts) (Figure 3) [20,26-30].

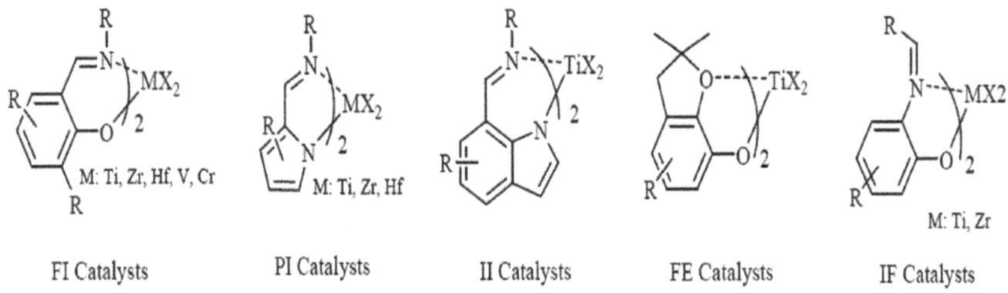

Figure 3. Development of highly active single-site ethylene polymerization catalysts [20].

In particular, bis(phenoxy-imine) group 4 metal catalysts, developed by Fujita [19–21] caused a new revolution in the field of catalytic olefin polymerization. These complex catalysts exhibit unique characteristics for production of new polymers that are not prepared by conventional Ziegler–Natta catalysts, as well as by ordinary metallocene-type catalysts [16]. The key feature of these complexes is the incorporation of nonsymmetric bidenate or tridentates ligands that possess electronically flexible properties (e.g., phenoxy–cyclopentadienyl, phenoxy–imine, phenoxy–ether, phenoxy–pyridine, pyrrolide–imine, and indolide–imine). These complexes can typically combine with appropriate cocatalysts to form highly active catalysts for the polymerization of ethylene. Among these new complexes, bis(phenoxy–imine) early transition metal complexes (named FI Catalysts) are particularly versatile for olefin polymerization when activated (Figure 4) [20,21].

Although a large number of families of high-performance single-site catalysts have been developed thus far, improvements in some aspects of catalytic performance (e.g., temperature stability, precise control of chain transfer, comonomer sequence distribution control, precise control of polymer stereochemistry, and the ability to incorporate sterically encumbered monomers and polar monomers) are still required to achieve both greater control over polymer microstructures and extension of generic polyolefinic materials by introducing new monomer combinations [31].

2. Characteristics of FI catalysts

In 1997, researchers at Mitsui Chemicals introduced phenoxy–imine [O-N] ligand early transition metal complexes (now known as FI catalysts) for the controlled polymerization and copolymerization of olefinic monomers [30]. Depending on the ligand design, the catalysts show different behaviors in ethylene and propylene polymerization, and the ligands strongly influence catalyst parameters such as activity, polymerization mechanism, and polymer properties including molecular weight. Extensively, FI catalysts, can polymerized ethylene with high efficiency, independent of the transition metals that are employed (Ti, Zr, Hf, etc.), showing the notable ability of phenoxy–imine ligands for efficient ethylene insertion.

The electronically flexible nature of the phenoxy–imine ligands may be responsible for these superior results. FI catalysts have the following structural and electronic features resulting in unique polymerization catalysis as well as formation of distinctive polymers [31-34].

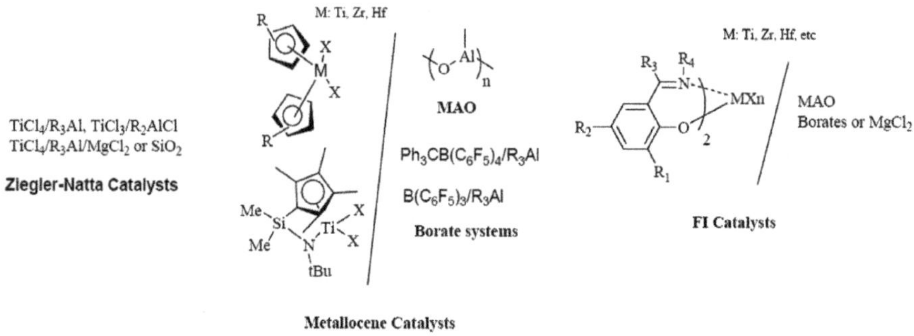

Figure 4. The progression of catalyst discovery, conventional olefin catalysts and FI catalyst [20,21].

2.1. Structure of FI catalysts

Since an FI catalyst contains two bidentate nonsymmetric phenoxy-imine ligands, it can potentially display five isomers from **a** to **e** (Figure 5) arising from the coordination modes of ligands in an octahedral configuration. X-ray analysis has established that, in the solid state, an FI catalyst normally exists as the isomer **a**, meaning, it has a *trans*-O, *cis*-N, and *cis*-Cl arrangement, and thus C_2 symmetry [35]. The C_2-symmetric nature implies that the catalyst may be an isospecific catalyst for propylene polymerization [35].

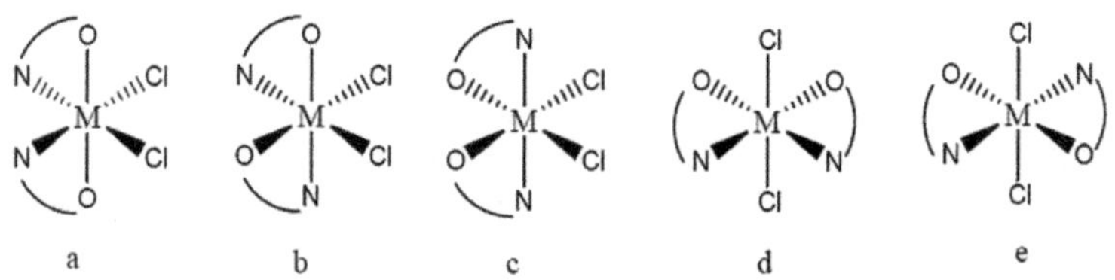

Figure 5. Possible isomers for FI catalysts. a) N-cis, O-trans, Cl-cis, C_2 symmetry b) N-cis, O-cis, Cl-cis, C_1 symmetry c) N-trans, O-cis, Cl-cis, C_2 symmetry d) N-cis, O-cis, Cl-trans, C_2 symmetry e) N-trans, O-trans, Cl-trans, C_2 symmetry [20,35].

When FI catalysts possess extremely bulky groups on the imine-Ns (*N*-C_6H_3-2,6-R^2, R: Me, *i*-Pr), they can adopt a *cis*-O, *trans*-N and *cis*-X (X= O-*i*-Pr) arrangement. Any further increase in the steric bulk of the R^1 substituent (*N*-C_6H_3-2,6-Ph_2 or *N*-$CHPh_2$) forces one of the *N*-Ti bonds to become dissociated, resulting in a five-coordinated complex [36]. However, FI catalysts generally exist as a mixture of isomer **a**, which is normally predominant (*trans*-O, *cis*-N, and *cis*-Cl arrangement: C_2 symmetry), and isomer **b** (*cis*-O, *cis*-N, and *cis*-Cl arrangement: C_1 symmetry [36,37]. These isomers are often fluxional and can mutually transform each other on a nuclear magnetic resonance (NMR) time scale [35].

In the other words, among the five possible isomers (Figure 5), the crystallographically determined structures of FI catalysts, adopt a configuration in which the shortest M-O bonds (M = group 4 transition metal) are positioned *trans* to each other, and the nitrogen atoms and Cl ligands are in *cis*-positions, displaying overall C_2 symmetry (Figure 4a) [34,38]. In this "ordinary" C_2 symmetric FI catalyst, the two imine-N's are positioned in a plane defined by a metal M and two Cl ligands, and therefore, substituents on the imine-N's (R^1) are on a plane at the backside of the X-M-X moiety, while substituents *ortho* to the phenoxy-O's (R^2) are located above and below the X-M-X moiety (Figure 6). Since the X-M-X moiety becomes a reaction site after activation, these R^1 and R^2 substituents have the most pronounced effects on polymerization reactions, but each of the substituents work in a different manner due to their particular spatial arrangements [38].

As discussed, an important feature of these complexes is that the chlorines occupy mutually *cis* coordination sites. This is potentially significant for generating efficient polymerization active centers since a crucial requirement for a highly-active catalyst is to have a pair of *cis* located sites for olefin polymerization [38].

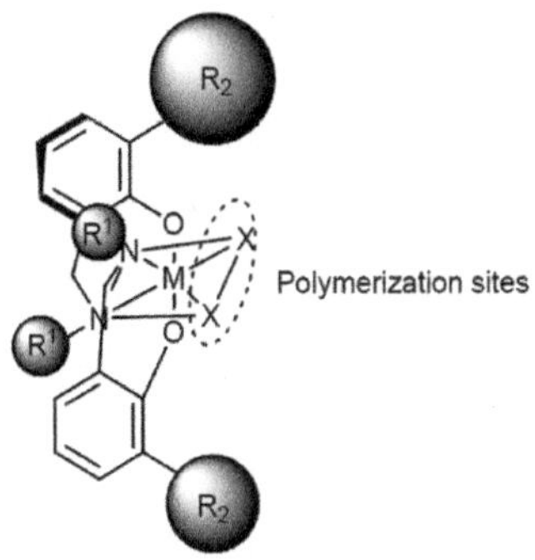

Figure 6. Molecular structures of the FI catalysts [39].

Nevertheless, the DFT calculations suggest that the Zr–N bonds that lie on the same plane as the polymerization sites expand and shrink according to the reaction coordinate of the ethylene insertion (2.23–2.34 Å), while the Zr–O bond length remains virtually unchanged (Figure 7). From studying these results, we believe that this variable Zr–N bond length (which facilitates a smooth and flexible electron exchange between the metal and the ligands) and the *cis* located active sites, as well as the electronically flexible nature of the phenoxy-imine ligands, are responsible for the high polymerization activities of FI catalysts [38,39].

In general, FI ligands can be obtained in practically quantitative yields by the Schiff-base condensation of *ortho*-hydroxy aromatic aldehydes or ketones and primary amines.

These phenol derivatives and amines are easily synthesized, and thus have a rich inventory of commercially available compounds. Therefore, FI catalysts have a wide range of catalyst design possibilities, which is the most important feature of FI catalysts.

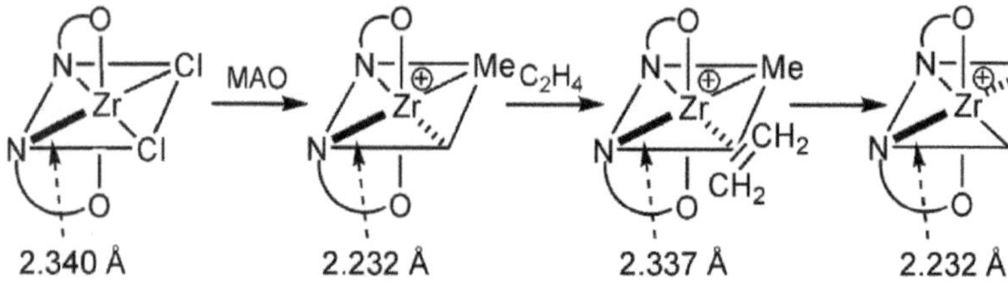

Figure 7. DFT calculation results for ethylene polymerization with FI catalyst/MAO [38].

As a result of this feature, FI catalysts have enormous and diverse ligand structures with a wide variety of substituents, including O, S, N, P and halogen-based functional groups. This enormous structural diversity of FI catalysts has given rise to unprecedented olefin polymerization catalysis and unique polymer formation [34,35,37].

2.2. Electronic characteristics of FI catalysts

In general, the discovery of new olefin polymerization catalysts seems to be a time-intensive, trial-and-error process, with no guarantee of success. However, theoretical calculations on ethylene polymerization with a metallocene catalyst have provided clues about how to design highly active catalysts. Polymerization olefin catalyst components has been depicted in Figure 8 including a metal, ligand(s), a growing polymer chain, a coordinated olefin, and a cocatalyst [40-42] .

As can bee seen in Figure 8, FI catalysts are heteroatom [O^-, N] coordinated early transition metal complexes, which makes FI catalysts different from group 4 transition metals metallocene catalysts that possess cyclopentadienyl (Cp) carbanion-based ligands. Because of the coordination of heteroatom-based [O^-,N] ligands that are more electron withdrawing than Cp carbanion-based ligands, the catalytically active species originating from FI catalysts possess a highly electrophilic nature relative to the active species derived from group 4 transition metals metallocene catalysts [39].

As explained before, DFT calculations using a cationic metallocene complex ($H_2SiCp_2ZrMe^+$) as a model suggest that olefin polymerization is a process that involves intense electron exchange between a ligand and a metal. Since all transition metals (even Mn and Fe) potentially possess the capability of olefin insertion, it is believed that ligands with an electronically flexible nature are a prime requisite for achieving high activity. Therefore, the combination of a transition metal and electronically flexible ligand(s) can yield a highly active olefin polymerization catalyst. Electronically flexible ligands typically possess well-balanced electron donating and withdrawing properties, indicated by a small energy gap between HOMO (the highest occupied molecular orbital) and LUMO (the lowest unoccupied molecular orbital). Therefore, they are capable of receiving electrons from the coordinating olefin through the metal atom and of releasing electrons whenever required to expedite the olefin insertion process. As a result, it can be realized that it is the ligands that play the predominant role in polymerization catalysis among the typical components of the catalyst, and that electronically flexible ligands combined with transition metals form high-activity catalysts when activated as long as the potentially active species possesses an appropriate electron deficiency (10- to 16-electron species) as well as a pair of available *cis*-located sites for polymerization. According to the fact, many researches have been devoted to ligand-oriented catalyst design in an attempt to discover highly active catalysts [38,39].

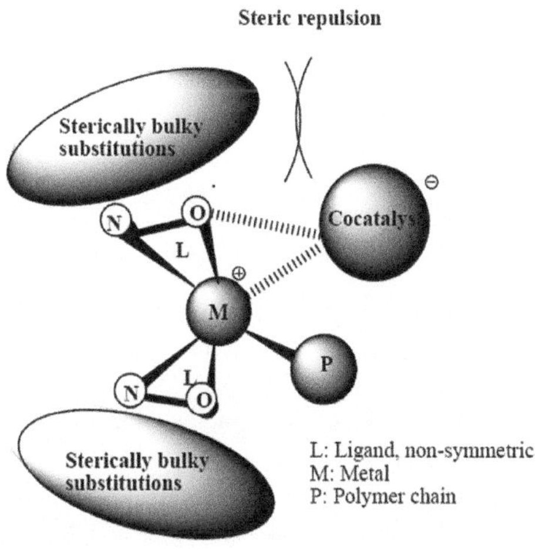

Figure 8. Polymerization components [42].

FI catalysts have another distinctive feature verses metallocenes, that is, M-ligand bonding characteristics. The presence of FI ligands with heteroatom donors renders the complex more electrophilic, a requirement for an active olefin polymerization catalyst, as supported by DFT calculations, which demonstrated that the Mulliken charge at the metal center (in au) in three cationic species increases in the following order, $(C_5H_5)_2TiMe^+$ < CGC $(Me_2Si(C_5Me_4)$-(tBu-N)$TiMe^+$) < (Ph-N=CH-C_6H_3-2-O-3-tBu$)_2TiMe^+$ (Figure 9) [40,41].

Furthermore, DFT calculation revealed that anionic phenoxy-imine chelate ligand possess a smaller energy gap between HOMO and LUMO than a Cp ligand ($C_5H_5^-$) and thus presumably display electronically more flexible properties than the Cp ligand (Figure 10) [42]. Small energy gap between HOMO and LUMO represents well-balanced electron-donating and withdrawing properties which are anticipated to expedite ethylene polymerization process [42].

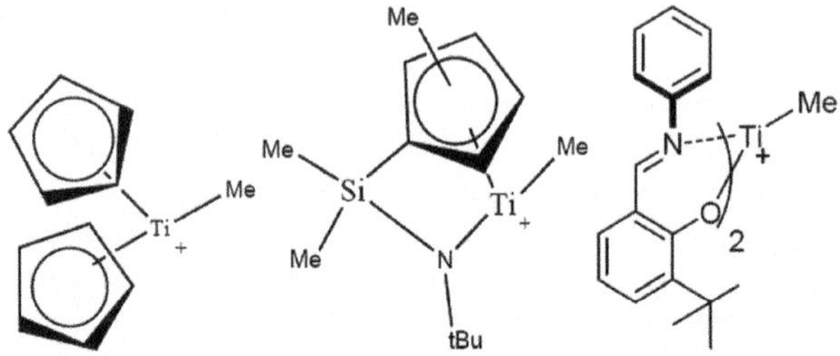

Mulliken charge: 1.417 1.599 1.741

Figure 9. Mulliken charges of the central metals of cationic active Ti species [40,41].

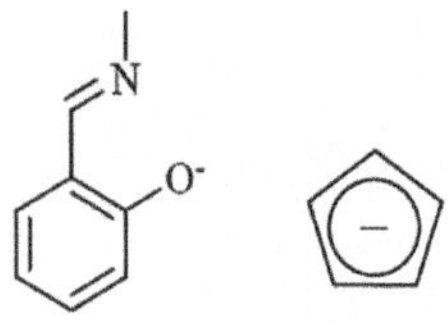

2.4 5.3

Figure 10. Energy gap between HOMO and LUMO (eV) [42].

According to the ligand-oriented catalyst design concept, Fujita and co-workers were designed transition metal complexes incorporating nonsymmetric and electronically flexible ligands as candidates for high-activity catalysts [39]. These complexes were examined as catalysts for the polymerization of ethylene with MAO as the cocatalyst at 25 °C under atmospheric pressure. As a result, a number of complexes were found to be highly active catalysts for the polymerization of ethylene. Polymerization results indicate that phenoxy–imine ligands can produce highly active catalysts when attached to a variety of transition metals, supporting the idea that ligands play the predominant role in polymerization catalysis and that electronically flexible ligands can engender highly active catalysts. A remarkable feature of phenoxy–imine ligated early transition metal complexes is that they possess ligands that can be readily tailored synthetically from both an electronic and steric point of view, and thus possess a wide range of possibilities in terms of catalyst design [39-41].

3. Catalytic properties of FI Catalysts for ethylene polymerization

One of the unique characteristics of FI catalysts, due to its flexible structure, is that the minor change in the structure of the FI catalysts leads to the major changes in activities as well as molecular weight and other characteristics of the resulting polymer. The flexibility refers to the simple Schiff base condensation of the aniline and salicylaldehyde derivatives which can produce a wide range of phenoxy–imine ligands. As depicted in Fig. 11, FI catalysts can be synthesized by treating the phenoxy-imine ligands and transition metal halids to furnish FI catalysts (Figure 11) [42].

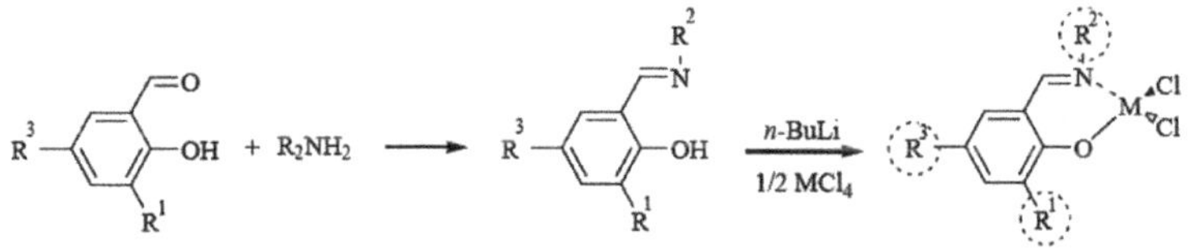

Figure 11. General procedure of FI catalysts [42].

3.1. Influence of alkyl substitutions on the catalyst behavior

Ethylene polymerization revealed that FI catalyst requires steric bulk ortho to the phenoxy-O in order to exhibit high ethylene polymerization activity [42,43]. The steric bulk is thought to afford steric protection toward the anionic phenoxy-O donors from coordination with Lewis acidic compounds such as MAO or from another molecule of the catalytically active cationic species, which are supposed to be highly electrophilic, and the inducement of effective ion separation between the cationic active species and an anionic cocatalyst, resulting in enhanced catalytic activity [44]. Moreover, Cavallo and co-workers [45] have proposed a site-inversion mechanism that can explain the unexpected syndiospecificity exhibited by the Ti-based FI complexes with C_2 symmetry. Additionally, their proposal can also explain that the steric bulk of the substituent *ortho* to the phenoxy–oxygen in bis(phenoxy–imine)Ti complexes plays a crucial role in determining the syndiospecificity of the polymerization [44,45].

Ethylene polymerization via FI catalysts has been shown that the increase in the steric bulk of the R^1 substituent resulted in the marked enhancement in both the catalytic activity and the product molecular weight [46]. The increase in the catalytic activity as a result of introducing sterically-hindered substituent at the R^1 position may be attributed to the fact that the steric bulk of the R^1 substituent plays an essential role in the ion separation between the cationic active species and the anionic cocatalyst. The effective ion separation will provide more space for polymerization and, in addition, enhances the degree of unsaturation associated with the catalytically active cationic species. On the other hand, the increase in the product molecular weight may be ascribed to the fact that the steric congestion exerted by the R^1 substituent diminishes the rate of chain termination [46].

For instance, FI catalyst bearing a hydrogen at the R^1 position ($R^2 = R^3 = H$) gave polyethylene with an *Mw* of 66,000, displaying a moderate activity of 38 kg-PE/mol-cat h under these conditions. Another FI catalyst having a sterically bulkier substituent, a methyl, at the R^1 position yielded much higher molecular weight polyethylene (*Mw* 402,000) with a considerably enhanced activity of 246 kg-PE/mol-cat h. The effects of the steric bulk of the R^1 substituent were more pronounced for FI catalyst possessing a *t*-butyl group (R1 = *t*-butyl) which displayed a higher activity of 3240 kg-PE/mol-cat h and provided polyethylene with a further enhanced molecular weight (*Mw* 1,281,000). Thus, the attachment of the methyl increased catalytic activity and product molecular weight of the produced polymer [46].

The steric bulk of the R^2 substituent also has a significant influence on the catalytic performance of the complexes but differently compared with the R^1 substituent [46]. A Ti-based FI catalyst bearing a methyl at the R^2 position demonstrated an activity of 301 kg-PE/mol-cat h and gave polyethylene with *Mw* of 355,000. The activity and molecular weight values obtained with this catalyst are much lower than complex having hydrogen substituent (R^2 = H). In addition, FI catalyst with *i*-propyl at the R^2 position exhibited more reduced activity (186 kg-PE/mol-cat h) and furnished polyethylene with further decreased molecular weight (*Mw* 296,000). Therefore, it can be concluded that the increase in the steric bulk of the R^2 substituent resulted in the quite deleterious effects on the catalytic activity and the product molecular weight. The decrease in the catalytic activity as a result of introducing the alkyl group into the R^2 position seems reasonable and is probably attributed to the increased steric congestion in close proximity to the active site, which reduces the catalytic activity by hindering access of ethylene to the active site and/or growth of the polymer chain. However, the trend for molecular weight for the Zr-based FI catalysts is inversed, i.e., the attachment of the methyl or the *i*-propyl to the R^2 position greatly enhances the product molecular weight [47,48].

Eventually, the substituent at the R^3 position exercised an effect on both the catalytic activity and the product molecular weight but not significantly, presumably because of the location of the R^3 position being far from the polymerization center [48,49].

Not only catalyst activity is affected by substitutions, but even thermal stability of the catalyst can be affected by electronic aspects of the substitutions. For instance, the introduction of an electrondonating methoxy group *para* to the phenoxy-*O* was found to increase the thermal stability of FI catalysts, making them available for polymerizations at industrially practical higher temperatures [30,39,49].

4. Polymerization of higher α-olefin with FI catalyst

Catalyst systems based on bis(phenoxy-imine) complexes of Ti(IV), Zr(IV) and MAO are effective catalysts for polymerization of ethylene and propylene [19,33,34,38,50,51-54]. Detailed ^{13}C NMR analysis of end-groups in polypropylene prepared with several catalysts of this type showed unusual chemical and regio-effects [52,55].

Phenoxy-imine ligands in complexes contain alkyl substituents in the 2nd and the 4th positions of their phenyl rings, they produce moderately syndiospecific catalysts [55,56]. Detailed investigations of the Ti and Zr complexes with the same ligands showed a surprising difference in the chemistry of polymerization reactions.

Active centers of the highest stereospecificity are produced when the aryl group attached to the nitrogen atom in the phenoxyimine ligand is C_6F_5 (Figure 3) and the phenyl group attached to the oxygen atom carries bulky t-Bu or SiMe substituents in the ortho position to the C–O bond. The [rrrr] content in polypropylene produced with these complexes at 0–20 ^{0}C can reach 0.95–0.96 corresponding to a very high probability of syndiotactic linking, ~0.99 [55-58]. However, a replacement of these substituents with two bromine or two iodin atoms changes the regiocontrol to primary and the stereocontrol to moderately isospecific, the [mm] values of polypropylene prepared at 0 ^{0}C increases from 0.17 for dialkyl-substituted complexes to 0.73 for diiodo-substituted complexes [58].

Bis(phenoxy-imine) complexes of Ti(IV) and Zr(IV) combined with MAO or with common organoaluminum compounds form effective catalysts for polymerization of ethylene, 1-alkenes, and for ethylene/1-alkene copolymerization. The productivity of catalyst systems containing the Ti complexes and MAO in ethylene polymerization reactions at 25–75 ^{0}C can reach 2,000–4,000 kg/mol Ti. h [34,52,55,59-61] .

Bis(phenoxy-imine) complexes of Ti (IV) containing C_6F_5 groups attached to the N atom in each bidentate ligand form catalyst systems with very low chain transfer rates. They copolymerize ethylene with propylene and higher 1-alkenes under living-chain conditions (Mw/Mn = 1.07–1.19) at temperatures as high as 50^{0}C [52,55,62,63] and they are suitable for the synthesis of alkene block-copolymers. These catalysts were employed for the synthesis of various alkene diblock- and triblock-copolymers [48,52,55,63].

Polypropylene produced with combinations of the aldimine Ti complexes and MAO or AlR3-$[Ph3C]^{+}$ $[B(C6F5)4]^{-}$ as cocatalysts under moderate conditions also has a very narrow molecular weight distribution [52,55,61,62]. The stereospecificity of catalyst systems based on bis(phenoxy-imine) complexes depends on the type of substituents in the ligands (substituents R in Fig. 11). When R = H or Me, the catalysts produce polypropylene with a predominantly syndiotactic structure. However, complexes of the same type bearing two bromine or two iodine atoms in the 2nd and the 4th positions of the phenyl groups produce moderately isotactic polypropylene with [mm] from ~0.5 to 0.73, depending on reaction temperature, and the dichloro-substituted complex produces atactic polypropylene [58].

Bis(phenoxy-imine) complexes of Ti(IV) activated with ion-forming cocatalysts instead of MAO, such as Ali-Bu3-[CPh3]+ [B(C6F5)4]- are efficient single-center catalysts for polymerization of higher 1-alkenes, 1-hexene, 1-octene, 1-decene, and 4-methyl-1-pentene. The catalysts have very high activity and produce regioand stereo-irregular polymers with very high molecular weights, $<1.4\times10^{6}$ [64].

The Zr bis(phenoxy-imine) complex was used for the synthesis of an ethylene/1-octene copolymer with a low 1-octene content, and a complex of Hf with a tridentate ligand was used for the synthesis of an ethylene/1-octene copolymer with a high 1-octene content.

The flexibility of the FI catalysts allows for the making of new polymers which are difficult or impossible to prepare using group 4 metallocene catalysts. For example, it is possible to prepare low molecular weight (Mv ~10^{3}) polyethylene or poly(ethylene-copropylene) with olefinic end groups, ultra-high molecular weight polyethylene or poly(ethylene-co-propylene), high molecular weight poly(1-hexene) with atactic structures including frequent regioerrors, monodisperse poly(ethylene-co-propylene) with various propylene contents, and a number of polyolefin block copolymers [e.g., polyethylene-b-poly(ethylene-co-propylene), syndiotactic polypropylene-b-poly-(ethylene-co-propylene), polyethylene-b-poly(ethylene-co-propylene)-b-syndiotactic polypropylene. These unique polymers are predictable to possess novel material properties and uses [65].

5. Living polymerization with FI catalyst

The synthesize polymers with completely defined structures is a goal. Living polymerization is known to control some elements like as degree of polymerization, chain-end structures, stereochemistry, especially molecular weight and chain-end structures of polymer. Although there have been a number of transition metal catalysts which can polymerize ethylene or α-olefins in a living fashion [66-68] there are few catalysts that are useful for both ethylene and α-olefins. Besides, most catalysts require a low polymerization temperature, usually below room temperature, to suppress chain termination and therefore exhibit low activities and insufficient polymer molecular weights. The titanium FI catalyst bearing a perfluorophenyl group as R (Figure 12) with MAO exhibited living polymerization with ethylene even at 50 ^{0}C to afford high molecular-weight polymer ($Mw=424\times10^3$) having a narrow molecular weight distribution (Mw/Mn=1.13). This catalyst exhibits living polymerization behavior over a wide range of temperature [65].

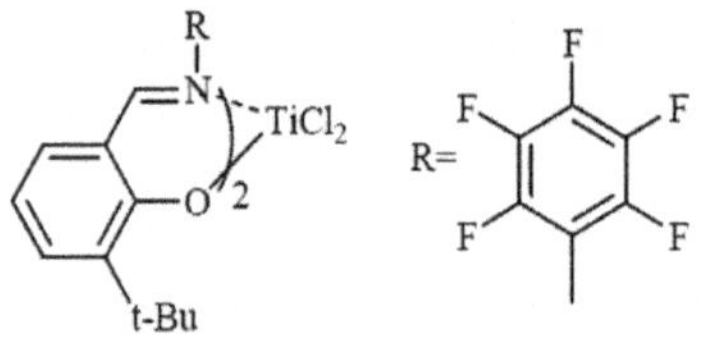

Figure 12. FI catalyst with perfluorophenyl group [65].

There is a clear difference in polymerization results depending on the fluorine substitution patterns. First of all, living polymerizations proceed only when at least one fluorine is located in the 2,6-positions of R-phenyl group (Figure 12). Second, the activity of living systems is considerably lower than that of non-living systems. Finally, the activity increases with the number of fluorine atoms in both living and non-living systems. It is reasonable that electron-withdrawing fluorines enhance the electrophilicity and consequently the reactivity of the active centers. The ortho-fluorine substituent effect represents a novel strategy for the design of a new transition metal complex for living olefin polymerization [65].

Interestingly, a titanium FI catalyst having a chlorine at the 2-position of the R-phenyl group instead of fluorine, bis[N-(3-t-butylsalicylidene)-2-chloroanilinato]titanium(IV) dichloride, also promoted ethylene polymerization at 25 ^{0}C to produce polyethylene having a narrow molecular weight distribution (Mw/Mn=1.23), implying that the interaction with β-hydrogen is potentially achieved by any substituent possessing lone-pair electrons. Propylene polymerization with the above complex (Figure 12) at room temperature also turned out to be living and produced polypropylene with a controlled molecular weight, a narrow molecular weight distribution (Mw/Mn=1.07-1.14), and surprisingly high syndiotacticity (rr=87%, 98%) [59].

Mitani et al, introduce a Ti-FI catalyst having fluorine- and trimethylsilyl-containing ligands, which polymerizes propylene above room temperature to form highly syndiotactic monodisperse PPs with extremely high *Tm*'s. The titanium complex employed in this study was bis[*N*-(3-trimethylsilylsalicylidene)-2,3,4,5,6-pentafluoroanilinato]Ti(IV) dichloride. Polymerization of propylene with this complex using methylalumoxane (MAO) cocatalyst under atmospheric pressure at 25 °C for 5 h yielded crystalline polypropylene having an extremely narrow polydispersity (*Mn*=47 000, *Mw*/*Mn* = 1.08). Interestingly, the syn-PP produced displayed an extremely high *Tm* (152 ^{0}C) [69].

Kinetic studies of ethylene polymerization reactions [70] and propylene polymerization reactions [62] with FI catalysts demonstrated that all features typical for living-chain reactions are present when these reactions are carried out at 25^{0}C. The catalyst activity remains constant for over 1 hour in the ethylene polymerization reactions and for 3 hours in the propylene polymerization reactions and it very slowly decreases after that. The molecular weight increases with reaction time in a nearly linear manner in the same time ranges, and the molecular weight distributions of the polymers remain narrow, Mw/Mn =1.07-1.10. The value of the propagation rate constant for ethylene polymerization reaction at 25^{0}C was ~1,400–1,900 $M^{-1}.s^{-1}$, it is comparable to the same value in metallocene catalysis [70].

The syndiospecific polymerization of propylene with this types of catalyst also exhibits kinetic features of living-chain polymerization reactions [52,55]. Both the polymer yield and the molecular weight of polypropylene produced at 0 ^{0}C increase nearly proportionally to the reaction time, and the molecular weight distribution of the polymer remains narrow (Mw/Mn =1.08–1.11) [52]. Kinetic parameters of the propylene polymerization reactions at 0-25 ^{0}C are: k_p = 0.05–0.06 $M^{-1}.s^{-1}$, [C]/[Ti] = 0.4–0.6% [52,71]. However, when the same complex contains 3,5-difluoro-substituted benzene rings, chain transfer reactions occur much more readily and the polymer has the typical molecular weight distribution characteristics of a material synthesized with a single-center catalyst, Mw/Mn ~2.5 [55].

Polymerization reactions with Ti diamide-based catalysts: Low-temperature propylene polymerization reactions with a Ti diamide complex $LTiMe_2$ activated with MMAO or with silica- and alumina-supported MMAO give an interesting example of long-term living-chain reactions. These catalysts are very stable kinetically at 0 ^{0}C and the Mn value of the produced polymers increases with time in a linear manner for over 30 minutes resulting in the formation of atactic polymers of a very high molecular weight, with Mn ~1.5×10^6. The k_p value for the homogeneous catalyst system is ~5 $M^{-1}.s^{-1}$ at 0 ^{0}C and when MMAO is supported on silica (this step increases the acidity of Al atoms in MMAO), the k_p value increases to ~20 $M^{-1}.s^{-1}$ [72,73].

6. Modification of the FI catalysts

Fujita group at Mitsui Chemicals discovered the fluorinated Ti–FI catalysts that can promote unprecedented living ethylene and propylene polymerization, resulting in the formation of functionalized polymers and block copolymers from ethylene, propylene, and higher α-olefins [59]. Coates [74] and Sakuma [75] reported that fluorinated Ti–FI catalysts are capable of mediating the highly controlled, thermally robust living polymerization of ethylene and propylene. Electronic attractive interaction between a fluorinated phenoxy-imine ligand and a growing polymer chain has significant effects on the catalytic properties of Ti–FI catalysts [59,74,75]. Additionally, the presence of electron withdrawing fluoro substituents on aniline is considered to be highly beneficial for increasing the catalyst activity. Ishii and coworkers described that introduction of electron-withdrawing F atoms on the ligand structure results in an increase in metal–carbon reactivity leading to reduced activation energy for ethylene insertion [76].

Although the catalytic behavior of ethylene and propylene polymerization, as well as ethylene–propylene copolymerization, using fluorinated Ti–FI complexes are extensively described in literature [77-79], but less attention has been paid to the fluorinated Zr–FI complexes. Recently, Zohuri and coworkers have reported the catalytic properties of a fluorinated Zr–FI catalyst in ethylene polymerization (Figure 13) [3].

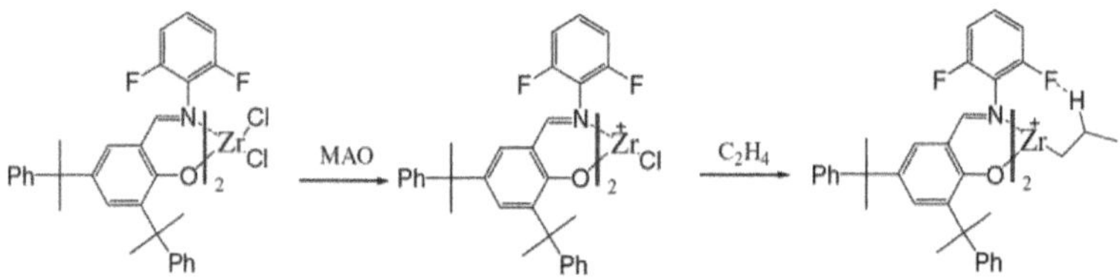

Figure 13. Structure of the new Zr-FI catalyst and the electronically interaction between H-b and ortho-F-atom-substituted phenyl ring on the N [3].

The synthesized catalyst could produce polyethylene with high molecular weight (M_v=1.39 ×10^6) which is surprisingly higher in comparison with the similar non-fluorinated FI catalysts [33,42] indicating the dramatic electronic effect of *ortho*-F substitution on the polymerization mechanism [3] (Fig. 13). The M_v values of 36 × 10^4, 10.4 × 10^4, and 1.84 × 10^4 have been reported already for the polyethylene obtained using the non-fluorinated bis[N-(3-t-butyl-5-methoxy salicylidene) anilinato] zirconium (IV) dichloride, bis[N-(3-cumyl-5-methoxy salicylidene) cyclohexy laminato] zirconium (IV) dichloride and bis[(3-cumyl-5-methyl salicylidene) anilinato] zirconium (IV) dichloride, respectively [33,74]. It has been proposed that attractive interaction between the ortho-F and the β-H on a growing polymer chain, which is expected to effectively curtail β–H transfer to the central metal and incoming monomer, is responsible for the unprecedented behavior [42]. Furthermore, it has been shown that *ortho*-F-substituted phenyl ring on the N electronically plays a key role in the suppression of chain transfer reactions resulted in increasing the molecular weight and moderation of the catalyst activity [3].

Additionally, a highly active Zr-based FI catalyst of bis[N-(3,5-dicumylsalicylidene)cyclohexyl-aminato]zirconium(IV) dichloride has been synthesized and used for polymerization of ethylene by Zohuri et al [80]. The synthesized catalyst by changing the phenyl group on the imine nitrogen of FI catalyst (Fig. 11, R_2) to a cyclohexyl group exhibited enhanced activity, may be due to the electron-donating effects of an aliphatic group at the R_2 position. The prepared FI catalyst displayed a very high activity of about 3.2×10^6 g PE/mmol Zr. h in ethylene polymerization at the monomer pressure of 3 bars. Despite such high activity, this catalyst showed a short lifetime [80].

Some complexes with ligands based on naphthalene carbaldehydes having a hydroxy group in the ortho position with respect to the aldehyde group were described [81,82], but no data on their catalytic activity were given. Ahmadjo and coworkers [53] used 2-hydroxynaphtaldehyde instead of salicylaldehyde for the preparation of naphthoxy-imines as ligands and three FI-like Zr-based catalysts, Bis[1-[(phenylimino)methyl]-2-naphtholato]zirconium(IV) dichloride (1), Bis[1-[(mesitylimino)methyl]-2-naphtholato] zirconium(IV) dichloride (2) and Bis[1-[(2,6-diisopropylphenyl)imino] methyl-2-naphtholato]zirconium(IV) dichloride (3) were prepared by changing the ligand from salicylaldehyde imine ligand, to 2-hydroxynaphthalene-1-carbaldehyde imine ligand and used for polymerization of ethylene (Figure 14) [53]. It has been reported that introducing the sterically bulky isopropyl groups enhances molecular weight of the resulting polymer through destabilization of β–agostic interaction due to the steric repulsion between β–hydrogen of the growing polymer and isopropyl leading to decrease the β–H elimination [34].

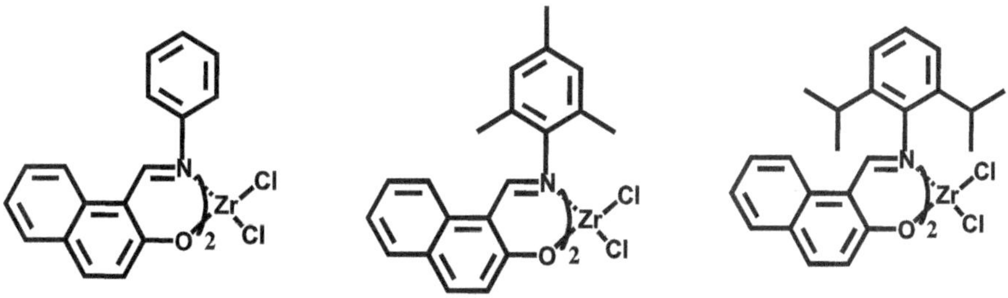

Figure 14. Structure of the new FI catalysts reported by Ahmadjo et al [53].

The transition state for β-hydride elimination requires the overlapping of a σ C-H orbital with an empty d orbital of the metal, and the two carbon adjacent to the metal and β-hydrogen must be on the same plane containing the metal between the two phenoxy-imine kelates. Therefore, the energy of the conformation needed for β-hydride elimination increases with increase of steric hindrance around the metal [83,84]. In the catalyst containing bulky isopropyl group phenyl ring on the N (Figure 14), it seems that β-carbon of polymer chain is not easily accommodated in the plane because of the steric interaction between naphthoxy-imine ligands of the catalyst and β-H of the growing polymer chain (Figure 15). It can be concluded that the sterically bulky substituents on the aniline ring of the naphthoxy-imine ligands cause a strong suppression of the both β-hydride elimination and β-hydride transfer to monomer leading to increase the molecular weight of the resulted polymer [53].

Moreover, Damavandi and coworkers have reported the synthesis of novel FI Zr-type catalysts for ethylene polymerization [85]. Figure 16 depicted the structure of the catalysts.

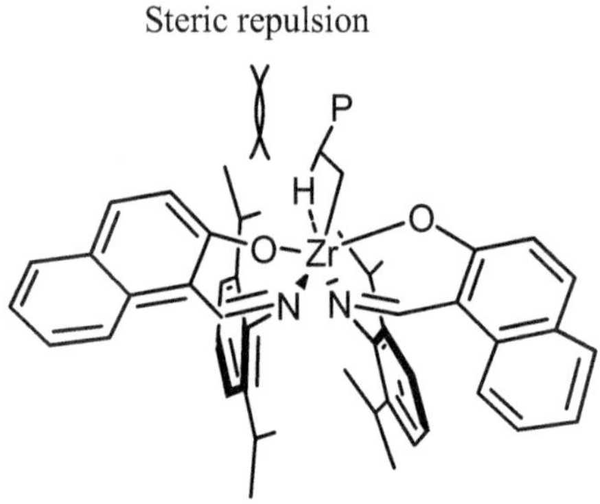

Figure 15. β–agostic interaction: steric repulsion between β–hydrogen of the growing polymer chain and isopropyl alkyl substitutions phenyl ring on the N [53].

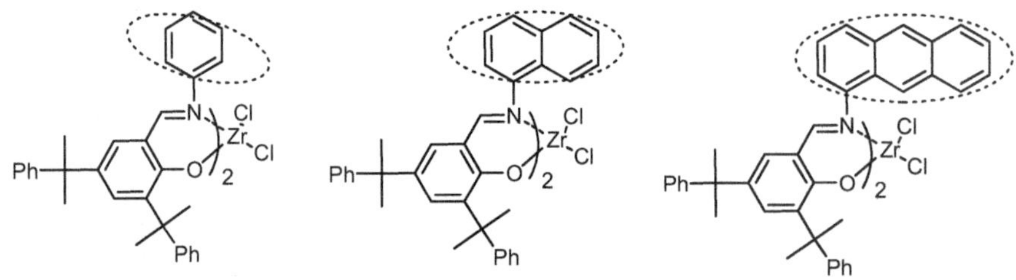

Figure 16. Structure of the FI Zr-type catalysts [85].

Plurality of the fused aromatic rings on the N atom of the imine in the catalyst structure affected the polymerization activity and molecular weight of the resulting polymer as well. It has been reported that changing of ligand from FI-type single aryl substituted on the N atom of imine to further aromatic fused rings, enhances the steric hindrance of the catalyst which resulted in diminishing the catalyst activity in comparison with FI catalysts including single aryl substituted on the N. However, this replacement could increase the molecular weight of the polymer obtained. It is presumable that rigidity of the catalyst structure, sterically causes low β-hydride transfer through destabilization of β–agostic interaction. The M_v values of the obtained polymer using the catalyst containing naphtholato group were higher than the obtained polymer using the catalyst containing anilinato group (Fig. 15). The increase in M_v values is referred to the steric repulsion between a β-hydrogen of growing polymer and aromatic fused rings on the imine-N which could diminish β-hydride elimination through destabilization of β–agostic interaction [53,85].

Furthermore, Damavandi et al. showed that polydispersity can remarkably increase with time which reveals that the system starts to deviate from pure living behavior [85]. Although a linear dependence between the polymerization time and the molecular weight was observed indicating for living behavior, but the polydispersity was broadened with the time (Figure 17). The reason can be that either the catalyst is not truly living over the full polymerization time or that the single-site system turns into a multi-site system due to heterogenization of the system. This phenomenon could be comparable with self-immobilization of single site catalysts that has been a subject of interest [86]. Similar result has been reported already by Ivanchev et. al [87]. The capture and blocking of active sites by the grown polymer after a certain polymerization time have been suggested.

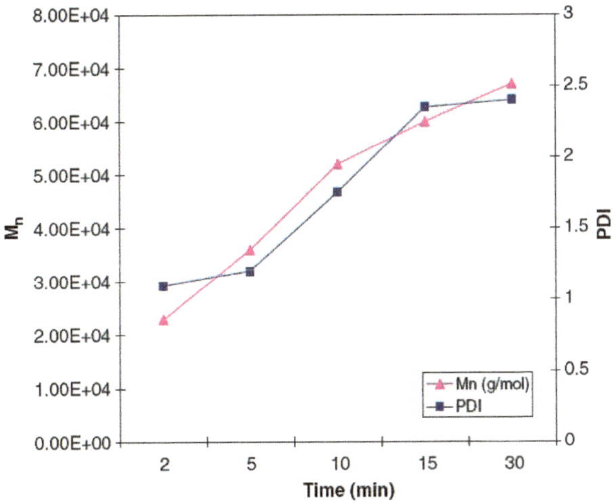

Figure 17. PDI and Mn versus time [85].

Also, Sandaroos and coworkers reported the synthesis and structure of novel Ti complex having a pair of chelating aminotropone [O–N] ligand [88]. Experimental as well as theoretical studies show that the active species derived from bis(aminotropone) Ti catalyst normally possess higher electrophilicity nature compared with those produced using bis(phenoxyimine) Ti complexes (Ti–FI catalysts) (Figure 18) [88].

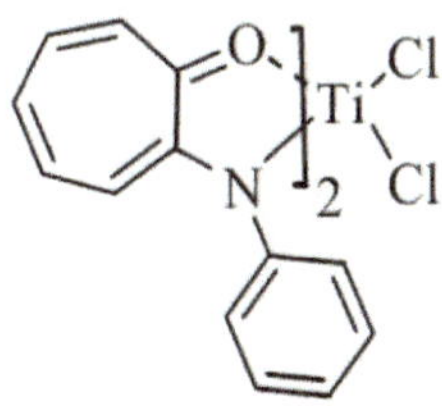

Figure 18. Structure of bis(aminotropone) Ti catalyst [88].

Ahmadjo et al. [53] Studied ethylene polymerization using FI catalysts in the presence of different amount of hydrogen. They reported that the polymerization activity can be increased with increasing the hydrogen concentration due to the fast hydrogenation of sterically more hindered and less reactive intermediates such as those resulting from 2,1-insertions (Figure 19) [53].

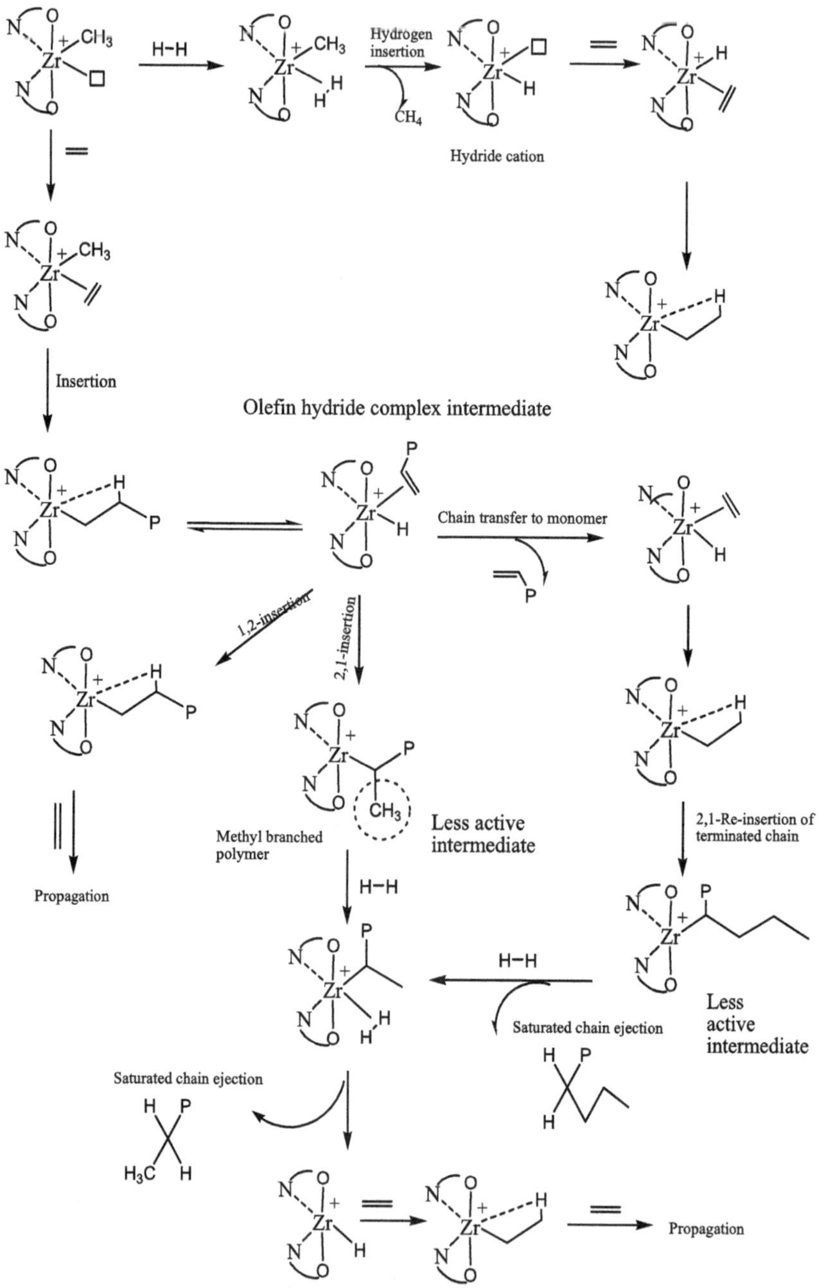

Figure 19. Proposed mechanisms for reactivation of catalyst active centers by hydrogen [53].

Isomerization is possible when the polymer chain at the Zr center undergoes β-hydride elimination forming an intermediate of olefin hydride complex and the olefin reinserts in a 2,1-manner into the Zr–hydride bond before insertion of the next monomer takes place. Consequently, methyl branches are formed (Figure 19). The new Zr-C bond is expected to be more hindered and less reactive for further monomer insertion which can be reactivated by hydrogen. As it is shown in Figure 19, in the presence of hydrogen, a dihydrogen complex is formed in the first step. This intermediate can subsequently form a hydride cation and a saturated alkyl chain, which is ejected from the active site. A new ethylene monomer can be inserted into the hydride cation structure and a new polymer chain starts to grow [53].

Additionally, 2,1-reinsertion of short olefin branches, terminated by β-H elimination which are still capable to be coordinated to the active centers, is probable. As it is shown in Figure 19, in the presence of hydrogen, the resulting sterically more hindered and less reactive intermediate undergoes fast hydrogenation to form hydrid cation which is susceptible to start new ethylene polymerization. However, it has been suggested that the catalyst structure limits the hydrogen role during the polymerization specifically in the latest ligand-oriented catalysts [53].

7. Immobilizing polymerization catalysts

The cocatalysts for bis(phenoxy-imine)Zr and Ti complexes are generally MAO (methylaluminoxane), *i*-$Bu_3Al/Ph_3CB(C_6F_5)_4$ [64,79,89], or $MgCl_2$-based activators [73]. There has been few report on common alkylaluminums alone as cocatalyst in the catalytic system of bis(phenoxy-imine)Zr and Ti complexes, they used some alkylaluminums, such as triethylaluminum (Et_3Al), trihexylaluminum (He_3Al), or trimethylaluminum (Me_3Al), could activate bis(phenoxy-imine)Zr and Ti complexes for ethylene polymerization without MAO or $Ph_3CB(C_6F_5)_6$, It is found that only the alkylaluminum with three linear alkyls can activate catalyst. However, no polymer was found using *i*-Bu_3Al or Et_2AlCl as cocatalyst for complex. The molecular weights of the polyethylenes produced by FI catalyst with Et_3Al, He_3Al, or Me_3Al are much higher than those produced with MAO. Also the polydispersities (Mw/Mn) of polyethylenes produced by FI catalyst with trialkylaluminums are higher than with MAO [54].

Up to now, several highly active transition metal complexes including metallocenes with $MgCl_2$-based activator catalyst systems, often introduced as $MgCl_2$-supported catalyst systems, have been reported [90,91]. These catalyst systems normally give polymers with broad molecular weight distributions.

$MgCl_2$ may work as an activator for the bis(phenoxy-imine)Ti complexes since these complexes possess O and N heteroatoms in the ligands, which are capable of electronically interacting with $MgCl_2$. Thus, researcher decided to investigate $MgCl_2$ as an activator for the bis(phenoxy-imine)Ti complexes in the hope of developing high performance catalysts based on these complexes [73,92].

At first dealcoholysis of a $MgCl_2$/alcohol adduct with *i*-Bu_3Al since this method produces highly porous small $MgCl_2$ particles, which should be suitable as an activator. Ethylene polymerization with complexes Bis[*N*-(3-*t*-butylsalicylidene)anilinato]titanium(IV) dichloride (1), Bis[*N*-(3-*t*-butylsalicylidene)cyclohexylaminato]titanium(IV) dichloride (2), Bis[*N*-(3-phenylsalicylidene)anilinato]titanium(IV) dichloride (3) were conducted under ethylene (0.9 MPa pressure) at 50 °C for 30 min using the above mentioned $MgCl_2$/*i*-$Bu_mAl(OR)_n$ as an activator [92].

i-Butylaluminum 2-ethyl-1-hexoxide, *i*-$Bu_mAl(OR)_n$, such as *i*-Bu_2Al(2-ethyl-1-hexoxide), are expected to work as in situ alkylating reagents for the Ti complexes as well as scavengers in a polymerization sysytem, and $MgCl_2$ may function as a Lewis acid to generate a cationic active species from an alkylated complex [93].

The complexes with $MgCl_2$/*i*-$Bu_mAl(OR)_n$ activator systems displayed high activities in the range of 20.8–36.3 kg-PE/mmol cat·h, with a steady uptake of ethylene over the 30 min duration of the run.

It is of great significance that the activities obtained using $MgCl_2$/*i*-$Bu_mAl(OR)_n$ as the activator are comparable to those seen with the MAO activator system. Notably, the activities obtained with complexes/$MgCl_2$/*i*-$Bu_mAl(OR)_n$ are comparable to or exceed that found for the Cp_2TiCl_2/MAO catalyst system under the same conditions. Alternatively, complexes **1–3** combined with *i*-$Bu_mAl(OR)n$ or *i*-Bu_3Al exhibited practically no activities under identical polymerization conditions. These results demonstrated that $MgCl_2$/*i*-$Bu_mAl(OR)_n$ is an effective activator for bis(phenoxy-imine)Ti complexes, and $MgCl_2$ is a pivotal component of the activator. The high performance exhibited by $MgCl_2$/*i*-$Bu_mAl(OR)_n$ as the activator towards the bis(phenoxy-imine)Ti complexes **1–3** may be related to the fact that the bis(phenoxy-imine) complexes possess O and N heteroatoms in the ligand, which are capable of electronically interacting with $MgCl_2$.Remarkably, the molecular weight distribution (*Mw*/*Mn*) values of the polyethylenes arising from complexes **1–3**/ $MgCl_2$/*i*-$Bu_mAl(OR)_n$ systems lie in the range of 2.40–2.67 (MAO activation; 2.07–2.74), suggesting that the polymers are produced by single active species [92].

Moreover, Bis(phenoxy-imine) V complex in association with $MgCl_2/RmAl(OR)_n$ demonstrate high activities at elevated temperatures (75 °C, 65.1 kg-PE/(mmol-cat h), atmospheric ethylene pressure) and a thermally robust single-site V-based olefin polymerization catalysts. All of the V compounds with $MgCl_2/Et_mAl(OR)_n$ provided very high molecular weight PEs (Mv > 5,000,000) for which we are unable to determine the molecular weights and molecular weight distributions using GPC analyses. All of the polyethylenes formed from the $MgCl_2/R_mAl(OR)_n$ systems display good polymer morphology, indicating that complexes are heterogenized on the surface of the $MgCl_2/R_mAl(OR)_n$. These results suggest that $MgCl_2/R_mAl(OR)_n$ not only works as an excellent cocatalyst, but is also a good support for phenoxy-imine ligated V complexes [73].

Xu et al reported Bis(phenoxy-imine) zirconium catalyst, bis-[*N*-(3-*tert*butylsalicylidence) cyclohexylaminato] zirconium(IV) dichloride, that directly impregnated onto MAO modified spherical $MgCl_2$ support . The resultant solid catalyst was proved to be very active in ethylene polymerization, leading to the formation of spherical polymer particles with high bulk density. The activities of supported catalysts and the bulk density of the resultant

PEs are strongly dependent on dealcoholization temperature, while dealcoholization temperature also influences the molecular weight and molecular weight distribution of the resultant polymers. When the support was thermally pretreated at lower temperature, the resultant solid catalysts, were inactive for ethylene polymerization. On the contrary, when the support was pretreated at higher temperature, the corresponding solid catalysts were all active for ethylene polymerization [94].

Ethylene polymerization by titanium complex having two phenoxy-imine with different supports and solvents has been investigated by Srijumnong et al [95]. The catalytic activity depended on supports used, and especially on the types of solvent medium. For the supported system, catalytic activities decreased in the following order: TiO_2 >TiO_2-SiO_2 > SiO_2. This can be attributed to the strong interaction of the TiO2 with dried MMAO (dMMAO) and the larger amount of dMMAO present on the TiO_2 than other supports.

Zhang et al. reported self-immobilized titanium and zirconium complexes with allyloxy substituted phenoxy-imine ligands as well as their catalytic performance for ethylene polymerization. The results of ethylene polymerization showed that the self-immobilized titanium (IV) and zirconium (IV) catalysts kept high activity for ethylene polymerization. SEM showed the immobilization effect could greatly improve the morphology of polymer particles to afford micron-granula polyolefin as supported catalysts. Polymer produced by self-immobilized catalysts had broader molecular weight distribution (3.5–19.2), while polymer produced by titanium and zirconium complexes free from allyloxy groups had very narrow molecular weight distribution (<3.0) in most case [96].

Author details

S. Damavandi
Department of Chemistry, Sarvestan Branch, Islamiz Azad University, Iran

S. Ahmadjo
Department of Catalyst, Iran Polymer and Petrochemical Institute, Tehran, Iran

R. Sandaroos
Department of Chemistry, Faculty of Science, Birjand University, Birjand, Iran

G.H. Zohuri*
Department of Chemistry, Ferdowsi University of Mashhad, Mashhad, Iran

8. References

[1] Brinzingger HH, Fisher, D, Mulhaupt R, Rieger R, Waymouth RM (1995) Stereospecific Olefin Polymerization with Chiral Metallocene Catalysts. Angew. Chem. Int. Eng. 34: 1143-1170.

* Corresponding Author

[2] Kaminsky W, (1999) Metalorganic Catalysts for Synthesis and Polymerization. Springer-Verlag, Heidelberg.

[3] Zohuri GH, Damavandi S, Sandaroos R, Ahmadjo S (2011) Ethylene Polymerization Using Fluorinated FI Zr-Based Catalyst. Polym. Bull. 66: 1051–1062.

[4] Kaminsky W, (2005) in Handbook of Polymer Synthesis, 2nd ed. Kricheldorf H, Nuyken O, Swift G, Eds. Marcel Dekker, New York, Pp 1-73.

[5] Kuran W, (2001) Principles of Coordination Polymerization, Wiley, New York.

[6] Moor EP, (2003) Polypropylene Handbook, Hanser; Munich, Germany.

[7] Resconi L, Cavallo L, Fait A, Piemontesi F, (2000) Selectivity in Propene Polymerization with Metallocene Catalysts. Chem. Rev. 100: 1253-1345.

[8] Wang B (2006) *Ansa*-Metallocene Polymerization Catalysts: Effects of the Bridges on the Catalytic Activities. Coord. Chem. Rev. 250: 242-258.

[9] Irwin LJ, Miller SA (2007) Catalyst System for High Activity and Stereoselectivity in the Homopolymerization and Copolymerization of Olefins. U.S. Patent 7214749.

[10] Yoshida Y, Matsui S, Fujita T (2010) FI Catalysts: Unique Olefin Polymerization Catalysis for Formation of Value-added Olefin-based Materials. Journal of the Japan Petroleum Institute, 53: (3), 111-129.

[11] Kaminsky W (2004) The Discovery of Metallocene Catalysts and Their Present State of the Art. J Polym. Sci. Part A. 42: 3911–3921.

[12] Gibson VC, Spitzmesser SK (2003) Advances in Non-Metallocene Olefin Polymerization Ccatalysis. Chem. Rev. 103: 283–315.

[13] Britovsek GJP, Gibson VC, Wass DF (1999) The Search for New-Generation Olefin Polymerization Catalysts: Life Beyond Metallocenes. Angew. Chem. Int. Ed. 38: 428–447.

[14] Baumann R, Davis WM, Schrock RR (1997) Synthesis of Titanium and Zirconium Complexes that Contain the Tridentate Diamido Ligand, $[((t\text{-}Bu\text{-}d_6)N\text{-}o\text{-}C_6H_4)_2O]_{2\text{-}}$ ($[NON]_{2\text{-}}$) and the Living Polymerization of 1-Hexene by Activated $[NON]ZrMe_2$. J. Am. Chem. Soc. 119: 3830–3831.

[15] Nomura K, Sagara A, Imanishi Y (2002) Olefin Polymerization and Ring-Opening Metathesis Polymerization of Norbornene by (Arylimido)(Aryloxo)Vanadium(V) Complexes of the Type VX_2(NAr)(OAr¢). Remarkable Effect of Aluminum Cocatalyst for the Coordination and Insertion and Ring-Opening Metathesis Polymerization. Macromolecules 35: 1583–1590.

[16] Johnson LK, Killian CM, Brookhart M (1995) New Pd(II)- and Ni(II)-Based Catalysts for Polymerization of Ethylene and α-Olefins. J. Am. Chem. Soc. 117:6414-6415.

[17] Quijada R, Rojas R, Bazan G, Komon ZJA, Mauler RS, Galland GB (2001) Synthesis of Branched Polyethylene from Ethylene by Tandem Action of Iron and Zirconium Single Site Catalysts. Macromolecules 34: 2411–2417.

[18] Azoulay JD, Bazan GC, Galland GB (2010) Microstructural Characterization of Poly(1-hexene) Obtained Using a Nickel R-Keto-β-diimine Initiator. Macromolecules 43:2794–2800.

[19] Terao H, Ishii S, Mitani M, Tanaka H, Fujita T (2008) Ethylene/Polar Monomer Copolymerization Behavior of Bis(phenoxy–imine)Ti Complexes: Formation of Polar Monomer Copolymers. J. Am. Chem. Soc. 130: 17636–17637.

[20] Suzuki Y, Terao H, Fujita T (2003) Recent Advances in Phenoxy-Based Catalysts for Olefin Polymerization. Bull. Chem. Soc. Jpn. 76: 1493–1517.

[21] Furuyama R, Saito J, Ishii SI, Mitani M, Tohi Y, Makio H, Matsukawa N, Tanaka H, Fujita T (2003) Ethylene and Propylene Polymerization Behavior of a Series of Bis(Phenoxy–imine)Titanium Complexes. J. Mol. Cat. A. 200: 31–42.

[22] Mecking S, Johnson LK, Wang L, Brookhart M (1998) Mechanistic Studies of the Palladium-Catalyzed Copolymerization of Ethylene and α-Olefins with Methyl acrylate. Journal of the American Chemical Society, 120(5): 888–899.

[23] Liu W, Malinoski JM, Brookhart M (2002) Ethylene Polymerization and Ethylene/Methyl 10-undecenoate Copolymerization Using nickel(II) and Palladium(II) Complexes Derived from a Bulky P,O Chelating Ligand. Organometallics, 21(14): 2836–2838.

[24] Sun J, Shan Y, Xu Y, Cui Y, Schumann H, Hummert M (2004) Novel Cyclohexyl-Substituted salicylaldiminato-Nickel(II) Complex as a Catalyst for Ethylene Homopolymerization and Copolymerization. J. Polym. Sci. Part A, Polym. Chem. 42 (23): 6071–6080.

[25] Matsugi T, Matsui S, Kojoh S, Takagi Y, Inoue Y, Nakano T, Fujita T, Kashiwa N (2002) New Titanium Complexes Bearing Two Indolide–Imine Chelate Ligands for the Polymerization of Ethylene. Macromolecules, 35: 4880-4887.

[26] Yoshida, Y, Mohri J, Ishii S, Mitani M, Saito J, Matsui S, Makio H, Nakano T, Tanaka H, Onda M, Yamamoto Y, Mizuno A, Fujita T (2004) Living Copolymerization of Ethylene with Norbornene Catalyzed by Bis(Pyrrolide–Imine) Titanium Complexes with MAO. J. Am. Chem. Soc. 126: 12023-12032.

[27] Suzuki Y, Inoue Y, Tanaka H, Fujita T (2004) Phenoxyether Ligated Ti Complexes for the Polymerization of Ethylene. Macromol. Rapid Commun. 25: 493-497.

[28] Suzuki, Y, Tanaka H, Oshiki T, Takai K, Fujita T (2006) Titanium and Zirconium Complexes with Non-Salicylaldimine-Type Imine–Phenoxy Chelate Ligands: Syntheses, Structures, and Ethylene-Polymerization Behavior. Chem. Asian J. 1: 878-887.

[29] Matsugi T, Fujita T, (2008) High-Performance Olefin Polymerization Catalysts Discovered on the Basis of a New Catalyst Design Concept. Chem. Soc. Rev. 37: 1264-1277.

[30] Fujita T, Tohi Y, Mitani M, Matsui S, Saito J, Nitabaru M, Sugi K, Makio H, Tsutsui T (1998) Olefin Polymerization Catalysts, Transition Metal Compounds, Processes for Olefin Polymerization, and Alpha-Olefin/Conjugated Diene Copolymers. EP Patent 0874005.

[31] Matsugi T, Fujita T (2008) High Performance Olefin Polymerization Catalysts Discovered on the Basis of a New Catalyst Design Concept, Chem. Soc. Rev. 37: 1266-1277.

[32] Tohi Y, Nakano T, Makio H, Matsui S, Fujita T, Yamaguchi T (2004) Polyethylenes Having Well-Defined Bimodal Molecular Weight Distributions Formed with Bis(phenoxy-imine) Zr Complexes, Macromol. Chem. Phys. 205: 1179-1183.

[33] Matsui S, Mitani M, Saito J, Tohi Y, Makio H, Matsukawa N, Takagi Y, Tsuru K, Nitabaru M, Nakano T, Tanaka H, Kashiwa N, Fujita T (2001) J. Am. Chem. Soc. A

Family of Zirconium Complexes Having Two Phenoxy–Imine Chelate Ligands for Olefin Polymerization, 123: 6847-6856.
[34] Tohi Y, Makio H, Matsui S, Onda M, Fujita T (2003) Polyethylenes with Uni-, Bi-, and Trimodal Molecular Weight Distributions Produced with a Single Bis(phenoxy–imine)zirconium Complex. Macromole. 36: 523-525.
[35] Strauch J, Warren TH, Erker G, Fröhlich R, Saarenketo P (2000) Formation and Structural Properties of Salicylaldiminato Complexes of Zirconium and Titanium. Inorg. Chim. Acta. 300-302: 810-821.
[36] Johnson, AL, Davidson MG, Lunn MD, Mahon MF (2006) Synthesis, Isolation and Structural Investigation of Schiff-Base Alkoxytitanium Complexes: Steric Limitations of Ligand Coordination. Eur. J. Inorg. Chem. 15: 3088-3098.
[37] Pärssinen A, Luhtanen T, Klinga M, Pakkanen T, Leskelä M, Repo T (2007) Alkylphenyl-Substituted Bis(salicylaldiminato) Titanium Catalysts in Ethene Polymerization. Organometallics, 26: 3690-3698.
[38] Mitani M, Saito J, Ishii S, Nakayama Y, Makio H, Matsukawa N, Matsui S, Mohri J, Furuyama R, Terao H, Bando H, Tanaka H, Fujita T (2004) FI Catalysts: New Olefin Polymerization Catalysts for the Creation of Value-Added Polymers. Chem. Rec. 4: 137–158.
[39] Makio H, Fujita T (2009) Development and Application of FI Catalysts for Olefin Polymerization: Unique Catalysis and Distinctive Polymer Formation. Acc. Chem. Res. 42: (10) 1532-1544.
[40] Chen E YX, Marks T J (2000) Cocatalysts for Metal-Catalyzed Olefin Polymerization: Activators, Activation Processes, and Structure-Activity Relationships. Chem. Rev. 100: 1391-1434.
[41] Matsui S, Mitani M, Saito J, Tohi Y, Makio H, Tanaka H, Fujita T (1999) Post-Metallocenes: a New Bis(Salicylaldiminato) Zirconium Complex for Ethylene Polymerization. Chem. Lett. 28: 1263-1264.
[42] M Mitani, T Nakano, T Fujita (2003) Unprecedented Living Olefin Polymerization Derived from an Attractive Interaction Between a Ligand and a Growing Polymer Chain. Chem. Eur. J. 9:(11) 2396-2403.
[43] Makio H, Fujita T (2005) Propene Polymerization with Bis(phenoxy–imine) Group 4 Transition Metal Complexes. Bull. Chem. Soc. Jpn. 78: 52-66.
[44] Yoshida Y, Matsui S, Fujita T (2005) Bis(pyrrolideimine) Ti Complexes with MAO: a New Family of High Performance Catalysts for Olefin Polymerization. J. Organometal. Chem. 690: 4382-4397.
[45] Milano G, Cavallo L, Guerra G, (2002) Site Chirality as a Messenger in Chain-End Stereocontrolled Propene Polymerization. J. Am. Chem. Soc. 124: 13368-13369.
[46] Furuyama R, Saito J, Ishii SI, Mitani M, Matsui S, Tohi Y, Makio H, Matsukawa N, Tanaka H, Fujita T (2003) Ethylene and Propylene Polymerization Behavior of a Series of Bis(phenoxy–imine)Titanium Complexes. J. Mol. Cat. A: Chem. 200: 31-42.
[47] Matsui S, Mitani M, Saito J, Matsukawa N, Tanaka H, Nakano T, Fujita T (2000) Post-Metallocenes: Catalytic Perfomance of New Bis(salicylaldiminato) Zirconium Complexes for Ethylene Polymerization. Chem. Lett. 5: 554-555.

[48] Fujita T, Coates GW (2002) Synthesis and Characterization of Alternating and Multiblock Copolymers From Ethylene and Cyclopentene Macromolecules, 35: 9640–9647.
[49] Matsui S, Fujita T (2001) FI Catalysts: Super Active New Ethylene Polymerization Catalysts. Catal Today, 66: 63-73.
[50] Ittel SD, Johnson LK, Brookhart M (2000) Late-Metal Catalysts for Ethylene Homo-and copolymerization. Chem. Rev. 100: 1169-1203.
[51] Rieger B, Saunders Baugh L, Kacker S, Striegler S (eds) (2003) Late Transition Metal Polymerization Catalysis. Wiley-VCH, Weinheim.
[52] Tian J, Hustad PD, Coates GW (2001) A New Catalyst for Highly Syndiospecific Living Olefin Polymerization Homopolymers and Block Copolymers from Ethylene and Propylene. J. Am. Chem. Soc. 123: 5134-5135.
[53] Ahmadjo S, Zohuri GH, Damavandi S, Sandaroos R (2010) Comparative Ethylene Polymerization Using FI-like Zirconium Based Catalysts, Reac. Kinet. Mech. Cat. 101: 429-442.
[54] Liu D, Wang S, Wang H, Chen W, (2006) Trialkylaluminums: Efficient Cocatalysts for Bis(phenoxy-imine)Zirconium Complexes in Ethylene Polymerization, J. Mol. Catal. A: Chem. 246: 53-58.
[55] Hustad PD, Tian J, Coates GW, (2002) J. Am. Chem. Soc. Mechanism of Propylene Insertion Using Bis(phenoxyimine)-Based Titanium Catalysts: An Unusual Secondary Insertion of Propylene in a Group IV Catalyst System. 124: 3614-3621.
[56] Busico V, Cipullo R, Cutillo F, Friederichs N, Ronca S, Wang B, (2003) Improving the Performance of Methylalumoxane: A Facile and Efficient Method to Trap "Free" Trimethylaluminum, J. Am. Chem. Soc. 125: 12402-12403.
[57] Lamberti M, Pappalardo D, Mazzeo M, Pellecchia C (2004) Effects of the Reaction Conditions on the Syndiospecific Polymerization of Propene Promoted by Bis(phenoxyimine) Titanium Catalysts Macromol. Chem. Phys. 205: 486-491.
[58] Mazzeo M, Strianese M, Santoriello I, Pellecchia C (2006) Phenoxyaldimine and Phenoxyketimine Titanium Complexes in Propene Polymerization. A Different Effect of o-Phenoxy Halide Substituents. Macromolecules, 39: 7812-7820.
[59] Saito J, Mitani M, Mohri J, Ishii S, Yoshida Y, Matsugi T, Kojoh S, Kashiwa N, Fujita T (2001) Highly Syndio Specific Living Polymerization of Propylene Using a Titanium Complex Having Two Phenoxy-imine Chelate Ligands. Chem. Lett. 30: 576–582.
[60] Arriola DJ, Carnahan EM, Hustad PD, Kuhlman RL, Wenzel TT (2006) Catalytic Production of Olefin Block Copolymers via Chain Shuttling Polymerization. Science, 312: 714-719.
[61] Lamberti M, Pappalardo D, Zambelli A, Pellecchia C (2002) Syndiospecific Polymerization of Propene Promoted by Bis(salicylaldiminato)Titanium Catalysts: Regiochemistry of Monomer Insertion and Polymerization Mechanism. Macromolecules, 35: 658-663.
[62] Mitani M, Furuyama R, Mohri, JI, Saito J, Ishii S, Terao H, Nakano T, Tanaka H, Fujita T (2003) Syndiospecific Living Propylene Polymerization Catalyzed by Titanium

Complexes Having Fluorine-Containing Phenoxy–imine Chelate Ligands. J. Am. Chem. Soc. 125: 4293-4305.

[63] Furuyama R, Mitani M, Mohri JI, Mori R, Tanaka H, Fujita T (2005) Ethylene/Higher α-Olefin Copolymerization Behavior of Fluorinated Bis(phenoxy–imine)titanium Complexes with Methylalumoxane: Synthesis of New Polyethylene-based Block Copolymers, Macromolecules, 38: 1546-1552.

[64] Saito J, Suzuki Y, Makio H, Tanaka H, Onda M, Fujita T (2006) Polymerization of Higher α-Olefins with a Bis(Phenoxyimine)Ti Complex/i-$Bu_3Al/Ph_3CB(C_6F_5)_4$: Formation of Stereo- and Regioirregular High Molecular Weight Polymers with High Efficiency. Macromolecules, 39: 4023-4031.

[65] Makio H, Kashiwa N, Fujita T, (2002) FI Catalysts: a New Family of High Performance Catalysts for Olefin Polymerization. Adv. Synth. Catal. 34: 477-493.

[66] Matsugi T, Matsui S, Kojoh S, Takagi Y, Inoue Y, Fujita T, Kashiwa N (2001) New Titanium Complexes Having Two Indolide-Imine Chelate Ligands for Living Ethylene Polymerization. Chem. Lett. 30: 566-577.

[67] Gottfried AC, Brookhart M, (2001) Living Polymerization of Ethylene Using Pd(II) α-diimine Catalysts, Macromolecules, 34: 1140-1142.

[68] Tshuva EY, Goldberg I, Kol M (2000) Isospecific Living Polymerization of 1-Hexene by a Readily Available Nonmetallocene C2 Symmetrical Zirconium Catalyst. J. Am. Chem. Soc. 122: 10706-10707.

[69] Mitani M, Furuyama R, Mohri J, Saito J, Ishii S, Terao H, Kashiwa N, Fujita T (2002) Fluorine- and Trimethylsilyl-Containing Phenoxy–Imine Ti Complex for Highly Syndiotactic Living Polypropylenes with Extremely High Melting Temperatures. J. Am. Chem. Soc. 124: 7888-7889.

[70] Mitani M, Mohri J, Yoshida Y, Saito J, Ishii S, Tsuru K, Matsui S, Furuyama R, Nakano T, Tanaka H, Kohoj S, Matsugi T, Kashiwa N, Fujita T (2002) Living Polymerization of Ethylene Catalyzed by Titanium Complexes Having Fluorine-Containing Phenoxy–imine Chelate Ligands. J. Am. Chem. Soc. 124: 3327-3366.

[71] Zambelli A, Longo P, Pellecchia C, Grassi A (1987) Beta-Hydrogen Aabstraction and Regiospecific Insertion in Syndiotactic Polymerization of Styrene. Macromolecules, 20: 2035-2037.

[72] Hagimoto H, Shiono T, Ikeda T (2004) Supporting Effects of Methylaluminoxane on the Living Polymerization of Propylene with a Chelating (diamide)dimethyltitanium Complex. Macromol. Chem. Phys. 205: 19-26.

[73] Nakayama Y, Bando H, Sonobe Y, Fujita T (2004) Olefin Polymerization Behavior of Bis(phenoxy-imine) Zr, Ti, and V Complexes with $MgCl_2$-Based Cocatalysts, J. Mol. Catal. A. 213: 141-150.

[74] Coates GW, Tian J, Hustad PD (2003) Bis(salicylaldiminato)Titanium Complex Catalysts, Highly Syndiotactic Polypropylene by a Chain-end Control Mechanism, Block Copolymers Containing this. US Patent 6562930.

[75] Sakuma A, Weiser MS, Fujita T (2007) Living Olefin Polymerization and Block Copolymer Formation with FI Catalyst. Polym. J. 39(3): 193–207.

[76] Ishii SI, Saito J, Mitani M, Mohri JI, Matsukawa N, Tohi Y, Matsui S, Kashiwa N, Fujita T (2002) Highly Active Ethylene Polymerization Catalysts Based on Titanium Complexes Having Two Phenoxyimine Chelate Ligand. J. Macromol. Catal. 179: 11-16.

[77] Nakayama Y, Saito J, Bando H, Fujita T (2005) Propylene Polymerization Behavior of Fluorinated Bis(Phenoxy-imine) Ti Complexes with an $MgCl_2$-Based Compound ($MgCl_2$-Supported Ti-Based Catalysts). Macromol. Chem. Phys. 206: 1847-1852.

[78] Ishii SI, Furuyama R, Matsukawa N, Saito J, Mitani M, Tanaka H, Fujita T (2003) Ethylene and Ethylene/Propylene Polymerization Behavior of Bis(Phenoxy-imine) Zr and Hf Complexes with Perfluorophenyl Substituents. Macromol. Rapid. Commun. 24: 452-456.

[79] Yasunori Y, Shigekazu M, Terunori F (2005) Bis(pyrrolide-imine) Ti Complexes with MAO: A New Family of High Performance Catalysts for Olefin Polymerization. J. Organometal. Chem. 690: 4382-4397.

[80] Zohuri GH, Damavandi S, Sandaroos R, Ahmadjo S (2010) Highly Active FI Catalyst of Bis[N-(3,5-dicumylsalicylidene)cyclohexylaminato] Zirconium(IV) Dichloride for Polymerization of Ethylene, Iran Polym. J. 19:(9) 679-687.

[81] Fujita T, Tohi Y, Mitani M, Matsui S, Saito J, Nitabaru M, Sugi K, Makio H, Tsutsui T (2005) Olefin Polymerization Catalysts, Transition Metal Compounds, Processes for Olefin Polymerization, and α-Olefin/Conjugated Diene Copolymers. US Patent 6875718.

[82] Oleinik II, Oleinik IV, Ivanchev SS (2008) Design of Schiff Base-Like Postmetallocene Catalytic Systems for Polymerization of Olefins: VIII. Synthesis of N-(o-cycloalkylphenyl) 2-Hydroxynaphthalene-1-Carbaldehyde imines. Russ. J. Org. Chem. 44:(1) 103-106.

[83] Burger BJ, Thompson ME, Cottor WD, Barcaw JE (1990) Ethylene Insertion and Beta.-hydrogen Elimination for Permethylscandocene Alkyl Complexes. A Study of the Chain Propagation and Termination Steps in Ziegler-Natta Polymerization of Ethylene. J. Am. Chem. Soc. 112: 1566-1577.

[84] Bordwell F, Bausch MJ (1983) Methyl Effects on the Basicities of Cyclopentadienide and Indenide Ions and on the Chemistry of Their Transition Metal Complexes. J. Am. Chem. Soc. 105: 6188-6189.

[85] Damavandi S, Galland GB, Zohuri GH, Sandaroos R (2011) FI Zr-type Catalysts for Ethylene Polymerization. J. Polym. Res. 18: 1059-1065.

[86] Zhang J, Wang X, Jin GX (2006) Polymerized Metallocene Catalysts and Late Transition Metal Catalysts for Ethylene Polymerization. Coord. Chem. Rev. 250: 95-109

[87] Ivanchev SS, Trunov VA, Rybakov VB, Albov DV, Rogozin DG (2005) Dik Phys Chem 404:165.

[88] Sandaroos R, Damavandi S, Nazif A, Goharjoo M, Mohammadi A (2011) Highly Efficient Bis(aminotropone) Ti Catalyst for Ethylene Polymerization. Chinese Chemical Letters, 22: 213–216.

[89] Nakayama Y, Bando H, Sonobe Y, Fujita T (2004) Development of Single-site New Olefin Polymerization Catalyst Ssystems Using $MgCl_2$-Based Activators: MAO-Free $MgCl_2$-Supported FI Catalyst Systems. Chem. Soc. Japan, 77: 617-625.

[90] Soga K, Uozumi T, Saito M, Shiono T (1994) Structure of Polypropylene and Poly(ethylene-co-propylene) Produced with an Alumina-Supported $CpTiCl_3$/Common Alkylaluminium Catalyst System. Macromol. Chem. Phys. 195: 1503-1515.

[91] Satyanarayana G, Sivaram S (1993) An Unusually Stable Supported Bis(cyclopentadienyl)titanium dichloride-trialkylaluminum Catalyst System for Ethylene Polymerization. Macromolecules, 26: 4712-4714.

[92] Nakayama Y, Bando H, Sonobe Y, Kaneko H, Kashiwa N, Fujita T (2003) New Olefin Polymerization Catalyst Systems Comprised of Bis(Phenoxy-imine) Titanium Complexes and $MgCl_2$-Based Activators. J. Cat. 21: 171-175.

[93] Hedden D, Marks TJ (1988), $[(CH_3)_5C_5]_2Th(CH_3)_2$ Surface Chemistry and Catalysis. Direct NMR Spectroscopic Observation of Surface Alkylation and Ethylene Insertion/Polymerization on $MgCl_2$, J. Am. Chem. Soc. 110: 1647-1649.

[94] Xu R, Liu D, Wang S, Wang N, Mao B (2007) Preparation of Spherical $MgCl_2$-Supported Bis(Phenoxy-imine) Zirconium Complex for Ethylene Polymerization. J. Mol. Catal. A. 263: 86-92.

[95] Srijumnonga S, Suttipitakwong P, Jongsomjita B, Praserthdama P (2008) Effect of Supports and Solvents on Ethylene Polymerization with Titanium Complex Consisting of Phenoxy-imine Ligands/dMMAO Catalytic System. J. Mol. Catal. A. 29: 1-7.

[96] Zhang D, Jin G (2004) Self-immobilized Titanium and Zirconium Catalysts with Phenoxy-imine Ligands for Ethylene Polymerization. X-ray Crystal Structure of Bis(N-(3-t-butylsalicylidene)-4'-allyloxyanilinato) Zirconium (IV) dichloride. Applied Catalysis A: General ,262: 85-91.

Chapter 7

High Affinity Polymers by Molecular Imprinting for Drug Delivery

Mara Soares da Silva and Teresa Casimiro

Additional information is available at the end of the chapter

1. Introduction

Molecular recognition processes found in nature have always inspired scientists to mimic these systems in synthetic materials such as molecular imprinted polymers. Selective receptors within a polymer have a huge range of applications such as in separation processes, analytical chemistry, sensors, catalysis, drug discovery and therapeutics. In this last area in particular, polymers are having a central role in recent developments and materials with molecular recognition ability can respond to new challenges and opportunities.

The synergy of traditional pharmaceutical formulations with new polymers and hybrid and conjugated materials is leading to new approaches of treating diseases in medicine. In 2003 Langer and Peppas have identified the combination of new polymers with molecular imprinting, as a new field in drug delivery but with applications not immediately forthcoming [1]. Since then the field has grown with increasing reported applications of molecular imprinted polymers for drug delivery and pharmaceutical applications, although not at the expected rate. From a literature search (Web of Knowledge database, April 2012) one can see that from the ca. 13000 published papers on molecular imprinting only less than 170 are related to drug delivery. We believe this relatively small number of contributions can be attributed to difficulties in the synthesis and especially in the optimization of the polymers. Traditional methods are time-consuming involving multi-steps and in general MIPs are prepared using organic solvents which eventually leave toxic traces, incompatible with pharmaceutical and biomedical applications. Also safety, in particular biodegradability and cytotoxicity issues, must be addressed in applications where MIPs are in contact with biological tissues. In addition there is an extra difficulty of preparing MIPs with performance in aqueous solutions which is crucial to *in vivo* applications.

It has already been demonstrated that MIPs are highly promissory drug delivery systems (DDS). Not just because of the high specificity of the active sites but also due to the possibility of introducing stimulus-responsiveness by choosing appropriated monomers. In addition the physicochemical properties of MIPs are able to protect the drug from enzymatic degradation during trafficking in the body [2]. However, only recently MIPs have been applied to drug release [3]. The possibility of preparing cheaper robust materials with molecular recognition capability and with tailored properties, is highly attractive to pharmaceutical applications and in particular for drug delivery.

This Chapter will focus on the use of supercritical fluid technology as a viable and greener alternative to the synthesis and processing of molecularly imprinted polymers for potential pharmaceutical applications in particular for drug delivery.

2. Molecular imprinting

Molecular recognition comprises the specific interactions that enable one chemical entity to selectively recognize its physical and chemically complementary target molecule. The recognition mechanism is mainly mediated by weak non-covalent interactions, such as hydrogen bonding, ion-pairing, hydrophobic interactions and dipolar associations that as a whole govern the structural conformation of macromolecules and influence their interaction with other molecules. In nature there are several examples of successful recognition mechanisms, such as DNA-protein, RNA-ribosome and antigen-antibody. Biomolecules however tend to lose their recognition properties in abiotic environments. Furthermore they are considerably expensive, which limits their applicability. For those reasons scientists have shown a great interest in the design of artificial supramolecular systems that exhibit molecular recognition. Examples of these materials are crown ethers and calixarenes, that selectively bind specific cations [4], derivatives of parental cyclodextrins [5], dendrimers [6] and molecularly imprinted polymers (MIPs).

Molecular imprinting is an emerging technique to create high affinity polymeric matrices, MIPs, for target molecules. The methodology relies on the stabilization of the monomers-template assembly in the pre-polymerization step and posterior copolymerization with a crosslinking agent, in the presence of a porogen [7]. Template removal from the imprinted polymer at the end of the reaction leaves accessible chemical and sterically complementary binding sites. The generally accepted mechanism for the formation of MIPs is schematically presented in Fig. 1.

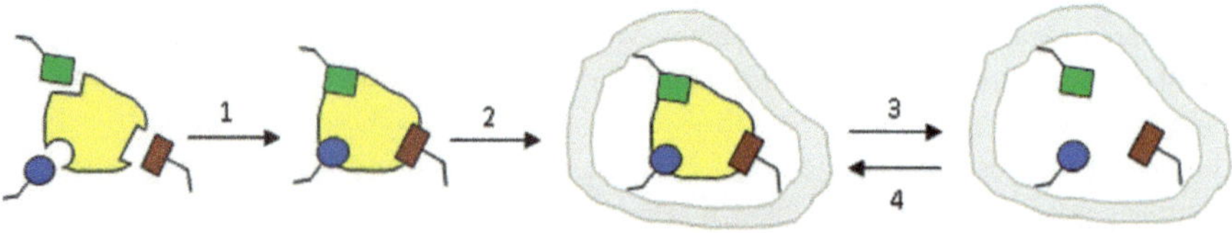

Figure 1. Molecular imprinting schematic mechanism: (1) Monomer and template pre-assembly; (2) Polymerization with crosslinking agent; (3) Polymer-template bonds disruption and subsequent template removal; (4) Template rebinding. [8]

Imprinted polymers are synthetic receptors with binding constants comparable to natural receptors [9,10], but capable of withstanding harsh conditions of pressure [11], temperature and extreme pH [12], and organic solvents [13]. They are cheap to synthesize and can be manufactured in large quantities with good reproducibility. These properties explain the growing interest in imprinted materials in the last decades, resulting in a significant increase in reported works in such diverse areas as in synthesis and catalysis [14,15], solid-phase extraction [16,17], chromatography [18,19], sensing [20,21] and in drug delivery [22,23].

In literature one can find many functional monomers and their combinations, typically used in molecular imprinting. Mayes and Whitcombe listed a significant part of monomers typically used in non-covalent molecular imprinting and corresponding literature [24]. The most common monomers are methacrylic acid (MAA), acrylic acid (AA), 4-vynilpyridine, diethylaminoethyl methacrylate (DEAMA), acrylamides, 2-hydroxyethyl methacrylate (HEMA), etc. Typical cross-linkers used in molecular imprinting are acrylated-based (e.g. ethyleneglycol dimethacrylate EGDMA, trimethylolpropane trimethacrylate TRIM, etc), styrenic-based (e.g. divinylbenzene DVB), water soluble cross-linkers (e.g. ethylene-*bis*-acrylamide), among many others. In 2004 Hilt and Byrne reported a quite complete list of molecularly imprinted biologically significant molecules, where is visible that methacrylic acid and ethylene glycol dimethacrylate are probably the most common monomer and cross-linker used in the synthesis of MIPs [25].

3. Synthetic approaches and design of MIPs

Molecularly imprinted polymers can be prepared according to a number of approaches that essentially differ in the way the template interact first with the functional monomer and later with the polymeric binding sites. Although the divisions among them are becoming somewhat blurred due to the emergence of more complex hybrid strategies, the main molecular imprinting approaches considered are covalent, non-covalent, semi-covalent and metal ion-mediated imprinting.

Covalent imprinting, first developed by Wulff [26], was the initial strategy to introduce molecular affinity in polymeric matrices. By means of this imprint methodology, templates with polymerizable groups are reacted to prepare the recognition matrix. At the end of the polymerization the template is cleaved from the matrix and the functionality left in the binding site is available for future covalent rebind of the target molecule. The great advantage of this approach is that the functional groups are only associated with template sites. However, difficulties in the cleavage of the template and limitations found in the type of compounds that can be imprinted by this way led to the development of other imprinting methodologies.

Non-covalent imprinting, pioneered by Mosbach [27], is currently the most used approach to produce molecular recognition matrices because of its simplicity and easy availability of the commercial monomers used. The efficiency of MIPs prepared by non-covalent methodology relies in the successful stabilization of template-monomer (T-M) complexes formed in the pre-polymerization step, through numerous non-covalent interactions such as hydrogen bonding, ion-paring and dipole-dipole interactions. After the matrix prepared and template-cleaved, the rebinding is achieved also via non-covalent interactions, providing higher kinetic constants than MIPs prepared by covalent imprinting. The simplicity of the method, allied with the easy and cheap preparation, as well as the high selectivity that the polymers can have, seem to overcome potential complex stabilization issues.

Semi-covalent imprinting route combines both covalent and non-covalent interactions during the synthesis and rebinding process, respectively. When using this approach, a template with covalently bound polymerizable groups is used to prepare the recognition matrix, affording polymers with narrower distribution of the binding sites. Also, the inexistence of other kinetic restrictions than those encountered by diffusion can be considered a big advantage of this methodology. Two different approaches have been developed in which the template and the monomer are directly connected or, hence, have a sacrificial spacer group between them. This latter has been intensively developed by the group of Whitcombe [28].

Metal ion-mediated imprinting uses metal ions to target a wide range of functional groups through the donation of electrons from heteroatoms of ligands to the unfilled orbitals of the outer coordination sphere of the metal. By means of this approach, a polymerizable ligand is used to complex the metal ion that coordinates to the template [29], yielding therefore matrices with highly specific active sites.

The development of molecular recognition polymers involves the optimization of a number of experimental variables that determine the thermodynamic balance of the system and consequently the specificity of the matrix towards the template molecule. The most important parameters that have to be carefully optimized are the choice of the functional monomers according to the template nature, monomer: template (M:T) ratio, the crosslinking degree and the solvent of the polymerization.

The non-covalent approach is the most widely adopted method due to the reasons already exposed above. M:T ratio has a major influence in non-covalent imprinting as it directs the equilibrium of formation of template-monomer complexes. Functional monomers are typically used in excess to shift the equilibrium towards complex formation, that results in some of functional groups being randomly distributed throughout the polymer, leading to non-specific binding [30]. Although there is no straightforward correlation between the amount of template and the formation of specific sites within the polymer, a slight decrease in the absolute number of high-affinity sites has been reported when low template concentrations are used while the relative yield in high-affinity sites increased [31].

This is extremely important in order to extend the application of non-covalent molecular imprinting to expensive or poorly soluble template molecules. On the other hand high concentrations of template lead to an increased number of random affinity binding sites, increasing non-specific interactions [31,32].

The cross-linker agent fulfils three main purposes in imprinting technique: 1) it controls the polymer morphology (whether the matrix formed is of gel type, macroporous or microgel powder); 2) it freezes the functional monomer-template complex within the matrix; and 3) imparts mechanical stability to the polymeric network. The integrity of the binding sites responsible for the recognition mechanism usually requires a rigid structure, with high crosslinking density, which minimizes the possible conformations of the matrix. However, when lower ratios of cross-linker are used, the matrix is more flexible, which can turn the equilibrium between drug release and uptake in the recognition site faster. Therefore, a compromise between the rigidity and flexibility of the network is necessary [23]. The imprinted cavities should be stable enough to maintain the conformation in the absence of the template, but also flexible enough to ease the fast equilibrium between the release and re-binding of the template in the cavity during its application.

The solvent or porogen used in the imprinting process has also a significant effect in surface morphology and molecular recognition properties of the polymers [33,34], particularly in the case of non-covalent imprinting. The success of imprinting relies on the relative amount of cross interaction between the solvent and the intended non-covalent interactions employed during T-M complex formation. If the solvent interferes or competes with any of these interactions, less effective recognition will occur. Furthermore it is well known that the performance of the MIP is usually optimized in the solvent used in the synthesis achieving better selectivity and rebinding results, i.e., there is a memory effect [35].

The influence of pressure in the imprinting process was studied by Sellergren and co-workers, by means of hydrostatic pressure, using three different solvents (dichlorometane, methanol and 2-propanol) and two different triazines as templates [36]. Their results support the idea that high-pressure could be used to stabilize template-monomer complexes and result in higher affinities.

The increasing restrictions in the use of organic solvents prompted the development of new greener methods for the preparation of MIPs. Some efforts have been made in the synthesis of imprinted polymers in aqueous environment; however it presents many problems since the high concentration of water molecules weakens the interactions between the monomer and the template, decreasing the selectivity of the MIP for the template molecule [35]. Attempts have been made by using functional monomers that may stabilize the complexes through hydrophobic interactions, such as cyclodextrins [37], metal coordination [38] or by multi-step polymerization procedure which is time-consuming and energy intensive drying steps are required.

4. Applications of MIPs

Imprinted polymers play an important role in separation processes. MIPs can be used as absorbents in solid phase extraction [39,40], extraction of contaminants from environmental samples [41], food analysis [42], analytical chemistry [43] and chromatography, which is the most traditional application of MIPs. In particular, MIPs can be used as chiral stationary phases, replacing expensive chiral columns and avoiding chiral mobile phase additives and derivatisation with chiral reagents. Successful imprinting systems are able to deliver polymeric networks with effective enantiomeric recognition properties when templating a chiral molecule. Higher affinity to the template enantiomer is translated in a predictable order of elution when using MIPs as chiral stationary phases, as the template establishes a higher number of interactions with the polymeric matrix and therefore is more retained. The use of MIPs as stationary phases has been object of intense research, with many papers published in the area [44,45,46,47,48].

Membrane technology is an active, growing field, which has gained a huge importance in the last forty years, as high-throughputs membranes were developed. New trends encounter the development of affinity membranes, with molecular recognition character inspired in natural mechanisms. Molecularly imprinted membranes (MIMs) emerge as potential affinity materials due to their low cost, ease of preparation and good molecular recognition performance [49]. Also the development of sensors and biosensors are gaining significant interest in the last years. MIPs have been used in transducers and integrated in sensors [2].

Another very promising area is the application of MIPs in catalysis. The robustness and high selectivity of molecular imprinted polymeric materials turn possible their use at harsh conditions of temperature and pressures, in a wide range of organic solvents and pH conditions, replacing for instance enzymes [50].

The use of molecular imprinted polymers in drug delivery systems is a new and attractive area of research. Developments in the field of polymer synthesis and polymerization mechanisms are being reflected in MIPs synthesis, where various mechanisms can be followed to obtain more controlled macromolecular architectures [51] and by exploring new intelligent analyte-sensitive hydrogel co-polymers [1]. We believe that the development of new biomimetic materials for controlled and targeted drug delivery will be for sure one of the most active areas of molecular imprinting in the next years, and where supercritical fluid technology can have a relevant role by delivering materials with the required features for pharmaceutical and biomedical applications.

5. Supercritical fluid technology

A supercritical fluid is a substance above its critical temperature and pressure but below the pressure required to condense it from the fluid to the solid state [52]. The properties of supercritical fluids are somewhat intermediate between those of liquids and gases (Table 1), they have gas-like diffusivities and liquid-like densities, in addition to an easily tuneable solvent power [53].

	Diffusivity (cm^2/s)	Density (g/mL)	Viscosity (Pa.s)
Gases	0.1	10^{-3}	10^{-5}
Supercritical fluids	10^{-3}	0.3	10^{-4}
Liquids	5×10^{-6}	1	10^{-3}

Table 1. Physical properties of gases, liquids and supercritical fluids (adapted from [53]).

Supercritical carbon dioxide ($scCO_2$), in particular, possesses numerous properties that made it emerge as the most extensively studied supercritical fluid. It has a low critical point, is inexpensive, readily available in high purity and is a gas under ambient conditions, meaning that by simple depressurization of the system no solvent residues will be found in the final product. It is a GRAS solvent (Generally Recognized As Safe) ([http://www.fda.gov) and considered a "green" alternative solvent.

In the last years, $scCO_2$ has already proved to be an excellent medium for the synthesis and processing of polymers [54,55,56], impregnation of active substances [57], formation of porous structures [58,59], textile cleaning [60], chemical reactions [61], extraction [62], etc.

Polymers and other high average molecular weight macromolecules and polar materials are typically not soluble in supercritical carbon dioxide, with exception of silicones and fluorinated polymers. On the other hand monomers are usually very soluble. This means that initial homogeneous phases are easily obtained at mild pressure conditions at the beginning of the reaction and when the polymer reaches a certain threshold molecular weight it precipitates being collected inside the reactor (Fig. 2b). This is a typical heterogeneous reaction by precipitation. Depending on the solubility of template and monomers in the $scCO_2$ medium, different polymerization approaches can be followed such as dispersion and emulsion [63].

Typically a polymer is synthesized in $scCO_2$ inside a stainless steel high-pressure vessel equipped with a pressure transducer. Several types of reactors can be used with magnetic or mechanically stirring or equipped with visual windows to allow full observation of the inside (Fig. 2a).

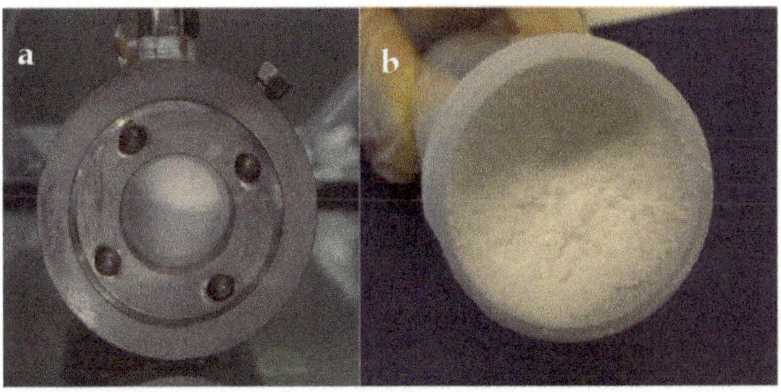

Figure 2. a) 33 mL high-pressure visual reactor used in our lab and b) ready-to-use dry MIP polymer powder synthesized in $scCO_2$.

Polymerization of the typical monomers used so far in $scCO_2$-assisted molecular imprinting is radical initiated, usually with AIBN. That is why typical reactions are carried out at around 65ºC, the optimal initiation temperature. Initial pressure of the reaction is chosen in order to have a one single phase at the beginning of the reaction, with all reactants, monomers, cross-linker and initiator soluble in the CO_2 continuous phase, as well the template molecule. There is the possibility of adding a stabilizer that can further control the morphology of the polymer by minimizing the surface area of the particles forming spherical particles with monodisperse particle size distribution [64,65].

Phase behaviour studies such as cloud point curve determination for the reaction mixture are important for the polymerization optimization [66]. In addition, as organic solvents are typically soluble in $scCO_2$, a small amount of an organic solvent co-solvent can be added in order to tune the polarity of the medium, and help dissolving more polar molecules.

6. Supercritical CO_2-assisted synthesis of MIPs

Supercritical CO_2-assisted molecular imprinting has recently demonstrated to be a clean and one-step synthetic route for the preparation of MIPs, with attested performance in chromatography [47], drug delivery [67,68,69], and extraction [70,71,72,73]. The advantages of using scCO2 as solvent in the molecular imprinting process go beyond environmental concerns. By performing the synthesis in an apolar aprotic porogen, such as scCO2, the T:M complexes can be highly stabilized giving rise to well-defined active sites and leading to materials with high affinity [74]. Furthermore the matrices can have a controlled morphology and are obtained as dry powders, with no solvent residues, avoiding posterior purification and drying steps, which are major advantages over conventional methods. Moreover the use of supercritical fluids in the removal of template molecules can increase the diffusion coefficient at least 10-fold [75], which can overcome this actual limitation of conventional molecular imprinting. Thus the use of $scCO_2$ can bring not just advantages in the synthesis of the polymers but also in the template desorption from the matrix at the end of the polymerization, and it can be further used in the impregnation of the bioactive molecules in the matrix by $scCO_2$-assisted impregnation.

Imprinted polymers have gained much importance over the last years, and their potential use in drug delivery applications, as unique delivery systems or incorporated into other drug delivery devices, are being widely studied. The use of MIPs could improve the control over the therapeutic release by sustaining the delivery of the agent and intelligently releasing the drug in response to small variations in the environment. Moreover the majority of the drugs act by a molecular recognition mechanism and molecular imprinting technique also enhances the loading capacity of the polymers towards the template drug.

The first work reporting the potential use of MIPs as sustained drug delivery carriers was conducted by the group of Nicholls [76] and studied the drug dissociation kinetic from theophylline-imprinted polymers, at different pH and loaded with different amounts of drug. The imprinted polymers synthesized demonstrated higher affinity towards theophylline than to caffeine and displayed a more sustained delivery than the corresponding control polymer. This work was followed by many others reporting the development of MIPs with potential use as DDS. Some recent developments include the use of molecular imprinting in the preparation of materials for ocular, transdermal and oral drug delivery. Templates such as theophylline, histamine, ephedrine [77], propranolol [77,78,79,80,81], citalolpram [82], H1-antihistamines [83], sulfasalazine [84], 5-Fluorouracil

[85], norfloxacin [86], hyaluronic acid [87], levofloxacin [88] among many others have been used with success. Propranolol a β-blocker widely used for hypertension treatment, is one of the most used templates, and is indeed the most successful imprinted molecule. Its choice is not only justified by its clinical relevance, inherent chirality and availability in enantiomerically pure and radio-labelled forms, but also because of its ability to stablish strong ion-pair interactions as well as hydrophobic [89]. The actual great challenge in the synthesis of MIPs is to imprint other interesting molecules such as macromolecules, charged or low functionality molecules [90].

Over the last years a great effort has been made to develop imprinted matrices for drug delivery applications with some chain flexibility. Although the majority of imprinted polymers reported possess high crosslinking degree and therefore limited chain mobility, a certain degree of flexibility can be found in biological systems, with recognition occurring in aqueous environment as the result of numerous non-covalent interactions. Much of the theory concerning conformational memory in biological macromolecules, such as proteins, is valid for molecularly imprinted polymers. Thus, the design of imprinted matrices with a certain mobility of the polymeric chains has been studied to improve the properties of common hydrogels.

Hydrogels are networks of hydrophilic polymers that have attracted wide research interest as controlled release devices due to their tunable chemical and three-dimensional physical structure, good mechanical properties, high water content and biocompatibility [1]. Hydrogels can be successfully synthesized in $scCO_2$ [91,92]. Poly(N-isopropylacrylamide) (PNIPAAm) hydrogel with controlled morphology was synthesised in $scCO_2$, with advantages when compared with conventional polymerization since there is no need for intensive drying of the material before further processing or characterization steps. Thus the combination of hydrogel development using supercritical fluid technology with molecular imprinting has a strong potential in the design of new targeted DDS therapies.

Supercritical CO_2-assisted synthesis of MIPs as potential DDS was first developed by Casimiro and collaborators [67]. The work reports the synthesis in supercritical medium of spherical particles of poly(diethylene glycol dimethacrylate), PDEGDMA, using a carboxylic acid end-capped perfluoropolyether oil as stabiliser. Polymerisations were carried out in the presence of different concentrations of salicylic acid and acetylsalicylic acid and results showed that MIPs were able to uptake higher amounts of template drug during $scCO_2$-assisted impregnation and release it in a more sustained way when compared with non-imprinted polymers (NIPs). In addition there was an evident correlation between the amount of template used in the synthesis and the loading capacity in the impregnation of the polymers using $scCO_2$. These promising results prompted the optimization of other imprinting systems, through the study of the influence of some experimental variables in the performance of MIPs as drug delivery carriers.

Copolymers of methacrylic acid (MAA) with ethylene glycol dimethacrylate (EGDMA) and N-isopropylacrylamide (NIPAAm) with EGDMA were designed in $scCO_2$ to have molecular recognition to flufenamic acid (FA) [68]. The synthesized polymeric matrices with different ratios between template and functional monomer and different crosslinking degrees were further impregnated with FA in $scCO_2$ and the drug release profiles evaluated *in vitro*. The results showed that by changing the experimental parameters above mentioned it is possible to tune the molecular recognition of MIPs and therefore the drug delivery. Although highly cross-linked matrices possess higher imprinting factor, all the imprinted polymers presented more sustained release of FA than control polymers, due to the affinity introduced by molecular imprinting. High-pressure NMR was used to study the interactions between the template and the monomers in the pre-polymerization mixture. The results confirm that there are hydrogen bond interactions between the carboxylic acid groups of FA and monomers, which are responsible for the specificity of the binding sites.

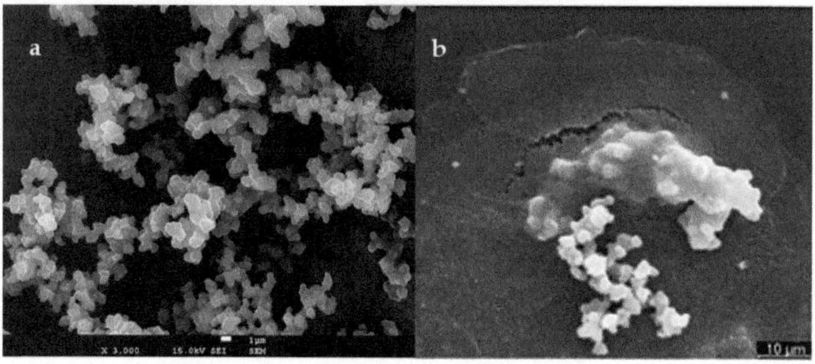

Figure 3. a) SEM image of MIP polymer and b) MIP particles in contact with Caco-2 cells (1500×, 20 kV) from [69].

Low cross-linked polymeric networks composed of 2-(dimethylamino)ethyl methacrylate (DMAEMA) and EGDMA with molecular recognition to ibuprofen were developed using supercritical fluid technology [69]. After impregnation of the matrices with the drug using $scCO_2$-assisted impregnation, the polymers were evaluated as potential DDS in two different *in vitro* situations. In the first the polymers were allowed to release the drug over several days and in the second an oral administration situation was simulated and the polymers were subjected to different pHs and the drug release profiles studied for a period of 8 hours. The results show that MIP had an extraordinary ability to uptake the template drug during $scCO_2$-assisted impregnation, when compared with the NIP (33.1 wt% versus 10.2 wt%). Further, cytotoxicity experiments performed with Caco-2 cells revealed that the polymers are biocompatible and could therefore be used in biomedical applications. Fig. 3a shows the MIP particles synthesized in $scCO_2$ and Fig. 3b the proliferation and the adhesion of Caco-2 to the particles which shows its compatibility.

Furthermore, the affinity structures were able to maintain their selective recognition properties in aqueous environment. Water-compatible molecular recognition is a real challenge as traditionally MIPs present better performance in hydrophobic organic solvents. Supercritical CO_2 can provide a potential alternative to the synthesis of MIPs with good performance in biological fluids, crucial for *in vivo* applications.

Though we believe the use of supercritical fluid technology is very promising and challenging in the development of MIPs, there are of course still some limitations. An example is the development of molecularly imprinted contact soft lenses where transparency of the polymer is a required parameter. Typically polymers are obtained as amorphous white powders not in accordance with this specific requirement. However even in these cases supercritical fluid technology can be beneficial for instance in the impregnation of drugs into the lenses taking advantages of the high diffusivity of $scCO_2$ [93].

Very recently a MIP was developed for the first time using the semi-covalent approach using supercritical fluid technology [71]. Molecularly imprinted polymeric particles with molecular recognition towards bisphenol A (BPA) were synthesized using bisphenol A dimethacrylate, a monomer containing the template. Bisphenol A was then cleaved from the polymeric matrix by hydrolysis with tetrabutylammonium hydroxide (n-Bu4OH) with $scCO_2$, taking advantage of its high diffusivity. Once again the MIP showed a good performance in aqueous solutions, due to the fact of CO_2 is an aprotic solvent, as explained above. This could open up a new possibility of molecular imprinting for drug molecules, which however needs the use of a template-containing molecule with polymerizable backbones. Still it could lead to a stricter control over the functional group location and uniform distribution binding sites, characteristic of covalent imprinting, and the reduced kinetic restriction during rebinding, characteristic of non-covalent imprinting approach.

Recently supercritical fluid technology has also been proposed as a viable alternative to prepare hybrid MIMs [70]. This methodology took advantage of the ability of $scCO_2$ to induce phase inversion of poly(methyl methacrylate), to prepare PMMA-based membrane with imprinted particles casted. Results showed that the immobilization of imprinted polymers by incorporation of MIPs into a casting solution for membrane production is a viable method to produce affinity membranes, which are able to be loaded with higher amounts of the template molecule than the non-imprinted membrane, even in dynamic conditions. This means that it is possible to confer molecular recognition to other types of structures and scaffolds by immobilizing MIP particles.

7. Final remarks

The continuous advances in materials science, in particular in the development of smart targeted DDS and materials with molecular recognition will for sure bring important progresses in biomaterials applications. The development of molecularly imprinted polymers as drug delivery systems although in an emerging stage, has already shown its high potential. The affinity to the target molecules, introduced by molecular imprinting, provides polymers with the ability to load higher amounts of the template molecule, compared to the non-imprinted polymers, either in aqueous solutions or in supercritical environment. In addition *in vitro* drug delivery experiments revealed that the macromemory effect yields polymeric DDSs able to release the drug in a more sustained way. Additional degrees of control in drug release can be introduced by tuning the MIP composition for instance with stimulus-responsive polymers. Furthermore the immobilization of imprinted particles within porous structures, such as membranes, seems to be a feasible option for the preparation of affinity devices with potential application as DDS.

Supercritical fluid technology is a very promising way to prepare MIPs. While it provides a clean route for the synthesis, since it avoids the use of organic solvents and multistep processes, it can also bring high control over the morphology of the polymers. We believe that this technology has still much to offer to molecular imprinting and that the combination of both technologies is a good synergy to design new materials and structures with a high purity level and tunable molecular recognition, with the needed requirements for biomedical and pharmaceutical applications.

Author details

Mara Soares da Silva and Teresa Casimiro *
REQUIMTE, Departamento de Química, Faculdade de Ciências e Tecnologia, FCT, Universidade Nova de Lisboa, Caparica, Portugal

Acknowledgement

The authors would like to thank Fundação para a Ciência e Tecnologia (FCT-Lisbon) for financial support through project PTDC/QUI-QUI/102460/2008 and contract PEst-C/EQB/LA0006/2011.

8. References

[1] Langer R, Peppas N A. Advances in Biomaterials, Drug Delivery, and Bionanotechnology. AIChE J. 2003; 49 (12): 2990-3006.
[2] Vasapollo G, Sole RD, Mergola L, Lazzoi MR, Scardino A, Scorrano S, Mele G. Molecularly imprinted polymers: Present and Future prospective. Int. J. Mol. Sci. 2011; 12: 5908-5945.
[3] Byrne M E, Hilt J Z, Peppas N A. Recognitive biomimetic networks with moiety imprinting for intelligent drug delivery. J. Biomed. Mat. Res. A, 2008; 84(1): 137-147.
[4] Ludwig R, Dzung NTK. Calixarene-based molecules for cation recognition. Sensors. 2002; 2: 397-416.
[5] Malta LFB, Cordeiro Y, Tinoco LW, Campos CC, Medeiros ME, Antunes OAC. Recognition mechanism of D- and L-tryptophan enantiomers using 2-hydroxypropyl-α- or β-cyclodextrins as chiral selectors. Tetrahedron: Asymmetry. 2008; 19 (10): 1182-1188.
[6] Pittelkow M, Nielsen CB, Kadziola A, Christensen J.B. Molecular recognition: minimizing the acid-base interaction of a tunable host–guest system changes the selectivity of binding. J. Incl. Phenom. Macrocycl. Chem. 2009; 63(3-4): 257-266.

* Corresponding Author

[7] Alexander C, Andersson HS, Andersson LI, Ansell RJ, Kirsch N, Nicholls IA, O'Mahony J, Whitcombe MJ. Molecular imprinting science and technology: a survey of the literature for the years up to and including. J. Mol. Recognit. 2003; 19(2): 106-180.
[8] Soares da Silva ML. Development of molecularly imprinted polymers using supercritical fluid technology, PhD Thesis, Universidade Nova de Lisboa, 2011a.
[9] Chianella I, Lotierzo M, Piletsky SA, Tothill IE, Chen B, Karim K, Turner APF. Rational Design of a polymer specific for microcystin-LR using a computational approach. Anal. Chem. 2002; 74(6): 1288-1293.
[10] Hoshino Y, Shea KJ. The evolution of plastic antibodies. J. Mater. Chem. 2011; 21: 3517-3521.
[11] Baggiani C, Giraudi G, Trotta F, Giovannoli C, Vanni A. Chromatographic characterization of a molecular imprinted polymer binding cortisol, Talanta. 2000; 51 (1): 71-75.
[12] Svenson J, Nicholls IA. On the thermal and chemical stability of molecularly imprinted polymers. Anal. Chim. Acta. 2001; 435(1): 19-24.
[13] Bossi A, Bonini F, Turner APF, Piletsky SA. Molecularly imprinted polymers for the recognition of proteins: The state of the art. Biosens. Bioelectron. 2007; 22 (6): 1131-1137.
[14] Wulff G. Enzyme-like catalysis by molecularly imprinted polymers. Chem. Rev. 2002; 102(1): 1-28.
[15] Ramström O, Mosbach K. Synthesis and catalysis by molecularly imprinted materials. Curr. Opin. Chem. Biol. 1999; 3(6): 759-764.
[16] Baggiani C, Anfossi L, Giovannoli C. Solid phase extraction of food contaminants using molecular imprinted polymers. Anal. Chim. 2007 Acta, 591 (1): 29-39.
[17] Andersson LI, Paprica A, Arvidsson T. A highly selective solid phase extraction sorbent for pre-concentration of sameridine made by molecular imprinting. Chromatographia. 1997; 46 (1-2) 57-62.
[18] Wei S, Mizaikoff B. Recent advances on noncovalent molecular imprints for affinity separations. J. Sep. Sci. 2007; 30 (11): 1794-1805.
[19] Kempe M. Antibody-mimicking polymers as chiral stationary phases. Anal. Chem.1996; 68(11) 1948-1953.
[20] Guan G, Liu B, Wang Z. Imprinting of molecular recognition sites on nanostructures and its applications in chemosensors. Sensors. 2008; 8(12): 8291-8320.
[21] Haupt K, Mosbach K. Molecularly imprinted polymers and their use in biomimetic sensors. Chem. Rev. 2000; 100 (7): 2495-2504.
[22] Hung C, Huang Y, Huang H, Hwang C. Preparation of (S)-Ibuprofen-Imprinted Polymer and Its Molecular Recognition Study. J. Appl. Polym. Sci. 2006; 102(3): 2972–2979.
[23] Alvarez-Lorenzo C, Concheiro A. Molecularly imprinted polymers for drug delivery. J. Chromatogr. B Analyt. Technol. Biomed. Life Sci. 2004; 804(1): 231-245.
[24] Mayes AG, Whitcombe MJ. Synthetic strategies for the generation of molecularly imprinted organic polymers. Adv. Drug Del. Rev. 2005; 57: 1742-1778.

[25] Hilt JZ, Byrne ME. Configurational biomimesis in drug delivery: molecular imprinting of biologically significant molecules. Adv. Drug del. Rev.2004; 56: 1599-1620.

[26] Wulff G, Schauhoff S. Enzyme-analog-built polymers. 27. Racemic resolution of free sugars with macroporous polymers prepared by molecular imprinting. Selectivity dependence on the arrangement of functional groups versus spatial requirements. J. Org. Chem. 1991; 56(1): 395-400.

[27] Siemann M, Andersson LI, Mosbach K. Selective Recognition of the Herbicide Atrazine by Noncovalent Molecularly Imprinted Polymers. J. Agric. Food Chem. 1996; 44 (1): 141-145.

[28] Kirsch N, Alexander C, Davies S, Whitcombe MJ. Sacrificial spacer and non-covalent routes toward the molecular imprinting of "poorly-functionalized" N-heterocycles. Anal. Chim. Acta. 2004; 504(1): 63–71.

[29] Qu S, Wang X, Tong C, Wu J. Metal ion mediated molecularly imprinted polymer for selective capturing antibiotics containing beta-diketone structure. J. Chromatogr. A. 2010; 1217(52): 8205–8211.

[30] Haupt K. Imprinted polymers-Tailor-made mimics of antibodies and receptors, Chem. Commun. 2003; 2: 171-178.

[31] Yilmaz E, Mosbach K., Haupt K. Influence of functional and cross-linking monomers and the amount of template on the performance of molecularly imprinted polymers in binding assays. Anal. Commun. 1999; 36(5): 167-170.

[32] Nicholls IA, Adbo K, Andersson HS, Andersson PO, Ankarloo J, Hedin-Dahlstrom J, Jokela P, Karlsson JG, Olofsson L, Rosegren J, Shoravi S, Svenson J, Wikman S. Can we rationally design molecularly imprinted polymers? Anal. Chim. Acta. 2001; 435(1): 9-18.

[33] Esfandyari-Manesh M, Javanbakht M, Atyabi F, Badiei A, Rassoul D. Effect of porogenic solvent on the morphology, recognition and release properties of carbamazepine-molecularly imprinted polymer nanospheres. J. Appl. Polym. Sci. 2011; 121 (2): 1118-1126.

[34] Yu C, Mosbach K. Influence of mobile phase composition and cross-linking density on the enantiomeric recognition properties of molecularly imprinted polymers. J. Chromatogr. A. 2000; 888(1-2): 63-72.

[35] Yu C, Mosbach K. Insights into the origins of binding and the recognition properties of molecularly imprinted polymers prepared using an amide as the hydrogen-bonding functional group. J. Mol. Recogn. 1998; 11(1-6): 69-74.

[36] Sellergren B, Dauwe C, Schneider T. Pressure-Induced Binding Sites in Molecularly Imprinted Network Polymers. Macromolecules. 1997; 30(8): 2454-2459.

[37] Asanuma H, Akiyama T, Kajiya K, Hishiya T, Komiyama M. Molecular imprinting of cyclodextrin in water for the recognition of nanometer-scaled guests, Anal. Chim. Acta. 2001; 435: 25-33.

[38] Piletsky SA, Andersson HS, Nicholls IA. Combined hydrophobic and electrostatic interaction-based recognition in molecular imprinted polymers. Macromolecules. 1999; 32(3): 633-636.

[39] Turiel E, Martin-Esteban A. Molecularly imprinted polymers for solid-phase microextraction. J. Sep. Sci. 2009; 32(19): 3278-3284.
[40] Qiao F, Sun H, Yan H, Row KH. Molecularly Imprinted Polymers for Solid Phase Extraction. Chromatography. 2006; 64: 625–634.
[41] Andersson LI. Selective solid-phase extraction of bio- and environmental samples using molecularly imprinted polymers. Bioseparation. 2001; 10(6): 353-364.
[42] Ramström O, Skudar K, Haines J, Patel P, Brüggemann O. Food analyses using molecularly imprinted polymers. J. Agric. Food Chem. 2001; 49 (5): 2105-2114.
[43] Haupt K. Molecularly imprinted polymers in analytical chemistry. The Analyst. 2001; 126: 747-756.
[44] Fu Q, Sanbe H, Kagawa C, Kunimoto KK, Haginaka J. Uniformly sized molecularly imprinted polymer for (S)-Nilvapidine. Comparison of chiral recognition ability with HPLC chiral stationary phases based on a protein. Anal. Chem. 2003; 75(2): 191-198.
[45] Huang BY, Chen YC, Wang GR, Liu CY. Preparation and evaluation of a monolithic molecularly imprinted polymer for the chiral separation of neurotransmitters and their analogues by capillary electrochromatography. J. Chromatogr. A. 2011; 1218(6): 849-855.
[46] Khan H, Khan T, Park JK. Separation of phenylalanine racemates using d-phenylalanine imprinted microbeads as HPLC stationary phases. Sep. Purif. Tech. 2008; 62(2): 363-369.
[47] Soares da Silva M, Vão ER, Temtem M, Mafra L, Caldeira J, Aguiar-Ricardo A, Casimiro T. Clean synthesis of molecular recognition polymeric materials with chiral sensing capability using supercritical fluid technology. Application as HPLC stationary phases. Biosens. Bioelectron. 2010; 25(7): 1742-1747.
[48] Yilmaz E, Ramström O, Möller P, Sanchez D, Mosbach K. A facile method for preparing molecularly imprinted spheres using spherical silica templates. J. Mater. Chem. 2002; 12(5): 1577-1581.
[49] Donato L, Tasselli F, Drioli E. Molecularly Imprinted Membranes with Affinity Properties for Folic Acid. Sep. Sci. Tech. 2010; 45(16): 2273-2279.
[50] Li S, Ge Y, Tiwari A, Wang S, Turner APF, Piletsky SA. "On/Off"-switchable catalysis by a smart enzyme-like imprinted polymer. J. Catal. 2011; 278: 173-180.
[51] Matyjaszewski K. Controlling polymer structures by atom transfer radical polymerization and other controlled/living radical polymerizations. Macrom. Symp. 2003; 195(1): 25-31.
[52] Jessop P, Leitner W. Chemical Synthesis using supercritical fluids. New York: Wiley-VCH; 1999.
[53] Leitner W, Jessop PG. Handbook of green chemistry, Vol. 4: Supercritical Solvents. Weinheim: Wiley-VCH; 2010.
[54] Kendall JL, Canelas DA, Young JL, DeSimone JM. Polymerizations in supercritical carbon dioxide. Chem. Rev.1999; 99: 543-563.
[55] Giles MR, Griffiths RMT, Aguiar-Ricardo A, Silva MMCG, Howdle SM. Fluorinated graft stabilizers for polymerization in supercritical carbon dioxide: the effect of stabilizer architecture. Macromolecules. 2001; 34: 20-35.

[56] Jung J, Perrut M. Particle design using supercritical fluids: Literature and patent survey. J. Supercrit. Fluids. 2001; 20(3): 179-219.

[57] Kikic I, Vecchione F. Supercritical impregnation of polymers. Curr. Opin. Solid State Mater. Sci. 2003; 7: 399-405.

[58] Reverchon E, Cardea S. Production of controlled polymeric foams by supercritical CO_2, J. Supercrit. Fluids. 2007; 40(1): 144-152.

[59] Temtem M, Casimiro T, Aguiar-Ricardo A. Solvent power and depressurization rate effects in the formation of polysulfone membranes with CO_2-assisted phase inversion method. J. Membr. Sci. 2006; 283(1-2): 244-252.

[60] Sousa M, Melo MJ, Casimiro T, Aguiar-Ricardo A. The art of CO_2 for art conservation: a green approach to antique textile cleaning. Green Chem. 2007; 9: 943- 947.

[61] Bogel-Łukasik E, Fonseca I, Bogel-Łukasik R, Tarasenko, YA, Nunes da Ponte M, Paiva A, Brunner G. Phase, equilibrium-driven selective hydrogenation of limonene in high-pressure carbon dioxide. Green Chem. 2007; 9: 427-430.

[62] Simões PC, Carmelo PJ, Pereira PJ, Lopes JA, Nunes da Ponte M, Brunner G. Quality assessment of refined olive oils by gas extraction. J. Supercrit. Fluids. 1998; 13(1-3): 337-341.

[63] Kazarian S G. Polymer Processing with Supercritical Fluids. Polymer Science C, 2000; 42(1): 78-101.

[64] Christian P, Howdle SM, Irvine D. Dispersion polymerization of methyl methacrylate in supercritical carbon dioxide with a monofunctional pseudo-graft stabilizer. Macromolecules. 2000; 33(2): 237–239.

[65] Casimiro T, Banet-Osuna A, Ramos A M, Nunes da Ponte M, Aguiar-Ricardo A. Synthesis of highly cross-linked poly(diethylene glycol dimethacrylate) microparticles in supercritical carbon dioxide. Eur. Polym. J. 2005; 41: 1947-1953.

[66] Lora M, McHugh MA. Phase behavior and modeling of the poly(methyl methacrylate)-CO_2-methyl methacrylate system. Fluid Phase Equil. 1999; 157(2): 285-297.

[67] Duarte ARC, Casimiro T, Aguiar-Ricardo A, Simplício AL, Duarte CM. Supercritical fluid polymerisation and impregnation of molecularly imprinted polymers for drug delivery. J. Supercrit. Fluids. 2006; 39(1): 102-106.

[68] Soares da Silva M, Nobrega FL, Aguiar-Ricardo A, Cabrita EJ, Casimiro T. Development of molecularly imprinted co-polymeric devices for controlled delivery of flufenamic acid using supercritical fluid technology, J. Supercrit. Fluids. 2011b; 58(1): 150-157.

[69] Soares da Silva M, Viveiros R, Morgado PI, Aguiar-Ricardo A, Correia IJ, Casimiro T. Development of 2-(dimethylamino)ethyl methacrylate-based molecular recognition devices for controlled drug delivery using supercritical fluid technology. Int. J. Pharm. 2011c; 416: 61-68.

[70] Soares da Silva M, Viveiros R, Coelho M, Aguiar-Ricardo A, Casimiro T. Supercritical CO_2-assisted preparation of a PMMA composite membrane for Bisphenol A recognition in aqueous environment. Chem. Eng. Sci. 2012a: 68(1): 94-100.

[71] Soares da Silva M, Viveiros R, Aguiar-Ricardo A, Bonifácio V, Casimiro T. Novel supercritical CO_2- based strategy to prepare semi-covalent molecularly imprinted

materials with water-compatible molecular recognition, RSC Advances. 2012b; doi: 10.1039/C2RA20426F.

[72] Ye L, Yoshimatsu K, Kolodziej D, Francisco JDC. Preparation of molecularly imprinted polymers in supercritical carbon dioxide. J. Appl. Polym. Sci. 2006; 102(3): 2863-2867.

[73] Kobayashi T, Leong SS, Zhang Q. Using polystyrene-co-maleic acid for molecularly imprinted membranes in supercritical carbon dioxide. J. Appl. Polym. Sci. 2008; 108(2): 757-768.

[74] Schweitz L, Andersson LI, Nilsson S. Capillary electrochromatography with predetermined selectivity obtained through molecular imprinting. Anal. Chem. 1997; 69(6): 1179–1183.

[75] Ellwanger A, Berggren C, Bayoudh S, Crecenzi C, Karlsson L, Owens P K, Ensing K, Cormack, P., Sherrington D, Sellergren B. Evaluation of methods aimed at complete removal of template from molecularly imprinted polymers. Analyst. 2001; 126 (6): 784-792.

[76] Norell MC, Andersson HS, Nicholls IA. Theophylline molecularly imprinted polymer dissociation kinetics: a novel sustained release drug dosage mechanism. J. Mol. Recognit. 1998; 11: 98-102.

[77] Allender C J, Richardson C, Woodhouse B, Heard C M, Brain K R. Pharmaceutical applications for molecularly imprinted polymers. Int. J. Pharm., 2000. 195: 39-43.

[78] Suedee R, Srichana T, Martin GP. Evaluation of matrices containing molecularly imprinted polymers in the enantioselectivity-controlled delivery of β-blockers. J. Control. Release. 2000; 66(2-3): 135-147.

[79] Suedee R, Srichana T, Chotivatesin R, Martin GP. Stereoselective release behaviors of imprinted bead matrices. Drug Dev. Ind. Pharm. 2002; 28: 545-554.

[80] Andersson LI. Application of molecular imprinting to the development of aqueous buffer and organic solvent based radioligand binding assays for (S)-propranolol, Anal. Chem. 1996; 68(1): 111–117.

[81] Bodhibukkana C, Srichana T, Kaewnopparat S, Tangthong N, Bouking P, Martin GP, Suedee R. Composite membrane of bacterially-derived cellulose and molecularly imprinted polymer for use as transdermal enantioselective controlled-release system of racemic propanolool. J. Control. Release. 2006; 113(1): 43-56.

[82] Abdouss M, Asadi E, Azodi-Deilami S, Beik-mohammadi N, Aslanzadeh S A. Development and characterization of molecularly imprinted polymers for controlled release of citalopram. J. Mater. Sci: Mat. Med. 2011; 22: 2273-2281.

[83] Venkatesh S, Sizemore SP, Byrne ME. Biomimetic hydrogels for enhanced loading and extended release of ocular therapeutics. Biomaterials. 2007; 28 (4): 717-724.

[84] Puoci F, Iemma F, Muzzalupo R, Spizzirri UG, Trombino S, Cassano R, Picci N. Spherical molecularly imprinted polymers (SMIPs) via a novel precipitation polymerization in the controller delivery of sulfasalazine. Macromol. Biosci. 2004; 4 (1): 22-26.

[85] Puoci F, Lemma F, Cirillo G, Picci N, Matricardi P, Alhaique F. Molecularly Imprinted Polymers for 5-Fluorouracil Release in Biological Fluids. Molecules. 2007; 12 (4): 805-814.

[86] Alvarez-Lorenzo C, Yañez F, Barreiro-Iglesias R, Concheiro A. Imprinted soft contact lenses as norfloxacin delivery systems. J. Control. Release. 2006; 113(3): 236-244.
[87] Ali M, Byrne ME. Controlled Release of High Molecular Weight Hyaluronic Acid from Molecularly Imprinted Hydrogel Contact Lenses. Pharm. Res. 2009; 26(3): 714-726.
[88] Mazzotta E, Malitesta C, Díaz-Álvarez M, Martín-Esteban A. Electrosynthesis of molecularly imprinted polypyrrole for the antibiotic levofloxacin. Thin Solid Films. 2012; 520 (6): 1938-1943.
[89] Mayes AG. Polymerisation techniques for the formation of imprinted beads. In: Sellergren B, editor. Molecular Imprinted Polymers. Man-made mimics of antibodies and their applications in analytical chemistry. Amsterdam: Elsevier Science BV; 2001.
[90] Luk Y, Allender CJ, Wirth T. Molecular imprinted polymers binding low functionality templates. Tetrahedron Letters. 2010; 51(45): 5883–5885.
[91] Temtem M, Casimiro T, Mano JF, Aguiar-Ricardo A. Green synthesis of a temperature sensitive hydrogel. Green Chem. 2007; 9(1): 75-79.
[92] Costa E, de-Carvalho J, Casimiro T, Lobato da Silva C, Cidade M T, Aguiar-Ricardo A. Tailoring thermoresponsive microbeads in supercritical carbon dioxide for biomedical applications. J. Supercrit. Fluids. 2011; 56(3): 292-298.
[93] Costa VP, Braga MEM, Duarte CMM, Alvarez-Lorenzo C, Concheiro A, Gil MH, de Sousa HC. Anti-glaucoma drug-loaded contact lenses prepared using supercritical solvent impregnation. J. Supercrit. Fluids. 2010; 53(1-3): 165-173.

Chapter 8

Recent Progress in Benzocyclobutene Related Polymers

Junxiao Yang, Yawen Huang and Ke Cao

Additional information is available at the end of the chapter

1. Introduction

The developing history of polymer materials tells that the invention of novel polymers with excellent performance is generally originated from the synthesis of new polymerizable monomers. Among kinds of monomers, benzocyclobutene (BCB) has attracted much interest because of its unique molecular structure and polymer properties (1-2). The structure of BCB is shown as following with the numbering system, the CA name of which is bicyclo[4,2,0]octa-1,3,5-triene. In terms of the four-membered ring structure, BCBs are able to transform into the reactive *o*-quinodimethane, also known as *o*-xylylene, upon heating at around 200 °C. This very reactive specie readily undergoes inter- and intramolecular Diels-Alder reaction with various dienophiles to produce polycyclic compounds with high stability, such as cyclooctadiene. On the other hand, the polymerization of *o*-quinodimethane similar to that of 1,3-diene would give linear polymers. Additionally, the reaction of BCB does not require any catalyst and release any small molecules, thus hardly introducing any impurities into materials. These features undoubtedly offer BCB based materials several superiorities in several specific applications, such as microelectronic applications, optic applications, electrical applications, etc.

Nowadays, hundreds of BCB based structures have been synthesized, which constituted abundant resource for producing high performance and functional materials. Varying the structures and properties of BCB building blocks and constructing methods would afford materials with tunable or attractive properties and performance. Early prepared BCB monomers were mainly BCB-bicapped structures. As described previously, these compounds can be utilized to generate reactive oligomers which are processable intermediates to prepare polymeric materials. In its most basic statement, the concept of reactive oligomers involves in-situ conversion of low molecular weight compounds into high molecular weight materials. Reactive oligomers are in fact a very old technology with the earliest examples being found in the preparation of organic coatings. This technology however is limited to inferior film-forming properties. To address this problem, more recently, new structures were introduced to make a modification of present structures. Of them, poly (vinyl-benzocyclobutene) has attracted increased attention (3). In different with the polymers prepared from BCB-dicapped monomers, the BCB unit were placed on the side chains in poly (vinyl-benzocyclobutene). Thanks to their well-controllable structures, properties and performance, these polymers show potential application in nanoparticles preparation, surface modification, apart from microelectronic. On the other hand, early polybenzocylobutenes also suffer from relatively low thermal stability. As a respond to the requirement of high thermal resistance, a new class of polybenzocylobutenes, benzocyclobutene-siloxane resins, has been extensively studied. Through introducing of thermostable polysiloxane chains, remarkable enhancement in thermal resistance has attained.

Overall, since Finkelstein reported the synthesis of a BCB derivative in 1909 for the first time, the research on BCB has experienced rapid development. The research field has extended from chemistry to materials even physics. Correspondingly, the research targets have also shown a transition from low molecular weight molecules to high molecular weight materials. Furthermore, the applications of BCB-based materials have extended from dielectrics to materials modification, nanostructure construction, and so on. All of these trends suggest a prospective future of BCB related materials. Previously, several excellent reviews have dealt with the development of BCB related science (1-2). More recently, although new reviews involving the BCB chemistry have been reported (4-5), the progress in BCB related polymers has not been reviewed until now. In this chapter, we will present a review of BCB from 1996, which will concern about two types of newly developed BCB-based materials, poly (vinyl-BCB) and BCB-siloxane polymers.

2. Vinyl-benzocyclobutene

From a closer look of the structures of the classical BCB resins prepared from BCB di-capped monomers, it can be noted that these cross-linked polymers were prepared by directly taking use of intramolecular Diels-Alder reaction and the functionality of these monomers was more than 2. As a result, cross-linking and polymerization take place simultaneously. Because they lack well-defined structures, conventional BCB-based polymers are difficult to provide controllable properties. Consequently, their applications in many areas are limited. In addition, the network structures are mainly constructed by the polycyclic structure. This highly rigid structure is in general unfavorable for generating higher cross-linking degree.

Polyolefins represent an important class of polymer whose structure could be conveniently controlled by sophisticated methods. More importantly, reactive functional groups are able to be introduced into polyolefins conveniently (6-7). Such built-in cross-linkable functionality leads to two-step crosslinking, offering substantial advantages over alternative two-component mixtures or one-step formation of cross-linked, network solids. First of all, chain growth and cross-linking reactions become independent of one another, either expanding the materials application range or improving the processability of thermally stable polymers. Second, this cross-linking protocol provides complete control over the chemical structure of the primary polymer, network structures and physical/chemical properties of the matrix. Meanwhile, the cross-linking agent is incorporated as a comonomer into the backbone of the relative high molecular weight polymers, allowing the cross-linking density to be directly related to the monomer composition. Hence, constructing new BCB polymers with polyolefin backbone are expected to open a new road towards controllable structure of BCB-based materials.

2.1. Preparation and properties of vinyl-BCB polymers

2.1.1. Poly (4-vinyl benzocyclobutene)

As a typical vinyl-BCB, 4-vinyl benzocyclobutene (4-VBCB) as shown in **Scheme 1**, was firstly synthesized by Endo et al. (8) and more recently by Hawker et al. (9). Of several well-established polymerization protocols, free radical and anionic methods have been so far demonstrated available for the polymerization of 4-VBCB. Copolymerization of vinyl-BCB with other vinyl monomers such as styrene, methyl methacrylate, and *n*-butyl acrylate, were proved to be a controllable procedure, leading to random incorporation of the reactive BCB units. For example, So et al. reported the copolymerization of 4-VBCB with styrene initiated by benzoyl peroxide (10). As-prepared polystyrene-*co*-4-VBCB random copolymer contains 26 mol % of 4-VBCB, with M_n of 23104 g/mol and PDI of 1.83. Harth et al (9, 11) employed nitroxide mediated living radical polymerization to prepare 4-VBCB/styrene random copolymers (Scheme 1). It was pointed out that at molecular weight less than 120,000 the PDIs for these random copolymers were 1.08-1.16, which slightly increased to 1.19-1.26 for molecular weights above 200,000. Using living anionic polymerization initiated by *sec*-butyllithium and *n*-butyllithium, Baskaran et al. have prepared poly (4-VBCB) successfully (3). The polymerization was allowed to proceed at 25 °C for 15 h. A linear first-order semilogarithmic time-conversion plot was obtained, which indicated the absence of termination reactions during the polymerization of 4-VBCB at room temperature. The molecular weight of resulted poly (4-VBCB) was tailored by varying the amount of initiators and the maximum M_n can reach as high as 58,000 in their experiments. In addition, all of the prepared poly (4-VBCB)s exhibited narrow molecular weight distributions (<1.1).

In the above examples, 4-VBCB was generally used as a reactive functional group, which acts as active sites for cross-linking reactions. Taking poly(styrene-r-4-VBCB) as an example, the crosslinking structure is composed of cyclic and linear structure (Scheme 2) as suggested in many reports (10, 12, 13). Moreover, the crosslinking of these polymers were fulfilled by the Diels-Alder reaction of BCB with high atom economy.

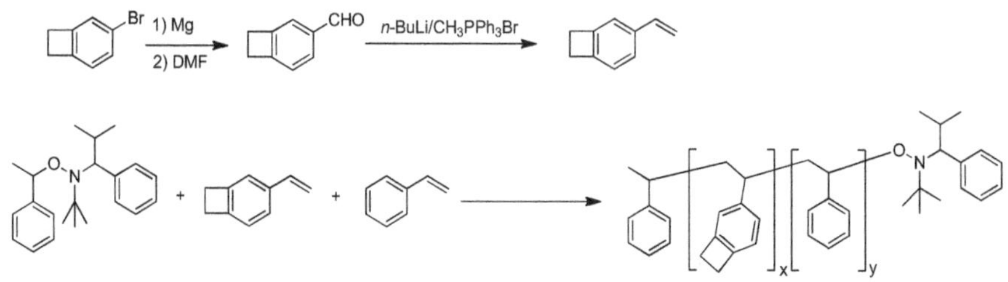

Scheme 1. Synthesis route to 1-VBCB and its nitroxide-mediate polymerization.

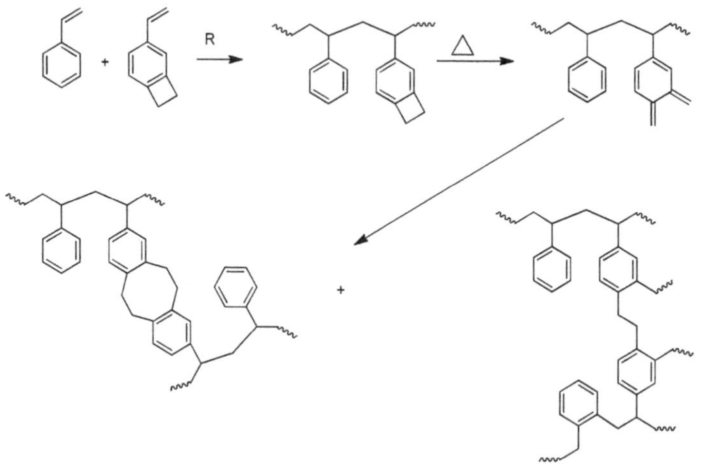

Scheme 2. Copolymerization of styrene and 4-VBCB and the cross-linking structure of poly (St-r-4-VBCB)

The cross-linking of polyolefins by means of BCB presents a considerable significance in stabilizing the polymeric matrix, including enhancing the glass transition temperature (T_g), thermal stability and mechanical stability. Resulting cross-linked materials can be potentially exploited in the design of high voltage cable insulation, biomedical joint replacement parts, hot-water piping, and etc (7). Recently, it has been reported for other applications including generation of cross-linked/monomolecular polystyrene nanoparticles (9, 14, 15), modification of solid surfaces (16-17), stabilization of nanostructures, thermally stable diblock copolymers in data storage and nanoscale templating (11).

On the basis of 4-VBCB and derivative poly (4-VBCB), Yang et al. have designed and synthesized a series of new vinyl-BCBs (Scheme 3), including 4-vinylsilylbenzocyclobutene (4-DMVSBCB) (18), 1-vinylsilylbenzocyclobutene (1-DMVSBCB), benzocyclobutene-4-yl acrylate (4-ABCB) (19-20), and benzocyclobutene-1-yl acrylate (1-ABCB). 4-DMVSBCB was synthesized conveniently by Grignard reaction of 4-bromobenzocyclobutene with vinyldimethylchlorosilane. With the same procedure, 1-DMVSBCB could be synthesized in good yield. 4-ABCB was synthesized through a facile esterification reaction. The detailed routes are outlined in **Scheme 3**. These new monomers, especially 4-DMVSBCB showed different polymerization behavior compared with 4-VBCB, probably caused by the incorporation of silicon containing group.

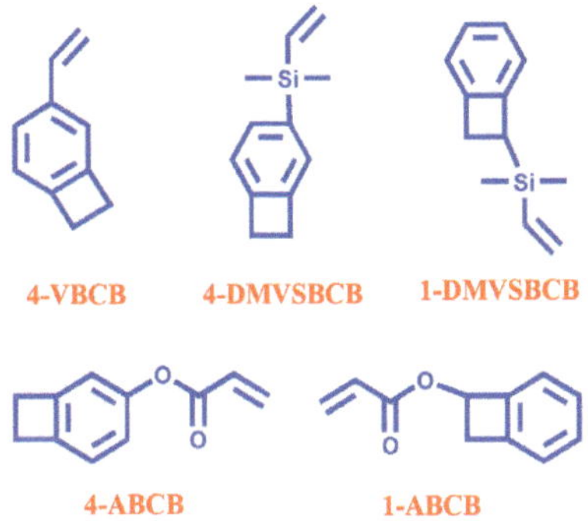

Scheme 3. Chemical structure of several new vinyl-BCBs

2.1.2. *Poly (4-vinylsilylbenzocyclobutene)*

As a vinylsilane, the polymerization of 4-DMVSBCB with different groups has been studied extensively. In fact, it exhibited almost the same behavior with another vinylsilane compound, 4-dimethylvinylphenylsilane (4-DMVSPh), which was reported to polymerize by the radical, anionic and coordination methods. The homo-polymerization of 4-DMVSBCB is sluggish when initiated by BPO or AIBN, only giving trace amount of poly(4-DMVSBCB) with M_n between 7500 to 8900 g/mol. Theoretically, it is probably attributed to the lack of conjugated structure and the electron-donating effect of silicon group. Contrastively, the copolymerization of 4-DMVSBCB with conventional α-olefins was successfully performed with low to moderate yield (< 50 %). However, the incorporation ratio of 4-DMVSBCB is still low (<10 mol %) and hard to increase even employing high concentration of 4-DMVSBCB (**Table 1**). Thus, alternative polymerization method to enhance polymerization activity and incorporation ratio in conventional polyolefins is required.

As compared with radical polymerization, the anionic polymerization of 4-DMVSBCB affords well-controlled and high molecular weight, and relatively low polydispersity. Anionic homo-polymerization of 4-DMVSBCB was performed using *n*-BuLi and *sec*-BuLi as initiators (Scheme 4). White solid polymers with molecular weight (M_n) of 9800–26,000 g/mol and PDI in the range of 1.6-1.8 were obtained in excellent yields (up to 94.5%). Structural studies have shown that dπ–pπ interactions in the vinyl-silicon bond give resonance contributions to the vinyl-silicon bond, which provides the polarization of the vinyl group necessary for anionic polymerization. Brooks et al. has reported that RLi adds to vinylsilanes so as to generate a secondary lithium compound. Furthermore, as noted before, the incorporation of aryl groups, especially phenyl, is more favorable to enhance anionic polymerization activity than that of alkyl groups. Related studies revealed that the aryl groups on silicon atom could provide silicon atom with enhanced positive charge and α-stabilizing effect due to the electron-withdrawing property of aryl groups.

Entry	Initiator (10^{-2} mmol)	4-DMVSBCB (mmol)	comonomer (mmol)	temp (°C)	time (h)	content of 4-DMVSBCB (mol.%)	yield (%)
1	AIBN 8.64	4.28		60-65	60		
2	BPO 5.08	2.64		75-80	60		
3	AIBN 6.35	2.65	St, 2.69	60	48	5.3	45%
4	AIBN 8.34	3.19	St, 1.61	60	48	7.3	36%
5	AIBN 8.74	4.26	St, 1.06	60	48	9.5	29%
6	AIBN 7.92	2.68	4-BrSt, 2.58	60	48	13	48%
7	BPO 7.38	3.19	St, 1.73	75	36		38%
8	BPO 8.68	2.16	MMA, 2.21	75	36		
9	BPO 11.46	3.22	VA, 3.16	75	36		

Table 1. Polymerization and Copolymerization of 4-DMVSBCB

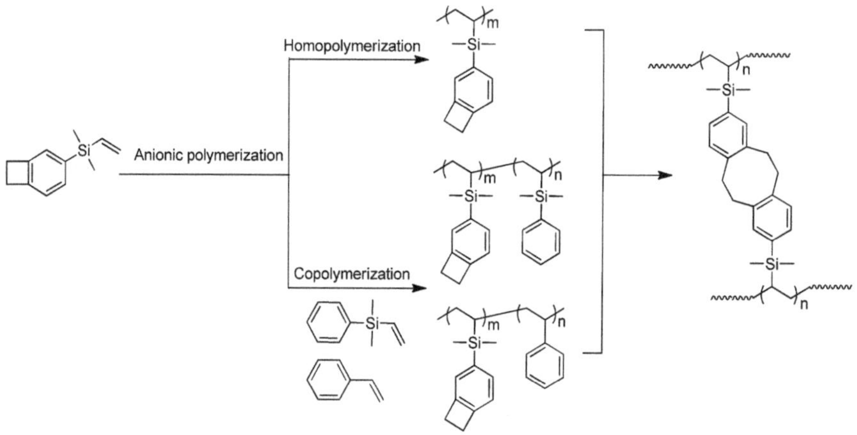

Scheme 4. Anionic homo and copolymerization of 4-DVMSBCB

Poly (4-DMVSBCB) shows several outstanding properties. First, it shows superior film forming property. A problem encountered in the film forming process for the low molecular weight polymer is its low viscosity. One solution to this issue was to pre-crosslink the polymer in prior to film forming. In the film forming process of the polymer derived from BCB bis-capped monomer, the pre-crosslinking process is hard to control as gel formation occurred immediately when gel point reached. In comparison, pre-crosslinking process of poly (4-DMVSBCB) is controllable because its polymerization and cross-linking processes were separated and the incorporating ratio of BCB cross-linkable functionality into polyolefins was controllable. Second, TGA in air shows that the initial decomposition temperature (T_0) and 5 wt % weight loss temperature ($T_{5\%}$) of poly (4-DMVSBCB) are about

318 °C and 400 °C, respectively. In particular, the equilibrium rate of mass loss for the cross-linked poly (4-DMVSBCB) was 0.95 wt %/h, which is lower than that of Cyclotene™ resins. The dielectric constant of subtle cross-linked poly (4-DMVSBCB) is in the range of 2.41~2.45 from 5 MHz to 20 MHz. This value is lower than that of most polymeric matrix. Third, a film fabricated from highly cross-linked poly (4-DMVSBCB) shows a greatly small roughness. These properties show that it could be applied in some interconnect fabrication processes which are implemented at temperature as high as 350 °C.

The copolymerization of 4-DMVSBCB with styrene and DMVSPh was also performed (Scheme 4). It is worth noting that the ratios of 4-DMVSBCB to 4-DMVSPh in copolymers showed well accordance with the feed ratios due to their similar structure and reactivity (Table 2). In contrast, when copolymerizing 4-DMVSBCB with styrene, the latter show higher reactivity and incorporation ratio than the former because of the conjugation effect of phenyl group. More importantly, the incorporation of styrene or DMVSPh into poly (4-DMVSBCB) did not lead to obvious reduction in thermal resistance, even when 1:1 of comonomers to 4-DMVSBCB was used.

Entry	initiator I	[I]/[M][a]	M_n[b]	M_w/M_n	yield[c] (%)
1	*n*-BuLi	1.9×10^{-3}	13 900	1.62	92
2	*n*-BuLi	1.2×10^{-3}	14 500	1.71	89
3	*n*-BuLi	9.4×10^{-4}	22 400	1.77	93
4	*n*-BuLi	5.7×10^{-4}	-	-	27
5	*n*-BuLi	1.0×10^{-3}	21 900	1.45	91

Table 2. Results of the anionic homo-polymerization of 4-DMVSBCB

2.1.3. *Poly (1-vinylsilylbenzocyclobutene)*

In comparison with 4-DMVSBCB, where the silicon atom bonds with benzene ring (Bz), 1-DMVSBCB exhibit a structure change that the silicon atom bonds with cyclobutene (Ct) (Scheme 5). This appeared slight structural change leads to a significantly alteration in polymerization behavior, especially for anionic polymerization. The anionic polymerization did occur for 1-DMVSBCB but required several days. In addition, it was noted that the polymerization did not proceed except using *sec*-BuLi with higher reactivity. The yield (60~70 wt.%) and molecular weight (around 10000 g/mol) of poly (1-DMVSBCB) are lower than that of poly (4-DMVSBCB). These results indicated a lower polymerization activity of 1-DMVSBCB, which is possibly attributed to that the electron donating property of cyclobutene reducing the stability of carbon anion or radicals. DSC analysis showed that the initial cross-linking temperature of poly (1-DMVSBCB) is about 161 °C, which is lower than that of poly (4-DMVSBCB), about 180 °C. It once again supports that the substituting of cyclobutene would enable the tuning of ring-opening temperature of benzocyclotuene. Unfortunately, poly (1-DMVSBCB) shows lower initial decomposition temperature and thermal stability relative to poly (4-DMVSBCB). Apparently, 1-subsitution would also lead to a reduction in the stability of derived dibenzocyclooctadiene except for that of cyclobutene. Hence, it may be a big challenge to simultaneously achieve high thermostability of cross-linked polymer and low cross-linking temperature.

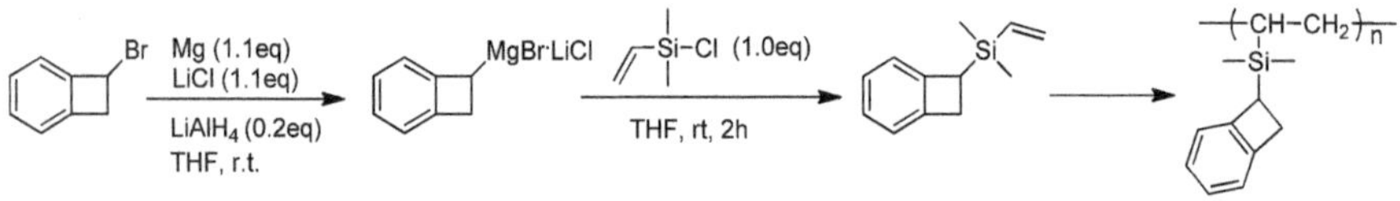

Scheme 5. Synthesis route of 1-DMVSBCB and poly (1-DMVSBCB)

Other vinyl-BCB polymers

Benzocyclobutene-4-yl acrylate exhibits effective homo-polymerization and co-polymerization by free radical methods as its similar property with methyl acrylate (Scheme 6). BCB units are able to be incorporated into a series of polyarylates or other polyolefins by copolymerization. Related studies (19) showed that the incorporation ratio of 4-ABCB in copolymer could reach 6.7 mol% when using 4.7 mol% feed ratio of 4-ABCB. This result indicates that 4-ABCB exhibit higher polymerization activity than MA.

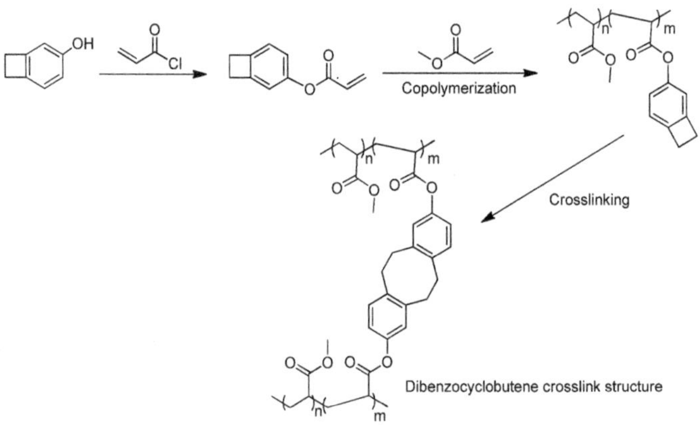

Scheme 6. Preparation and cross-linking of 4-ABCB/MA copolymer.

In recent years, apart from its conventional application in coating, adhesive, etc, polyarylates have been studied for use in a wide range of other areas such as gas separation membranes, hydrogels, tissue engineering, drug delivery, dentistry, and shape-memory materials for cardiovascular applications. Particularly, the potential applications of polyarylates as shape memory materials, which are capable of memorizing temporary shapes and recovering their permanent shape upon thermal activation, have attained much attention (21). For polyarylates to possess shape memory properties, it is necessary to have a permanent network which can be achieved via chemical or physical cross-linking. Interestingly, the cross-linking of polyarylates by the reaction of BCB generates nanoparticles which serve as physical cross-linkers (19). As a result, physical cross-links were generated, apart from chemical cross-links. This unique cross-linking architecture would be beneficial to broaden the list of cross-linking methods used in the production of SMPs.

Benzocyclobutene-1-yl acrylate (1-ABCB) exhibits similar radical polymerization behavior with 4-ABCB (Scheme 7). 1-ABCB also shows higher activity than MA as 1-ABCB/MA copolymerization results indicate that 1-ABCB/MA ratio in copolymer is higher than 1-ABCB/MA feed ratio. Interestingly, the initial ring-opening temperature of poly (1-ABCB) is approximately 170 °C, which is lower than that of 4-subsituted BCB.

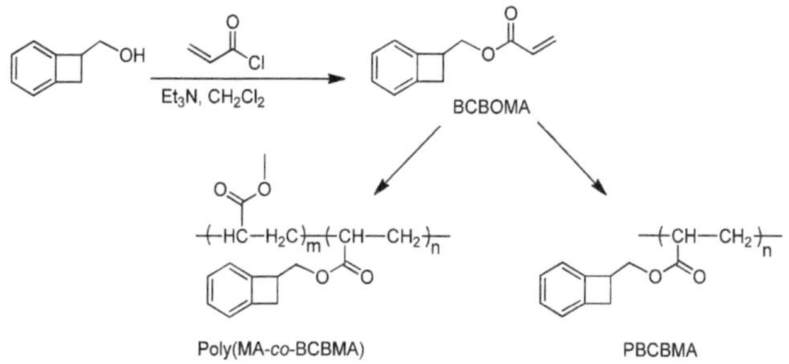

Scheme 7. Synthesis route of 1-ABCB and homo/copolymers of 1-DMVSBCB.

Benzocyclobutene-siloxane resins

Benzocyclobutene-linear siloxane

The purpose of designing benzocyclobutene-siloxane structure is to introduce the excellent thermostability of siloxane materials into BCB structure. The benzocyclobutene-siloxane materials are expected to possess integrated properties of polysiloxane and BCB related materials. As described above, one representative resin developed earlier is poly (DVS-BCB). However, poly (DVS-BCB) shows poor controllability of the property of pre-crosslinked polymer used for film-forming. To solve this problem, Yang et al. have prepared a mixture of poly (4-VBCB-*co*-St) and poly (DVS-BCB) (22). Cross-linking of this blend generates a complicated network (Scheme 8). By incorporating poly (4-VBCB-*co*-St), the viscosity was improved and slight cross-linking would afford polymers with adequate viscosity for film-forming. In this work, two poly (4-VBCB-*co*-St)s with 4-VBCB/St ratios of 1:3 and 1:5 were used. Correspondingly, the 5 % weight loss temperature of the resulted cross-linked polymers is around 410 °C and 400 °C, respectively.

On the other hand, the application of poly (DVS-BCB) have been limited largely in several standard interconnects due to their lower thermal stability and higher fragility relative to polysiloxanes. It implies that a simple combination of siloxane and BCB structure scarcely affords materials with high performance as expected. Structural analysis shows that the relatively low thermostability of poly (DVS-BCB) may be due to the incomplete crosslinking of DVS-BCB, the presence of residue ethylene moieties which are prone to take oxidation reaction, or the BCB-siloxane alternating copolymerization structure which destroys the thermally stable Si-O-Si backbone. Yang et al. have prepared a new siloxane-BCB monomer,

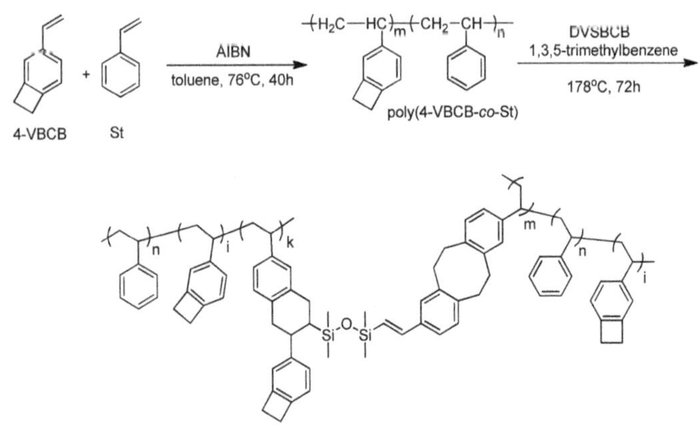

Scheme 8. Preparation of poly (4-VBCB-co-St) and the conetwork structure with DVSBCB.

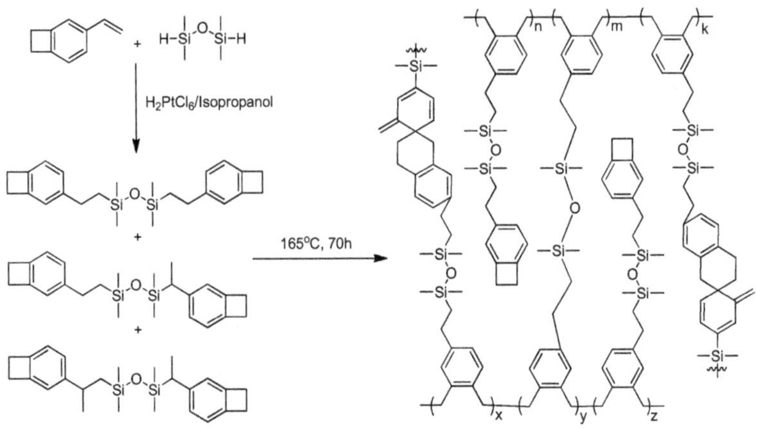

Scheme 9. Synthesis route to DES-BCB and the cross-linked poly (DES-BCB)

1,1,3,3-tetramethyl-1,3-bis[2′-(4′-benzocyclobutenyl)]vinyldisiloxane (DES-BCB) (**Scheme 9**). Ethyl was introduced instead of vinyl in DVS-BCB, which might enhance the thermal oxidation stability and decrease the fragility of cross-linked polymers simultaneously. However, TGA of poly (DES-BCB) shows a relatively inferior thermal stability with 5 % weight loss temperature of 380 °C, indicating that alkene oxidation would not be responsible for the inferior thermal stability of poly (DVS-BCB).

In their further work, a new BCB-siloxane polymer with polysiloxane backbone and BCB pendant groups was reported. This polymer was prepared by Pt/C or H_2PtCl_6 catalyzed hydrosilylation between poly (methylhydrosiloxane) and 4-DMVSBCB or poly (methylvinylsiloxane) and 4-dimethylhydrosilylbenzocyclobutene (4-DMHSBCB) (**Scheme 10**). The incorporation ratio of 4-DMVSBCB pendant groups is tunable, and the maximum ratio is up to 72 %. The corresponding cross-linked BCB-siloxane resins exhibit an initial decomposition temperature of 429 °C, showing a superior thermal resistance over poly (DVS-BCB) resins and other common polysiloxanes.

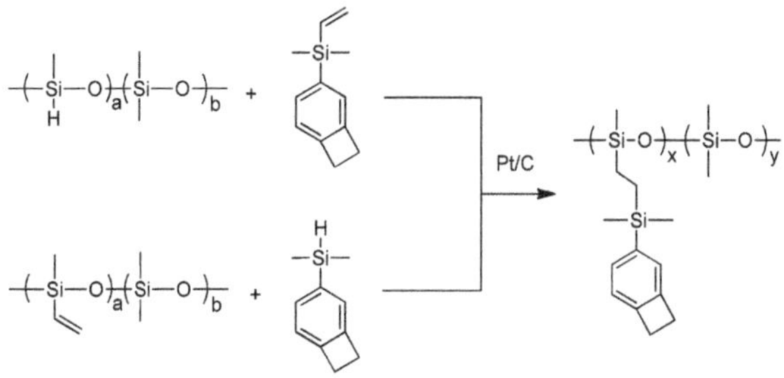

Scheme 10. Preparation of polysiloxane with BCB containing pendant groups.

Fan et al (23) have prepared two novel siphenylene-siloxane copolymers bearing latent reactive BCB groups by polycondensation procedure of 1,4-bis (hydroxydimethylsilyl)benzene with bis (dimethylamino) methyl (4′-benzocyclobutene)silane and bis(dimethylamino)methyl[2′-(4′-benzocyclobutenyl)vinyl]silane (Scheme 11). The BCB-based monomers were synthesized from Grignard and Heck reaction, respectively. Copolymers with high molecular weight (M_w) of 1.96 and 1.61×10^5 g/mol have been obtained which were readily converted to cross-linked films upon heating. The copolymer originated from SiViBu showed a big exotherm peak starting at about 230 °C. This temperature is similar with DVS-BCB but is higher than that of copolymer originated from SiBu. It may be attributed to their different cure mechanism. The cross-linking of SiBu proceeds via a Diels-Alder reaction of two *o*-quinodimethanes, while that of SiViBu and SiBu proceeds via a Diels-Alder reaction of *o*-quinodimethane as a diene and vinyl group as a dienophile. The final resins exhibited excellent thermal stability with high temperatures at 5 % weight loss (553 and 526 °C in N_2, 530 and 508 °C in air) owing to the incorporation of the benzene unit into the mainchains of the copolymers. The films of cured copolymers also demonstrated low dielectric constants of 2.66 and 2.64 and low dissipation factors of 2.23 and 2.12×10^{-3}.

In their further work (24), two new methylphenylsiloxane monomers terminated with cross-linkable benzocyclobutene functionalities, 1,1,5,5-dimethyldiphenyl-1,1,5,5-di[2′-(4′-benzocyclobutenyl)vinyl]-3,3-diphenyltrisiloxane (BCB-1) and 1,1,3,3-dimethyl-diphenyl-1,1,3,3-di[20-(40-benzocyclobutenyl)vinyl]disiloxane (BCB-2), were prepared (Scheme 12). These two monomers possess similar structure with DVS-BCB. Correspondingly, they show a simultaneous polymerizing and cross-linking process. Cured BCB-1 and BCB-2 exhibited good thermal stability ($T_{5\%} > 472$ °C both under N_2 and air), which is better than that of DVS-BCB resins and poly (4-DMVSBCB). From the comparison among DVS-BCB, BCB-1 and BCB-polysiloxane resins, it can be seen that elongating the length of polysiloxane chains and incorporating rigid functional groups on Si are beneficial for enhancing thermal stability. BCB-1 and BCB-2 resins also demonstrated low dielectric constants (2.69 and 2.66) and low dissipation factors of 2.23 and 2.12×10^{-3}.

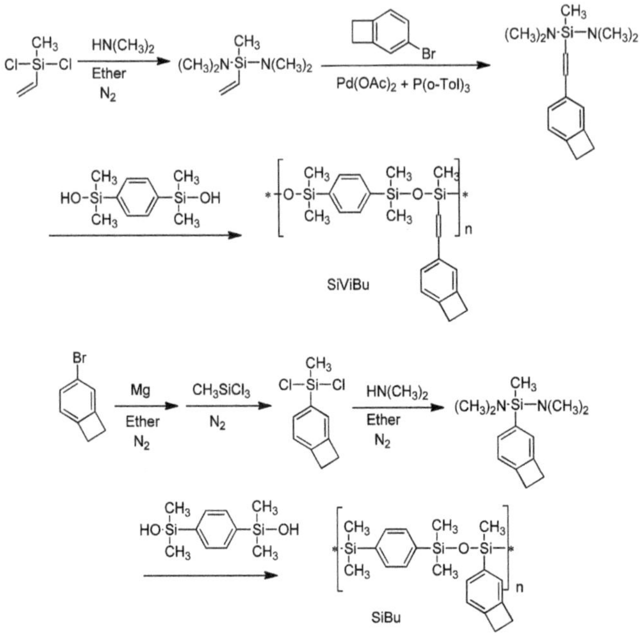

Scheme 11. Synthesis route to copolymers originated from SiViBu and SiBu, respectively.

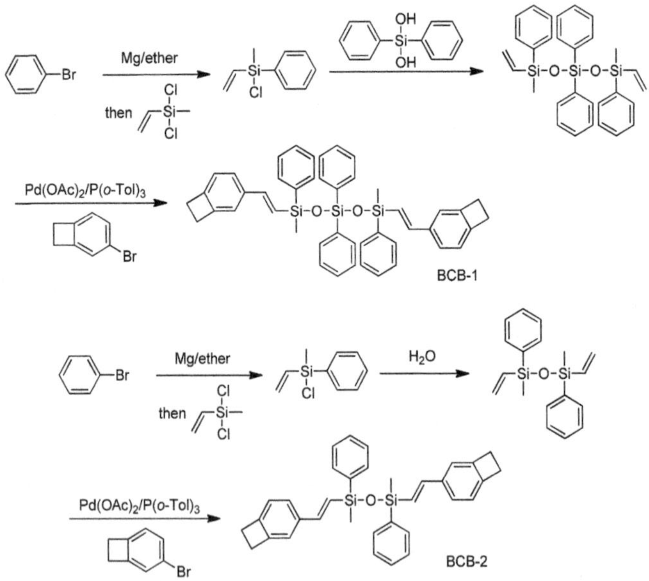

Scheme 12. Synthesis route to the monomers BCB-1 and BCB-2.

3. The application of BCB-based materials

3.1. Low dielectric materials

The performance of integrated circuit devices, historically limited by characteristics of the transistors, has become limited by the back-of-end-of-line (BEOL) signal delay of interconnect caused by wire capacitance, crosstalk and power dissipation. As one of the major technological strategy to decrease signal delay, developing new generation of low dielectric materials as alternatives of SiO_2 have received much attention in recent years. Besides low dielectric constant, a qualified dielectric materials must satisfy other requirements, including high thermal stability, high mechanical strength, high breakdown field above 2 MV cm^{-1}, low moisture absorption, low thermal swelling coefficient or film stress, good adhesion to various matrix, ease of damascence processing or etching, reactive inertness, good film forming property, etc. Among kinds of the accessible alternatives, some polymers with outstanding and promising mechanical and physicochemical properties have been widely studied (25). In the past few years, the studies were mainly concentrated on the following polymers: polyimides, BCB resins, poly (binaphthylene ether), polydiphenyl, polyquinolines, polynorbornene, silicon-containing polymers and fluorinated polymers.

Although several potential applications were discovered for BCB-based materials, only the use as low dielectric materials was accessible for application until now. One of the reasons for the intense research on BCB-based low dielectric materials is that the thermal activation would generate a highly reactive intermediate to form clean products, which are beneficial for designation of high precision curing processes. The application of BCB resins in microelectronic should start from Cyclotene (26). So far, several formulations of these BCB-based resins have been commercialized for thin-film dielectric applications and are currently used in the fabrication of liquid crystal displays, GaAs integrated circuit interconnect, interlayer connection and chip bumps. Although BCB-based resins displayed several advantages over many other polymeric materials, their use in standard interconnect is limited by their relatively low thermal stability than the required process temperatures.

Table 2 lists the thermal and dielectric performances of BCB-based materials reported in the past few years. Most of them show low dielectric property with dielectric constant below 2.65, it is worth noting that poly (4-DMVSBCB) possesses the lowest dielectric constant probably due to the absence of polar atoms in this carbosilane structure. Moreover, these new BCB-based materials show higher thermal stability except poly (4-vinylBCB). Most of them could be considered as potential alternatives of poly (DVS-BCB).

3.2. Vinyl-BCB used for the functionalization of Carbon Nanotube

The methods that have been explored for the functionalization of the sp^2 network of the carbon nanotubes (CNTs) include addition reactions of azomethine ylides, bromomalonates, nitrenes, nucleophilic carbenes, and free radicals. In 2007, Baskaran et al. prepared functionalized CNTs by [4+2] cycloaddition of BCB for the first time (27). The stained ring of

	T_0 (°C)	$T_{5\%}$ (°C)	Dielectric constant
Poly (DVSBCB)	370		2.65
Poly (4-vinylBCB)	~350		2.58
Poly (4-DMVSBCB)	380	428	< 2.45
Poly (DES-BCB)		380	
Trisiloxane-BCB		> 472	2.66~2.69
poly (SiViBu)		553	2.66
poly (SiBu)		526	2.64
Polysiloxane-BCB	427	493	

Table 3. Thermal and dielectric performance of several BCB-based materials.

the BCB is susceptible to thermal ring opening to form *o*-quinodimethane, which can undergo [4+2] cycloaddition exclusively with 6-6 bonds of fused aromatic rings. In an attempt to add reactive organic groups on the surface of CNTs, several substituted BCBs were reacted with CNTs (28-29). Vinyl substituted BCBs, typically1-phenyl-1-benzocyclobutene ethylene (BCB-DPE), are of particular useful because they can be attached on the surface of CNTs serving as anionic initiators for olefins polymerization. In a further work of Baskaran et al (28), the surface of multi walls carbon nanotubes (MWNTs) was modified via covalent attachment of BCB-polyethylene (BCB-PE) through Diels-Alder reaction at 235 °C (Scheme 13). TGA shows that MWNTs-*g*-$(BCB\text{-}PE)_n$ has 54 wt % precursor initiator. Anionic surface-initiated polymerization of styrene has been performed using such MWNTs functionalized with anionic initiators. Polymer grafted nanocomposites, MWNTs-*g*-$(BCB\text{-}PS)_n$, contains a very high percentage of hairy polymer with a small fraction of MWNTs (< 1 wt %). Size exclusion chromatography revealed a broad molecular weight distributions (1.3~1.8) of the polymer grafted MWNTs.

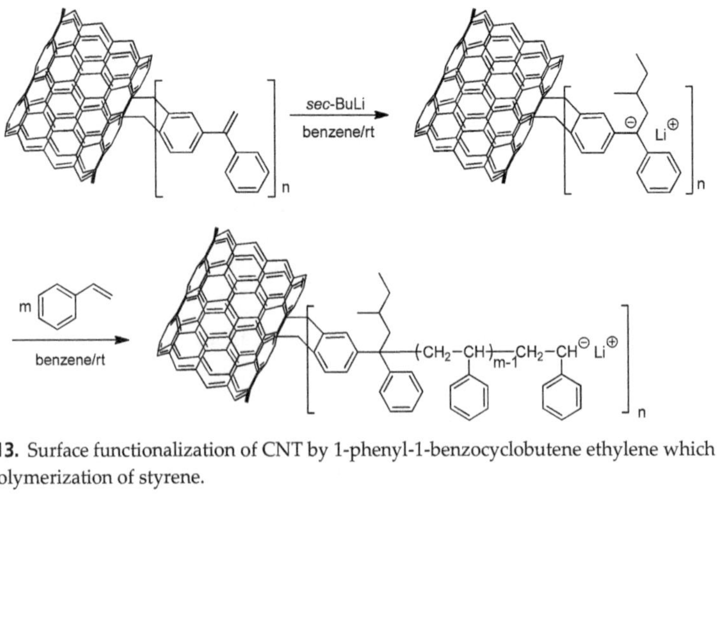

Scheme 13. Surface functionalization of CNT by 1-phenyl-1-benzocyclobutene ethylene which initiates anionic polymerization of styrene.

3.3. Applications in Organic Light Emitting Diodes (OLED)

OLEDs have attained much research interest in the past few decades due to their promising applications in full color displays and solid-state lighting. Device fabrication, involving the construction of hole-transporting layer (HTL), phosphorescent emitting layer (EL), and electron-transporting layer (ETL) multilayers structure, shows great impact on the ultimate performance. There are usually two methods to fabricate the layers. One is high vacuum vapor deposition of small molecules, and another is solution processing of polymers or dendrimers. The solution processing technique presents advantages in thermostability. However, it requires that each deposited layer must resistant to the solvent that used to deposit subsequent layers. Perhaps the most elegant strategy to address these issues involves the development of cross-linkable materials. To date, most of the reported thermally cross-linkable hole-transport materials are based on perfluorocyclobutane (PFCB) cross-linking unit. Recently, a novel thermally cross-linkable hole-transporting polymer for solution processible multilayer OLEDs was designed (30). This amorphous polymer was prepared from the radical copolymerization of vinyl-BCB with 4-[N-(4-Vinylphenyl)-N-(4-methylphenyl) amino]-4′-[N-phenyl-N-(4-methylphenyl) amino]biphenyl initiated by AIBN (Scheme 14). It contains ~9 mol % of BCB with the T_g of near 175 °C. Cross-linking of the copolymer at 200 °C led to insoluble polymer films with a smooth surface. Solution processed multilayer light emitting diodes were prepared using this new BCB-inducing cross-linked polymer as a hole-transporting layer. These light emitting diodes exhibited high performance with 10.4 % external quantum efficiency at a brightness of 350 cd/m².

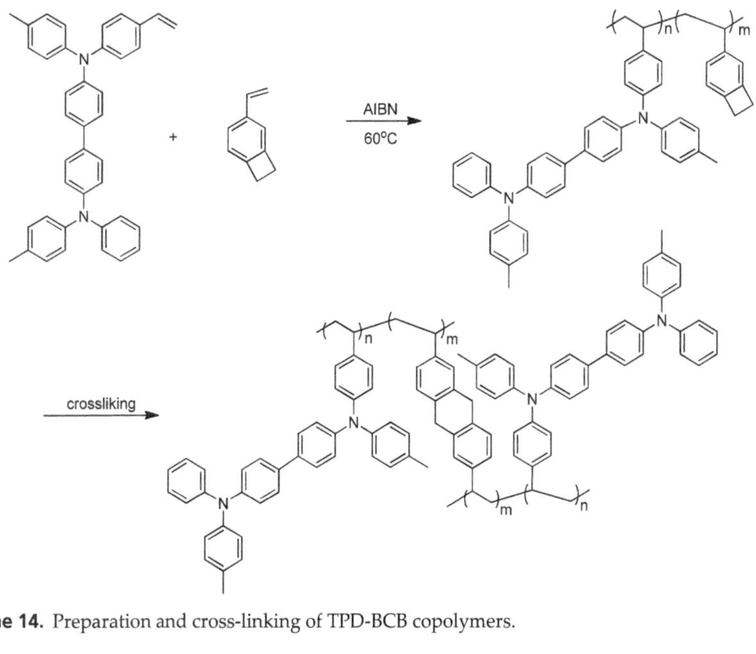

Scheme 14. Preparation and cross-linking of TPD-BCB copolymers.

3.4. Applications in organic field-effect transistors

Organic field-effect transistors (FETs) are promising in enabling disposable electronics and flexible electronics technologies. Fabricating ultrathin defect-free gate dielectric layer with a high quality interface to the semiconductor layer is crucial for polymeric FETs. In recent years, several polymer dielectric materials including spin-on silesquioxanes, photoresist, polyimide and poly (vinylphenol) (PVP) were investigated. However, these polymers are not wholly satisfactory because they still cannot give conformal and pinhole-free films in the sub-100-nm regime. Consequently, present spin-on polymer dielectrics are typically limited to more than 300 nm in thickness, thus demanding a large gate voltage to operate. Recent works show that DVS-BCB possesses most of the requisite gate dielectric properties (31). Furthermore, defect-free films down to a few tens of nanomaters in thickness are attainable by a simple solution casting technique. The fabricated devices can operate at a low voltage with a field-effect mobility of few 10^{-4} cm^2/Vs. They can be continuously operated at 120 °C, showing a good operating stability.

4. Perspective of BCB related chemistry and materials

4.1. Potential polymerization methods

Although most of the vinyl-BCBs can be polymerized by anionic and radical polymerization methods, their copolymerization with styrene, MMA, ethylene and so on encountered a big challenge as the great difference polymerization activity between them. Coordination polymerization method as a powerful tool has promoted a great development of polymer chemistry in the last decades. However, the coordination polymerization of vinyl-BCB monomers has not been reported yet. In fact, most of these monomers mentioned above are capable of polymerizing through coordination polymerization in terms of its similar structure with styrene and acrylate. Of course, the coordination polymerization of vinyl-silylBCB, such as 4-DMVSBCB and 1-DMVSBCB, is more difficult due to the Lewis acid property of silicon element and the large steric hindrance. The coordination polymerization of several simple vinylsilane compounds, indeed, has been studied. The firstly used catalysts were Ziegler-Natta catalysts, which however failed to induce accepted polymerization performance. An acceptable result was obtained till 2008 when Nomura et al has demonstrated a successful copolymerization of vinyltrialkylsilanes with ethylene catalyzed by nonbridged half-titanocenes (32). Thus, in this regard, a large number of works including new high-efficiency catalysts developing and in-depth polymerization behavior and mechanism studying are highly desired in the field of vinyl-BCB related polymer chemistry.

Earlier work demonstrated that free radical method could not initiate the polymerization of vinyl-silylBCB. Nevertheless, it does not mean that the radical polymerization definitely fails to give products with high molecular weight and poor yields. Theoretically, the living free radical method could stabilize the initiated monomer free radicals, thus extending the lifetime of them and consequently enhancing polymerization activity. In addition, living radical polymerization of silyl-functionalized vinyl monomers with large steric hindrance, for example vinylcyclicsilazane, has been indeed reported (33). The use of RAFT agents with anionic characteristic displayed excellent activity, giving the product with molecular weight up to 11300 g/mol. Thus, further studies regarding the radical polymerization of vinyl-silylBCB should include the investigation of living polymerization, such as RAFT and ATRP. Moreover, these methods can also be used for the polymerization of vinyl-BCB to afford improved and controlled molecular weight, enabling the generation of random even block copolymers.

4.2. Low temperature cross-linking

Although the cross-linking by BCB ring-opening reaction presents several superiorities over common cross-linking ways, its high ring-opening temperature would induce the deformation of films due to the generally low T_g of the linear BCB precursors, thus limiting its use in some extent. Previous work demonstrated that the ring-opening temperature of cyclobutene is tunable by incorporating variety of substituents on cyclobutene (34). In particular, when utilizing the electron-donating substituents, the stability of cyclobutene and the ring-opening temperature was pronounced decreased. Recent work (35) even reveals that when placed within long polymer strands, the *trans* and *cis* isomers of a 1, 2-disubstitued benzocyclobutene undergo an ultrasound-induced electrocyclic ring opening at 6-9 °C (Scheme 15). Overall, developing new BCB structures, which enable lower opening temperature, is great important for improving the performance and expanding the application range of the vinyl containing BCB polymeric resins and also other BCB resins. Moreover, the cross-linking structure ascribed from the ring-opening reaction of 1-substitued or 1, 2-disubstituted benzocyclobutene is still unclear and needs to be further studied.

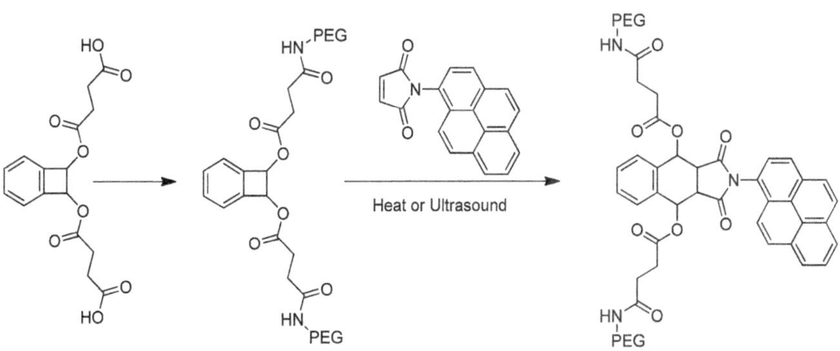

Scheme 15. Preparation and reaction of mechanosensitive polymers.

4.3. Nanomaterials

Currently, the nanomaterials with unique chemical structure and corresponding high performance and attractive functions have been attracted increased attention. Hence, exploiting vinyl-BCB based materials with nanostructure would be an interesting research topic because of the cross-linkable characteristic upon heating and the unique cross-linking behavior of BCB. Based on the well-established in-situ methods to construct nanostructures from polymerization of olefins, similar nanostructures based on poly (vinyl-BCBs) could be easily constructed. These BCB-based nanomaterials can be potentially used as adsorption materials, optic-electronic materials, coating materials, and so on.

More recently, Harth et al. present a new BCB based cross-linking structure that allows a 100 °C lower ring-opening temperature in comparison with conventional ones (36). They found that the incorporation of BCB based cross-linking unit by copolymerization shows a limited molecular weight control. Thus, as an alternative, they explored a grafting-on method by grafting a low-temperature cross-linking unit, 2-(1,2-dihydrocyclobutabenzen-1-yloxy)ethanol, onto a linear acrylate polymer backbone (Scheme 16). To afford the required end group functionality, a convenient method was adopted to convert the alcohol to an amine for grafting onto the polymer. The polyacrylate was prepared via reversible addition-fragmentation transfer (RAFT) polymerization. A polymeric precursor with a 75,000 M_w polyacrylate with a PDI of 1.21 was obtained using AIBN catalyst towards the RAFT initiator. The amine was grafted to poly (acrylic acid) through chloroformate activation chemistry and the 5 % conversion ratio of the carboxylic acid functional groups to functionalized benzocyclobutene units was obtained. Monitoring by ^{1}H NMR show that the ring-opening was initiated at the temperature higher than 100 °C and the optimal

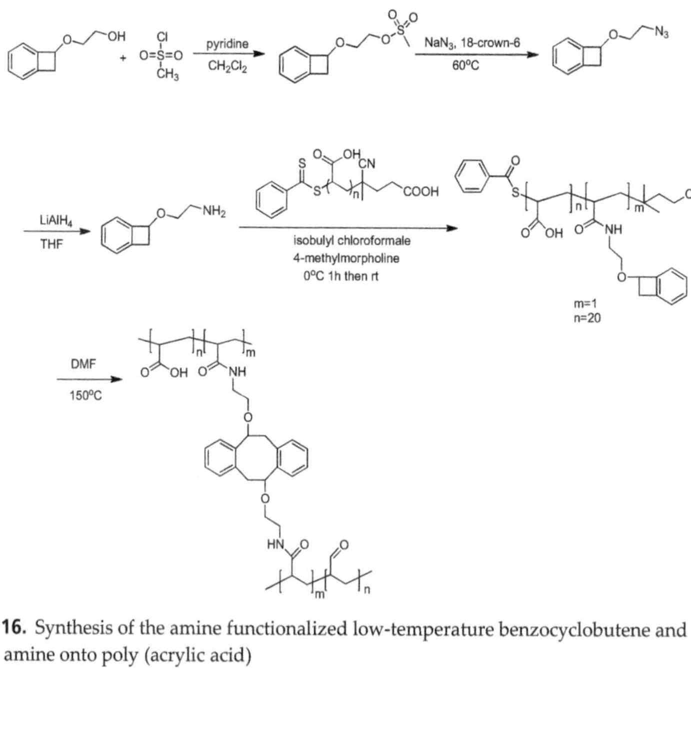

Scheme 16. Synthesis of the amine functionalized low-temperature benzocyclobutene and Grafting of this BCB amine onto poly (acrylic acid)

temperature is above 130 °C at which the polymeric precursor could be fully converted. Dynamic Light Scattering (DLS) and TEM confirmed the formation of nanoparticles from the cross-linking of linear copolymers. TEM and DLS suggest that the mean diameter of nanoparticles is approximately 4.3 ± 0.8 and 7.3 ± 0.5 nm, respectively.

5. Conclusions

In the past few years, the approach to the utilization of benzocyclobutenes to construct polymers has experienced significant change. BCB-related polymers based on two-step cross-linking technology, represented by poly (vinyl-BCB)s, has brought remarkable improvement in controllability of materials structures and properties. Apart from this, their excellent polymerization compatibility with traditional olefins allows the convenient incorporation of BCB functionalities, significantly improving properties of original polymers and extending the potential application of BCB-related polymers. Besides, a series of BCB-siloxane polymers with novel structures exhibiting excellent thermal stability and low dielectric properties have been prepared. Indeed, some of these polymers are promising alternatives of poly (DVSBCB) and would be widely used in microelectronic devices. These exciting works are just a beginning. A new window has been opened for BCB related science and technology. The future development will mainly focus on the following areas: 1) searching for more suitable BCB-related dielectric materials; 2) utilizing BCB to construct novel structures and improve performance of original materials; 3) exploiting more applications suitable for BCB-related polymers.

Author details

Junxiao Yang, Yawen Huang and Ke Cao
School of Material Science and Engineering, Southwest University of Science and Technology, P. R. China

Acknowledgement

The authors would like to thank the financial support through the National Natural Science Foundation of China-NSAF and China-NSFC, and Open Project of State Key Laboratory Cultivation Base for Nonmetal Composites and Functional Materials for Southwest University of Science and Technology. Prof. Yang would like to thank students from his research for their contributions.

6. References

[1] Kirchhoff, R.A.; Bruza, K.J. (1993) Benzocyclobutenes in polymer synthesis. *Progress in Polymer Science*, 18, 85-185.
[2] Farona, M.F. (1996) Benzocyclobutenes in Polymer Chemistry. *Progress in Polymer Science*, 21, 505-555.

[3] Sakellariou, G.; Baskaran D.; Hadjichristidis, N.; Mays, J.W. (2006) Well-defined poly (4-vinylbenzocyclobutene): synthesis by living anionic polymerization and characterization. *Macromolecules,* 39, 3525-3530.

[4] Mehta, G.; Kotha, S. (2001) Recent chemistry of benzocyclobutenes. *Tetrahedron*. 57, 625-659.

[5] Sadana, A.K.; Saini, R.K.; Billups, W.E. (2003) Cyclobutenes and related compounds. *Chemistry of Review*. 103, 1539-1602.

[6] Boaen, N.K.; Hillmyer, M.A. (2005) Post-polymerization functionalization of polyolefins. *Chemical Society Reviews*. 34, 267-275.

[7] Leibfarth, F.A.; Schneider, Y.; Lynd, N.A.; Schultz, A.; Moon, B.; Kramer, E.J.; Bazan, G.C.; Hawker, C.J. (2010) Ketene functionalized polyethylene: control of cross-link density material properties. *Journal of the American Chemical Society*. 132, 14706-14709.

[8] Endo. T. Koizumi T. Takata T. Chino K. (1995) Synthesis of poly(4-vinylbenzocyclobutene) and its reaction with dienophiles. *Journal of polymer science, Part A*. 33, 707-715.

[9] Harth, E.; Horn, B. V.; Lee, V. Y.; Germack, D. S.; Gonzales, C. P.; Miller, R. D.; Hawker, C. J. (2002) A facile approach to architecturally defined nanoparticles. *Journal of the American Chemical Society*. 124, 8653–8660.

[10] So, Y.H.; Hahn, S.F.; Li, Y.; Reinhard, M.T. J. (2008) Styrene 4-vinylbenzocyclobutene copolymer for microelectronic application. *Journal of polymer science, Part A*. 33, 707-715.

[11] Leiston-Belanger, J.M.; Russell, T.P.; Drockenmuller, E.; Hawker, C.J. (2005) A thermal and manufacturable approach to stabilized diblock copolymer templates. *Macromolecules*, 38, 7676-7683.

[12] Chino, K.; Takata, T.; Endo, T. (1998) Polymerization of o-Quinodimethanes. III. Polymerization of o-Quinodimethanes Bearing Electron-Withdrawing Groups Formed in Situ by Thermal Ring-Opening Isomerization of Corresponding Benzocyclobutenes. *Journal of polymer science, Part A*. 37, 1555-1563.

[13] Hahn, S.F.; Martin, S.J.; McKelvy, M.L. (1992) Thermally Induced Polymerization of an Arylvinylbenzocyclobutene Monomer. *Macromolecules*. 25, 1539-1545.

[14] Pyun, J.; Tang, C.B.; Kowalewski, T.; Frechet, J.M.J.; Hawker, C.J. (2005) Synthesis and Direct Visualization of Block Copolymers Composed of Different Macromolecular Architectures. *Macromolecules*, 38, 2674-2685.

[15] Kim, Y.; Pyun, J.; Frechet, J.M.J.; Hawker, C.J.; Frank, C.W. (2005) The Dramatic Effect of Architecture on the Self-Assembly of Block Copolymers at Interfaces. *Langmuir*. 21, 10444-10458.

[16] Ryu, D.Y.; Shin, K.; Drockenmuller, E.; Hawker, C.J.; Russell, T.P. (2005) A generalized approach to the modification of solid surfaces, *Macromolecules*. 308, 236-239.

[17] Ryu, D.Y.; Wang, J.Y.; Lavery, K.A.; Drockenmuller, E.; Satija, S.K.; Hawker, C.J.; Russell, T.P. (2007) Surface Modification with Cross-Linked Random Copolymers: Minimum Effective Thickness. *Macromolecules*. 40, 4296-4300.

[18] Yang, J.X.; Liu, S.C.; Zhu, F.H.; Huang, Y.W.; Li, B.; Zhang, L. (2011) New Polymers Derived from 4-Vinylsilylbenzocyclobutene Monomer with Good Thermal Stability,

Excellent Film-Forming Property, and Low-Dielectric Constant, *Journal of Polymer Science, Part A*, 49, 381-391.
[19] Yang, J.X.; Xie, L.Q.; Zhu, F.H.; Sui, H.B.; Li, H.L.; Huang, Y.W. (2011) Incorporation of Benzocyclobutene Cross-Linkable Moieties in Poly(Methyl Acrylate): A Novel Approach to Shape-Memory Polymers Accompanied with Microphase Separation, *Journal of Macromolecular Science, Part B*. 50, 2129-2139.
[20] Xie, L.Q.; Yang, J.X.; Zhu, F.H.; Yang, H.J.; Liu, C.L.; Zhang, L. (2010) Preparation and Polymerization of Benzocyclobuten-4-yl Acrylate Monomer, *Chinese Journal of Polymer Science*, 28, 877-885.
[21] Smith, K.E.; Parks, S.S.; Hyjek, M.A.; Downey, S.E.; Gall, K. (2009) The effect of the glass transition temperature on the toughness of photopolymerizable (meth)acrylate networks under physiological conditions. *Polymer*. 50, 5112-5123.
[22] Xu, Y.W.; Zhu, F.H.; Xie, L.Q.; Yang, J.X.; Zhang, L.; Xie R.G. (2010) Synthesis and property studies of oligomer obtained from the reaction of 4-vinylbenzocyclobutene and styrene with divinyl tetramethyl disiloxane-bisbenzocyclobutene, *e-Polymers*, 013.
[23] Zuo, X.B.; Chen, J.S.; Zhao, X.J.; Yang, S.Y.; Fan, L. (2008) Synthesis and Characterization of Siloxane Resins Derived from Silphenylene-Siloxane Copolymers Bearing Benzocyclobutene Pendant Groups, *Journal of Polymer Science, Part A*, 46, 7868-7881.
[24] Zuo, X.B.; Yu, R.L.; Shi, S.; Feng, Z.H.; Li, Z.P.; Yang, S.Y.; Fan, L. (2009) Synthesis and Characterization of Photosensitive Benzocyclobutene-Functionalized Siloxane Thermosets, *Journal of Polymer Science, Part A*, 47, 6246-6258.
[25] Zhao X.Y. Liu H.J. (2010) Review of polymer materials with low dielectric constant. *Polymer International*, 59, 597-606.
[26] Martin S.J. Godschalx J.P. Mills M.E. Shaffer E.O. Townsend P.H. (2000) Development of a Low-Dielectric-Constant Polymer for the Fabrication of Integrated Circuit Interconnect. *Advanced Material*. 12, 1769-1778.
[27] Sakellariou, G.; Ji, H.N.; Mays, J.W.; Hadjichristidis, N.; Baskaran, D. (2007) Controlled Covalent Functionalization of Multiwalled Carbon Nanotubes using [4 + 2] Cycloaddition of Benzocyclobutenes. *Chemistry of Materials*. 19, 6370-6372.
[28] Sakellariou, G.; Ji, H.N.; Mays, J.W.; Baskaran, D. (2008) Enhanced Polymer Grafting from Multiwalled Carbon Nanotubes through Living Anionic Surface-Initiated Polymerization. *Chemistry of Materials*. 20, 6217-6230.
[29] Priftis, D.; Petzetakis, N.; Sakellariou, G.; Pitsikalis, M.; Baskaran, D.; Mays, J.W.; Hadjichristidis, N. (2009) Surface-Initiated Titanium-Mediated Coordination Polymerization from Catalyst-Functionalized Single and Multiwalled Carbon Nanotubes. *Macromolecules*. 42, 3340-3346.
[30] Ma, B.W.; Lauterwasser, F.; Deng, L.; Zonte, C.S.; Kim, B.J.; Frechet, J.M.J. (2007) New thermally cross-linkable polymer and its application as a hole-transporting layer for solution processed multilayer organic light emitting diodes. *Chemistry of Materials*. 19, 4827-4832.

[31] Chua, L.L.; Ho, P.K.H.; Sirringhaus, H.N.; Friend, R.H. (2004) High-stability ultrathin spin-on benzocyclobutene gate dielectric for polymer field-effect transistors. *Applied Physic Letters*, 84, 3400-3402.

[32] Nomura, K.; Kakinuki, K.; Fujiki, M.; Itagaki, K. (2008) Direct precise functional group introduction into polyolefins: efficient incorporation of vinyltrialkylsilanes in ethylene copolymerizations by nonbridged half-titanocenes. *Macromolecules*. 41, 8974-8976.

[33] Nghiem, Q.D.; Nguyen, C.T.; Kim, D.P. (2008) Controlled/Living Radical Polymerization of Vinylcyclicsilazane by RAFT Process and Their Block Copolymers. *Journal of Polymer Science. Part A*. 46, 4594-4601.

[34] Matsuya, Y.; Ohsawa, N.; Nemoto, H. (2006) Accelerated Electrocyclic Ring-Opening of Benzocyclobutenes under the Influence of a α-Silicon Atom. *Journal of the American Chemical Society*, 128, 412-413.

[35] Hickenboth, C.R.; Moore, J.S.; White, S.R.; Sottos, N.R.; Baudry, J.; Wilson, S.R. (2007) Biasing reaction pathways with mechanical force. *Nature*, 446, 423-427.

[36] Dobish, J.N.; Hamilton, S.K.; Harth, E. (2012) Synthesis of low-temperature benzocyclobutene cross-linker and utilization. *Polymer Chemistry*. 3, 857-860.

Chapter 9

An Advanced Approach on the Study of Emulsion Polymerization: Effect of the Initial Dispersion State of the System on the Reaction Mechanism, Polymerization Rate, and Size Distribution of Polymer-Monomer Particles

M. Khaddazh, I.A. Gritskova and G.I. Litvinenko

Additional information is available at the end of the chapter

1. Introduction

Emulsion polymerization (EP) is a complex heterophase process which involves many components and also runs sequentially and in parallel physicochemical processes that determine the composition of the emulsion, the formation mechanism of polymer-monomeric particles (PMP), conditions for the formation of interfacial layer, etc. There are a number of factors that collectively affect the kinetics of emulsion polymerization, among them:

- initiator decomposition,
- dispersion of the monomer,
- emulsifier redistribution between phases,
- microemulsification,
- formation of interfacial adsorption layer,
- initiation of polymerization,
- PMPs formation,
- diffusion of monomer into PMPs.

The initial stage of emulsion polymerization defines the basic parameters of the process: the rate of polymerization and the formation of interfacial layer, the number of particles, their size distribution, molecular masses of polymers and molecular mass distribution. The mechanism of formation of PMPs has been widely discussed in the literature and the researchers can not possibly come to a consensus.

This can be explained by the fact that in the original system, different types of particles may be present depending on the nature of system components and their concentrations:

- individual molecules of surfactant
- micelles of surfactant
- monomer swollen surfactant micelles
- surface-active oligomers formed during the initiation of polymerization in the aqueous phase
- mixed micelles (surfactant + oligomers)
- macro- and microdroplets of the monomer.

In addition to these particles, the emulsion of the monomer may contain specially added low-molecular-weight substances necessary for the preparation of polymeric suspensions with desired properties.

2. Mechanisms of polymer-monomer particles formation in emulsion polymerization

In the literature there are various hypotheses about the mechanism of particle formation in emulsion polymerization, the main ones assume the formation of PMPs from:

- micelles of emulsifier,
- macromolecules of the polymer dropped out in water (the mechanism of homogeneous nucleation),
- by the mechanism of nucleation aggregative which combines elements of different modes of particle formation
- microdroplets of monomer.

None of the hypotheses proposed to date is confirmed by reliable experimental data. Hereinafter we briefly recall these hypotheses.

2.1. Micellar mechanism

Micellar mechanism of particle formation is based on a qualitative model proposed simultaneously and independently by Harkins [1,2] and Yurzhenko [3,4]. The basic points of such mechanism are as follows:

1. Emulsion polymerization is a usual process of a radical polymerization, its features are explained by the fact that the main site of reactions - micelles and PMP - have discrete volumes.
2. At the initial stage of the polymerization the reaction system consists of emulsifier micelles, PMP and large monomer droplets.
3. Polymerization starts only in the emulsifier micelles, which when attacked by a radical convert to PMP, monomer droplets being only the source of monomer. Diffusion of monomer from droplets through the aqueous phase and into the PMPs does not limit the process that leads to the establishment of equilibrium monomer concentration PMP, which persists as long as monomer droplets are present in the system.

According to these ideas, in the presence of surfactants, the initial emulsion consists of two kinds of particles of different sizes: the monomer droplets with diameters of 5-20 microns and colloidal degree of dispersion and monomer-swollen surfactant micelles (5-10 nm).

The mechanism of Harkins-Yurzhenko lied in the base of Smith and Ewart quantitative theory [5-7], with further refinements in the works of other authors [8-70]. The main aspect of this theory is that radicals formed in the aqueous phase are trapped by the monomer swollen surfactant micelles and turn to PMP. It is assumed that only one out of every 100-1000 micelles captures a radical and becomes PMP, and the rest of the micelles are spent to stabilize the growing PMPs. Polymer-monomer particles formation ends with the disappearance of micelles of the emulsifier in the aqueous phase, after which the number of particles remains constant.

Initial system contains monomer droplets with a diameter of 5-15 microns, their concentration being 10^{12}-10^{14} droplets/l, the monomer-swollen micelles with diameters of 5-10 nm and the number of micelles 10^{18}-10^{21} l^{-1}, and a water-soluble initiator (usually potassium persulfate) at a concentration of 1% per monomer [8]. Only a small fraction of the molecules of the monomer is located in the interior of the micelles (1-2%) and is dissolved in the aqueous phase (0.03% for styrene). In the aqueous phase, 10^{14}-10^{16} PMP / l are formed with a diameter in the range of 20-200 nm. Monomer droplets due to their relatively small surface area hardly compete with micelles in capturing radicals.

In the polymerization process PMPs increase their size due to the diffusion of the monomer from droplets and monomer-swollen micelles which contains no radicals [1-3, 5, 9-15].

Three limiting cases were considered:

1. The number of free radicals in the particles is small compared with the total number of radicals, thus the average number of radicals per particle is much less that unity.
2. The average number of radicals per particle is equal to 0.5. This case is realized under following conditions: the activity of the radical in the particle persists as long as the second radical enter the particle, and the time of chain-breaking is small as compared with the average time interval of successive absorption of radicals by the particle.
   The authors believe that case 2 corresponds to the emulsion polymerization of styrene in the presence of potassium persulfate [5]. The rate of formation of primary radicals is equal to 10^{13} radicals• $ml^{-1}s^{-1}$. The average number of polymer particles is of the order of 10^{14} -10^{16} in 1 ml of the system. If all the radicals formed by the decay of the initiator enter the polymer particles, the average frequency at which a radical enters a particle is once per 10 - 100 sec.
   At any time, the particle contains either one radical or does not contain radicals at all, since it is assumed that chain termination occurs immediately in contact with the second radical in the particle. The particle is inactive until the next radical does not enter the particle, i.e. by 10-100 sec. Consequently, half of time each particle contains a radical and another half of time does not contain radicals, i.e. one half of the particles are active and each of them contains one polymer radical. Thus, the polymerization rate, referred to 1 ml of latex, is expressed by the equation:

$$R_p = k_p \cdot \frac{N}{2}[M] \quad (1)$$

where N is the number of PMPs in 1 ml of latex, k_p is the propagation rate constant,[M] is the concentration of monomer in the PMP.

The rate of polymerization is determined by the number of particles and it does not depend on particle size if it is not very large.

3. If the particle size is large, they may contain several radicals at the same time, as in the case of suspension polymerization. This condition corresponds to the third case: the number of free radicals in the polymer particle is large, each particle has a certain steady-state concentration of radicals.

After completion of the formation of PMPs their concentration in water remains relatively constant until the end of polymerization. The size of particles becomes larger due to the diffusion of monomers from monomer droplets which serve as reservoirs for the growing particles. Most of the monomer is consumed in the growth stage of particles (approximately 10 to 60% monomer conversion). The stage of particle growth (interval II) ends with the disappearance of monomer droplets in the system. For case 2 the following assumptions were implanted:

1. The number of particles per unit volume of water remains constant throughout the polymerization;
2. The particle size distribution is relatively narrow;
3. No desorption of free radicals from the particles;
4. Bimolecular termination of polymer radicals, located in a particle diffusing with a radical from the aqueous phase is instantaneous.

Figure 1 illustrates the evolution of the rate of emulsion polymerization. The kinetic curve contains three phases: a relatively short stage I, characterized by the growing polymerization rate, stage II of a constant rate of the process, and stage III, where the polymerization rate decreases.

Stage I refers to a micellar stage of emulsion polymerization. It describes the formation of PMP and the increase in their number. In stage II, the formation of new particles does not occur, and the main process parameters: growth rate constant, k_p, the number of radicals in the particle, n; number of particles, N, and monomer concentration [M], respectively, and the constant changes in the rate of polymerization not observed. Stage III begins after the disappearance of monomer droplets. In this case, all the monomer is contained in the volume of PMP and the rate of polymerization of the consumption of monomer decreases.

Shortly after the publication of the Smith-Ewart theory, many other interpretations that discuss deviations from the theory were issued [8-33]. There have been many recent indications of change in the number of particles in the emulsion polymerization. In stage II, the Smith-Ewart assumption on the discreteness of the latex particles is not supported.

Nevertheless, the consideration of the number of particles in the emulsion system as the main kinetic factor remains the basic idea of the Smith-Ewart theory.

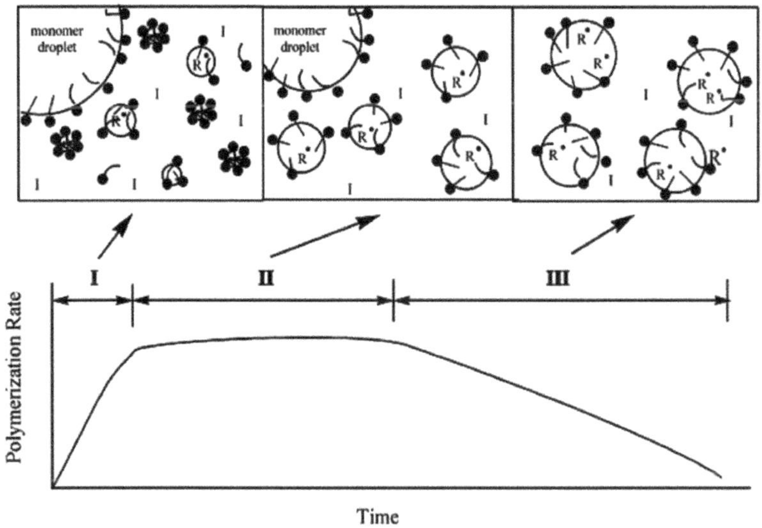

Figure 1. Three intervals of the dependence of polymerization rate on monomer conversion (from Prog. Polym. Sci. 26, 2001, p. 2094) [16]

Today it is obvious that the observed phenomena in emulsion polymerization are extremely diverse and can not yet be explained by a single theory.

Kinetic description of emulsion polymerization is complicated due to the difficulty of establishing a place where elementary reactions proceed, and concentrations of reagents in these places.

The classical theory of emulsion polymerization of the Smith-Ewart, establishing a link between the synthesis conditions and the basic parameters of the process, proved to be applicable only to a limited number of objects. It describes EP of nonpolar monomers in the presence of surfactants insoluble in monomer.

Later on, these ideas have been modified by Gardon [17-23], Harada [24], Stokmayer [25], O'Toole [26], Ugelstad [27-29], Kuchanov [30-32], etc.

A more general review of topochemistry and kinetics of polymerization of the latex were presented by Medvedev et al. [8-11, 14]. Agreeing with Harkins-Yurzhenko's ideas concerning particles formation, Medvedev, nevertheless, has proposed that micelles have properties of swarms, and an exchange of molecules and radicals between (a) micelles themselves, and (b) micelles and polymer particles takes place. This assumption considers the processes occurring with the participation of micelles as homogeneous ones averaged over the main parameter - the concentration of emulsifier.

Key topochemical and kinetic features of emulsion polymerization according to Medvedev may be formulated as follows:

I. With water-soluble peroxide compounds which form initiating radicals in aqueous solution polymerization, depending on the nature of monomers, starts in water or micelles. When using oil-soluble polymerization initiators, regardless of the nature of the monomer, it begins in micelles since they contain both monomer and initiator.

The process proceeds further in 10-100 nm PMP, in which monomer concentration during the reaction remains constant (for polymers which are soluble in their monomers, this concentration is 40-50%).

It is essential that the polymerization does not start simultaneously in all the micelles, but only in a small part of them. This is because the concentration of initiating radicals is usually much lower than the number of micelles. The emulsifier contained in the polymer-free micelles is used to stabilize the increase in volume of PMP. Since initiating radicals are formed in the molecular aqueous solution or in adsorption layers of the emulsifier (which means in the micelles or at the surface of polymer particles), the polymer radicals occur near these adsorbed layers. Because of the low rate of diffusion of polymer radicals in viscous phase of PMP the reaction growth and chain termination take place not in the whole volume of the particles (especially in their relatively large volume), but in some zone near the surface. The volume of this zone is determined by the concentration of emulsifier in the process, and in some cases remains approximately constant.

II. The total rate of emulsion polymerization is 10^2-10^3 higher than the polymerization rate in homogeneous systems with the same initiators (peroxides, azo compounds, radiation). The average molecular mass of the polymers is also much higher in emulsion polymerization.

Therefore increasing the total rate of the process due to the decrease of the reaction rate of chain termination, leads to an increase in both rate and average length of the polymer molecular chains.

Reducing rate of chain termination reactions may be due to two reasons: 1) a low rate of diffusion of polymer radicals in a viscous medium, as in emulsion polymerization, even in early stages, the process takes place in concentrated solutions of polymer in monomer, 2) the radical separation to individual particles. Qualitatively, these two assumptions explain the simultaneous increase in the speed of the process and the molecular weight of polymer in latex polymerization.

Medvedev theory satisfactorily describes the quantitative relationships emulsion polymerization of many monomers.

Ivanchev and Pavlyuchenko [33, 34] attempted to study the elementary reactions of emulsion polymerization and use the results for the synthesis of polymers with properties fundamentally different from those obtained in homogeneous systems. In their study of initiation reaction in emulsion polymerization of styrene, it was suggested that the adsorption layer of PMP has concentrating and orienting effect on the initiator molecule.

2.2. The mechanism of homogeneous nucleation

According to the hypothesis of the formation mechanism of PMP for homogeneous nucleation [27, 35-42], the formation of particles or macromolecules derived from the radicals who have reached the critical length (j_c), where they lose solubility and precipitate in the aqueous phase. Further growth of the polymer particles dropped in the water is treated differently. Some authors believe that these particles grow by diffusion of monomer from the monomer droplets, i.e. as well as in the case of micellar mechanism, other authors suggest that there exist their limited flocculation and, consequently, the formation of PMPs.

First who suggested the formation of the PMP precipitated from an aqueous solution of polymer molecules was Khomikovsky in 1948 in his study of emulsion polymerization of methyl methacrylate and vinylcyanide [43]. He found that the dependence of the rate of polymerization of the two monomers on the emulsifier concentration is different: the rate of vinylcyanide polymerization decreases with increasing concentration of emulsifier and the rate of MMA polymerization increases, the concentration of emulsifier in these experiments being above the critical micelle concentration (CMC). Khomikovsky explained the results in such a way that vinylcyanide polymerization, initiated by potassium persulfate, begins in water.

Upon reaching the length of polymer chain when oligomers became insoluble in water they precipitate in the aqueous phase to form particles that are stabilized by the adsorption emulsifier molecules on their surface. Initiation of polymerization transforms them to PMP. Here, the polymerization proceeds similarly to that in PMP formed from micelles during polymerization of monomers poor soluble in water.

Later on, these ideas found their development in the investigations of Priest [35], Roe [36], Fitch and Tsai [39, 40], Yeliseyeva [37, 38], Christiansen [44], Ugelstad [27], Pepard [45] Wilkinson [46-48], Oganesyan [49-51], Tauer [52-57], etc.

A quantitative description of the processes of polymerization of the monomers, partially soluble in water, was given by Fitch and Barrett [58].

Fitch proposed the theory called the theory of homogeneous nucleation. Quantitative analysis of this theory was based on the process of self-exhaustion of oligomeric radicals that have reached a critical degree of polymerization.

Considering the theory of Fitch, Barrett has identified two possible distribution of radicals between the aqueous phase and particles:

1. The equilibrium distribution of radicals between the particle and the volume of the aqueous phase, the radical capture rate here is proportional to the volume of the particle.
2. The equilibrium distribution of radicals between the surface of the particle and the aqueous phase, the radical capture rate in this case is proportional to the square of the particle surface.

He showed that the second case is realized more likely, explanation being that small particles are more effective for the capture of oligomeric radicals than larger particles.

The number of particles, according to the theory of Barrett, depends mainly on the distribution coefficient of oligomeric radicals between the surface and volume of the particles and the aqueous phase.

As the number of particles increases during polymerization, the rate of radical absorption by particles increases, as long as all the newly formed radicals will not be absorbed by the particles without having to reach a critical length.

Two variants of the flow of polymerization in the absence of coagulation of particles were considered:

a. a short nucleation time (due to a high rate of initiator decomposition), here all the particles start to grow almost simultaneously at the same average rate, and reach finally approximately the same size. The result of this reaction is the large number of particles of small diameter and narrow particle size distribution.
b. a long time of nucleation (due to a slow initiation), the particles are formed longer, which leads to a broader particle size distribution.

The rate of flocculation of the particles can be controlled by varying the concentration of emulsifier, which is adsorbed on the surface of primary particles and prevents flocculation due to the formation of the electric double layer or due to steric stabilization factor.

In the absence of emulsifier or at its very low concentration and a high rate of initiation,a rapid increase in the number of particles due to the high rate of decomposition of the initiator and the formation of free radicals is noticed during the first stage of polymerization.

Further, due to flocculation, the number of particles drops sharply at the end of the nucleation, particles reach a certain stability by increasing the density of the surface electric charge. If the period of the particles nucleation is short, then, the polymer suspension is characterized by a narrow particle size distribution, on the contrary, the distribution is broad if this period is long.

Ugelstadt and Hansen [27] consider that EP model with partially water soluble monomers should take into account the different stages of the diffusion of oligomeric radicals, as well as the reaction of oligomeric radicals with monomer in the particles.

Using the basic assumption of Fitch that the oligomeric radicals have to reach a certain critical degree of polymerization at which they spontaneously precipitate from the aqueous phase, Ugelstadt and Hansen examined the effect of diffusion of oligomeric radicals in the formation of PMP on the number of particles formed by the mechanism of homogeneous nucleation.

The main provisions of the model are similar to those observed in the Fitch theory and are as follows:

1. Oligomeric radicals spontaneously precipitate when a critical degree of polymerization is reached;
2. Absorption of oligomeric radicals by particles depends on the molecular weight of the radicals;
3. There is a desorption of monomeric radicals from particles;
4. The role of the emulsifier (at a concentration below CMC) is to stabilize the primary particles.

According to this model, in the absence of emulsifier radicals which are formed by the decay of the initiator can:

- add monomer dissolved in the aqueous phase;
- be captured by existing polymer-monomeric particles or adsorbed on the surface of the PMP;
- recombine in the aqueous phase with the other radicals;
- associates to form micelles with other types of radicals dissolved;
- precipitate when the oligomeric radicals critical chain length (j_c).

The tendency of dissolved oligomeric radicals to the association is small because of their low concentrations (<10^{-7} mol/l).

Termination of oligomeric radicals in the aqueous phase may, or may not lead to the formation of new particles. Loss due to recombination leads to a doubling of the chain length of oligomeric radicals, but in this case oligomeric product contains two ionic end groups, resulting in a critical chain length required will increase. Therefore it is not clear whether the loss be oligomeric product from the solution.

According to Fitch model, Ugelstadt determined the number of PMPs in the reaction system is determined by a limited coagulation process of the primary particles.

To explain the influence of emulsifier on the nature of the stabilization process of PMPs Yeliseyeva et al. [12, 13] proposed to use the emulsifier adsorption isotherms at PMPs as a main characteristic.

These authors believe that in EP the rate of emulsifier adsorption on the surface of PMP is a function of adsorption energy, which, in turn, depends on the nature of the emulsifier, and the nature of the surface of the PMP. Later, in the publications of Yeliseyeva et al. adsorption of emulsifier on the particle surface was considered as the main factor determining the kinetics of EP of polar monomers.

Because of the high flow rate of the polymerization process it is difficult experimentally to determine the number of particles during their formation.

Published data show that existing models of EP, based on ideas about the formation mechanism of PMPs homogeneous nucleation satisfactorily describe only the polymerization with a low content of monomer in the initial system (5-10% wt.) in the absence of an emulsifier, or at very low (below the CMC) concentrations.

2.3. Aggregative nucleation mechanism

Along with the micellar and homogeneous mechanisms, aggregative nucleation of particles in EP is widely recognized. It is carefully designed by Lichti [59-61] and Feeney [62-64] on the basis of the study of the size distribution of polystyrene latex particles stabilized with sodium dodecyl sulfate. According to them, precursor particles are formed as a result of the growth of oligomeric radicals in the surfactant micelles. Precursor particles are highly unstable and aggregate, reducing the interfacial tension at the oil / water boundary. Thus, there are germs of the particles.

The rate of nucleation time is extremely high, the authors explain this fact as a result of coagulative (aggregative) nucleation. To describe aggregate nucleation a mathematical model has been used which is a combination of the kinetic theory of coagulation, and Muller-Smoluchowski theory of DLVO (Derjaguin, Landau, Verwey and Overbeek) [65], which determines the strength of the electrostatic repulsion opposing the dispersion forces of attraction as the main factor responsible for the stability of the colloidal particles.

A further development of the mechanism of nucleation of aggregate particles is reflected in the publications of Tauer, Kuhn et al. [52-57, 66]. This theory is indeed a combination of ideas proposed by Fitch et al. [39-41] and Oganesyan [51] with the difference that the authors consider the surface energy of the particles as an adjustable parameter of the model.

The basic assumption of aggregative mechanism is the formation of water-soluble oligomers of PMP with a certain chain length. The proposed mechanism can be regarded as a refinement of the mechanism of homogeneous nucleation of particles and act as an independent theory. This mechanism is similar to the mechanism of homogeneous nucleation, but in this case the particles are not dropped from the aqueous phase oligomeric radicals that have reached a critical chain length, and from particles formed by aggregation of several water-soluble oligomeric radicals which are called clusters [66].

This theory is based on the results of a study of the polymerization of styrene, methyl methacrylate, and vinyl acetate in the absence of surfactant [58]. Selected monomers suitable for this model as well studied in the literature are presented all the necessary constants and experimental data on the critical value (j_c) the chain length of oligomeric radicals, in which they precipitate in the aqueous phase, in addition, these monomers differ significantly in their solubility in water [22, 67-70].

As part of this mechanism depends significantly on the over-saturation of the oligomeric chain length (j), which in turn depends on the polymerization conditions. The higher values of j_c, the lower the solubility of oligomers and the lower concentration required to achieve saturation and the phase formation.

For different monomers nucleation occurs at different times and for different values of j: the higher solubility of the monomer in water, the higher the j_c and the longer time required for the appearance of the particles. For the three monomers selected above the calculated j_c agree well with the experimental values.

The theory of Hansen and Ugelstad of homogeneous nucleation of particles is based on several assumptions whose validity has not yet been proved. As a main place of particle formation the authors consider a saturated aqueous solution of monomer and initiator and do not take into account the processes occurring at the monomer/water interface.

The formation of the polystyrene suspension during emulsifier-free polymerization initiated by potassium persulfate and the detection of styrene oligomers with the 8-9 degree of polymerization in the aqueous phase are considered as an experimental proof of this theory [27].

Identifying the formation of particles according to the homogeneous nucleation process that takes place in a perfectly pure supersaturated solutions, these authors did not find the original monomer/water system emulsifier micelles or microdroplets of monomer.

In these studies it is assumed that ion-radicals falls out of solution as a new phase at attaching a certain number of monomer molecules (up to 10). However, the size of oligomeric radicals is widely debated by many authors.

Detailed studies of physico-chemical processes occurring at the interface as well as in separate bulk phases in monomer-water solution of potassium persulfate static systems were performed and described by Oganesyan et al. [49-51].

The authors studied the polymerization of styrene in static conditions in the absence of an emulsifier, using potassium persulfate as initiator at 50°C in temperature-controlled reactors. The system was kept in the oven for 2 hrs where decomposition of initiator and initiation of polymerization took place, the aqueous phase became turbid. After standing for more than two days, the aqueous phase transformed into stable latex. These studies were carried out at a volume ratio styrene/water equal to 1:7.

The authors explained these results via the formation of particles at the droplets of the monomer / water interface.

To confirm this conclusion, the authors have found conditions which allow to extend the residence time of particles at the interface and to enable them to build-out there. They increased the density of the aqueous phase by increasing the concentration of initiator. The experiments require a high purity and it was necessary to exclude the influence of possible impurities on the stability of the system under study and eventual variations in temperature. Experiments were carried out in batch reactors, which are attached to the cover of the crystallization apparatus designed for growing crystals under isothermal conditions and having an electronic device to provide a constant temperature with high accuracy.

Crystals of potassium persulfate are grown up in the crystallization apparatus. They are used to initiate polymerization at concentrations in the range of 0.5 - 3% by weight. Aqueous solutions of potassium persulfate and styrene separately thermostated at a temperature of 50°C, and carefully layered on the aqueous phase styrene. Painting turbidity of the aqueous phase was dependent on initiator concentration (>2%), turbidity appeared at high concentrations of initiator in the narrow boundary layer from the aqueous phase, and then distributed throughout the volume. For the remaining initiator concentrations area of the initial turbidity of water phase was extended.

Authors believe that the formation of monomer droplets in static conditions is necessary to find the source of energy to perform work on the dispersion of the monomer, which is determined by temperature and chemical potential of the contacting phases. It is believed that if the heat of polymerization is able to transfer a certain amount of monomer in the aqueous phase, then the polymerization reaction at the interface can deform the surface of the interface and disperse system. Further, they assume that at an initial equilibrium state of the system the transfer of certain amount of monomer in the monomer-rich aqueous phase is equivalent to supersaturation of water molecules of the monomer, and nucleation of monomer droplets in the aqueous phase near the interface can be expected.

2.4. Mechanism of particle formation from microdroplets of monomer

According to the representations of micellar [1-7] and homogeneous [27, 35-40] theories of particles nucleation, the monomer droplets play the role of a reservoir from which the monomer in the polymerization process goes into growing PMP by the diffusion through the aqueous phase. Subsequent studies performed by Pravednikov, Gritskova, Taubman, Nikitina, Ugelstad and co-workers have shown that the monomer droplets may be involved in the formation of PMPs if their size may be reduced to the size of the order of 1 micron or less [28, 49-51, 71-111].

Droplet size of emulsions obtained by emulsification of the monomer in the aqueous solution with stirring may be evaluated by the equation:

$$D \approx \left(\frac{\sigma_{12}}{d_{wp}} \right)^{3/5} E^{-2/5} \tag{2}$$

where σ_{12} is interfacial tension, d_{wp} is the density of the dispersion medium, and E is the energy expended to move a unit mass of the medium [80].

The dispersion of the monomer depends on the power consumed for mixing of the reactor design, the type of the mixer and mixing rate. The presence of the surfactant in the system has a significant influence on the dispersion of the monomer emulsion : adsorption of surfactant at the interface alters the interfacial tension, which facilitates the process of fragmentation and prevents the coalescence of droplets of the monomer when they collide. On the other hand, the irregular distribution of surfactant between monomer and aqueous phases in the initial system provokes transfer of surfactant across the interface, which can lead to the destruction of the border and microemulsification. Thus, in addition to fragmentation and coalescence processes that determine the size of the emulsion droplets in the presence of surfactants, microemulsification may take place. Assignment of microemulsification as a separate factor influencing the formation and composition of the initial emulsion system relates to the fact that the size of droplets produced here is significantly less than those determined by the direct effect of mechanical agitation (less than 0.2 microns). Therefore, the investigation of microemulsification goes beyond traditional ideas about the formation of disperse systems under mixing.

A significant attention in the literature was devoted to the investigation of the properties of microemulsions (ME) and conditions of their formation [81-94], however, the thermodynamic equilibrium of lyophilic systems [81, 82] was mainly considered.

Rebinder was one of the pioneers who studied and later developed the process of spontaneous microemulsification [89, 90]. According to his ideas, this process results in the formation of lyophilic colloidal phase corresponding to such a state when an increase of free energy in the formation of the colloidal phase is close to the energy of thermal motion. This process is accompanied by an increase in entropy of the system due to the formation of a large number of colloidal particles. Entropy factor will compensate for the increase in free energy associated with an increase in the interface area.

In contrary to the real (true) spontaneous emulsification when dispersion occurs in the total volume of the dispersed phase with the formation of thermodynamically stable lyophilic system [79, 95-100], in this case the dispersion may occur in one or both phases simultaneously with interfacial surface tension, far from the critical (σ_{min}), as a result of hydrodynamic instability of the interfacial layer, and leading to mass transfer of one phase to another. As a result, only part of the dispersed phase the in the layer adjacent to the interface can be involved in emulsification [91-94].

The mechanism of interphase mass transfer of surfactant with microemulsification remains open. Various hypotheses are put forward, including those which are based on interfacial instability in the development of interfacial turbulence caused by Marangoni effect. Besides, local fluctuations of the interfacial tension, which lead to the movement of the surface layers of liquid, which, in turn, can increase the interfacial tension gradients in the presence of phase transfer surfactant may be also took into consideration [101].

Thermodynamically stable ME are translucent, lyophilic systems containing spherical aggregates whose size is in the range of 10 to 20 nm. The formation and the type of microemulsion depend on the ratio of the components in the system and the interaction between the surfactant molecules, hydrocarbon and water, the length of the alkyl chain in the molecules of surfactant and co-surfactant (if the later exists), as well as on the nature of the hydrocarbon [111-116].

It should be noted that microemulsions formed as a result of surfactant phase transfer are only kinetically stable and significantly different from those traditionally considered as thermodynamically stable ME. Again we note that the size of monomer microdroplets exceeds the characteristic size of the emulsifier micelles, but is much smaller than droplet size which can be obtained with the corresponding value of interfacial tension by mechanical fragmentation of the monomer.

Mass transfer at the interface caused by the diffusion of surfactant, soluble in both phases, provided greater solubility of the surfactant in one phase, without stirring, bringing with him through the phase boundary, and enjoys a certain amount of the solvent (monomer) that is emulsified in the other phase.

The intensity of microemulsification is slowed over time and can be completely terminated with prolonged contact of phases. This is explained by the fact that the interface formed a dense layer of ME, retarding the process of mass transfer. Microemulsification observed in systems with sufficiently high values of interfacial tension σ_{12} (about 1-10 mN / m). The increase in interfacial surface at high values of σ_{12} can not be explained by thermodynamic factors, because the corresponding increase in free energy can not be offset by an increase in entropy of microdroplets. This means that the stability of such a dispersed system has a kinetic rather than thermodynamic in nature.

In this case, when the surface energy of the droplets is much higher than *KT*, the formation of droplets can not be explained only by thermal fluctuations. Note that the ratio of the surface energy of a *KT* scan, depending on the size of droplets can be large, even for low values of σ_{12}. So for a drop having a diameter D = 100 nm and $\sigma_{12} = 10^{-2}$ mN / m, for this ratio, we obtain

$$\frac{\pi D^2 \sigma_{12}}{KT} \approx 75 \quad (3)$$

This process is a kind of "quasi-spontaneous" emulsification, and once again we recall that in contrast to the spontaneous emulsification of the true droplets can form various degrees of dispersity, from colloidal particles to the droplets, significantly exceeding the size, for example, microscopic investigation of emulsion obtained in a glass capillary at the interface of 10% aqueous solution of sodium butylnaphthalenesulfonate / xylene, clearly showed that the formation of direct and inverse emulsions. Unstable inverse emulsion of water in coarse xylene - the diameter of water droplets is in the range 1-5 microns. The emulsion is xylene in the aqueous phase (direct emulsion) is visually observable in the form of a milky-white layer below the interface and is extremely stable and formed droplets whose size is beyond the interval of an optical microscope. Therefore, the determination of the dispersion ultra ME is carried out by electron microscopy [79]. The diameters of the particles according to electron microscopy were in 20-40 nm intervals.

There is a proportional relationship which was noticed between the resulting number of ME microdroplets and the quantity of surfactant have passed through the interface. The effectiveness of microemulsification is associated not only with surfactant phase transfer caused by initial non-equilibrium distribution of surfactants between the phases. Parameters determining the condition of interfacial instabilities are also: a sign of the derivative $d\sigma_{12}/dG$ (G - surface concentration of surfactant), the relationship of diffusion coefficients D_1/D_2 and coefficients of the kinetic viscosity γ_1/γ_2 in different phases [77]. It should be noted that the surface concentration of surfactant uniquely determines the surface tension only in the absence of local gradients of G, leading to a surface instability. Dynamic surface tension measured in the presence of interfacial instability, depends on the distribution of local values of G, and may be substantially lower than the equilibrium value. Numerous experimental determination of σ_{12} under non-equilibrium conditions confirms this fact.

Interphacial transfer causing the destruction of the interface can also lead to the fragmentation of large monomer droplets, whose dimensions are determined mainly by conditions of mechanical stirring. By analogy with the effect of interphase mass transfer direction that depends on the surfactant distribution coefficient on the rate of microemulsification, we can assume that the fragmentation of large droplets of the monomer will depend on the way the surfactant is introduced to the system:

- surfactant is dissolved initially in aqueous phase (traditional way);
- surfactant is dissolved initially in monomer;
- surfactant is formed on the interface layer (when acid and alkaline components of surfactant are dissolved in monomer and aqueous phase respectively)

Investigation of polymerization of vinyl and diene monomers in the presence of a nonionic emulsifier, performed in [79, 95, 100] led the authors to conclude that monomer microemulsification is the first stage of PMP formation in EP.

When free radicals are injected into the microdroplets of monomer, they become PMP. Polymer-monomer particles are microdroplets of polymer solution in monomer, on the surface of which the polymer sedimentate, forming a polymer film. This film is the site of fixation of polymer radicals and, therefore, it is the place of formation of high molecular mass polymer.

Monomer to the reaction site (surface layer) of a particle diffuses as from inside this particle as from the outside monomer droplets, if later are present in the system and if the monomer concentration in microdroplets reached critical values lower than monomer/polymer equilibrium concentration. The authors believe that, along with the diffusion of monomer through the water, the transfer of the monomer can be carried out through direct contact of PMP with a monomer droplets.

Thermodynamic substantiation of formation of monomer microdroplets was investigated by Oganesyan [49, 51, 109]. In his study of the static monomer/water system the author tried to find the source of energy needed to perform work on the formation of droplets. His analysis came from the following considerations: the minimum work necessary to create a unit of the interface (the specific surface free energy) is the interfacial tension. New surface can be created by elastic deformation of the interface, the transfer of some amount of a substance from one phase to another and the creation of the surface bumps or depressions, as well as the division of each of the phases in small particles. If both phases are liquid, the minimum work to create a unit surface area for all modes is the same because it is defined only by temperature and chemical potential of the contacting phases [79]. If the released heat of polymerization at the interface is able to transfer a certain amount of monomer in the aqueous phase, it can be assumed that the polymerization reaction can also deform the interface and disperse system. If we start from an initial equilibrium state of the system, the transfer of certain amount of monomer in the monomer-rich aqueous phase of water is equivalent to supersaturation with respect to monomer and nucleation can be expected in the monomer droplets in the aqueous phase in the vicinity of the interface.

Oganesyan et al. [49-51] believe that the formation of droplets of the monomer due to polymerization at the interface monomer - water, can also be explained on the basis of their dependence on the specific surface energy, γ, and temperature.

$$\gamma = \gamma_0 (1 - T / T_c)^n \tag{4}$$

where n is a constant depending on the nature of the substance (for organic liquids it is equal to 11/9 [51]). T_c is the critical temperature (at $T = T_c$ there is a mixture of phases). From equation (4) it follows that with increasing T, γ strongly decreases. Thus, for every act of the reaction heat release in certain areas of the interface performs a partial mixing of fluids, and if stabilizing agents, such as surfactants or oligomeric growing radicals, are present in the system this process will lead to the formation of microdroplets of the monomer.

2.5. Mixed mechanism of PMP nucleation

In [34, 35, 117, 118] Hansen and Ugelstad proposed to consider all three mechanisms of nucleation of PMP (micellar, homogeneous, and from microdroplets) when investigating EP at a concentration of emulsifier above CMC, assuming that the micelles containing solubilized monomer and monomer microdroplets compete with one another in the capture of oligomeric radicals from the aqueous medium. This causes that the rate of formation is the sum of the rate of the particle nucleation by all three mechanisms.

$$\frac{dN_p}{dt} = \rho_i (P_m + P_h + P_d) \tag{5}$$

where P_m and P_d - the probability of absorption of oligomeric radicals by solubilized micelles and microdroplets of monomer respectively, and P_h is the probability of homogeneous nucleation in the aqueous phase. The sum of all probabilities is equal to unity:

$$P_m + P_h + P_d + P_{p0} + P_{p1} = 1 \tag{6}$$

where P_{p1} and P_{p0} are the probability of capture of oligomeric radicals by PMP containing one and zero free radical.

On the basis of this model, several extreme cases have been investigated, such as

a. homogeneous nucleation of particles, taking into account the possible absorption of oligomeric radicals by particles and nuclei of small flocculation [34];
b. the limited flocculation of particles and nuclei, and small micellar nucleation PMP microdrop [34 , 35];
c. competing homogeneous and micellar nucleation, with the possibility of desorption and reabsorption of free radicals with little flocculation of particles [117];
d. homogeneous nucleation of microdrop and low flocculation of particles [118].

Song and Poeleyn suggested a scheme (Figure 2) of particles nucleation at the initiation of EP monomer and potassium persulfate developed general kinetic model that takes into account all the mentioned above three possible mechanisms of particle nucleation [119-120].

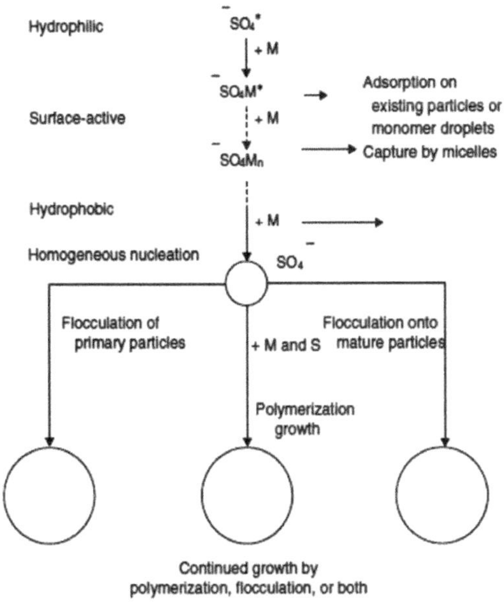

Figure 2. Paths for the formation of particle nuclei starting from persulfate initiator radicals generated in the continuous aqueous phase. The symbols M and S represent monomer and surfactant species, respectively (from Principles and applications of emulsion polymerization / by Chorng-Shyan Chern. John Wiley & Sons, Inc., Hoboken, New Jersey 2008) [121].

These published data suggest that current theoretical ideas on the mechanism of PMP varied. This complicates the establishment of methods of synthesis of polymeric suspensions with regulated properties and makes it relevant and important research aimed at solving this problem.

3. Model of styrene emulsion polymerization. The effect of the initial dispersion of the system on the polymerization rate and size distribution of polymer–monomer particles

The models of latex particle nucleation described in the literature were generalized in [122]. These models are the "collision process" [123, 124], "diffusion process" [27, 29], "diffusion/propagational" model (Maxwell et al. [125]), "collision/empirical" model (Dougherty [126] and Penlidis et al. [127]), "surface coverage" model (Yeliseyeva and Zuikov [128]), and "colloidal" model (Penboss et al. [129]).

The authors of [74] assume that the original monomer emulsion contains, in addition to surfactant micelles, monomer microdroplets with a size of 50–150 nm that are formed owing to fragmentation of monomer droplets in the initiation of polymerization and owing to the mass transfer of the emulsifier at the monomer/water interface. It was shown that the ratio of the number of micelles and microemulsion droplets in the system depends on both the type of surfactant and the method of introducing it into the system [101]. The number of microdroplets affects the pattern of the particle size distribution and the molecular mass distribution of the polymer. According to the authors of [101], PMPs are formed from both micelles and microemulsion droplets.

The simulation of emulsion polymerization of styrene is the subject of many publications, in which the authors discuss the dependence of the kinetics of the process on the emulsifier and initiator concentration [130], the particle size distribution [127, 131-134], and the molecular mass distribution [135-139] and perform a complete simulation of the kinetic characteristics of the process [24, 38, 130, 140-145]. However, all these studies are based on the Harkins–Yurzhenko qualitative theory and disregard the participation of microdroplets in the formation of PMPs.

In this study, we attempt to develop a mathematical model of emulsion polymerization for a partially water-soluble monomer (styrene) that takes into consideration that the original emulsion contains microdroplets that participate in the formation of PMPs.

The calculations are performed with disregard for homogeneous nucleation, polymerization in emulsion macrodroplets, and radical desorption from PMPs. In our opinion, these assumptions are fully justified for the polymerization of partially water-soluble monomers, such as styrene, and in the presence of a fairly high amount of emulsifier.

Let us introduce the following notations: The concentration of PMPs with volume ν with volume fraction of the monomer in them containing i growing radicals at time t is $f_i(\nu, \phi, t)$, the total concentration of PMPs in the system is expressed through $f(\nu, \phi, t)$ as , and their average diameter

$$\bar{D} = \frac{\sum_i \int D f_i(v,\phi,t)dv}{\sum_i \int f_i(v,\phi,t)dv} \tag{7}$$

To determine the effect of the dispersion state of the original emulsion system on the characteristics of emulsion polymerization, we performed a model calculation of the polymerization rate and the PMP size distribution in systems with different initial states.

Consider two limiting cases:

i. A system that contains, in addition to the aqueous phase and monomer emulsion droplets, microdroplets of the same size (here micelles can be regarded also as a limiting case of microdroplets). With this aim in view, we calculate the dependence of the polymerization rate and the PMP size distribution on the diameter of microdroplets.
ii. A system that contains the aqueous phase, monomer emulsion droplets, and micelles and microdroplets of the same size. In this case, we estimate the effect of the ratio between the number and size of micelles and microemulsion droplets on the polymerization rate and the PMP size distribution.

3.1. A monodisperse system consisting of monomer microdroplets

Consider an emulsion system that contains emulsion droplets dispersed in the aqueous phase and microdroplets of diameter D_0 (accordingly, of volume V_0), the concentration of which is M_0 particles per cubic centimeter. Monomer microdroplets will be present in the system if $M_0V_0 < (1+r)^{-1}$, where r is the volume ratio of the aqueous and monomer phases. We assume that the monomer diffusion from droplets towards PMPs through the aqueous phase does not limit the process of emulsion polymerization.

If the emulsifier concentration in the system is high, that is, much higher than the critical concentration of micelle formation (it is this case that will be discussed below), we can disregard the fraction of a surfactant adsorbed on droplets. In this case, M_0 and D_0 are related as follows:

$$\pi M_0 D_0^2 = a_{m(D_0)} S_E N_A$$

where S_E is the emulsifier concentration in the system (mol•cm^{-3}), a_m is the area occupied by one surfactant molecule on the microdroplet surface, and N_A is Avogadro's number.

In the absence of PMP coalescence and radical desorption from particles, the system of equations for the PMP distribution function $f_i(v)$ for volume v; the volume fraction of the monomer in them, ϕ ; and number i of growing radicals has the form

$$\frac{\partial f_i}{\partial t} + i\frac{\partial}{\partial v}(f_i\theta_v) + i\frac{\partial}{\partial \phi}(f_i\theta_\phi) = \\ = j(v)(f_{i-1} - f_i) + \frac{k_t}{2v}\left[(i+1)(i+2)f_{i+2} - i(i-1)f_i\right] + M(t)\delta(v - V_0)j(v)\delta(\phi - 1) \quad (8)$$

In Eq. (8), the first term in the right-hand side describes the arrival of a radical to a particle from the aqueous phase, the second term defines the change in the number of radicals during termination, and the last term describes the formation of new particles after the entry of a radical into a microdroplet. In the left-hand side of Eq. (8), θ_v and θ_ϕ define, respectively, the volume growth rate and the rate of change in the volume fraction of the monomer for a PMP that contains one radical. These rates can vary with the particle size. In addition, $j(v)$ is the diffusion flux of radicals towards particles; k_t is the chain termination rate constant; and $\delta(v - V_0)$ is the generalized Dirac function,

$$\delta(x) = \begin{cases} 0, & x \neq 0 \\ \infty, & x = 0 \end{cases}$$

From the balance for the emulsifier, we derive the time dependence of the microdroplet concentration:

$$\frac{\pi M(t) D_0^2}{a_{m(D_0)}} + \sum_i \iint dv d\phi \frac{f_i v^{2/3} \sqrt[3]{36\pi}}{a_{m(v)}} = S_E N_A \quad (9)$$

Depending on the time of PMP formation, the volume fraction of the monomer, ϕ, of some particles will be higher than equilibrium volume fraction ϕ_M, which is determined from the thermodynamic equilibrium; for some particles, $\phi = \phi_M$. We divide f_i into two parts: PMPs with $\phi > \phi_M$ $(Y_i(\phi,t))$ and particles with $\phi = \phi_M$ $(X_i(v,t))$; that is,

$$f_i(v,\phi,t) = X_i(v,t)\delta(\phi - \phi_M) + Y_i(\phi,t)\delta(v - V_0)$$

The total concentration of PMPs in the system is

$$N = \sum_i \left(\int_{\phi_M}^{1} Y_i d\phi + \int_{V_0}^{\infty} X_i dv \right) \tag{10}$$

The polymerization rate W is expressed in terms of the functions Xi and Yi according to the equation

$$W = \frac{k_p}{v_M} \sum_i i \left(\int_{\phi_M}^{1} Y_i \phi d\phi + \phi_M \int_{V_0}^{\infty} X_i dv \right) \tag{11}$$

Equations (12)–(14) completely describe the first and second stages of emulsion polymerization. To calculate the distribution function fi, we introduce dimensionless variables and parameters:

$$X_i = \frac{M_0}{V_0}\tilde{X}_i \quad v = V_0 y \quad \gamma = \frac{k_p M_0}{V_0 \rho}$$

$$Y_i = M_0 \tilde{Y}_i \quad t = M_0 x / \rho \quad \kappa = \frac{M_0 \theta_v}{V_0 \rho} \equiv \frac{d_M}{d_P} \cdot \frac{\phi_M}{1-\phi_M} \cdot \gamma$$

$$M = M_0 \xi \quad \lambda = \frac{k_t M_0}{2 V_0 \rho}$$

If $\lambda >> 1$, then all $\tilde{X}_i$ and $\tilde{Y}_i$ at $i \geq 2$ are equal to zero (i.e., the approximation of fast termination is valid for the system), and the system with an infinite number of equations will transform into a system of equations for f_i with $i = 0$ and 1. The appropriateness of the approximation of fast termination, other things being equal, is determined by the value of termination constant k_t. If we take the termination constant value for in bulk polymerization as k_t, then, for typical values of the initiation rate ρ = 10^{11}–10^{13} cm^{-3} s^{-1} and the emulsifier concentration $S_E \approx 10^{-4}$ mol/cm^3, the value of λ will be $\lambda \approx 10^{10}$ for particles with a size of D_0 = 10 nm and $\lambda \approx 10^3$ for particles with D_0 = 150 nm. The termination constant for the emulsion polymerization in particles with a volume fraction of the monomer of $\phi = \phi_M = 0.5$ is two to four orders of magnitude lower than that for bulk polymerization; therefore, the value of λ will be lower as well. Nevertheless, at the first stage of emulsion polymerization, the approximation of fast termination is valid for small particles owing to $\lambda >> 1$ that is due to the low value of D_0; for large particles (up to $D_0 \approx 300$–500 nm), for which $\phi > \phi_M$ throughout the first stage, the approximation of fast termination is appropriate because of the high value of the termination constant. However, note that, at the second stage of emulsion polymerization, when the volume fraction of the polymer is high in large PMPs, the approximation of fast termination may be inapplicable to these large particles.

In the approximation of fast termination we obtain the following system

$$\begin{cases} \dfrac{\partial \tilde{X}_0}{\partial x} = J(y)(\tilde{X}_1 - \tilde{X}_0) \\ \dfrac{\partial \tilde{X}_1}{\partial x} + \kappa \dfrac{\partial \tilde{X}_1}{\partial y} = J(y)(\tilde{X}_0 - \tilde{X}_1) + \gamma\phi_M \tilde{Y}_1(\phi_M)\delta(y-1) \end{cases} \tag{12}$$

$$\begin{cases} \dfrac{\partial \tilde{Y}_0}{\partial x} = J(1)(\tilde{Y}_1 - \tilde{Y}_0) \\ \dfrac{\partial \tilde{X}_1}{\partial x} - \gamma \dfrac{\partial (\tilde{Y}_1 \phi)}{\partial \phi} = J(1)(\tilde{Y}_0 - \tilde{Y}_1) + J(1)\xi\delta(\phi - 1) - \gamma\phi_M \tilde{Y}_1(\phi - \phi_M) \end{cases}$$

where κ is the dimensionless parameter in equations

3.1.1. *A system containing small microdroplets.*

The criterion of a small size of microdroplets is short time t_M of settling of the equilibrium volume fraction of the

Monomer ϕ_M in PMPs relative to the duration of the first stage t_1; that is, $t_M < t_1$.

Figure 3 shows the PMP size distribution by the end of the first stage and the time dependence of microdroplet concentration in the system. It is evident that the PMP concentration increases almost linearly with time nearly to the end the first stage ($t < t_1$), and the microdroplet concentration decreases also almost linearly; only near the end of the first stage, a sharp decrease in the microdroplet concentration down to zero occurs.

These results are derived under the assumption that the time of settling of the equilibrium volume fraction of the monomer, ϕ_M, in a PMP is much shorter than the duration of the first stage. Using the expression for the dependence of the first-stage duration on parameters of the system, we find that the above calculations are valid for systems whose parameters satisfy the inequality

$$t_M = \frac{V_0}{k_p} \ln \frac{1}{\phi_M} << \frac{M_0}{\rho} \kappa^{-2/5} \approx t_1 \tag{13}$$

or, since $\phi_M \approx 0.5$ and $d_P/d_M \approx 1$, this inequality is equivalent to inequality $\kappa >> 1$.

Thus, if inequality (13) is fulfilled (the system consists of small microdroplets), then the PMP concentration will depend only on the initiator and emulsifier concentrations and will not depend on the size of microdroplets.

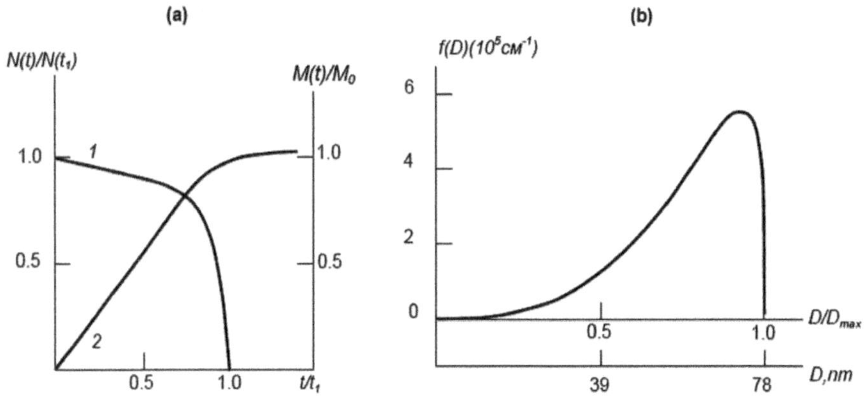

Figure 3. (a) Time dependences of the concentrations of (*1*) microdroplets and (*2*) PMPs and (b) PMP size distribution by the end of the first stage: $D_{max} = D_0(1+\theta t_1 / V_0)$, and t_1 is the time of exhaustion of microdroplets.

3.1.2. A system containing large microdroplets

If the original emulsion system contains microdroplets with a large diameter ($\kappa \leq 1$), throughout the first stage (at certain sizes, including a part or the entire second stage), the volume fraction of the monomer in PMPs formed from these microdroplets will be higher than equilibrium, $\phi > \phi_M$. In this case, PMPs will not absorb the monomer from emulsion droplets and, consequently, change their volume. Thus, the surfactant adsorbed on the surface of microdroplets is not consumed for the stabilization of PMPs, and the finite number of PMPs will be equal to the number of microdroplets in the original system.

In this system, the concentration of PMPs with $\phi = \phi_M$ is zero, $X_i = 0$, and the polymerization process is described by equations for functions $Y_i(\phi, t)$, which, in the approximation of fast termination, in dimensionless variables, have the form

$$\begin{cases} \frac{\partial \tilde{Y}_0}{\partial x} = \tilde{Y}_1 - \tilde{Y}_0 \\ \frac{\partial \tilde{Y}_1}{\partial x} - \gamma \frac{\partial}{\partial \phi}(\phi \tilde{Y}_1) = \tilde{Y}_0 - \tilde{Y}_1 + \xi\delta(\phi - 1) \end{cases} \tag{14}$$

Let us use $n_0(x) = \int \tilde{Y}_0 d\phi$ and $n_1(x) = \int \tilde{Y}_1 d\phi$ to denote the fractions of dead and alive PMPs in the system, respectively (relative to the total number of PMPs and microdroplets equal to M_0); $W_0 = \int \phi \tilde{Y}_0 d\phi$ and $W_1 = \int \phi \tilde{Y}_1 d\phi$. The PMP concentration in the system and the polymerization rate are expressed in terms of functions n_0, n_1, W_0, W_1.

For the time dependence of conversion, from the condition of balance for the monomer, we have

$$p(t) = \frac{d_P}{d_M} \frac{M_0 V_0}{(r+1)^{-1}} \left((n_0 + n_1) - (W_0 + W_1) \right) \tag{15}$$

Figure 4 shows the time dependence of conversion for the original system of large microdroplets, which is calculated through formula (15), and, for comparison, the time dependence of conversion for the original system of small microdroplets. It is evident that, in the approximation of fast termination, the polymerization occurs faster in the system with small microdroplets.

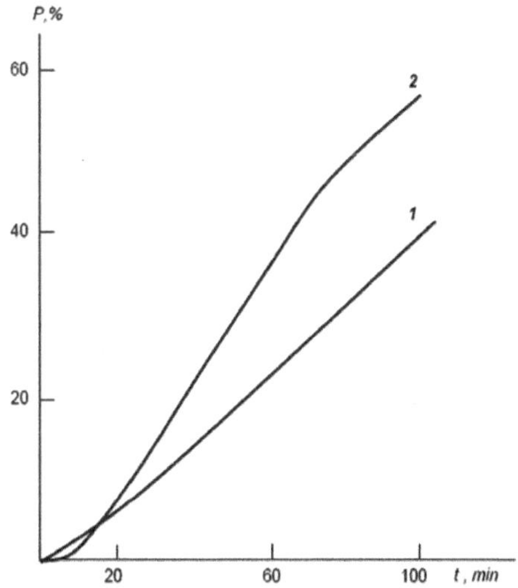

Figure 4. Time dependence of monomer conversion in systems with microdroplets of different sizes: D0 = (*1*) 150 and (*2*) 10 nm, a_m = 0.4 nm^2, $S_E = 2 \bullet 10^{-4}$ mol cm^{-3}, and $\varrho = 10^{12}$ cm^{-3} s^{-1}.

Figure 5 shows the calculated dependence of the final concentration of PMPs in the system on the size of microdroplets for the case in which the rate of radical entry into particles is proportional to their surface. The characteristic size of microdroplets, D_C, starting from which the PMP concentration decreases, is determined by the value of parameter $\kappa = 1$; therefore, it depends not only on D_0 but also on the initiation rate and the emulsifier concentration. For typical values of concentrations of the components, this size is $D_C \approx$ 40–60 nm.

To determine the effect of microdroplet size on the PMP size distribution through solution of system of equations (12)–(13), we calculated also the PMP size distribution by the end of the first stage for different values of D_0 and changes in the distributions during polymerization. Figure 6 represents the PMP diameter distribution by the end of the first stage; Fig. 6 depicts a change in the distribution during polymerization, it shows that the PMP size distribution becomes narrower as the size of microdroplets increases.

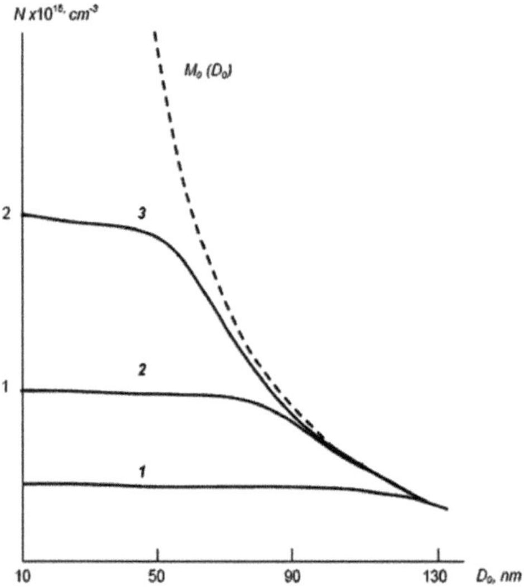

Figure 5. Dependence of PMP concentration in the system on the initial size of microdroplets: $\varrho = 10^{11}$ *(1)*, 10^{12} *(2)* and $5x10^{12}$ cm^{-3} s^{-1} (3) , $a_m = 0.4$ nm^2, $S_E = 2 \bullet 10^{-4}$ mol cm^{-3}.

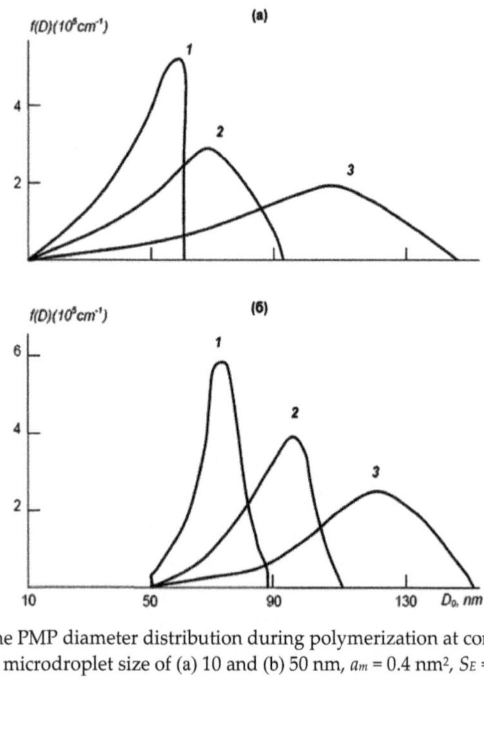

Figure 6. Change in the PMP diameter distribution during polymerization at conversions *P* = (*1*) 5, (*2*) 15, and (*3*) 50%. Initial microdroplet size of (a) 10 and (b) 50 nm, $a_m = 0.4$ nm^2, $S_E = 2 \bullet 10^{-4}$ mol cm^{-3}, and $\varrho = 10^{12}$ cm^{-3} s^{-1}.

3.2. A bidisperse system of microdroplets

To determine the mechanisms of polymerization in the case of comparable rates of PMP formation from micelles and microemulsion droplets, we study a bidisperse system that contains M_0 microemulsion droplets with diameter D_0 and μ_0 micelles with diameter d_0 and calculate the polymerization process in this system.

Depending on the method of introducing the emulsifier into the system, the formation of new droplets of the microemulsion can occur during polymerization; however, owing to the absence of experimental data on the rate of microemulsification in these systems and its dependence on the concentration of the free emulsifier and its distribution between phases, we restrict ourselves to the discussion of a system with a fixed number of microemulsion droplets. Consideration for the finite rate of formation of the microemulsion will only lead to a change in the duration of the stage of PMP formation and slightly affect other characteristics of the process.

To describe the rate of polymerization and the concentration of PMPs in the system under discussion, we introduce parameter Γ that characterizes the fraction of the surfactant adsorbed on the microemulsion surface:

$$\Gamma = \frac{\pi M_0 D_0^2}{N_A S_E a_m(D_0)} \tag{16}$$

where M_0, D_0, μ_0, and d_0 are related as follows

$$\frac{\pi M_0 D_0^2}{a_m(D_0)} + \frac{\pi \mu_0 d_0^2}{a_m(d_0)} = S_E N_A$$

We shall divide all PMPs in the system into two types: PMPs formed from micelles (with subscript μ) and PMPs formed from the microemulsion droplets (with subscript M):

$$f_i(v,\phi,t) = f_{\mu,i}\delta(\phi - \phi_M) + f_{M,i}(v,\phi,t)$$

Here, we take into account that, in PMPs formed from micelles, the equilibrium volume fraction of the monomer, ϕ_M, is settled almost immediately. By time t, the volume fraction of the monomer of a portion of the PMPs formed from microemulsion droplets will be equal to ϕ_M and that of another portion will be higher than ϕ_M; that is, $f_{M,i}$ can be represented as the sum

$$f_{M,i} = X_i(v,t)\delta(\phi - \phi_M) + Y_i(\phi,t)\delta(v - V_0)$$

the total number of PMPs formed from the microemulsion droplets, N_M, is equal to the initial number of droplets, M_0, and the number of PMPs formed from micelles, N_μ, is $N_\mu = \mu\kappa^{-2/5}$ at $\kappa >> 1$. The total number of PMPs per unit volume of the system by the end of the first stage is

$$N = N_{\mu} + N_{M} = \mu_0\kappa^{-2/5} + M_0 = N_{\mu}^{0}\left[1-\Gamma(1-\frac{d_0}{D_0}\kappa^{-2/5})\right] \quad (17)$$

where

$$N_{\mu}^{0} = \frac{S_E a_m N_A}{\pi d_0^2}\kappa^{-2/5}$$

is the total number of PMPs formed in the system if it contained only micelles. Since for large particles, for which the above formulas are valid, $M_0\theta/V_0\varrho < 1$, then the total number of PMPs is less than N_{μ}^{0}.

The PMP size distribution is the sum of two distributions: the distribution corresponding to PMPs formed from micelles and the distribution corresponding to PMPs formed from microemulsion droplets. The index of diameter polydispersity by the end of the first stage is derived through calculation of the average diameter and the square of the average diameter of PMPs:

$$\begin{cases} \bar{D} = \frac{3}{4}d_0\kappa^{1/5}\dfrac{1+\frac{4}{3}\frac{M_0 D}{\mu_0 d}\kappa^{1/5}}{1+\frac{M_0}{\mu_0}\kappa^{2/5}} \\ \overline{D^2} = \frac{3}{5}d_0^2\kappa^{2/5}\dfrac{1+\frac{5}{3}\frac{M_0}{\mu_0}\left(\frac{D_0}{d_0}\right)^2}{1+\frac{M_0}{\mu_0}\kappa^{2/5}} \\ K_D = \left(\dfrac{\overline{D^2}}{\bar{D}^2}\right)^{1/2} = \dfrac{\left(1+\frac{5}{3}\frac{\Gamma}{1-\Gamma}\right)\left(1+\frac{\Gamma}{1-\Gamma}\left(\frac{d_0}{D_0}\right)^2\kappa^{2/5}\right)}{\left(1+\frac{4}{3}\frac{\Gamma}{1-\Gamma}\frac{d_0}{D_0}\kappa^{1/5}\right)^2} \end{cases} \quad (18)$$

Figure 7 shows the dependence of K_D on Γ for $\kappa = 10^4$. It is evident that index K_D by the end of the first stage depends on the d_0/D_0 ratio and has maximum.

Figure 8 shows the dependence of the finite number of PMPs formed in the system on parameter Γ at different values of d_0/D_0. It is evident that, during an insignificant difference between the sizes of the two fractions, the PMP concentration hardly changes at all; however, when the size of large microdroplets is five (or more) times larger than the size of small microdroplets (solubilized micelles), the PMP concentration sharply decreases with an increase in Γ. Figure 9 depicts the size distributions of PMPs by the end of the first stage for various values of Γ and d_0/D_0. It is evident that the distribution is bimodal. Figure 10 shows that the PMP size distribution by the end of the first stage is bimodal in the general case. The peak relating to a smaller diameter corresponds to PMPs formed from micelles; that relating to a larger diameter corresponds to PMPs formed from microemulsion droplets. The transition from the bimodal size distribution of PMPs to the unimodal size distribution of PMPs by the end of the first stage occurs at a ratio of the initial sizes of micelles and microemulsion droplets of $d_0/D_0 \approx \kappa^{-1/5}$.

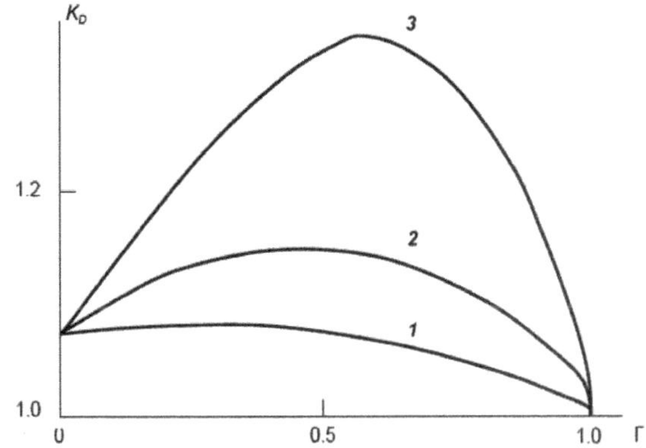

Figure 7. Dependence of the polydispersity index of the PMP size distribution on the amount of the surfactant adsorbed on the microemulsion surface (parameter Γ): a_m = 0.4 nm², $S_E = 2 \bullet 10^{-4}$ mol cm⁻³, and $\varrho = 10^{12}$ cm⁻³ s⁻¹. d_0 = 10 nm; D_0 = (1) 80, (2) 100, and (3) 150 nm.

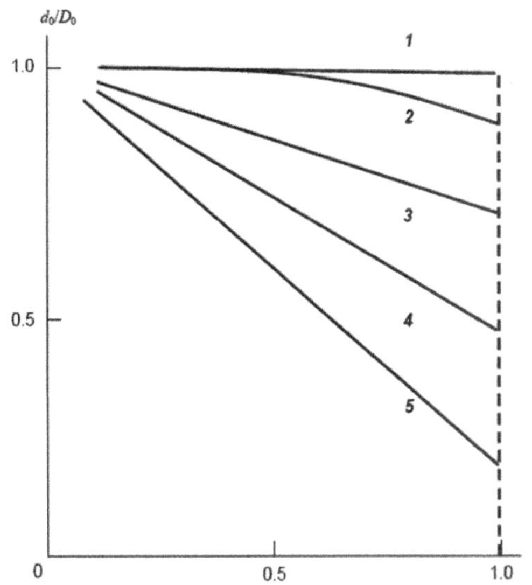

Figure 8. Dependence of PMP concentration on parameter Γ: d_0 = 10 nm; D_0 = 20 (1), 50 (2), 80 (3), 100 (4) and (5) 150 nm

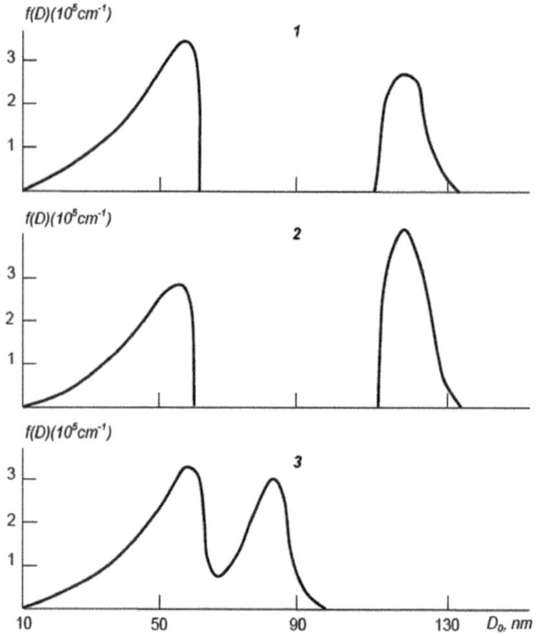

Figure 9. PMP size distribution in systems with simultaneous PMP formation from micelles and microemulsion droplets. The end of the first stage. Here, d_0 = 10 nm, D_0 = (*1*, *2*) 100 and (*3*) 50 nm, and Γ = (*1*) 0.25 and (*2*, *3*) 0.5.

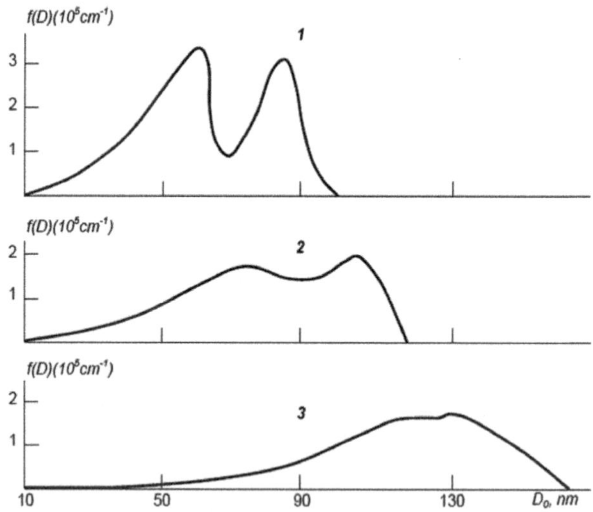

Figure 10. Change in the PMP diameter distribution during polymerization: d_0 = 10 nm, D_0 = 50 nm, and Γ = 0.5. Monomer conversion: (*1*) 3, (*2*) 15, and (*3*) 50%.

So a mathematical model of emulsion polymerization of poor soluble monomers has been developed. It takes into account the basic physio-chemical characteristics of the system and enables simultaneous calculating important parameters of the system such as polymerization rate, the time dependence of conversion, the concentration and size distribution of PMPs and molecular mass distribution of polymer[105]. The models predicts that with increasing microdroplet size D_0 above the critical value $D_C \sim$ 40–60 nm, the PMP concentration and the polymerization rate will decrease; at $D_0 < D_C$, both the polymerization rate and the PMP concentration do not depend on D_0. The PMP distribution becomes narrower with an increase in D_0. In the case of simultaneous formation of PMPs from micelles and monomer microdroplets, the PMP size distribution is broad or, under certain conditions, even bimodal.

Analysis of experimental data shows that monomer microdroplets play an important role which should not be ignored neither in studying mechanism of polymer-monomer particles nucleation in emulsion polymerization nor in modeling this complicated process.

Author details

M. Khaddazh*
Kurnakov Institute of General and Inorganic Chemistry Russian Academy of Sciences, Moscow, Russia

I. A. Gritskova
Lomonosov State University of Fine Chemical Technologies, Moscow, Russia

G. I. Litvinenko
Karpov Institute of Physical Chemistry, Moscow, Russia

4. References

[1] Harkins W D (1947) Theory of the Mechanism of the Emulsion Polymerization. J.Amer.Chem.Soc. 69: 1428-1444.
[2] Harkins W D (1950) General Theory of Mechanism of the Emulsion Polymerization. J. Polym. Sci. 5: 217-251.
[3] Yurjenko A I (1946) Physico-chemical investigation of hydrocarbons polymerization in emulsions (in Russian). J.Obschei Himii 16: 1171-1188.
[4] Yurjenko A I, Kolechkova M S (1947) Physico-chemical investigation of hydrocarbons polymerization in emulsions (in Russian). J.Obschei Himii17: 354-357.
[5] Smith W V, Ewart R M (1948) Kinetics of Emulsion. J.Chem.Phys.16: 592-599.
[6] Smith W V (1948) The Kinetics of Styrene Emulsion Polymerization. J.Am.Chem.Soc.70: 3695-3702.

* Corresponding Author

[7] Smith W V (1949) Chain initiation in Styrene Emulsion Polymerization. J.Am.Chem.Soc.70: 4077-4082.
[8] Khomikovsky P M (1958) Kinetics and topochemical particularities of emulsion polymerization (in Russian). Uspehi Himii 27: 1025-1055.
[9] Khomikovsky P M (1959) Elementary reactions in emulsion polymerization (in Russian). Uspehi Himii 287: 545-575.
[10] Sheinker A P, Medvedev S S (1956).Rubber Chem.Technol. 29: 121
[11] Berezhnoi G D, Khomikovsky P M Medvedev S S (1960) Kinetics of emulsion polymerization(in Russian). Vysokomol.Soed. 2: 5-17.
[12] Yeliseeva V I, Aslamazova T R (1991) (in Russian). Uspehi Himii. 2: 398
[13] Yeliseeva V I, Zharkova N G, Lukyanovich V M (1968) (in Russian). Dokl. AN SSSR 178: 1113-1116.
[14] Medvedev S S (1968) Emulsion polymerization. Kinetics, Mechanism and Transformation of Macromolecules (in Russian). Ed. KorshakV V. Moscow.
[15] Bovey F A, Kolthoff I M, Medalsa A J, Meehan E I (1955) Emulsion Polymerization. New York – London. Interscience Publ. 455.
[16] Qiu J, Charleux B, Matyjaszewski K (2001) Controlled/living radical polymerization in aqueous media: homogeneous and heterogeneous systems. Prog.Polym.Sci. Elsevier. 2083-2134
[17] Gardon J L (1968) Emulsion polymerization. I Recalculation and extension of the Smith-Ewart theory. J.Polym.Sci. Chem. 6: 623-641.
[18] Gardon J L (1968) Emulsion polymerization. II Review of experimental data in the context of the revised Smith-Ewart theory. J.Polym.Sci. Chem. 6: 643-664.
[19] Gardon J L (1968) Emulsion polymerization. III Theoretical prediction of the effects of slow termination rate within latex particles.J.Polym.Sci.Chem. 6: P. 665-685.
[20] Gardon J L (1968) Emulsion polymerization. IV Experimental verification of the theory based of latex particles. J.Polym.Sci. Chem. 6: 687-710.
[21] Gardon J L (1968) Emulsion polymerization. V. Lowest theoretical limits of the ratio k_t/k_p J.Polym.Sci. Chem. 6: 2853-2857.
[22] Gardon J L (1968) Emulsion polymerization. . VI. Concentration of monomer in latex particles J.Polym.Sci. Chem. 6: 2859-2879.
[23] Gardon J L (1977) Emulsion Polymerization.– In: Polym/ Process. New York. 143-197.
[24] Harada M., Nomura M, Kojima H, Eguchi W, Nagata S (1972) Rate of emulsion polymerization of styrene. J.Appl.Polym.Sci. 16: 811-833.
[25] Stockmayer W H (1957) Note on the kinetics of emulsion polymerization. J.Polym.Sci. 24: 313 – 317.
[26] O'Toole J T (1965) Kinetics of emulsion polymerization. J.Appl.Polym.Sci. 9:1291 – 1297.
[27] Hansen F K, Ugelstad J (1978) Particle nucleation in emulsion polymerization. I. A theory for homogeneous nucleation. J.Polym.Sci.Polym.Chem. 16: 1953.

[28] Hansen F K, Ugelstad J (1979) Particle nucleation in emulsion polymerization. IV. Nucleation in monomer droplets. J.Polym.Sci.Polym.Chem. 17: 3069-3078.

[29] Hansen F K, Ugelstad J (1982) Particle formation mechanism. In: Piirma I, editor. Emulsion polymerization. New York: Academic Press. 51-92.

[30] Kuchanov S I. (1975) Digest of Chemistry and Technology of High-Molecular Compounds Moscow. 7: 167.

[31] Kuchanov S I. (1978) Methods of Kinetic Calculations in Polymer Chemistry (in Russian). Moscow-Leningrad. Himiya. 368.

[32] Pismen L M, Kuchanov S I (1971) Qualitative theory of emulsion polymerization (in Russian). Vysokomol.Soed. A. 13: 1055-1065.

[33] Pavlyuchenko V N, Ivanchev S S (1981) Emulsion polymerization of non-polar monomers (Development of representation about kinetics and topochemistry) (in Russian). Uspehi Himii. 50: 715-745.

[34] Ivanchev S S, Pavlyuchenko V N, Rozhkova D A (1973) Surfactants as alkamons as components of redox initiating systems in emulsion polymerization of styrene (in Russian). DAN SSSR. 211: 885-888

[35] Priest W J (1952) Particle growth in the aqueous polymerization of vinyl acetate. 56: 1077–1083.

[36] Roe C P (1968) Surface chemistry aspects of emulsion polymerization. Ind.Eng.Chem. 60: 20 – 33.

[37] Yeliseyeva V I (1977) New Theory of Emulsion Polymerization. Sunthesis and Modification of Latexes (in Russian). Moscow, Himiya 3-9.

[38] Yeliseyeva V I, Ivanchev S S, Kuchanov S.I, Lebedev A I (1977) Emulsion Polymerization and its Applications in Industry (in Russian). Moscow, Himiya. 240.

[39] Fitch R M (1971) Homogeneous nucleation of polymer colloids. IV. The role of soluble oligomeric radicals. Polymer colloids, Plenum Press. New York.103 – 116.

[40] Fitch R M The homogeneous nucleation of polymer colloids. Br.Polym.J. -1973. - 5. – P. 467 – 483.

[41] Fitch R M (1973) Particle formation in polymer colloids. III. Prediction of the number of particles by a homogeneous nucleation theory. In: Fitch RM, editor. Polymer colloids. New York: Plenum Press.

[42] Fitch R M, Tsai C H (1970) Polymer colloids: particle formation in nonmicellar systems theory. J.Polym.Sci., Polym.Chem. B. 8: 703-710.

[43] Khomikovsky P M (1948) About mechanism of emulsion polymerization of vinyl acetate and methyl methacrylate in water and in soap solutions (in Russian). DAN SSSR. 60: 615-618.

[44] Christiansen J A, Nielsen A E (1952) The interplay between nucleus formation and crystal growth. Z. Electrochem. 56: 465-467.

[45] Pepard B (1974) Particle nucleation phenomena in emulsion polymerization of polystyrene. University Microfilms, – Ann Arbor.

[46] Goodal A R, Wilkinson M C, Hearn J (1975) Formation of anomalous particles during the emulsifier-free polymerization f styrene. J.Coll.Interface Sci. 53: 327.
[47] Cox R.A. , Wilkinson M C, Greasy J M (1977) Study of the anomalous particles formed during the surfactant-free emulsion polymerization of styrene. J.Polym.Sci., Polym.Chem.Ed. 15: 2311.
[48] Goodal A R, Wilkinson M C, Hearn J (1977) Mechanism of emulsion polymerization of styrene in soap-free systems. J.Polym.Sci., Polym.Chem.Ed. 15: 2193.
[49] Oganesyan A, Gukasyan A, Bayagyan V, Matsoyan S (1985) The generation place of growing radicals and their behavior in styrene – sodium persulate aqueous system (in Russian). Armenian Chem.J. 38: 331-333.
[50] Oganesyan A, Airapetyan K, Gukasyan A, Kinoyan (1986) About the chemistry of initiation reaction of monomer radicals in sodium persulate aqueous solution saturated with styrene (in Russian). DAN Armenian SSR. 82: 134-136.
[51] Oganesyan A (1986) Оганесян, A.A. Radical polymerization and phase formation in heterogeneous monomer/water systems. Dr. Sc. Thesis. Moscow. MITHT. 201.
[52] Tauer K , Kuhn I (1996) Modeling Particle Formation in Emulsion Polymerization: An Approach by Means of the Classical Nucleation Theory. Macromolecules. 28: 2236 - 2239.
[53] Tauer K (1998) Comment on the Development of Particle Surface Charge Density during Surfactant-Free Emulsion Polymerization with Ionic Initiators. Macromolecules. 31: 9390 - 9391.
[54] Tauer K , Kuhn I, Schellenberg C (1999) Tauer, K. A comprehensive experimental study of surfactant-free emulsion polymerization of styrene. Colloid Polym.Sci. 277: 607. – 626.
[55] Tauer K, Oz N (2004) Interfacial Energy Promotes Radical Heterophase Polymerization. Macromolecules. 37: 5880 - 5888.
[56] Tauer K , Schellenberg C, Zimmermann A (2000) Nucleation in heterophase polymerizations. Macromol. Symp. 150: 1-12.
[57] Kozempel S , Tauer K, Rother G (2005) Aqueous heterophase polymerization of styrene — a study by means of multi-angle laser light scattering. Polymer. 46: 1169 – 1179.
[58] Barrett K E (1975) Dispersion polymerization in organicmedia. Wiley-Interscience: Bristol.
[59] Lichti G, Hawkett B S, Gilbert R G, Napper D H, Sangster D F (1981) Styrene emulsion polymerization: particle-size distributions. J.Polym.Sci.Chem. 19: 925-938.
[60] Lichti G, Gilbert R G, Napper D H (1977) The growth of polymer colloids. J.Polym.Sci.Polym.Chem.Ed. 15: 1957-1971.
[61] Lichti G, Gilbert R G, Napper D H (1983) Mechanisms of latex particle formation and growth in the emulsion polymerization of styrene using the surfactant sodium dodecyl sulfate. J.Polym.Sci.Polym.Chem.Ed. 21: 269-291.

[62] Feeney P J, Napper D H, Gilbert R G (1984) Coagulative nucleation and particle size distributions in emulsion polymerization. Macromolecules. 17: 2520-2529.

[63] Feeney P J, Napper D H, Gilbert R G (1987) Surfactant-free emulsion polymerization: predictions of the coagulative nucleation theory.Macromolecules. 20: 2922-2930.

[64] Feeney P J, Napper D H, Gilbert R G (1985) Surfactant-free emulsion polymerization: predictions of the coagulative nucleation theory. J.Colloid Interface Sci. 107: 159-173.

[65] Deryagin B V, Churaev N V, Muller V M (1987) Surface Forces (in Russian). Moscow.Nauka.

[66] Adamson A W (1990) In Physical Chemistry of surfaces . J. Wileyand Sons. Inc. New York.

[67] Nielsen, A. E. Kinetics of Precipitation // Pergamon Press:Oxford. U.K.- 1964.

[68] Elias H G (1972) In Mdmmolekiile. Huthig & WepfVerlag:Basel,Heidelberg.

[69] Gunderl F, Wolf B A (1989) In Polymer Handbook. Wiley-Interscience. New York,.

[70] Barton A F M (1990) In CRC H of Polymer-Liquid Interaction Parameters and Solubility Parameters. CRCPRESS Inc. - Boca Raton.

[71] Hansen F K, Ofstad E B, Ugelstad J (1975) Emulsification of styrene with mixtures of anionic emulsifier and long chain fatty alkohols. Emulsion polymerization with initiation in monomer droplets. Theory and Practic of Emulsion Technology. Ed.by Smith London, Acad. Press. 13-25.

[72] El-Aasser M S, Lack C D, Chou Y T, Min T I, Vanderhoff J W, Fowkes F J (1984) Interfacial aspects of microemulsion polymers. Coll.S. 12: 79 - 97.

[73] El-Aasser M S, Miller C M (1997) Preparation of Latexes Using Microemulsions, in Polymeric Dispersions — Principles and Applications , J. M. Asua (Ed.). Kluwer, Dordrecht. 109–126.

[74] Gritskova I A, Sedakova L I, Muradyan G S, Pravednikov A N (1978) About topochemistry of emulsion polymerization (in Russian). DAN SSSR. 238: 607-610.

[75] Pravednikov A N, Gritskova I A, Muradyan G S, Margolin D N, Vassilev L A, Sedakova L I (1981) A new method of latex dispersion investigation (in Russian). Kolloid.J. 41: 595 - 597.

[76] Pravednikov A N, Simakova G A, Gritskova I A, Prokopov N I (1985) The formation of surfactant at the interface in emulsion polymerization (in Russian). Kolloid.J. 47: 192 - 194.

[77] Simakova G A, Kaminsky V A, Gritskova I A, Pravednikov A N (1984) Microemulsification in emulsion polymerization (in Russian). DAN SSSR. 276: 151 - 153.

[78] Gritskova I A (1979) Emulsion polymerization of monomers slightly soluble in water (in Russian). Dr. Sc. Thesis. Moscow. MITHT 304 C.

[79] Gritskova I, Marchenko S, Zaitsev S (2002) Polymer surfactant interfacial layer of polystyrene particles modeling by monolayer techniques. Coll.S. 198-200: 501-508.

[80] Hinze J O (1955) Fundamentals of the hydrodynamic mechanism of splitting in dispersion processes. Amer.Inst.Chem. Eng. J. 1: 289-295.

[81] Micellization, Solubilization, and Microemulsions/ Ed. by K.L. Mittal. N.Y.; L.: Plenum Press, 1977. 597.

[82] Yentov V M, Kaminsky V A, Lapiga Ye Ya (1976) Calculation of coalescence rate of emulsion in turbulent flow (in Russian). Izv. AN SSSR. Ser.Mech.Zhidk.Gas. 3: 47-55.

[83] Hicke H F, Rehark J (1976) On the formation of water/oil microemulsion. Helv.Chim.Acta. 59: 2883-2891.

[84] Rose S, Nishioka G (1975) Foaminees of binary and ternary solutions. J.Phys.Chem. 79: 1561-1565.

[85] Gierbacia W, Rosano H L (1973) Microemulsion formation and stabilization. J. Colloid Interface Sci. n 44: 242-248.

[86] Smith G D, Donelan C R, Barden R E (1977) Oil-continuous microemulsions composed of hexane, water and 2-propanol. J. Colloid Interface Sci. 60: 488-496.

[87] Bourrel M, Grasia A, Schecher R S (1979) The relation of emulsion stability to phase behavior and interfacial tension of surfactant systems. J. Colloid Interface Sci. 72: 161-163.

[88] Hall A C (1980) Interfacial tension and phase behavior in systems of petroleum sulfonate n-alkane. Coll.S. 1: 209-228.

[89] Rebinder P A (1946) About theory of emulsions (in Russian). Koll.J. 8: 157-174.

[90] Rebinder P A, Taubman A B (1961) Notes on the question about aggregative stability of disperse systems (in Russian). Koll.J. 23: 359-361.

[91] Simakova G A, Chalykh A Ye, Vasenin R M, Nikitina S A, Taubman A B (1969) Investigation of nonionic surfactants in water and xylene (in Russian). Koll.J. 31: 893-896.

[92] Chalykh A Ye (1987) Diffusion in Polymer Systems (in Russian). Moscow:Himiya. 312.

[93] Taubman A B, Nikitina S A (1962) Structural mechanical properties of surfactant surface layers and stabilization mechanism of concentrated emulsions (in Russian). Koll.J. 24: 633-635.

[94] Taubman A B, Nikitina S A (1965) Role of quasi-spontaneous emulsification in stabilation of emulsions (in Russian). Koll.J. 27: 291 - 295.

[95] Dudukin V V, Medvedev S S, Gritskova I A (1967) Styrene polymerization in the presence of nonionic surfactant polyoxyethylene polypropylene glycol (in Russian). DAN SSSR. 172: 1125 - 1127.

[96] Ishkov A, Cherkasov V, Prokopov N, Gritskova I (1996) Synthesis of polymer suspensions for immuno-chemical studies. J.Appl.Polym.Sci. 61: 1465-1471

[97] Gritskova I, Adebayu G, Krasheninnikova I, Kaminsky V (1998) The synthesis of polymer suspensions with narrow size distribution for immunochemistry. Coll.Polym.Sci. 276: 1068-1077.

[98] Lovell P A, El - Aasser M S (Eds.) (1997) Emulsion Polymerization and Emulsion Polymers. John Wiley & Sons , West Sussex , 1997 .

[99] Haward R N (1949) Polymerization in a system of discrete particles. J.Polym.Sci. 4: 273 – 287.

[100] Zhachenkov S V, Litvinenko G I, Kaminsky V A, Ilmenev P Ye, Pavlov A V, Guryanova V V, Gritskova I A, Pravednikov A N (1985) Particularities of styrene emulsion polymerization with various methods of initial emulsion preparation (in Russian). Vysokomol.Soed. 27: 1249 - 1253.

[101] Kaminsky V A, Gritskova I A (1996) Interface phenomena and formation of particles in emulsion polymerization. J.Phys.Khim. 70: 1516-1520.

[102] Gan L M, Lee K C, Chew C H, Ng S C (1995) Effects of Surfactant Concentration on Polymerizations of Methyl Methacrylate and Styrene in Emulsions and Microemulsions. Langmuir. 11: 449 - 454.

[103] Prokopov N I, Gritskova I A, Kiryutina O P, Khaddazh M, Tauer K, Kozempel S (2010) The mechanism of surfactant-free emulsion polymerization of styrene. Polym.Sci. Ser. B. 52: 339-345.

[104] Khaddazh M, Litvinenko G I, Gritskova I A (2011) A Theoretical Study of the Emulsion Polymerization of Styrene: The Effect of the Initial Dispersion of the System on the Polymerization Rate and Size Distribution of Polymer–Monomer Particles. Polym.Sci. Ser. B. 53: 209-222.

[105] Khaddazh M, Litvinenko G I, Gritskova I A (2011) A Theoretical Study of the Emulsion Polymerization of Styrene: The Effect of the Initial Dispersion of the System on the molecular-mass distribution of polystyrene. Polym.Sci. Ser. B. 53: 283-291.

[106] Gritskova I A, Zhachenkov S V, Tsar'kova M S, Levachev S M, Simakova G A, Khaddazh M, Prokopov N I (2011) Styrene polymerization in the presence of different initiators and properties of polymer suspensions. Polym.Sci. Ser. B. 53: 568-577.

[107] Khaddazh M, Zhachenkov S V, Gritskova I A (2011) The influence of method of preparation on kinetic characteristics of styrene emulsion polymerization (in Russian). Vestnik MITHT. 6: 97-101.

[108] Zhachenkov S V, Gritskova I A, Levachev S M, Khaddazh M, (2011) Particularities of styrene heterogeneous polymerization under statistic conditions in the presence of emulsifiers and co-surfactants. Vestnik MITHT. 6: 85-88.

[109] Oganesyan A A, Melkonyan L G, Gritskova I A, Bagdasaryan R V (1971) Dependence of rate of initiator-free chloroprene emulsion polymerization on the concentration of emulsifier (in Russian). Armenian Khim.J. 3: 284-288.

[110] Oganesyan A A, Melkonyan L G, Gritskova I A, Bagdasaryan R V (1971) About mechanism and Topochemical particularities of chloroprene emulsion polymerization (in Russian). Armenian Khim.J. 3: 660-667.

[111] Gilbert R G, Napper D H (1983) The direct determination of kinetic parameters in emulsion polymerization system. J.Macromol.Sci.Rev.Macromol.Chem.Phys. 23: 127-186.

[112] Candau F (1987) in Encyclopedia of Polymer Science and Engineering. Mark F, Bikales N M, Overberger C G, Mengers G (Eds.). Wiley Interscience New York. 718-724 .

[113] Chern C S (2002) in Encyclopedia of Polymer Science and Technology.. Kroschwitz J I (Ed.), John Wiley & Sons. New York.

[114] Guo J S, Sudol E D, Vanderhoff J W, El Aasser M S (1992) Kinetics and Mechanism of Styrene Microemulsion Polymerization /in Polymer Latexes: Preparation, Characterization, and Applications, Daniels E S, Sudol E D and El-Aasser M S (Editors), ACS Symposium Series. 492: 99-113.

[115] Tang P L, Sudol E D, Adams M E, Silebi C A, El Aasser M S (1992) Miniemulsion Polymerization / P.L. Tang, E.D. Sudol, M.S. El Aasser /in Polymer Latexes: Preparation, Characterization, and Applications, Daniels E S, Sudol E D and El-Aasser M S (Editors), ACS Symposium Series. 492: 72-98.

[116] Guo J S, El Aasser M S, Vanderhoff J W (1989) Microemulsion Polymerization of Styrene. J.Polym.Sci.:Part A: Polym.Chem. 27: 691-710.

[117] Hansen F K, Ugelstad J (1978) Particle nucleation in emulsion polymerization. II. Nucleation in Emul-sifier-free Systems Investigated by Seed Polymeriza-tion. J.Polym.Sci.Polym.Chem. 16: 3033-3046.

[118] Hansen F K, Ugelstad J (1978) Particle nucleation in emulsion polymerization. III. Investigation of Nucleation in Systems with Anionic Emul-sifier by Seeded and Unseeded Polymer-ization. J.Polym.Sci.Polym.Chem. 16: 3047-3068.

[119] Song Z, Poehlein G W (1989) Particle nucleation in emulsifier-free aqueous-phase polymerization : Stage 1. J. Coll.Interf.Sci. 78: 486-500.

[120] Song Z, Poehlein G W (1990) Kinetics of emulsifier free emulsion polymerization of styrene. J.Polym.Sci. 28: 2359-2392.

[121] Chern C S (2008) Principles and applications of emulsion polymerization. John Wiley & Sons, Inc., Hoboken, New Jersey. 252.

[122] Gao J., Pendilis A (2002) Mathematical modeling and computer simulator/database for emulsion polymerizations. Prog. Polym.Sci. 27: 403-535.

[123] Fitch R M, Tsai C H (1971) Particle formation in polymer colloids. III. Prediction of the number of particles by a homogeneous nucleation theory. In: Fitch RM, editor. Polymer colloids. New York: Plenum Press, 73-102

[124] Fitch R M, Tsai C H (1971) Homogeneous nucleation of polymer colloids. IV. The role of soluble oligomeric radicals. In: Fitch R M, editor. Polymer colloids. New York: Plenum Press. 103-116.

[125] Maxwell I A, Morrisson B R, Napper D H, Gilber R G (1991) Entry of free radicals into latex particles in emulsion polymerization. Macromolecules.;24:1629.

[126] Dougherty E P (1986) The SCOPE dynamic model for emulsion polymerization I. Theory. J.Appl.Polym.Sci. 32: 3051 -3078.

[127] Penlidis A, MacGregor J F, Hamielec A E. (1986) Mathematical modeling of emulsion polymerization reactors: a population balance approach and its applications. ACS Symp Ser. 313: 219—240.

[128] Yeliseyeva V I, Zuikov AV (1982) Interfacial phenomena in of polar monomers. In:Priima I, GardonJL, editors. Emulsion polymerization. ACS Symp Ser. 24: 62-81

[129] Penboss I A, Napper D H,. Gilbert R G (1983) Styrene emulsion polymerization. J.Chem.Soc.Faraday Trans. 79: 1257-1271.

[130] Bataille P, Gonzalez A (1984) Emulsion polymerization of styrene. III. Effect of Ti(III) and Cl(I) on the polymerization of styrene initiated by potassium persulfate. J.Polym.Sci.Chem.Ed. 22: I409-1417.

[131] Penlidis, A. Mathematical modeling of emulsion polymerization reactors: a population balance approach and its applications / A. Penlidis, J.F. MacGregor, A.E. Hamielec // ACS Symp. Ser. – 1986. – 313. – P. 219-240.

[132] Sudol E D, El-Aasser M S, Vanderhoff J W (1986) Kinetics of successive seeding of monodisperse polystyrene latexes. I. Initiation via potassium persulfate. J.Polym.Sci.Chem.Ed. 24: 3499-3513.

[133] Sudol E D, El-Aasser M S, Vanderhoff J W (1986) Kinetics of successive seeding of monodisperse polystyrene latexes. II. Azo initiators with and without inhibitors. J.Polym.Sci.Chem.Ed. 24: 3515-3527.

[134] Giannetti E (1993) Nucleation mechanisms and particle size distributions of polymer colloids. AIChE J. 39: 1210—1227.

[135] James Jr. H L, Piirma I (1975) Molecular weight development in styrene and methyl methacrylate emulsion polymerization. Polym.Prepr. 16: 198-204.

[136] Piirma I, Kamath V R, Morton M (1975) The kinetics of styrene emulsion polymerization. Polym.Prepr. 16: 187-191.

[137] Miller C M, Clay P A, Gilbert R G, El-Aasser M S (1997) An experimental investigation on the evolution of the molecular weight distribution in styrene emulsion polymerization. J.Polym.Sci.Chem.Ed. 35: 989-1006.

[138] Said Z F M. (1991) Molecular-weight distributions in the thermally initiated emulsion polymerization. Makromol.Chem. 192: 405-414

[139] Salazar A, Gugliotta L M, Vega J R, Meira G R (1998) Molecular weight control in a starved emulsion polymerization of styrene. Ind.Eng.Chem.Res. 37: 3582-3591.

[140] Morbidclli M, Storti G, Carra S (1983) Role of micellar equilibria on modeling of batch emulsion polymerization reactors. J.Appl.Polym.Sci. 28: 901-919.

[141] Bataille F., Bataille P, Fortin R (1988) Emulsion polymerization of styrene. IV. Effect of Co(II), Ni(II), and Pb(ll) on the polymerization of styrene initiated by potassium persulfate. J.Polym.Sci.Chem.Ed. 26: 1471-1477.

[142] Thickett S C, Gilbert R G (2006) Mechanism of radical entry in electrosterically stabilized emulsion polymerization systems. Macromolecules. 39: 6495.

[143] Kusters J M H, Napper D H, Gilbert R G, German A L (1992) Kinetics of particle growth in emulsion polymerization systems with surface-active initiators. Macromolecules. 25: 7043-7050.

[144] Anderson C D, Sudol E D, El-Aasser M S (2003) Elucidation of the miniemulsion stabilization mechanism and polymerization kinetics. J.Appl.Polym.Sci. 90: 3987-3993.

[145] Min KW, Ray W H (1974) On the mathematical modeling of emulsion polymerization reactors. J.Macromol.Sci,Rev.Macromol. C11: 177-255.

Chapter 10

Cationic Polymerization of Vinyl Monomers Under the Action of Metalorganic Compounds

Alexey Lyapkov, Elena Ionova, Vladimir Bondaletov and Alexey Pestryakov

Additional information is available at the end of the chapter

1. Introduction

The most part of known metalorganic compounds with Metal-carbon bond of σ-p type belongs to Lewis acid class. These compounds are the most effective catalysts of cationic polymerization of metalorganic monomers. Explanation of initiative action of such systems is very complicated because of their complex structure as well as contaminations in the reaction mixture.

Formation of active site is the main process in all initialization reactions (including metalorganic Lewis acid action). The process of interaction of vinyl monomers with metalorganic Lewis acid (MLA) consists of several equilibrium stages [1], such as formation of active charged particles and their counter ions and subsequent addition of the charged pair to the monomer. The initial equilibrium is usually slow process while the further attack of the monomer (initialization) is very fast and is comparable with the rate of the reaction of chain growth. Moreover, presence of the cationogenic contaminations in the reaction mixture causes a lot of confusion to chemistry of cationic processes. Now we can consider more or less confidently only two possible ways of initialization of polymerization of vinyl monomers by MLA: direct initialization and co-initialization [2-5].

2. Initiators and co-initiators

During co-initialization the charged particles are formed as e result of bimolecular reaction or heterolytic decomposition of a molecule:

$$SR + AlR'_2Cl \rightleftarrows R^+AlR'_2ClS^-$$

where SR – cationogen (H_2O, HHal, R-Hal, etc.).

After injection of cationogen into the last stage the complexing between the monomer and MLA is completing; and initialization in this case consists in the cationogen attack on the complex:

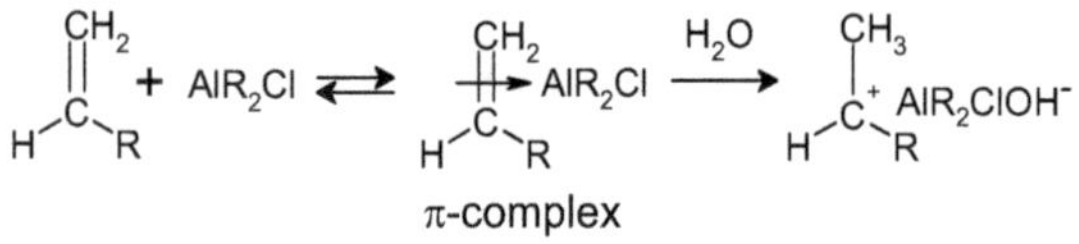

The π-complex can cause the initialization if it is transformed to isomeric carbenium ion:

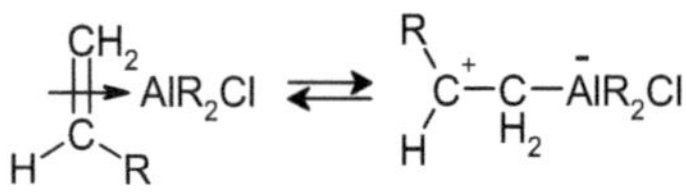

Such regrouping is possible if the electronic density of the double bond is small. Therefore, the heightened electronic density of the double bond of asymmetric substituted alkenes has to decrease significantly at interaction with MLA.

MLA can be divided into 2 types. The first type is presented by MLA of B and Al who has no d-electrons in the valence sphere of the central atom. The second one consists of compounds containing heavier atoms with d-electrons. Usage of MLA of the 2nd type makes possible formation of the π-complex from monomer and its further interaction due to overlapping of full d-orbitals of MLA and empty antibonding orbitals of vinyl monomer. Such interactions strengthen the complex and complicate transformation into the carbenium ions because of increasing the electron density of the double bond.

Thus, to start the initialization process it is usually necessary presence of cation donor. Role of MLA consists in assistance to cationogen to form the initiating particle by the way of complexing [6]. According to Ref. [7] MLA allows stabilizing and removing the anion; and initialization is caused by cationogen who is a principal initializer while MLA is only co-initializer.

A mechanism of initialization by alkylhalogenides is the most studied now. Role of R-Hal was cleared on the example of styrene polymerization because intermediate carbenium ion of styrene is rather stable [8]:

$$RCl + AlEt_2Cl \rightleftarrows R^+AlEt_2Cl_2^- \qquad 1.1$$

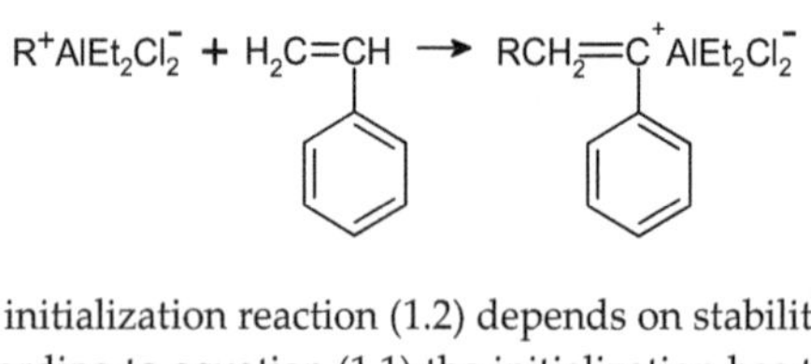

1.2

Efficiency of the initialization reaction (1.2) depends on stability of the carbenium ion and its availability. According to equation (1.1) the initialization has to be favored by R-Hal (or Ar-Hal) having low bond dissociation energies and low ionization potentials of alkyl groups

[9,10] as well as by solvatation of the formed ions [11]. In the case of R-Hal producing primary or secondary carbonium ions (e.g., n-butylchloride or isopropylchloride) the direct reaction is not observed. If such very active ions can be formed they have to initiate styrene polymerization and produce comparatively stable styrene cations:

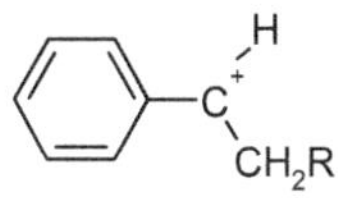

Tret-butylchloride forms rather stable tertiary carbonium cation. However, it is less stable as compared with styrene cation; that is why fast initialization takes place. In contrast, triphenylchloromethane forms very stable tret-cation which cannot cause styrene polymerization [9]. Thus, carbonium ion has to be less stable as compared with styrene cation, but not very unstable to exclude its formation from R-Hal and MLA.

Application of suitable R-Hal and Ar-Hal (together with MLA co-initiators) gives opportunity to control the initialization process [12,13]. In this case a type of alkylhalogenide influences structure of head group of the polymer:

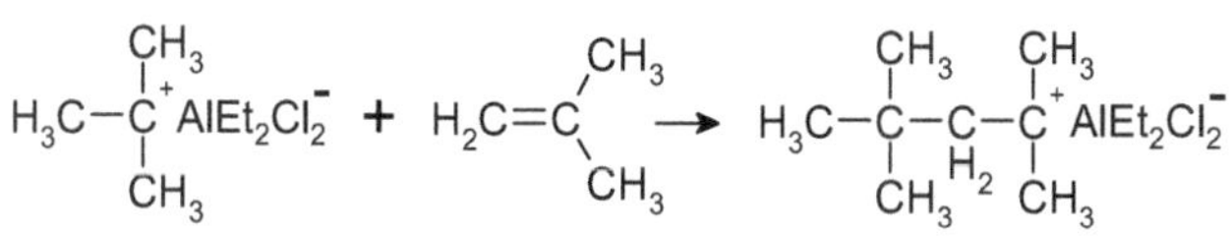

Now it is known a great number of cationogens which can efficiently initiate cationic polymerization of vinyl monomers and ensure production of polymers with certain head group.

In the case of absence of cation donors (e.g., at high pure conditions) other mechanisms of initialization of polymerization of vinyl monomers may be realizes under the action of MLA. For example, if the monomer contains allyl hydrogen tearing off hydride-ion from the monomer may be one of possible ways of initialization [14]:

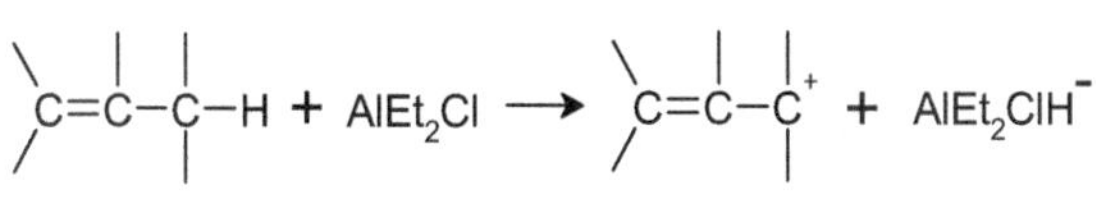

In this case the monomer is a donor of cations that is one of self-initialization mechanisms.

3. Direct initiation

The formation of the metal-carbon bonds in the polymer can occur in various ways:

1. Direct conjunction of MLA to the double bond of monomer followed ejection of cationogen and the formation of complex counter ion [15]

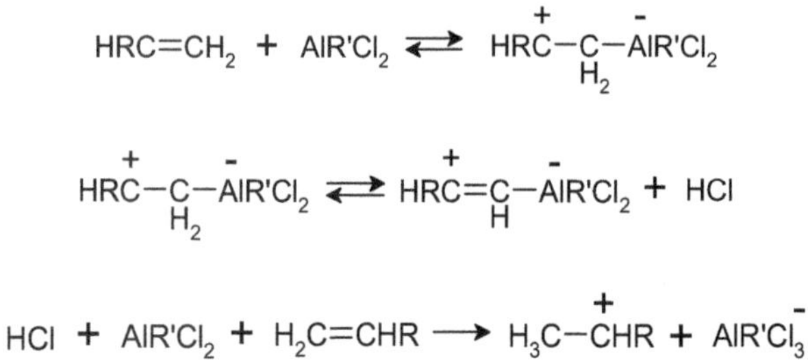

2. Electrophilic addition to the monomer cation, formed as a result of MLA ionization itself [16–18]

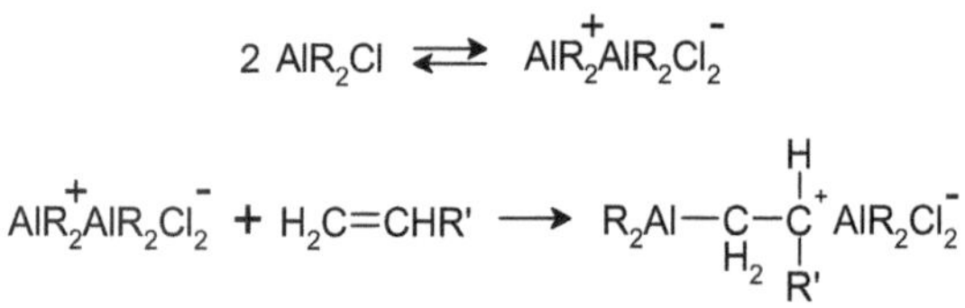

The requirement of initiation according to the scheme 1 is a possibility of formation of cationogen at zwitterionic decay; according to scheme 2 – the presence of a polar solvent or electron-donating compounds such as ethers, thioethers, tertiary amines, etc. MLA easily adds such compounds forming a rather stable complexes [19]. Such complexes usually have a composition of 1:1. The NMR study of complexation of organoaluminum compounds with methylpyridine (D) in solutions of 1,2-dichloroethane and dichloromethane showed, that the solution may contain the following complex compounds [20]:

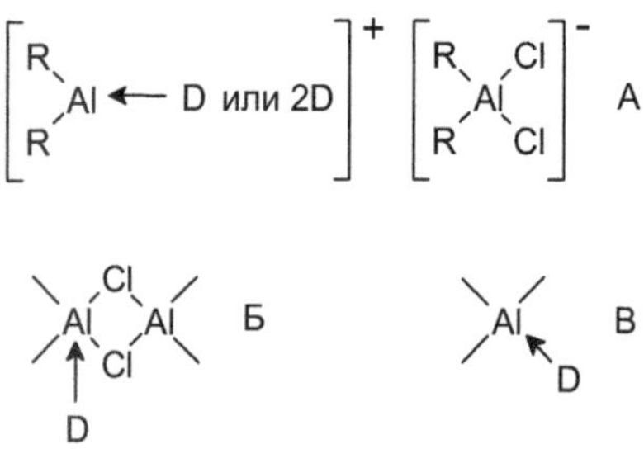

The presence in the solution structures of type A is also confirmed by determination of electrical conductivity.

By Wittig [21], one of the possible states of the MCL in the solutions are structures of $MeR^+_{n-1}MeR^-_{n+1}$. Hypothesis about the possibility of ionization of the MLA to form complex ions, including metal atom, is consistent with a number of electrochemical studies. The corresponding ionization schemes are proposed, for example, magnesium-organic compounds [22]:

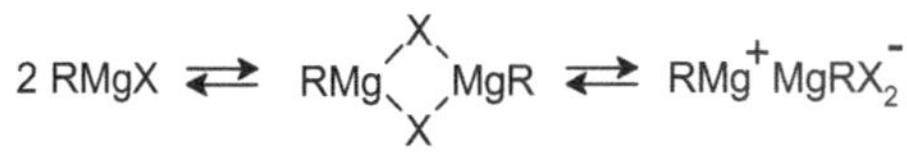

or

$$2\ RMgX \rightleftarrows Mg^{2+}\ MgR_2X_2^{-}$$

Such MLA ionization is characteristic not only for the individual MLA, but also for their mixtures. It was found that polymerization of butadiene in the presence of alkylaluminum-cobalt initiators is the most affective only in the presence of ion-pair $Et_2AI^{+}EtCI_3AI^{-}$. This ion pair is formed by mixing either $AIEt_2CI$ with $AIEtCl_2$, or $AlEt_3$ with $AlCl_3$ [6]:

$$AlEt_2Cl + AlEtCl_2 \rightleftarrows Et_2Al^{+}AlEtCl_3^{-}$$

$$AlEt_3 + AlCl_3 \rightleftarrows Et_2Al^{+}AlEtCl_3^{-}$$

At present, the mechanism of the formation of MCL self-ionization ion pair is widespread [4, 23]. However, the equilibrium constant for the reaction

$$2\ AlEt_2Cl \rightleftarrows Et_2Al^{+}AlEt_2Cl_2^{-}$$

Is much lower than in the case of interaction of MLA with a suitable donor of cations. For example:

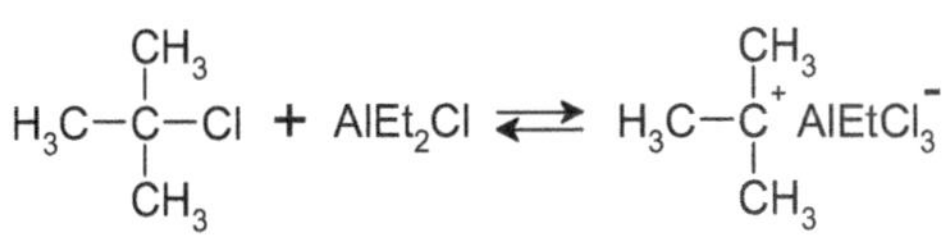

In this case a more stable carbonium ion is formed. Therefore, in order to a direct initiation we need to create conditions for guarantee the complete absence of cationogen impurities in the system. Only in this case MLA themselves can play a role of initiators.

Nevertheless, the polymerization of highly active monomers by initiation of the MCL self-ionization is possible. One of these monomers is a 9-vinylcarbazole (VC), it is very easy polymerizing for cationic mechanism [24–26]. Authors of Ref. [27] showed that the rate of chain growth in cationic polymerization of VC under the action of stable organic cations in more than 100 times higher than the corresponding value for vinyl ethers.

Authors [28] studied kinetics of polymerization of VC in toluene solution under the action of titanium tetrachloride by thermometric method [29-33] and found that the polymerization process begins with the formation of complex compounds $TiCl_4$ with 1-6 solvent molecules forming the solvation shell. It was established that complex formation is accompanied by a large thermal effect, which is reflected in the curve recorded a sharp jump in temperature at the initial time of polymerization. Since the system contains monomer (VC) with a developed system of π-electrons, the latter one competes with toluene and partially displaces it from the solvation shell of $TiCl_4$. As a result of new resolvation a complex compound of a rather complex structure is formed where the vinyl bond of the monomer is partially polarized. This facilitates the subsequent process of opening the double bond and leads to increasing reactivity of the monomer.

The active particle is formed by the further polarization of the monomer vinyl bond in the solvation shell of titanium tetrachloride. This is followed by the accession of the last one to the monomer double bond forming the corresponding ion (direct initiation). Direct connection $TiCl_4$ to monomer is confirmed by data of X-ray fluorescence analysis of the polymer samples. Diffractogram shows that the polymer contains a small amount of titanium chloride as terminal groups (Fig. 1).

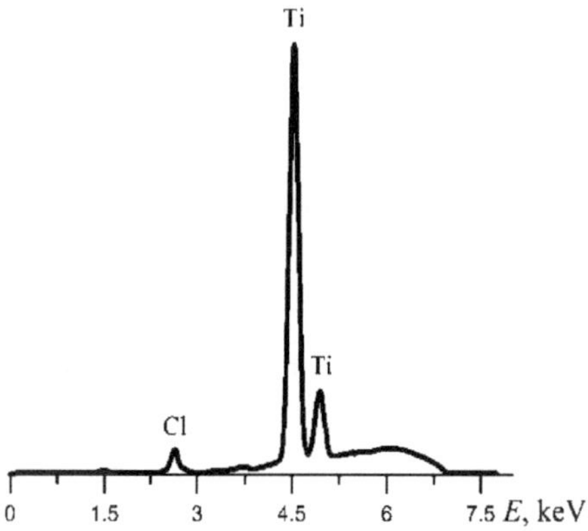

Figure 1. X-ray Fluorescence Spectrum of the sample Polyvinylcarbazole obtained under the action of $TiCl_4$ in toluene solution

Concentration of active sites may increase during the slower stage; it causes S-shaped kinetic curves. The chain growth on contact or separated ionic pairs is the most probable mechanism of the polymerization in the studied system. It is confirmed by permanency of the chain growth constant depending on the initial concentration of the initiator.

Initialization of the monomer polymerization by the way of ionization of complex Lewis acids can be illustrated by oligomerization of dicyclopentadiene (DCPD) under the action of catalytic system $TiCl_4$: Et_2AlCl [34–35]. This catalyst (similar to individual $TiCl_4$ [30]) causes process of cationic polymerization of DCPD. Lower value of k_P as compared with the process based on individual $TiCl_4$ can be explained by the presence of less hard counter ion on the base of $AlEt_2Cl$.

In both cases dicyclopentadiene polymerization goes according to the following mechanism:

Firstly, active complex (solvated ion pair) is formed as a result of self-ionization of Lewis acids:

$$Et_2Al{-}Cl + TiCl_4 \longrightarrow TiCl_3^{+} + AlEt_2Cl_2^{-} \xrightarrow{+M} TiCl_3M^{+}AlEt_2Cl_2^{-}$$

Also, we can suppose formation of π-complex (I) MLA with cyclopentadiene which is present at the system in small amount:

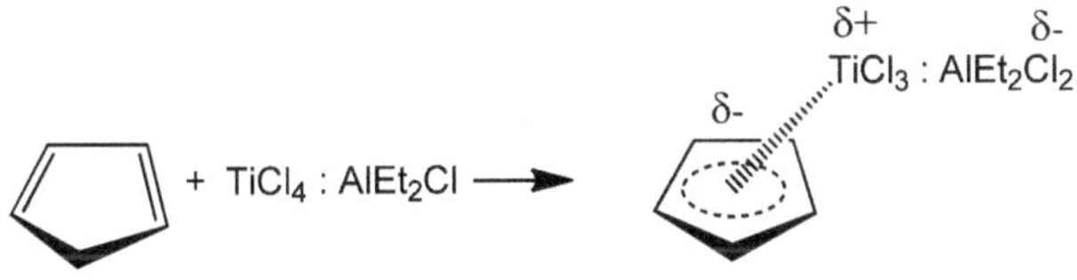

(I)

Then sandwich-complex is formed (II):

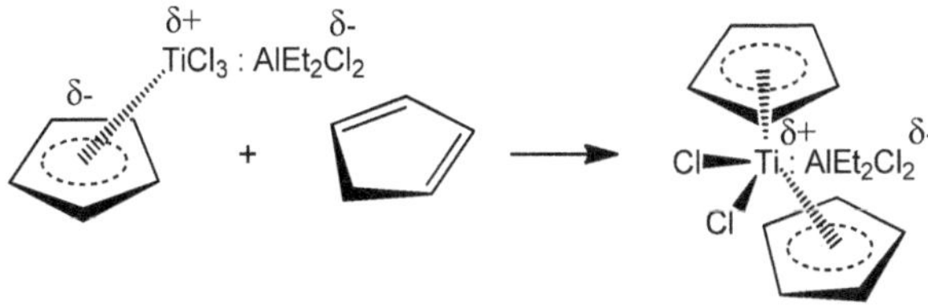

(II)

Complex equilibrium during initialization may be completed by regrouping the π-complex to σ-complex and isomer carbonium ion or by transformation of the sandwich-complex to semi-sandwich-complex (I):

$$Cp_2TiCl_2 : AlEt_2Cl_2 + TiCl_4 + AlEt_2Cl \longrightarrow 2CpTiCl_3 : AlEt_2Cl_2.$$

Then the reaction is also completed by regrouping the π-complex to σ-complex and isomer carbonium ion. Further chain growth occurs using one of two unsaturated bonds of DCPD – norbornene or cyclopentene [36].

In the case of initiation of olygomerization of vinyltoluene by $TiCl_4$ the active particle is formed due to further polarization of vinyl bond of the monomer in solvate shell of $TiCl_4$. Finally it adds to the double bond forming the respective ions (direct initialization). The obtained carbocation transfers to indane dimer and oligomeric products. At room temperature of the reaction mixture the indane dimer finally decomposes into toluene and indanyl cation [37, 38]:

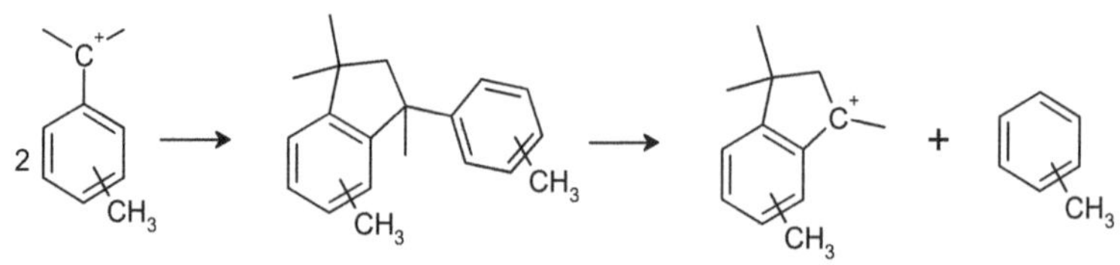

Then the cationic oligomerization of vinyltoluene takes place. In the case of use of catalytic system $TiCl_4$: Et_2AlCl (1:1) oligomerization of vinyltoluene occurs under similar conditions and is characterized by the presence of counter-ion of more complex composition.

4. Initiation of polymerization by electron transfer

Initiation of the polymerization by complete or partial electron transfer from the monomer to the initiator is a special case of interaction of the initiator with the monomer. This process is usually accompanied by the formation of intermediate charge-transfer complexes (CTC):

$$\text{Monomer} + \text{Initiator} \rightleftarrows \left[\text{CTC}\right] \rightleftarrows \text{Monomer}^{\overset{+}{\bullet}} + \text{Initiator}^{\overset{-}{\bullet}}$$

The last scheme is the most universal and attracts increasing attention in the field of polymerization of the polar electron-donor monomers. The phenomenon of charge transfer is evident not only in initiating the polymerization of such monomers with organic electron acceptors, but also occurs at the interaction with typical cationic polymerization initiators such as Lewis acids (including MLA) and stable organic cations, which act as acceptors relative to the electron-donating monomers [39, 40]. For example, in Ref. [41] it was shown that the initiation of polymerization of 9-vinylcarbazole under the action of $Ph_3C^+AsF_6^-$ proceeds through the stage of formation of cation-radical:

Then dication 9-vinylcarbazole is formed; and it is responsible for the polymerization:

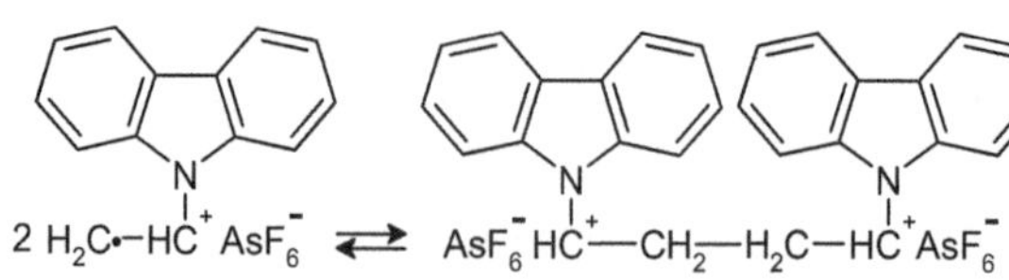

As MCL are very strong v-acceptors, their lowest molecular orbital is vacant v-valence orbitals of the metal atom. We can assume that stage of electron transfer from the monomer to a v-orbital of the MLA precedes the initiation process. Till now there is no adequate attention to the question of which kind of orbital electron-donating monomer delivers its electron acceptor. It is usually assumed that π-complex is formed as a result of this interaction [42, 43]. However, data concerning formation of π-complexes of vinyl monomers with MCL are practically absent, that is probably associated with their rapid transition to the σ-complexes, ionic and radical products.

In [44] VC polymerization was studied in the presence of $Ph_3C^+AlEt_2Cl_2^-$ in $CHCl_3$ solution at 20 °C. Process kinetics was studied using the stopped flow method with the registration in the IR range [2, 45–47]. In this case, the active site of polymerization is a stable organic cation Ph_3C^+, which is formed from the reaction of equivalent amounts of Ph_3CCl и $AlEt_2Cl$. However, counter-ion $AlEt_2Cl_2^-$, in contrast with hexa-fluoro-arsenate, hexa-chloro-antimonate and the like inorganic counter-ions of low nucleophilicity [27, 41, 47–50], associates with the organic cation. It allows concluding that the most likely type of active sites are separated solvate-ion pairs or even complex between VC and the contact ion pair.

The first reaction order was observed until complete consumption of the monomer. It indicates the absence of the chain break reactions. Moreover, when new portion of the monomer is added the reaction rate is not changed. It indicates the presence of the "live chains" in this system [51–54]. The behavior of this system is similar to the system VC – DEAC – chloroform studied earlier [46]. However, kinetic correlations have some differences which mean other mechanism of the active site formation. Kinetic measurements revealed that the most probable polymerization mechanism in the studied system $CHCl_3...Ph_3C+Et_2AlCl_2^-...VC$ is the chain growth on contact or solvate-separated ion pairs. It is confirmed by permanency of the kP value depending on the initial concentration of the initiator as well as its similarity to the ones known from literature.

For assessing reactivity of VC in the cationic polymerization the kinetics of its polymerization in solution under the action of diethylaluminum chloride was studied [46]. It was shown that the reaction is greatly influenced by the complexation among the monomer and

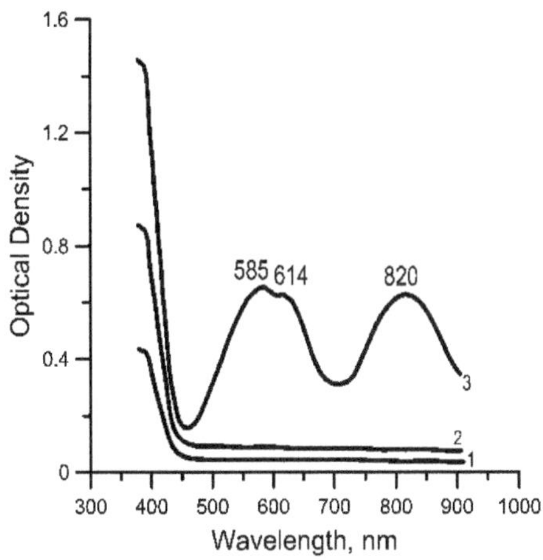

Figure 2. The absorption spectra of solutions in chloroform BK (1) $AlEt_2Cl$ (2) and complex of BK–$AlEt_2Cl$ (3)

initiator. In particular, this leads to the fact that the polymerization process in general is limited by the formation of active particles that are supposed to be dication of VC. Appearance of new charge-transfer band in the electronic spectrum of the products of their interaction is clear evidence of the formation of donor-acceptor complexes between $AIEt_2CI$ and VC (Fig. 1, curve 3).

Figure 2 shows that the spectra of the initial VC (curve 1) or $AIEt_2CI$ (curve 2) similar band is absent, and the observed charge-transfer band actually consists of two bands with maxima at $\lambda = 584$ and 614 nm. One of them, apparently, is associated with the transition of π-electron VC to the d-orbital $AIEt_2CI$, and another one - with the transition of n-electron. The obtained data reveal that the interaction of VC with a molecule of the initiator in the initial phase of the formation of donor-acceptor complex of medium strength; and degree of charge transfer from donor to acceptor is 0.29 of electron charge. The resulting donor-acceptor complex between VC and $AIEt_2CI$ has $\lambda_{max} = 584$ and 614 nm. It is close to the form shown in the literature for the complex value of λ_{max} VC with tetra-cyano-ethylene (590 nm, [55]). The interaction between VC and AIEt2CI leads to the fact that the effective charge on the β-carbon atom of the vinyl group BK significantly decreases; the length of the vinyl connection increases and its order is reduced. This greatly facilitates the subsequent process of formation of the active particles.

Formation of VC cation-radical followed by merge of two these cation-radicals to the dication may be such process. In this case evidence of formation of the latter one is the appearance in the electron spectrum of products of interaction and VC $AIEt_2CI$ a band at $\lambda_{max} = 820$ nm attributable to the long-wavelength absorption maximum dication VC (curve 3, Fig. 2). Authors [56] gave a value of $\lambda_{max} = 850$ nm for the cation formed by the interaction of ethyl-carbazole with $SnCl_4$. Further cationic polymerization of VC mechanism takes place.

In all cases the polymerization of VC under the action of DEAC occurred until complete consumption of the monomer and accompanied by formation of colored intermediate. This color did not disappear after complete monomer consumption.

Such polymerization character indicates the absence of the chain break reaction in the system. It is confirmed by the fact that after contact of the reaction mixture with moisture in air the color disappeared and did not appear after addition of new portion of the monomer. However, stopping the growth of the molar weight of the polymer at repeated addition of the monomer proofs availability of the chain transfer reactions in the system.

In some cases, the problem of donor site in the monomer molecule requires special study. For example, vinyl compounds containing a hetero atom in the structure (such as vinyl ethers, N-vinyl heterocyclic monomers, etc.) can function as both n-donors and π-donors, giving unshared electron pair of heteroatoms or electrons of the highest π-level for the intermolecular bond [57, 58]. It is largely dependent on the properties of an electron acceptor. When interacting with stronger acceptors (e.g., MCL metals of the third group of the Periodic table) sufficiently strong high polar nv-complexes are formed [59]:

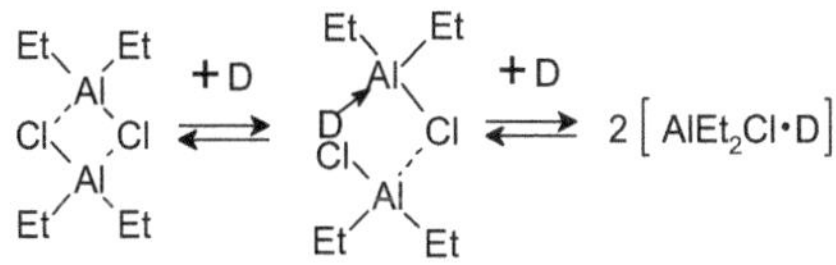

This will undoubtedly influence the formation of active particles in the system.

Thus, the generation of primary cations can pass through a series of relatively slow equilibrium stages, especially in the case of polymerization of monomers containing hetero atoms in the structure with an unshared electron pair. So often the formation of active sites becomes the limiting stage of the polymerization process as a whole, exerting a strong influence on the rest of the process.

5. Effect of reaction medium on the polymerization process

It is known that the influence of the reaction medium is a decisive factor, which determines the rate constants of individual stages of ionic polymerization [60]. We can distinguish two effects of medium influence: change of reactivity of the active sites and the stabilization of the formed ionized particles.

Process rate and reactivity of the active sites in various media will be determined by a number of factors: the influence of the polarity of the medium, co-catalytic action of solvents, specific solvation, and the formation of complexes with components of the reaction system. Experimental results show that the major factor is the polarity of the medium. It plays particularly noticeable role at the stage preceding the initiation and growth of the polymer chain. This is expected, since the processes occurring at the same time are due to the formation of intermediate active particles: free ions, contact and solvate-separated ion pairs, as well as other more complex aggregates:

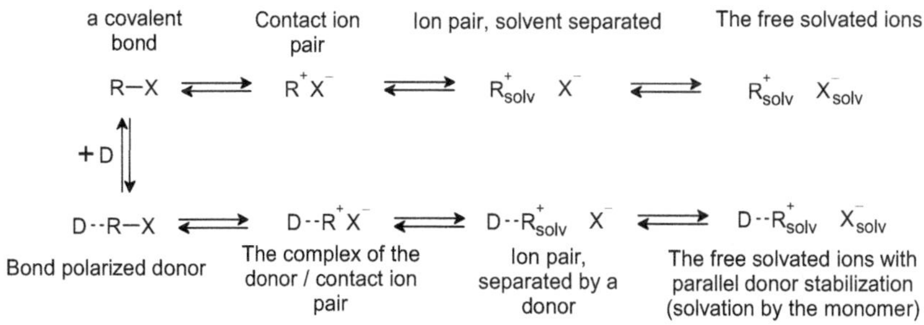

The main problems associated with these particles in the initiation of polymerization were discussed at 6th International Symposium on cationic polymerization [61–67].

Large dipole moment of the ion pairs leads to a strong interaction with polar molecules including molecules of polar solvents. Solvate separated ion pairs exist only in mediums where at least one of the ions in a free state is coordinated with solvent molecules. In the case of a slightly solvating solvents their function can be performed by the monomers, changing the dielectric constant of the medium, or other components that are directly involved in the polymerization. It results in a change in the kinetic order of reaction with respect to these components. This is explained by the fact that in media not providing the necessary solvation energy the ion pairs are stabilized by most polar or polarizable molecules from those ones presented in the system, i.e. monomer and initiator. They can be incorporated into the kinetic equation corresponding to the reaction, although they do not take direct part in it [8].

The existence of free ions in organic media because of their low stability is possible only under the following conditions [7]:

- ionization equilibrium is shifted toward preferably a covalent bond, i.e. there is a rapid reverse stabilization of charge;
- active formation of stabilized specific solvation of cations and anions with suitable solvents;
- nucleophility of the counter-ion is strongly reduced by acceptors in solvents with the prevailing acceptor character (H_2CCl_2, 1,2-diloretan, etc.); and there is no exchange with the cation except its electrostatic compensation.

The last condition, which would correspond to an almost "naked" cation takes place only at a strong intramolecular stabilization by delocalization of the charge (exchange interaction with the aromatic rings with a high density of π-electron, oxo-carbonium acids, allyl cations). Otherwise inductive effect of the ion carbenes on adjacent carbon atoms is so strong that the isomerization and cleavage for example, β-H-atoms will go faster than the reaction of chain growth:

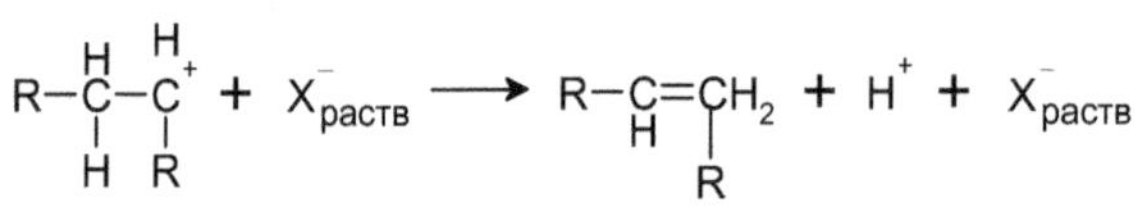

In such hard conditions the monomer itself is the strongest π-donor in the reaction system, which can reduce the high electrophilicity of carbonium ion. Therefore, the literature points to the special role the monomer solvation as a stabilizing factor for the active particles, helping them to implement the reaction growth [68].

Obviously, the ion pairs and free ions may have a different activity, and ion pairs are usually less active as compared with free ions. For example, for the polymerization of 9-vinylcarbazole in H_2CCl_2 solution under the $Ph_3C^+SbCl_6^-$ action at 20 °C a growth rate constants of the free ion and ion pair are equal $6 \cdot 10^5$ и $5 \cdot 10^4$ $L \cdot mol^{-1} \cdot s^{-1}$ [69]. According to data of other authors [49] the corresponding values can differ by several orders of magnitude. This is due to the higher effective charge density of the free cation in comparison with the corresponding ion pairs [1].

In real systems the situation is complicated by the fact that the reaction of chain growth can be represented as the competition of the monomer, solvent, and counter-ions around the electrophilic center [7]. The exchange interaction of solvent separated ion pair with the monomer leads to resolvation of the ion pair and its expansion in order to the counter-ion:

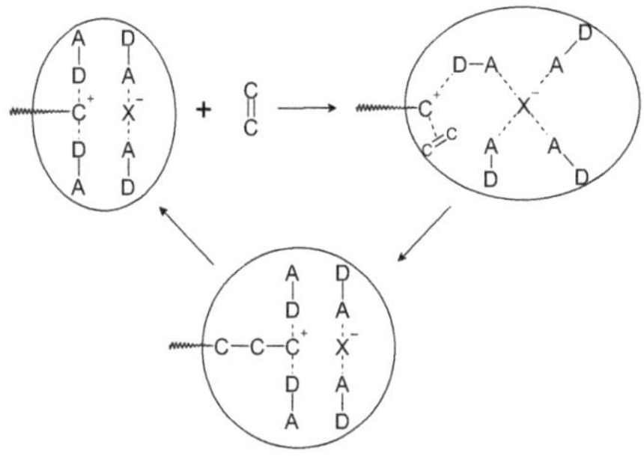

1.3

The exchange interaction between the counter-ion and the growing carbonium ion can be controlled by charge and must depend on the effective electrophilicity of the latter one.

According to current ideas about the mechanism of chain growth in cationic polymerization at high electrophilicity of the cation (aliphatic vinyl monomers) and the high nucleophilicity of the counter-ion $\left(CF_6^- \gg Cl^- \gg Br^- > I^-\right)$ polymerization does not take place. Monomers forming a resonance-stabilized cations with low electrophilicity (e.g., vinyl esters, 9-vinylcarbazole, dienes, p-methoxy-styrene) with counter-ions having low nucleophilicity (e.g., AsF_6^-, $SbCl_6^-$) at least at low temperatures are able polymerizing with a relatively small transfer chain. The growth occurs mainly through the free solvated ions. In the presence of counter-ions with the relatively high nucleophilicity (e.g., Br^-, I_{n+1}^-, $F_3CCO_2^-$ and others) these monomers are also capable for polymerization. Probably, the overall structure of the solution (scheme 1.3) reduces the potential energy [53], and wedging the monomer into an ion pair becomes favorable as compared with the monomer transfer to the chain.

Author details

Alexey Lyapkov, Elena Ionova, Vladimir Bondaletov and Alexey Pestryakov
National Research Tomsk Polytechnic University, Russia

6. References

[1] Reactivity, Mechanism and Structure in Polymer Chemistry; Jenkins, A.D.; Ledwith, A., Eds.; Wiley: London, 1974.
[2] Kennedy J.P. // Indian J. Phys. – 1982. V. A56. – № 4. – PP. 41–50.
[3] Miangos F. // Rev. Plast. Mod. – 1985. – V. 36. – № 346. – PP. 450–452.
[4] Cheradame H., Gandiny A. // Amer. Chem. Soc: Polym. Prepr.–1985. – V. 26. – № 1. – PP. 42–43.
[5] Sigwalt P. // Polym. J. – 1985. – V. 17. – № 1. – PP. 57–71.

[6] Kennedy J.P. Cationic polymerization of olefins: a critical inventory. Wiley: London, 1975.
[7] Reactions in polymer systems. / Ed. S.S. Ivanchev. – Chemistry, Leningrad. – 1987. – 304 p.
[8] Cationic Polymerization. P.Plesh, Ed., Mir, Moscow, 1966, p. 584.
[9] Kennedy J.P. // J. Macromol. Sci. – 1969. – V. A3. – № 5. –PP. 861–884.
[10] Kennedy J.P. // J. Macromol. Sci. – 1969. – V. A3. – № 5. –PP. 885–895.
[11] Heublein G. // Plaste und Kautsch. – 1981. – V. 28. – № 7.–PP. 361–364.
[12] Kennedy J.P. // «Int. Conf. Struct. – Prop. Relat. Rubber, 29–31 Dec., 1980». – Kharagpur. – 1981. – PP. 5–8.
[13] Kennedy J.P. // Rubber Chem. and Technol. – 1983. – V. 56. – № 3. – PP. 639–663.
[14] Kennedy J.P., Langer A.V. // Progress of Chemistry. – 1967. – T.35. – № 1. – C. 65–141.
[15] Sigwalt P. // Makromol. Chem. – 1974. – V. 175. – № 4. – PP. 1017–1034.
[16] Chmelir M., Marek M., Wichterle 0. // J. Polym. Sci. – 1965. – V. C16. – № 3. – PP. 833–839.
[17] Plesch P.H. // «Macromol. Chem. 8 Plenary and Main.Lect. Int. Symp. Macromol., Helsinki, 1972». – London. – 1973. – PP. 305–318.
[18] Plesch P.H. // Makromol. Chem. – 1974. – V. 175. – № 4. – PP. 1065–1076.
[19] Tolstikov G.A. Aluminum-organic synthesis// Moscow, USSR Acedemy of Sciences Publishing House, 1978, 286 p.
[20] Matsubayshi G., Wakatsuki K., Tanaka T. // Org. Magn. Reson. – 1971. – V. 3. – № 3. – PP. 703–711.
[21] Rokhov Yu., Kherd D., Lewis R., Chemistry of Metalorganic Compounds, Moscow: Publishing House of Foreign Literature, 1963.
[22] Ashby E.C., Smith M.B. // J. Amer. Chem. Soc. – 1964. – V. 86. – № 20. – PP. 4363–4370.
[23] Chmelir M. // J. Polym. Sci.: Polym. Symp. – 1976. – V. 56. – № 2. – PP. 311–322.
[24] Gorbachev S.G. // Abstract dissertation ... of candidate of chemical science. – Tomsk. – 1976. – 23 p.
[25] Lyapkov A.A., Sutyagin V.M., Lopatinskii V.P., Kubits Z.G. // Polymer Science U.S.S.R. – 1987. – V. 29. – № 12. – PP. 2939-2942.
[26] Sutyagin V.M., Lyapkov A.A. // Izvestiya Vuzov. Khimiya i Khimicheskaya Tekhnologiya. – 2000. – V. 43. – Issue 3. – PP. 87–91.
[27] Bowyer P.M., Ledwith A., Sherrington D. // Polymer. – 1971. – V. 12. – № 8. – PP. 509–520.
[28] Lyapkov A.A., Ionova E.I., Sutyagin V.M., Nikonova N.A. // Bulletin of the Tomsk Polytechnic University. – 2008. – V. 313. – №3. – PP. 59-64.
[29] Ionova E.I., Lyapkov A.A., Bondaletov V.G., Shipilova N.S. // Bulletin of the Tomsk Polytechnic University. – 2009. – V. 314. – №3. – PP. 100-105.
[30] Lyapkov A.A., Ionova E.I., Bondaletov V.G., Romanova A.A. // Bulletin of the Tomsk Polytechnic University. – 2009. – V. 314. – №3. – PP. 106-111.
[31] Ionova E.I., Lyapkov A.A., Bondaletov V.G., Bondaletova L.I., Petrenko T.V. // Coke and Chemistry. – 2009. – V. 52. – № 11. – PP. 496–500.

[32] Lyapkov A.A., Ionova E.I., Bondaletov V.G. // Bulletin of the Томsк Polyтechnic University. – 2011. – V. 318. – №3. – PP. 97-101.

[33] Ionova E.I., Lyapkov A.A., Bondaletov V.G., Karmanova O.I. // Bulletin of the Томsк Polyтechnic University. – 2011. – V. 319. – №3. – PP. 147-152.

[34] Kuznetsov N.N., Ionova E.I., Lyapkov A.A., Bondaletov V.G., Manankova A.A. // Chemical Industry. - 2009. - V. 86. - № 7. - PP. 367-378. 22 V. 311. - № 3. - PP. 124-129. 25

[35] Ionova E.I., Lyapkov A.A., Bondaletov V.G., Fiterer E.P. // Polzunov bulletin, 2009. – No 3. – PP. 163–167.

[36] Peng Y.X., Liu J.L., Cun L.F. // J. Polym. Sci. – 1996. – V. 34. – № 17. – pp. 3527–3530.

[37] M. Moreau, K. Matyjaszewski, P. Sigwalt. // Macromolecules. – 1987. – V. 20. – № 7. – PP. 1456–1464.

[38] K. Matyjaszewski, P. Sigwalt. // Macromolecules. – 1987. – V. 20. – № 11. – PP. 2679–2689.

[39] Shirota Y., Mikawa H. // Macromol. Sci.: Rev. Macromol. –Chem. – 1977. – V. C16. – № 2. – PP. 129–196.

[40] Geylord J. // Uspekhi Khimii – 1972. – T. 41. – No 6. – pp. 1067–1110.

[41] Rodrigues M., Leon L.M. // Eur. Polym. J. – 1983. – V. 19. – № 7. – PP. 585–588.

[42] Viberg N. // Angev. Chem. – 1968. – V.80. – № 20.–PP.809–822.

[43] Tinyakova E.I., Zhuravleva T.G., Kurenkina T.I., Kirilova I., Dolgoplosk B.A., Reports of USSR Academy of Sciences – 1962. – T. 144. – No 3. –pp. 592–595.

[44] Lyapkov A.A., Sutyagin V.M. // Bulletin of the Томsк Polytechnic University. – 2007. – V. 311. – № 3. – C. 124-129.

[45] Novikov V.T., Lyapkov A.A., Kubits V.V. // Polymer Science U.S.S.R. – 1987. – V. 29. – № 12. – PP. 2942-2944.

[46] Lyapkov A.A., Sutyagin V.M., Lopatinskii V.P. // Bulletin of the Tomsk Polytechnic University. - 2004. - V. 307. - № 4. - PP. 108-113.

[47] Peper D.C. // Eur. Polym. J. – 1979. – V. 16. – № 5. – PP. 407–411.

[48] Rooney J.M. // Makromol. Chem. – 1978. – V. 179. – PP. 165–171.

[49] Rodrigues M., Leon L.M. // Bur. Polym. J. – 1983. – V. 19. – № 7. – PP. 589–591.

[50] Bilbao E., Rodrigues M., Leon L.M. // Polym. Bull. – 1984. – V. 12. – № 4. – PP. 359–362.

[51] Sigwalt P., Moreau M. // Progress Polym. Sci. – 2006. – V. 31. – № 1. – PP. 44–120.

[52] Goethals E.J., Du Prez F. // Progress Polym. Sci. – 2007. – V. 32. – № 2. – PP. 220–246.

[53] Puskas J.E., Shaikh S., Yao K.Z., McAuley K.B., Kaszas G. // Eur. Polym. J. – 2005. – V. 41. – № 1. – PP. 1–14.

[54] Puskas J.E., Kaszas G. // Progress Polym. Sci. – 2000. – V. 25. – № 3. – PP. 403–452.

[55] Wojciechowski P., Kryszewski M. // Potsdam. Forsch. – 1979. – V. B. – № 20. – PP. 141–145.

[56] Partridge R.H. // Polymer. – 1983. – V. 24. – № 6. – PP. 733–738.

[57] Kearns D.R., Gardner P., Carmody 1. // J. Phys. Chem. –1967. – V. 71. – № 4. – PP. 931–937.

[58] Gurianova E.N., Goltsshtein I.P., Romm I.P.,Donor-acceptor bond, Moscow: Khimia – 1973. – 398 p.

[59] Heublein G., Kumpfel W., Schutz H. // Z. Chem. – 1978. –V. 18. – № 1. – PP. 34–36.

[60] Amis E. Influence of solvent on rate and mechanism of chemical reactions, Moscow: Mir, 1977

[61] Pask S.D., Nuyken 0., Visher A., Walter M. // «Cation Polym. and Relat. Process. Proc. 6th Int. Symp., Ghent, 30 Aug. – 2 Sept., 1983». – London. – 1984. – PP. 25–30.

[62] Heublein G., Bauerfeind D., Spange S., Knoppel G., Wondra–czek R. // ibid. – PP. 69–75.

[63] Plesch P.H. // ibid. – PP. 1–16.

[64] Matyjaszewski K., Szymanski R., Kubisa P., Penczek St. // Acta Polym. – 1984. – V. 35. – № 1. – PP. 14–22.

[65] Kunitake T., Takarada K. // J. Polym. Sci.: Polym. Symp. –1976. – № 56. – PP. 33–38.

[66] Puskas J., Kaszas G., Kennedy J.P., Tudos F. // J. Macromol. Sci. – 1982. – 1983. – V. A18. – № 9. – PP. 1353–1366.

[67] Higashimura T. // J. Polym. Sci.: Polym. Symp. – 1976. – № 56. – PP. 71–76.

[68] Heublein G. // J. Macromol. Sci. – 1981. – V. A16. – № 2. – PP. 563–577.

[69] Rooney J.M. // J. Polym. Sci.: Polym. Symp. – 1977. – № 56.– PP. 47–56.

Chapter 11

Synthesis of $Ba_{1-x}Ca_xTiO_3$ by Complex Polymerization Method (CPM)

Fabiana V. Motta, Ana Paula A. Marques, Carlos A. Paskocimas, Mauricio R. D. Bomio, Amélia S. F. Santos, Edson R. Leite, José A. Varela and Elson Longo

Additional information is available at the end of the chapter

1. Introduction

Oxide powders have been synthesized by solid-state reaction, using binary oxides or carbonates as precursors have been reported. The reaction with necessity of high temperatures and longer reaction times are obstacles in materials synthesis process because of yours high cost and because the use of high temperatures [1,2].

In the Complex Polymerization Method (CPM) theses problems are reduced because the synthesis occurs at low temperature and reduced reaction times are required, and, the immobilization of the metal complexes in such rigid organic polymeric networks can reduce the metal segregation, thus ensuring the compositional homogeneity at the molecular scale [3].

Organic polymer precursors are often utilized in the synthesis of multicomponent oxide powders. In the Pechini process, an aqueous solution of ethylene glycol, citric acid, and metals ions is polymerized to form a polyester resin. The compositional homogeneity is of vital importance for the synthesis of multicomponent oxides with complicated compositions, since the chemical homogeneity, with respect to the cation distribution throughout the entire gel system, often determines the compositional homogeneity of the final multicomponent oxides [4]. This method has been used to synthesize nanoparticles and thin films of several ferroelectric materials. The development of new semiconductor materials with wide band gaps (2.0–4.0 eV) may give rise to newoptoelectronic devices, particularly to materials for application in the development of green or blue light emission diodes (LED) or laser diodes. In many optoelectronic devices, amorphous semiconductors can replace single crystal semiconductors, particularly when cost is an important factor [5].

The CPM achieves molecular mixing of the starting materials and has been successfully used to synthesize several compounds, including $BaTiO_3$, $Pb(Zr,Ti)O_3$ and $BaMoO_4$ [6-10]. The barium titanate ($BaTiO_3$) perovskite, that is one ferroelectrics material, have been extensively studied due several possible applications such as electronic and optical devices [11-15]. The substitution of Ba by Ca in the $BaTiO_3$ perovskite results in an improvement of the stability of the piezoelectric properties, consequently the barium calcium titanate ($Ba_{1-x}Ca_xTiO_3$) solid solution has attracted great attention for use in the laser systems, eletro-optic material for various photorefractive and holographic applications [16-18]. The production of Ca-doped $BaTiO_3$ compounds by solid state reaction method the mixture of $BaCO_3$, $CaCO_3$ and TiO_2 is heat treated at high temperatures for long times and requires two successive calcinations to get high solubility of Ca^{2+} in the $BaTiO_3$ matrix [19-21]. It has been observed that Ca^{2+} replaces Ba^{2+} in $Ba_{1-x}Ca_xTiO_3$ to form tetragonal BCT solid solutions when x is less than ~0.23 [22, 23]. Moreover, Cheng et al. [23] obtained pure tetragonal phase for $x \leq 0.25$ and mixture of phases for x in the range of 0.3-0.85.

Photoluminescence (PL) property at room temperature occurs due to structural disorder existing in the perovskite system [24, 25]. However, the system can not be fully disordered owing to present a minimal order in the structure. This means that there is an order-disorder rate which favors the PL phenomenon in the system. Disorder in materials can be manifested in many ways: some examples are vibrational, spin and orientation disorder (all corresponding to a periodic lattice) and topological disorder. We will concentrate principally on the latter, which is the type of disorder associated with the structure of glassy and amorphous solids, a structure that cannot be defined in terms of a periodic lattice. Photoluminescence (PL) is a powerful probe of certain aspects of short-range order in the range of 2–5 Å and medium range of 5–20 Å, such as clusters, where the degree of local order is such that structurally inequitable sites can be distinguished because of their different types of electronic transitions and are linked to a specific structural arrangement [26]. The order was related to the presence of $[TiO_6]$ clusters, whereas the disorder was related to the presence of $[TiO_5]$ clusters. It is believed that PL occurs due to interaction of TiO_5-TiO_6 [27]. This PL emission was attributed to localized levels above the valence band (VB) and below the conduction band (CB). It has been demonstrated that a series of structurally disorder titanates synthesized by a soft chemical process have shown intense photoluminescence (PL) at room temperature [28-30]. In such cases, a minimal order in the system is necessary for the material exhibit PL property at room temperature.

Considering that the CPM is efficient to promote produce structural order-disorder complexes oxides powders and few have has been reported about the property PL of the barium calcium titanate (BCT). In this chapter, we present a detailed the synthesis and the effect structural of Ca^{2+} in the $BaTiO_3$ matrix powders of the conditions that favor the PL emission at room temperature.

2. Experimental section

2.1. Materials

The starting chemicals included titanium (IV) isopropoxide $[Ti[OCH(CH_3)_2]_4$ (Aldrich 97%), citric acid ($H_3C_6H_5O_7$) (Merck 99.5%), $CaCO_3$ (Aldrich 99 +%), $BaCO_3$ (Aldrich 99 +%), NH_4OH (Merck 99%) and ethylene glycol ($HOCH_2CH_2OH$) (J. T. Baker 99%). All of the chemicals were used without further purification.

2.2. Synthesis

$Ba_{1-x}Ca_xTiO_3$ (x = 20 % mol) was prepared by the Complex Polymerization Method (CPM) using barium carbonate ($BaCO_3$), calcium carbonate ($CaCO_3$) and titanium (IV) isopropoxide $[Ti[OCH(CH_3)_2]_4$ as starting materials. A flow chart representing of the synthesis by the CPM of BCT20 is outlined in Figure. 1. In this synthesis, the titanium citrate (TC) was formed by dissolution of titanium (IV) isopropoxide in aqueous solutions of citric acid (CA). The molar ratio of (CA) to titanium (IV) isopropoxide was 4:1. The citrate solution was well homogenized under constant stirring at a temperature of ~ 60-70°C, pH ~ 1.5. After achieving complete dissolution was add stoichiometric amounts of barium carbonate and calcium. The complex was well stirred for several hours at 60–70°C to produce a clear, homogeneous solution. After the solution was homogenized, ethylene glycol was added to promote the polyesterification. The citric acid/ethylene glycol mass ratio was set at 60:40. With continued heating at 100–120°C, the viscosity of the solution increased, albeit devoid or any visible phase separation. After partial evaporation of the water, the resin was heat

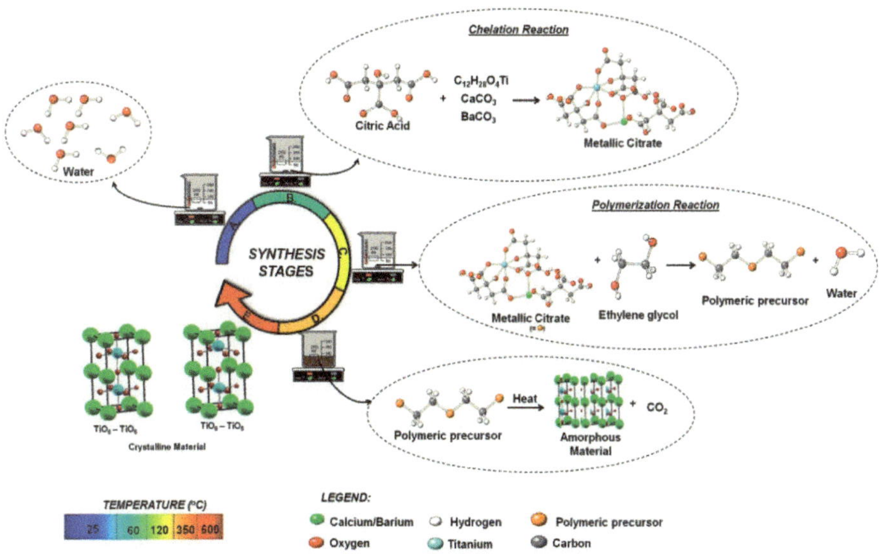

Figure 1. Flowchart representing the procedure for the synthesis of BCT20 powders.

treated at 350°C for 2 h, in a static oxidizing atmosphere, leading to the partial decomposition of the polymeric gel, forming an expanded resin, constituted of partially pyrolyzed material. This pre-pyrolyzed material (dark-brown powders) was removed from the beaker and deagglomerated using mortar. The powders were annealed at 400, 450, 500, 550, 575 and 600°C for 2 h.

2.3. Characterizations

FTIR transmittance spectroscopy and Liquid-state ^{13}C NMR at room temperature were performed on chemical solutions of TC (Ti-CA) and BCT (Ti-CA-Ba-Ca) at the initial reaction stage, before the polyesterification reaction by (EG). This study was carried out in order to compare the structures of complex species in these initial phase. The ^{13}C NMR characterization was conducted with a ^{13}C resonance at room temperature on a Bruker DRX 100Mz, 9.4T. These initial solutions and the calcined powder of BCT compounds (400°C $\leq$ T $\leq$ 900°C) were characterized by Fourier transformed infrared (FTIR) spectroscopy. The FTIR transmittance spectra were taken in the frequency range of 400–4000 cm^{-1} using an Equinox/55 (Bruker) spectrometer after disk samples was formed with KBr.

The thermal properties of the calcined dark-brown powders were investigated by simultaneous thermogravimetric analysis (DTA and TG) using a TA Instruments machine under oxygen flow (100 cm^3/min) and subjected to heating rate of 5°C/min at room temperature (~ 20°C) to 900°C. Samples were placed in platinum crucibles, weighing typically 8 mg, and a calcined alumina crucible was used as the reference material.

Powder X-ray Diffraction (XRD) patterns were collected from the heat-treated samples using a Rigaku diffractometer, model D/max-2500/PC with Cu Kα radiation. The X-ray diffraction data for the 1200°C sample were analyzed by the Rietveld method as implicated in the FULLPROF program, taking tetragonal perovskite (*P4mm*) as the starting model. A pseudo-Voigt function was fitted by refinement of the scale factor, background, peak width, peak asymmetry, atomic positions and lattice constants.

The Raman spectra for powders were obtained at room temperature using a RFS/100/S Bruker FT-Raman equipment over the scan range of 100-1400 cm^{-1}, using the 1064 nm exciting wavelength of a Nd:YAG laser. The infrared (IR) spectra were taken to characterize the lattice vibration of powders calcined at the different temperatures.

The PL spectra were collected with a digital monochromator internally integrated to a CCD with optical resolution of 1 nm and accuracy of 0.1 nm (Newport, OSM-400UV/VIS-U), using a time integration of 4 s (2.9 x 10-17 W per count/s) coupled to a optical fiber. The 355 nm exciting wavelength of a third harmonic of a Nd:YAG Q-switched laser (Brilliant B from Quantel) with a pulse duration of 4 ns and a repetition rate of 10 Hz was used, with an average energy of 4 mJ per pulse. The 488 nm exciting wavelength of a tunable optical parametric oscillator (OPO) pumped by 355 nm (3W) of a Q-switched Nd-YAG laser was also used, with an average energy of 4 mJ per pulse. All the measurements were taken at room temperature.

3. Results and discussion

3.1. Characterization of the precursor solution

CA-metal chelation in the precursor was accompanied by IR and NMR spectroscopy, in order to analyze the barium-calcium-titanium-citric acid complexes. Figure 2 shows the FTIR absorption spectra of TC and precursor solution of BCT. In the pH range 1–2, two peaks of asymmetric carboxyl groups are evident for the Ti-CA resin (TC) spectra; one at 1724 and the other at 1635 cm^{-1} referent to υ_{as}(COOH) and $\upsilon_{as}(COO^-)$, respectively. Symmetric stretching frequencies occurred at 1445 and 1389 cm^{-1} for the same carboxyl groups [31, 32]. BCT resin spectra do not display a unionized carboxyl group in the region close to 1720 cm^{-1}, indicating that all carboxyl groups were ionized in ph ~8.5 [33]. In the BCT spectra, two bands belonging to asymmetric and symmetric stretching are attributed, according to Rajendran et al. [31] and Yasodhai et al. [32], to the presence of a mixed-metal CA complex in BCT, which creates different types of carboxyl groups. In all complexes, the carbonyl asymmetric and symmetric stretching frequencies were in the range of 1665-1565 cm^{-1} and 1445-1389 cm^{-1}, respectively. The difference between the asymmetric (ν_{as}) and symmetric (ν_s) COO stretching modes ($\triangle(\nu_{as}-\nu_s)$) of the coordinated carboxyl groups varied from 220 to 176 cm^{-1}, suggesting that the carboxyl groups present an unidentate coordination in these resins [31, 32]. If the carboxyl groups were involved in bridging in the chelation of CA complex, the value for $\triangle\nu$ would be expected to be lower than $\triangle(\nu_{as}-\nu_s)$ ~ 170 cm^{-1} [32]. In the region below 1000 cm^{-1} a broad band that was attributed to stretching of the bond M-O appears, where M is Ti, Ba or Ca [34]. Similar bonds were verified by 13C NMR measurements performed on this solution, as discussed below.

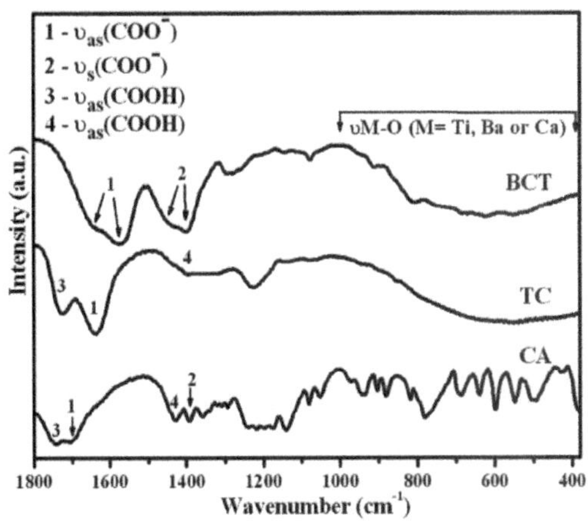

Figure 2. FTIR absorption spectra of free citric acid (CA), titanium citrate (TC), and precursor resin (BCT).

Figure 3(a) illustrates the ^{13}C NMR spectra for the Ti-CA solution with pH value close to 1.5. This figure revealed the presence of eight resonance signals, which were assigned to the free

CA and Ti-CA complex. The free CA is characterized by the four typical signals at 45.5, 75.5, 175.6, and 178.9 ppm. These four peaks represent the carbon centers in the methylene groups (CH_2 – C2 and C3 carbons), the alcoholic groups (COH – C1), the terminal carboxylic acid groups (t-COOH – C4 and C5), and the middle carboxylic acid groups (m-COOH – C6) of free CA, respectively [4, 33]. The other four signals are weak and the broad absorption signal is compared to the same peak of the free CA, this is a characteristic of coordination compounds with transition metals. The binding of Ti with CA can occur at different points of the CA chain, -COH, -mCOOH and -tCOOH. As the Ti ion is electron deficient, it can promote dislocations of the electronic density, possibly causing strengthening in bindings near Ti (C1′, C5′ and C6′). The resonance signal referent to C1′, C5′ and C6′ are 90.4, 175.9 and 187.8 ppm, respectively [4, 33]. The shifts of the alcoholic groups ($\Delta\delta$ 14.9 ppm) at 75.5 ppm and the middle carboxylic acid group ($\Delta\delta$ 8.9 ppm) at 178.9 ppm indicate that the –COH and –mCOOH groups in citric acid simultaneously coordinate the titanium atom. A small shift ($\Delta\delta$ 0.3 ppm) is observed for the uncoordinated –tCOOH, according to Deng et al. [35]. The resonance signal of the C2′, C3′ and C4′ in Ti-CA complex is not affected, consequently resonance signals of these carbons with the metals are similar to the free CA. Figure 3(b) shows a typical ^{13}C NMR spectrum of BCT solution, with the pH value at around 8.5. This figure revealed that all resonance signals belonging to BCT appear in larger ppm values (deshielding) than Ti-CA. This is related to the pH value of the BCT and Ti-CA solutions that are close to 8.5 and 1.5, respectively. Higher pH values (8-9) promote the deprotonation of the CA, in which the carbons of the acid groups are strongly linked with oxygen and come back more unprotected, the outcome is that more energy is necessary to make these carbon resonates and resonance signals to be shifted to larger ppm values (unshielded) compared to the systems with lower pH values [33]. The -tCOOH carbon group exhibits signals at around 180.9 and 180 ppm for Ba-C5″ and Ti-C5″, respectively [4, 33]. Resonance signals at around 77.7, 86.4, 91.5, 183.9, 189.3 and 190.3 belong to COH (C1), -COHBa (C1″), -COHTi (C1′), -mCOOH (C6), -mCOOBa (C6″) and -mCOOTi (C6′), respectively [4, 33]. The 13C NMR spectra of BCT resembled that of Ti-CA complex. Large downfield of the –COH ($\Delta\delta$ 8.7 and 13.8 ppm) and -mCOOH ($\Delta\delta$ 5.4 and 6.4 ppm) carbons groups related to the free citrate are observed for the complexation of CA with Ba^{2+} and Ti^{4+}, respectively [35]. The resonance signal around 47.8 ppm is attributed to the carbon center in the methylene groups (CH_2 – C2 and C3 for the free CA and C3 for the BCT). The resonance signal of the carbon center C2 in BCT is close to 47.1 ppm [4, 33].

3.2. Characterization of the calcined powders (BCT20)

The thermal decomposition of the pre-pyrolyzed (300°C for 2 h) BCT sample in synthetic air atmosphere is shown in Figure 4. The thermogravimetric analysis (TGA curve) shows four decomposition stages. First, TGA indicated a weight loss of ~ 15 % in the range of 25-120°C, which is related to the elimination of water and ethylene glycol. Second, it denoted a weight loss of ~ 62 %, as can be observed in the temperature range of 120 to 480°C, which is probably related to the decomposition of polymeric chains and the elimination of CO_2 and H_2O. A third weight loss of ~ 5 % occurs in the temperature range of 480-558°C and is attributed to the formation of $(Ba_{0.8}Ca_{0.2})_2Ti_2O_5.CO_3$ [36]. Lastly, from 558-675°C onward, the decomposition of intermediate phases occurs with a ~ 3 % weight loss during crystallization of pure BCT20 as the final reaction product. Above 675°C, no obvious weight loss was observed. Kumar and Messing [36] reported that the following two-stage process could represent such transformation.

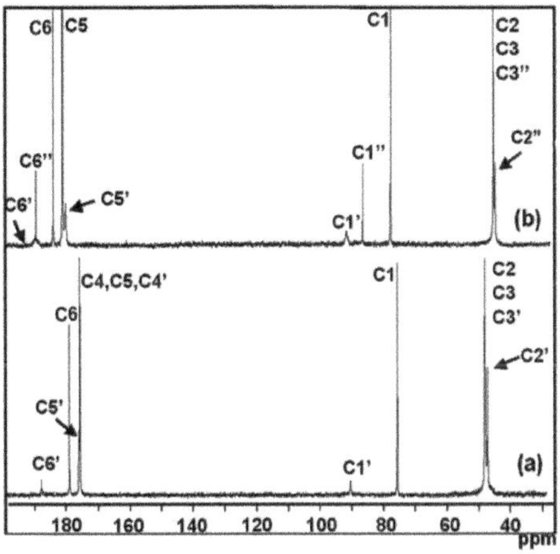

Figure 3. ^{13}C NMR spectra of: Ti-CA solution with pH ~1.5 (a); and BCT solution with pH ~ 8.5 (b). Here C = carbon centers in a citric acid free; C′ = carbon centers in a citric acid affected by Ti, and C″ = carbon centers in a citric acid affected by Ba.

$$\text{Ba, Ca, Ti precursor}_{(s)} + \text{air}_{(g)} \rightarrow \text{Ba, Ca,Ti oxycarbonate}_{(s)} + CO_{2(g)} \quad (1)$$

$$\text{Ba, Ca, Ti oxycarbonate}_{(s)} \rightarrow 2\ Ba_{0.8}Ca_{0.2}TiO_{3(s)} + CO_{2(g)} \quad (2)$$

DTA revealed the presence of a peak at temperatures close to 450 °C (Figure 4). This exothermic peak is assigned to organic species pyrolysis, during rupture of the polymeric chain. The low loss mass of the TG/DTA curves above T ~ 500°C suggests the beginning of the crystallization process. Moreover, it is important to notice the presence of two low intense exothermic peaks at 520 and 650 °C, which could indicate the formation and decomposition of intermediate phases as, for example, complex Ba, Ca and Ti carbonates.

Figure 5 shows the XRD patterns of the BCT20 powders heat treated at different temperatures, from 400 to 600°C. In this figure it is observed that the BCT20 heat treated at 400°C is disordered. This result is in agreement with the previous TG/DTA analysis and with those described in literature, indicating that the crystallization process occurs at temperatures higher than 450°C. The materials heat treated between 450-550ºC present only broad diffraction reflections at 2θ ~ 24.3°, 26.7º and 34.7º referent to the barium titanium oxycarbonate, $(Ba_{0.8}Ca_{0.2})_2Ti_2O_5.CO_3$ [37].

The X-ray patterns of heat-treated samples at 575°C display the Bragg reflections of this intermediate phase and two peaks at 2θ ~ 22º and 31.6º, which can be indexed as being characteristic of the perovskite BCT phase.

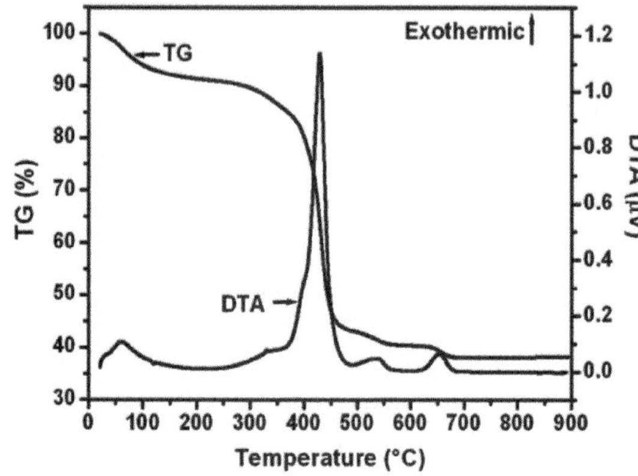

Figure 4. TGA/DTA range 25 - 900°C in synthetic air atmosphere of BCT pre-pyrolyzed at 300°C for 2 h.

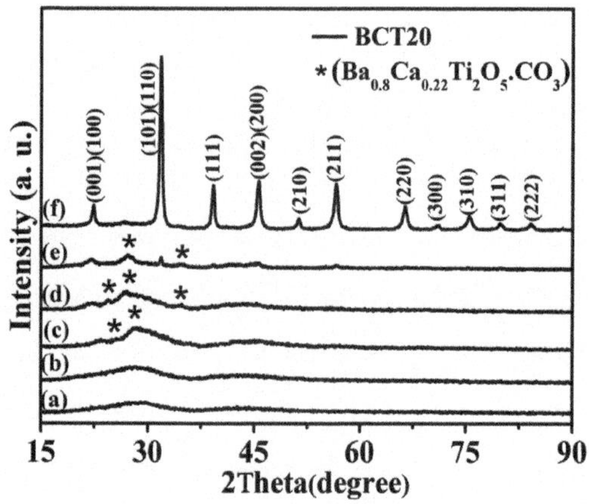

Figure 5. XRD patterns of the BCT powders calcined at temperatures: (a) 400°C; (b) 450°C; (c) 500°C; (d) 550°C; (e) 575°C and (f) 600°C.

The crystallization and decomposition process of this intermediate phase can be related to the presence of both exothermic peaks in the DTA curve. All samples heat-treated at 600°C are single-phased, with all Bragg reflections belonging to the perovskite BCT20. The pure BaTiO3 (BT) ceramic shows a tetragonal phase as identified and indexed using the standard XRD data of the corresponding BCT20 powders [19, 22]. The lattice parameters and the mean crystallite sizes were calculated from the peak positions displayed in Figure 5. The ordered BCT20 calculated crystallite sizes were around of 17 nm. The lattice parameters *a* and *c* were obtained using the least square refinement from the REDE93 program. The lattice parameters a and c were around of 3.9607(3) Å and 3.9941(3) Å, respectively; similar to the values reported for $Ba_{0.773}Ca_{0.227}TiO_3$ crystals (a = 3.962 Å and c = 3.999 (Å) [22]. The absence of the CT phase in the X-ray patterns of the BCT powders also suggests that samples prepared by this method present effective complexation of the Ca^{2+} ions in the matrix of $BaTiO_3$.

The Figure 6 depicts spontaneous Raman spectra of BCT20 powders recorded at the room temperature for samples calcined from 400-600°C. The samples heat treated at 400 and 450°C no present well-resolved sharp peaks in the Raman spectra, indicating that the material is structurally disordered at short range. The Raman spectra of powders heat treated between 500 and 575°C exhibited a fluorescent-like background. The band around of 1061 cm^{-1} (marked with a • in Figure 6) were attributed to the $(Ba_{0.8}Ca_{0.2})_2Ti_2O_5.CO_3$ intermediate phase [34]. Bands of the intermediate phase disappear completely in the powders heat treated at 600°C, agreeing with XRD data. These alterations were accompanied by appearance of five distinct broad bands referent to the $BaTiO_3$ *P4mm* tetragonal phase: 726, 526, 304, 256 and 186 cm^{-1} with its respective vibration modes $A_1(LO_3)$, $A_1(TO_3)$, E(TO), $A_1(TO_2$ and $A_1(TO_1)$ (marked with a ❖ in Figure 6) [2, 34]. The Raman peaks at 153 and 640 cm^{-1} (marked with a ◊ in Figure 6) were attributed to the satellite peaks, according to Cho et al. [2], however these peaks can be attributed too the $BaTiO_3$ hexagonal phase.

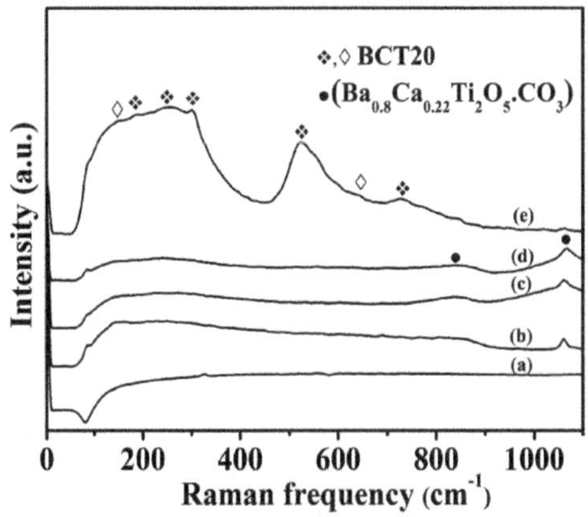

Figure 6. Spontaneous Raman spectra of BCT20 powders heat treated at (a) 400°C, (b) 500°C, (c) 550°C, (d) 575°C and (e) 600°C

The morphology of the BCT20 powders was investigated using HR-SEM. Figure 7 presents micrograph of powders heat treated at (a) 500°C and (b) 600°C. The powders calcined in the 400 - 575°C range for 2 h show free particles agglomerate with irregularly shaped Figure 7(a). The Figure 7(b) shows an agglomerate grain growth into of individual particles with irregularly forms developed as a result of the massive organic burn-off and consequent gas evolution, these characteristic were analogous to the powders treated at 600°C.

Figure 7. HR-SEM micrograph of BCT20 powders heat treated at (A) 500 °C and (B) 600 °C for 2 h.

Additional structural characterization was performed by infrared spectroscopy on the BCT20 sample. Figure 8 shows the FTIR spectra of the BCT20 treated at 400-600°C. From the spectrum of BCT annealed at 400-575°C, it is clear that undecomposed organic ligands are still present in the powders, both display a band at 1758 cm^{-1} attributed to C=O stretching mode [34, 38]. The broad absorption band corresponding to the O-H stretching modes in crystallization water occurs close to 3448 cm^{-1} to the BCT. The disordered compounds treated at 400-575°C showed a peak at ~ 1385 cm^{-1}, which was associated to the carboxylate group stretching mode [34, 38]. The spectra of compound heat-treated at 400-575°C present peaks at around 1060, 863, and 690 cm^{-1} that were attributed to carbonate ions ($BaCO_3$) [34]. These peaks seem to have lower intensity as the temperature increases, and in compounds treated at 600°C these peaks are absent. The band related to the carbonate ions were clearly observed for heat treatment up to 575°C (band at around 1450 cm^{-1}) reducing at higher temperatures and almost disappearing at 600°C [34]. In the BCT treated at 500 to 575°C, the peak related to the antisymmetric COO^- stretching mode of unidentate complex is observed at 1631 cm^{-1}, although for the samples heated treated at over 575°C, this peak is absent [34, 38]. The spectrum of the BCT heat-treated at 575-600°C, displays a very broad absorption band at around 522 cm^{-1}, referent to $F_2(\nu_3)$ antisymmetric stretch vibrations, ascribed to the Ti–O stretching vibration in $BaTiO_3$, as reported in literature [34, 38]. By increasing the heat treatment temperature, the crystalline structure seems to be more ordered and the band attributed to the Ti-O bonds, which are more defined. Also, the band at 2350 cm^{-1} refers to the adsorbed CO_2 from atmosphere [8, 9]. Such features are in good agreement with XRD and Raman Spectroscopy data and also indicate that CPM is effective in the synthesis of BCT powders.

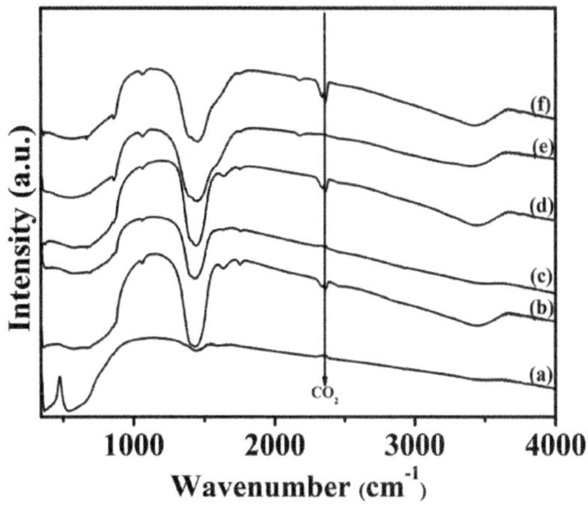

Figure 8. FTIR absorption spectra of dehydrated BCT powders heat treated at (a) 400°C, (b) 450°C (c) 500°C, (d) 550°C, (e) 575°C and (f) 600°C

The photoluminescence spectra recorded at room temperature for BCT20 powders heat treated at 400-600°C in two different excitation wavelength: the 355 nm (Figure 9a) and 488 nm (Fig. 9b) of an Nd:YAG laser system. BCT20 heat treated at 400-575°C presented PL emission. The BCT20 powder annealed at 500°C displays the most significant PL intensity when excited with these two wavelengths laser beam. The system has a totally ordered structure (samples annealed at 600°C), the PL emission is not observed.

This profile of the emission band is typical of a multiphonon process, i.e. a system in which relaxation occurs by several paths, involving the participation of numerous states within the band gap of the material [27]. This behavior is related to the structural disorder of BCT20 and indicates the presence of additional electronic levels in the forbidden band gap of the material.

The complex PL band spectra may be due to several components and generally can be decomposed in individual components. The spectrum is decomposed based on the nature of the process governing each component. The luminescence process is generally described by a Gaussian line broadening mechanism in which case, the luminescence intensity can be expressed in terms of a Gaussian (amplitude version) line-shape function and can be written by the following equation:

$$\mathrm{I}(h\nu) = \mathrm{I}_0 + \sum_{i=1}^{n} A_i \exp\left[-\frac{(h\nu) - E_{0i}}{2\sigma_i^2}\right] \quad (3)$$

where hv is the energy of the radiation emitted, I_0 is the offset intensity, A_i is the amplitude of each component, E_{0i} is the energy of each component in which the intensity is maximum, and σ_i is the standard derivation for each component (FWHM) [39].

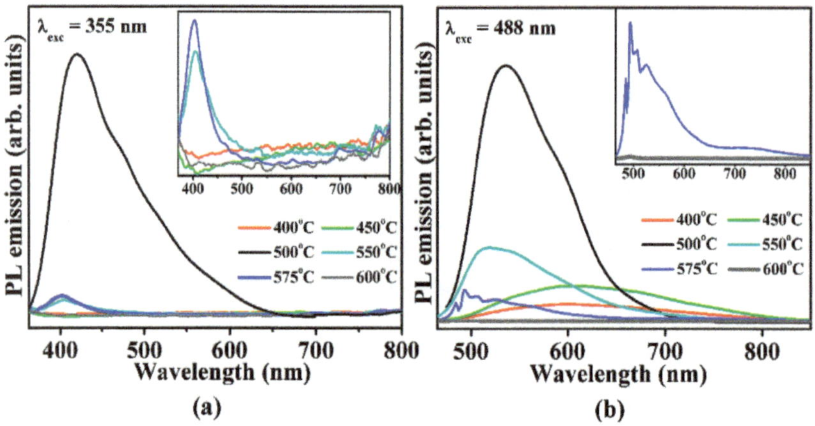

Figure 9. Room-temperature photoluminescence spectra of BCT powder samples annealed at 400, 450, 500, 550, 575 and 600oC for 2 h in an oxygen flow: (a) excitation with 355 nm wavelength of an Nd:YAG laser and (b) excitation with 488 nm wavelength of an Nd:YAG laser

The BCT20 powder annealed at 500ºC displays the most significant PL intensity when excited with these two laser beam wavelengths. Using the Gaussian method, the PL curves of the BCT20 sample annealed at 500ºC were decomposed into components, each of which refers to the region in the visible spectrum where its maximum peak intensity appears. Each color represents a different type of electronic transition and can be linked to a specific structural arrangement. To gain a better understanding of the properties of PL and its dependence on the structural order-disorder of the lattice, the PL curves were analyzed using the PeakFit deconvolution program [40]. Based on the Gaussian line broadening mechanism for luminescence processes, the fine features in the PL spectra of samples annealed at 500ºC were deconvoluted and extracted from the deconvolution curves. Figure 10a and 10b illustrates such decompositions, while Table 1 lists the areas under the curve of the respective transitions.

The decomposition band indicated the presence of different intermediate levels in the band gap. More energetic excitation wavelengths favor the transition of more energetic levels in the band gap. Thus, broad band PL emission consists of the sum of individual emissions. Such emissions arise from a radiative recombination between electrons and holes trapped in the gap states. The transitions of disordered titanates therefore occur at energies far below the band gap of these materials. The violet component is associated with more energetic transitions and is observed only after excitation with a laser beam wavelength of 355 nm. Otherwise, the green component is associated with a less energetic transition and is observed after excitation with a laser beam wavelength of 488 nm [16, 26].

In the structure perovskite-type the lattice former titanium is at the center of the cube, surrounded by six oxygens that occupy the middle of the faces, in a regular octahedral configuration. However, the structure before of get its ideal configuration (totally ordered)

is a mixture of TiO5-TiO6 clusters intercalated by Ba and Ca atoms. The higher the heat treatment temperature, the more frequent the TiO6 conformation and the more ordered the structure [10, 41]. According XRD results the BCT20 powders are ordered annealed at 600°C. The ordered powders where only TiO_6 clusters tend to exist does not allow the creation of point defects and do not presents PL emission at room temperature [10]. Similar to the CT:Sm studied by de Figueiredo et al. [10], the compound with intermediary range disorder (500°C) presented intense PL emission while compared to the compound disordered (400 and 450°C) and the ordered compound no PL property was showed.

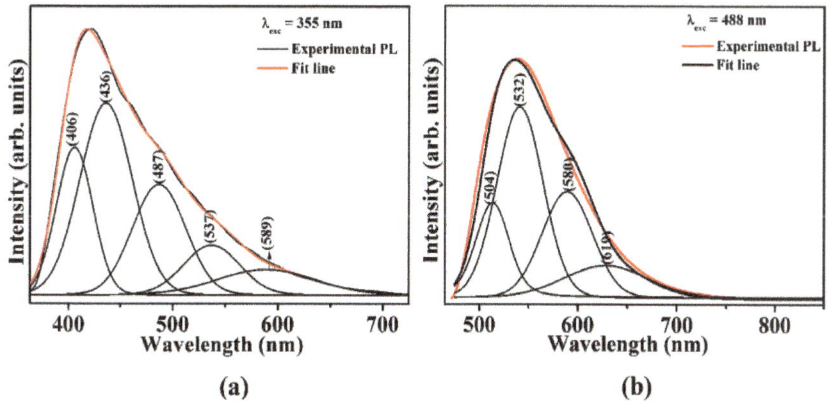

Figure 10. Deconvolution of PL curve fitted for the sample annealed at 500°C: (a) excitation with 355 nm wavelength and (b) excitation with 488 nm wavelength.

BCT20 annealed at 500°C					
Peak center (nm)					
λ (nm) 355 nm	599 %a	537 %a	487 %a	436 %a	406 %a
	9	10	22	38	20
Peak center (nm)					
λ (nm) 488 nm	619 %a	580 %a	532 %a	504 %a	
	17	27	45	11	

599 nm= orange component of PL; 580 = yellow component of PL; 532 and 537 nm = green component of PL; 487 nm = blue component; 436 and 406 = violet component of PL. % [a] obtained by dividing the area of each decomposed PL curves by the total PL area.

Table 1. Fitting parameters of Gaussian peaks.

X-ray Absorption Near Edge Structure (XANES) experimental results [42] pointed that the oxygen vacancies in titanates can occur in three different charge states: the [TiO_5. V_0^x] complex states, which is neutral relative and presents two electrons paired, the singly ionized [TiO_5. $V_0^{\bullet}$] complex state, that has one electron despaired, and the doubly positively charged [TiO_5. $V_0^{\bullet\bullet}$] complex state, which did not trap any electrons. Before donor excitation, a hole in the acceptor and an electron in a donor are created, according according to equations (2-4) using Kröger-Vink notation [43]:

$$\left[TiO_6\right]_x + [TiO_5.V_O^x] \rightarrow \left[TiO_6\right]' + [TiO_5.V_O^{\bullet}] \quad (4)$$

$$\left[TiO_6\right]_x + [TiO_5.V_O^{\bullet}] \rightarrow \left[TiO_6\right]' + [TiO_5.V_O^{\bullet\bullet}] \quad (5)$$

$$[TiO_5.V_O^{\bullet\bullet}] + 1/2\left(O_2\right) \rightarrow \left[TiO_6\right] \quad (6)$$

where [TiO_6]' is donor, [TiO_5. $V_O^{\bullet}$] is donors-acceptor and [TiO_5. $V_O^{\bullet\bullet}$] is acceptor.

These equations suggest that to the transition of a valence-band hole in the conduction band is a necessary requirement the oxygen-vacancy-trapped electron in the valence band. This means that most of electrons around oxygen vacancies are released and, therefore, such oxygen vacancy complex site is relatively positive charged. Moreover, oxygen vacancies tend to trap photo-generated electrons. The charge transfer occurring as proposed in the equations (2-4) create electrons and hole polarons that can be designed as bipolarons. After the photon excitation, the recombination and decay process follow the many valid hypotheses presented in the literature [44]. The present work shows that the emission process leading to PL is facilitated by previous existence of these complex clusters in the ground state. The order-disorder BCT20 powders thus intrinsically possess the necessary condition for creating PL at room temperature.

4. Conclusion

$Ba_{0.8}Ca_{0.2}TiO_3$ (BCT) compounds prepared by Complex Polymerization Method (CPM) showed effective chelation processes of the Ba-Ca-Ti-CA in initial precursor. The CPM shown is one advantageous method because in your simple procedure is used lower temperature and smaller reaction time. The other advantage of CPM is the low cost for the production of BCT powders.

The ordering system at short and long range was accompanied for the $Ba_{0.8}Ca_{0.2}TiO_3$ using Raman, PL band and XRD characterization. Raman data related to short-range and XRD results at long-range order shows that the material is organized at 600°C. However, the phenomenon of PL at room temperature is not observed in the order structure at 600°C. The introduction of Ca^{2+} ion in $BaTiO_3$ proportioned intrinsic defects. These intrinsic defects, linked to structural disorder, facilitate the main emission process to PL. Disorder in solids provokes degeneracy and destabilization in the localized states of the atoms acting as electron-hole pairs and supporting the broad PL band phenomena and electronic levels are fundamental to understanding the order-disorder process in the solid state. These optical properties exhibited by disordered BCT20 suggest that this material is a highly promising candidate for photoluminescent applications.

Author details

Fabiana V. Motta*, Carlos A. Paskocimas, Mauricio R. D. Bomio and Amélia S. F. Santos
Department of Materials Engineering, Federal University of Rio Grande do Norte, UFRN, Natal, RN, Brazil

Ana Paula A. Marques
Department of Exact and Earth Sciences, Federal University of São Paulo, UNIFESP, São Paulo, SP, Brazil

Edson R. Leite
Interdisciplinary Laboratory of Electrochemistry and Ceramics, Department of Chemistry, Federal University of São Carlos, UFSCar, São Carlos, SP, Brazil

José A. Varela and Elson Longo
Interdisciplinary Laboratory of Electrochemistry and Ceramics, Chemistry Institute, São Paulo State University, UNESP, Araraquara, SP, Brazil

Acknowledgement

This work was supported by Rede de Pesquisa em Catalisadores Ambientais – RECAM (grant no 564913/2010-3/CNPq) and FAPESP-CEPID (grant no 98/14324-8).

5. References

[1] Kakihana M., Yoshimura, M. (1999) Synthesis and Characteristics of Complex Multicomponent Oxides Prepared by Polymer Complex Method. Bull.Chem.Soc.Jpn 72:(1427-1443.

[2] Cho, W. S. (1998) Structural evolution and characterization of $BaTiO_3$ nanoparticles synthesized from polymeric precursor. J. Phys. Chem. Solids 59:(5) 659-666.

[3] Maurera, M., Souza, A. G., Soledade, L. E. B., Pontes, F. M., Longo, E., Leite, E. R.Varela, J. A. (2004) Microstructural and optical characterization of $CaWO_4$ and $SrWO_4$ thin films prepared by a chemical solution method. Mater. Lett. 58:(5) 727-732.

[4] Kakihana, M., Arima, M., Nakamura, Y., Yashima, M.Yoshimura, M. (1999) Spectroscopic characterization of precursors used in the Pechini-type polymerizable complex processing of barium titanate. Chem. Mat. 11:(2) 438-450.

* Corresponding Author

[5] Leite, E. R., Paris, E. C., Pontes, F. M., Paskocimas, C. A., Longo, E., Sensato, F., Pinheiro, C. D., Varela, J. A., Pizani, S. S., Campos, C. E. M. Lanciotti, F. (2003) The origin of photoluminescence in amorphous lead titanate. J. Mater. Sci. 38:1175-1179.

[6] Pontes, F. M. (2004) Characterization of $BaTi_{1-x}Zr_xO_3$ thin films obtained by a soft chemical spin-coating technique. J. Appl. Phys. 96:(8) 4386-4391.

[7] Pontes, F. M. (2003) Theoretical and experimental study on the photoluminescence in BaTiO3 amorphous thin films prepared by the chemical route. J. Lumin 104:(3) 175-185.

[8] Marques, A. P. D., de Melo, D. M. A., Paskocimas, C. A., Pizani, P. S., Joya, M. R., Leite, E. R. Longo, E. (2006) Photoluminescent $BaMoO_4$ nanopowders prepared by complex polymerization method (CPM). J. Solid State Chem. 179:(3) 671-678.

[9] Marques, A. P. A., de Melo, D. M. A., Longo, E., Paskocimas, C. A., Pizani, P. S.Leite, E. R. (2005) Photoluminescence properties of $BaMoO_4$ amorphous thin films. J. Solid State Chem. 178:(7) 2346-2353.

[10] de Figueiredo, A. T., Longo, V. M., de Lazaro, S., Mastelaro, V. R., De Vicente, F. S., Hernandes, A. C., Siu Li, M., Varela, J. A.Longo, E. (2007) Blue-green and red photoluminescence in $CaTiO_3$:Sm. J. Lumin. 126:(2) 403-407.

[11] Zhang, Q. S., Cagin, T.Goddard, W. A. (2006) The ferroelectric and cubic phases in $BaTiO_3$ ferroelectrics are also antiferroelectric. Proc. Natl. Acad. Sci. U. S. A. 103:(40) 14695-14700.

[12] Ahadi, K., Mahdavi, S.-M., Nemati, A., Tabesh, M.Ranjbar, M. Electronic structure and morphological study of BaTiO3 film grown by pulsed-laser deposition. Mater. Lett. 72:107-109.

[13] Takahashi, H. Development of lead-free $BaTiO_3$ ceramics possessing enhanced piezoelectric properties. Elect. Comm. Jp 95:(4) 20-26.

[14] Sonia, Patel, R. K., Kumar, P., Prakash, C.Agrawal, D. K. Low temperature synthesis and dielectric, ferroelectric and piezoelectric study of microwave sintered $BaTiO_3$ ceramics. Ceram. Int. 38:(2) 1585-1589.

[15] Duan, Y., Tang, G., Chen, C., Lu, T.Wu, Z. First-principles investigations of ferroelectricity and piezoelectricity in $BaTiO_3/PbTiO_3$ superlattices. Phys. Rev. B 85:(5)

[16] Motta, F. V., Marques, A. P. A., Espinosa, J. W. M., Pizani, P. S., Longo, E.Varela, J. A. (2010) Room temperature photoluminescence of BCT prepared by Complex Polymerization Method. Curr. Appl. Phys. 10:(1) 16-20.

[17] Kumar, P., Singh, S., Juneja, J. K., Prakash, C.Raina, K. K. Influence of calcium substitution on structural and electrical properties of substituted barium titanate. Ceram. Int. 37:(5) 1697-1700.

[18] Sonia, Patel, R. K., Prakash, C.Kumar, P. Effect of microwave processing on structural, dielectric and ferroelectric properties of calcium-doped $BaTiO_3$ ceramics. J. Ceram. Process. Res. 12:(6) 634-639.

[19] Chen, X. M., Wang, I.Li, J. (2004) Dielectric characteristics and their field dependence of (Ba, Ca)TiO_3 ceramics. Mater. Sci. Eng. B-Solid State Mater. Adv. Technol. 113:(2) 117-120.

[20] Chang, M. C.Yu, S. C. (2000) Raman study for $(Ba_{1-x}Ca_x)TiO_3$ and $Ba(Ti_{1-y}Ca_y)O_3$ crystalline ceramics. J. Mater. Sci. Lett. 19:(15) 1323-1325.

[21] Suzuki, T., Ueno, M., Nishi, Y. Fujimoto, M. (2001) Dislocation loop formation in nonstoichiometric (Ba,Ca)TiO_3 and $BaTiO_3$ ceramics. J. Am. Ceram. Soc. 84:(1) 200-206.

[22] Kuper, C., Pankrath, R.Hesse, H. (1997) Growth and dielectric properties of congruently melting $Ba_{1-x}Ca_xTiO_3$ crystals. Appl. Phys. A-Mater. Sci. Process. 65:(3) 301-305.

[23] Cheng, X.Shen, M. (2007) Different microstructure and dielectric properties of Ba1-xCaxTiO3 ceramics and pulsed-laser-ablated films. Materials Research Bulletin 42:(1662.

[24] Pontes, F. M., Longo, E., Leite, E. R., Lee, E. J. H., Varela, J. A., Pizani, P. S., Campos, C. E. M., Lanciotti, F., Mastelaro, V.Pinheiro, C. D. (2003) Photoluminescence at room temperature in amorphous $SrTiO_3$ thin films obtained by chemical solution deposition. Mater. Chem. Phys. 77:(2) 598-602.

[25] Orhan, E., Varela, J. A., Zenatti, A., Gurgel, M. F. C., Pontes, F. M., Leite, E. R., Longo, E., Pizani, P. S., Beltran, A. Andres, J. (2005) Room-temperature photoluminescence of BaTiO3: Joint experimental and theoretical study. Phys. Rev. B 71:(8) 085113.

[26] Motta, F. V., de Figueiredo, A. T., Longo, V. M., Mastelaro, V. R., Freitas, A. Z., Gomes, L., Vieira, N. D., Jr., Longo, E. Varela, J. A. (2009) Disorder-dependent photoluminescence in $Ba_{0.8}Ca_{0.2}TiO_3$ at room temperature. J. Lumin. 129:(7) 686-690.

[27] Pizani, P. S., Leite, E. R., Pontes, F. M., Paris, E. C., Rangel, J. H., Lee, E. J. H., Longo, E., Delega, P. Varela, J. A. (2000) Photoluminescence of disordered ABO_3 perovskites. Appl. Phys. Lett. 77:(6) 824-826.

[28] Pizani, P. S., Basso, H. C., Lanciotti, F., Boschi, T. M., Pontes, F. M., Longo, E. Leite, E. R. (2002) Visible photoluminescence in amorphous ABO_3 perovskites. Appl. Phys. Lett. 81:(2) 253-255.

[29] Orhan, E., Albarici, V. C., Escote, M. T., Machado, M. A. C., Pizani, P. S., Leite, E. R., Sambrano, J. R., Varela, J. A. Longo, E. (2004) A DFT rationalization of the room temperature photoluminescence of Li_2TiSiO_5. Chem. Phys. Lett. 398:(4-6) 330-335.

[30] Longo, V. M., Cavalcante, L. S., de Figueiredo, A. T., Santos, L. P. S., Longo, E., Varela, J. A., Sambrano, J. R., Paskocimas, C. A., De Vicente, F. S. Hernandes, A. C. (2007) Highly intense violet-blue light emission at room temperature in structurally disordered $SrZrO_3$ powders. Appl. Phys. Lett. 90:(9) 091906.

[31] Rajendran, M.Rao, M. S. (1994) Formation of $BaTiO_3$ from Citrate Precursor. J. Solid State Chem. 113:(2) 239-247.

[32] Yasodhai, S., Sivakumar, T.Govindarajan, S. (1999) Preparation, characterisation and thermal reactivity of transition metal complexes of hydrazine with citric acid. Thermochim. Acta 338:(1-2) 57-65.

[33] Fang, T. T. Tsay, J. D. (2001) Effect of pH on the chemistry of the barium titanium citrate gel and its thermal decomposition behavior. J. Am. Ceram. Soc. 84:(11) 2475-2478.

[34] Duran, P., Gutierrez, D., Tartaj, J., Banares, M. A. Moure, C. (2002) On the formation of an oxycarbonate intermediate phase in the synthesis of BaTO3 from (Ba,Ti)-polymeric organic precursors. J. European Ceram. Soc. 22:(6) 797-807.

[35] Deng, Y. F., Zhou, Z. H. Wan, H. L. (2004) pH-dependent isolations and spectroscopic, structural, and thermal studies of titanium citrate complexes. Inorg. Chem. 43:(20) 6266-6273.

[36] Kumar, S. Messing, G. L. (1994) Metal-Organic Resin Derived Barium-Titanate .2. Kinetics of $BaTiO_3$ Formation. J. Am. Ceram. Soc. 77:(11) 2940-2948.
[37] Kumar, S., Messing, G. L.White, W. B. (1993) Metal-Organic Resin Derived Barium-Titanate .1. Formation of Barium Titanium Oxycarbonate Intermediate. J. Am. Ceram. Soc. 76:(3) 617-624.
[38] Duran, P., Capel, F., Tartaj, J., Gutierrez, D. Moure, C. (2001) Heating-rate effect on the BaTiO3 formation by thermal decomposition of metal citrate polymeric precursors. Solid State Ion. 141:(529-539.
[39] Street, R. A. (1976) Luminescence in Amorphous-Semiconductors. Adv. Phys. 25:(4) 397-453.
[40] Ding, T., Zheng, W. T., Tian, H. W., Zang, J. F., Zhao, Z. D., Yu, S. S., Li, X. T., Meng, F. L., Wang, Y. M.Kong, X. G. (2004) Temperature-dependent photoluminescence in $La_{2/3}Ca_{1/3}MnO_3$. Sol. St. Commun. 132:(12) 815-819.
[41] Gurgel, M. F. C., Espinosa, J. W. M., Campos, A. B., Rosa, I. L. V., Joya, M. R., Souza, A. G., Zaghete, M. A., Pizani, P. S., Leite, E. R., Varela, J. A. Longo, E. (2007) Photoluminescence of crystalline and disordered BTO : Mn powder: Experimental and theoretical modeling. J. Lumin. 126:(2) 771-778.
[42] Asokan, K., Jan, J. C., Chiou, J. W., Pong, W. F., Tseng, P. K. Lin, I. N. (2001) X-ray absorption spectroscopy studies of $Ba_{1-x}Ca_xTiO_3$. J. Synchrot. Radiat. 8: 839-841.
[43] Kröger, F. A., Vink, H. J. Frederick Seitz, David, T. (1956). Relations between the Concentrations of Imperfections in Crystalline Solids, Vol. 3, Sol. St. Phys. 3: 307-435. Academic Press.
[44] Leonelli, R. Brebner, J. L. (1986) Time-Resolved Spectroscopy of the Visible Emission Band in Strontium-Titanate. Phys Rev B. 33:8649-8656.

Chapter 12

Titanium (IV) and Nickel (II) Catalysts Based on Anilinotropone Ligands

Reza Sandaroos, Tomás Cuenca, Gholam Hossein Zohuri, Saman Damavandi and Saeid Ahmadjo

Additional information is available at the end of the chapter

1. Introduction

Historically, the development of highly active olefin polymerization catalysts has been a trigger for creating new polymers which impact on our daily lives in countless beneficial ways [1-5]. A recent instance is the development of group 4 metallocene catalysts that exhibit very high ethylene polymerization activities [2,3]. Based on the highly active group 4 metallocene catalysts, high performance linear low-density polyethylene (LLDPE), isotactic poly-(propylene) (iPP) and syndiotactic polystyrene (sPS) etc., have been developed [6]. Therefore, much effort has been directed towards the development of highly active catalysts, following the group 4 metallocene catalysts. In consequence, quite a few highly active catalysts based on both early and late transition metal complexes have been developed [7-14]. There are, however, only a few examples of titanium complexes displaying high ethylene polymerization activities [15-19], though titanium metal is the major player in highly active heterogeneous Ziegler-Natta catalysts. Accordingly, further researches have been conducted on titanium catalysts with the intention of developing the highly active titanium catalysts and applying them to the polymerization of ethylene.

As a result of ligand-oriented catalyst design research, Sandaroos and coworkers [19-21] described the catalytic performance of new titanium complexes 1–9 containing aminotropone chelate ligands for ethylene polymerization (Scheme 1). The following subsections detail the results observed in these works.

2. Search for acquiring the fundamentally active ligand

Among the typical catalyst components, ligands play a predominate role in the polymerization process. During electron exchange between metal and monomer, ligands help the metal to balance its electron density with receiving electrons from the coordinated ethylene through the metal and releasing electrons whenever required to facilitate the ethylene insertion process (Figure 1). Accordingly, to obtain a highly active catalyst, the existence of ligands with a notable balance between their electron donating and withdrawing, evidenced by calculation of energy gap between the HOMO (the highest occupied molecular orbital) and LUMO (the lowest unoccupied molecular orbital) of them, is a predominate requirement [22]. For comparison, the energy gap between the HOMO and LUMO of three well-known ligands, namely phenoxy-imine [15,23-29], pyrrolide-imine [15,27-29], indolide-imine [30] and the new aminotropone chelate ligand [19] was studied using density-functional theory (DFT). Because of the reasonable energy gap between the HOMO and LUMO of aminotropone (2.6 eV), it was theoretically offered as a fundamentally active ligand (Scheme 2).

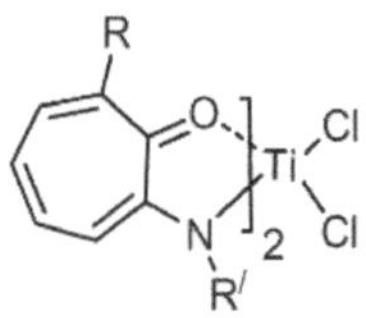

Catalyst	R'	R	Refrence(s)
1	phenyl	H	19-21
2	2,6-dimethylphenyl	H	19
3	cyclohexyl	H	19
4	4-(t-Buyl)cylohexyl	H	19
5	cyclooctyl	H	19
6	Ethyl	H	19
7	isoporpyl	H	19
8	phenyl	methyl	19,20
9	phenyl	t-Butyl	19,20

Scheme 1. Titanium (IV) complexes based on anilinotropone ligands 1–9.

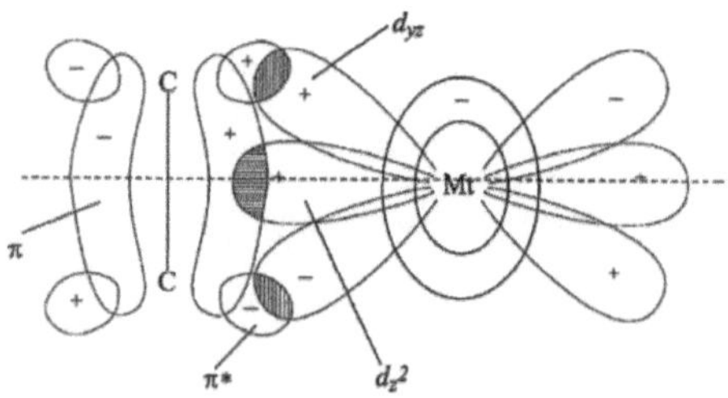

Figure 1. Presentation of the olefin coordination on the transition metal [19].

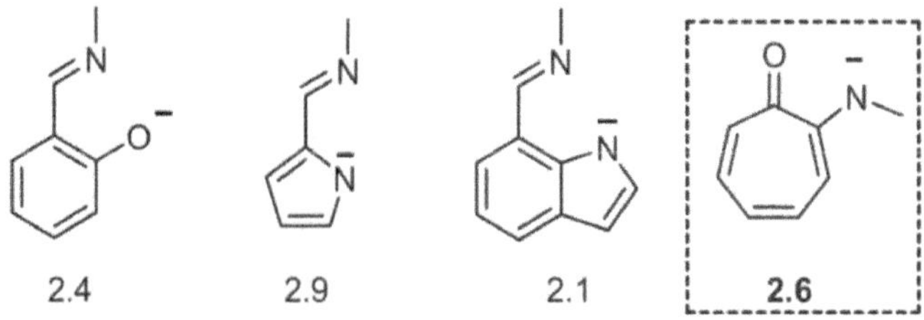

Scheme 2. Energy gap (eV) between HOMO and LUMO of aminotropone compared to those related to phenoxyimine, pyrrolide-imine and indolideimine [16,19].

3. Substituent effects

This section concentrates on the steric features of bis(aminotropone)Ti complexes compared with those of bis(phenoxyimine) Ti complexes (Ti-FI catalysts). Unlike Ti-FI catalysts, which require sterically demanding substituents in ortho position to the phenoxy oxygen atom in order to exhibit high ethylene polymerization activity [23-26], the activity of bis(pyrrolide-imine) Ti catalysts (Ti-PI) decreases with more congestion adjacent to the anionic N of the PI ligand because there is not enough space for monomer insertion [16]. The steric bulk in Ti-FI catalysts is thought to afford effective ion separation between the cationic active species and an anionic co-catalyst, resulting in enhancement of the catalytic activity. Accordingly, bis(aminotropone) Ti complexes 1, 8 and 9 with a series of subtituents (H, methyl and t-Butyl) adjacent to the carbonyl were prepared and examined as ethylene polymerization catalysts (Figure 2; Table 1, entries 1, 37 and 40) [19]. The polymerization results provided information on the potential of bis(aminotropone) Ti complexes for ethylene polymerization, and additional information about the effect of the substituent adjacent to the carbonyl on catalytic performance.

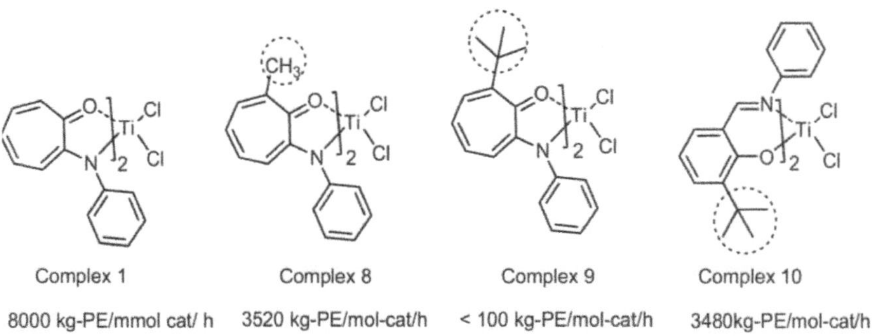

Figure 2. Investigation of substituent effects in bis(aminotropone) Ti complexes in comparison with a typical Ti-FI catalyst 10 [19].

The bis(aminotropone) Ti complex 9, which possesses the t-Butyl group, proved to be a poor catalyst for ethylene polymerization under the conditions employed (activity < 100 Kg PE mol cat^{-1} h^{-1}). Instead, the bis(aminotropone)Ti complex 1 with an H atom adjacent to the carbonyl formed polyethylene (PE) with M_w 98000 and a high activity of 8000 Kg PE mol cat^{-1} h^{-1}. Additionally, catalyst 8, which possesses the methyl group, exhibited almost moderate activity (activity 3520 kg PE mol cat^{-1} h^{-1}). These facts show that, for bis(aminotropone) Ti complexes 8 and 9, the methyl and t-Butyl groups in close proximity to the carbonyl evidently provide unfavorable steric hindrance to ethylene polymerization. This is in sharp contrast to Ti-FI catalysts, which require steric congestion near the polymerization center in order to display high activity. The activity obtained with complex 1 (8000 Kg PE mol cat^{-1} h^{-1}) is extremely high for a Ti complex with no cyclopentadienyl (Cp) or PI ligand(s) [16,23-26]. In fact, the activity exceeds that displayed by Ti-FI catalyst 10 under analogous conditions (3480 Kg PE mol cat^{-1} h^{-1}) (Figure 2) [22]. Since the substituent on the amine-N is situated near the active site, the bis(aminotropone)Ti complexes 1–7, containing three types of substituents, aromatic ring (catalyst 1-2), aliphatic ring (catalyst 3-5) and alkyl groups (catalyst 6-7) on amine-N were prepared and evaluated, with methylaluminoxane (MAO) activation, for the polymerization of ethylene to find a correlation between the N-aryl substituent and ethylene polymerization behavior. The polymerization results are presented in Table 1 which also includes the results for Ti-FI catalyst 10, Cp_2TiCl_2 and Cp_2ZrCl_2 as references (compare entries 1, 7, 12, 17, 35-36 and 40-43). The basic trend observed is that an increase in the steric bulk of the substituent in each of the three types of catalyst enhances catalytic activity. A reasonable hypothesis is that the sterically more encumbered substituent on the amine-N part of bis(aminotropone) Ti complexes induces more effective ion separation between the cationic active species and an anionic cocatalyst. In addition, it gives better steric protection to the anionic N from Lewis acidic compounds present in a polymerization medium, leading to increased catalytic activity.

Comparison of catalysts 1-2 with 3-5 revealed dramatically increased activities for 3-5, which is thought to be due to a good match between electronic and steric effects of them. Therefore, it can be concluded that the catalytic activity is influenced by both electronic and steric effects. In other words, a good integration between the steric and electronic properties of the catalyst is needed to reach the highest activities. It should be noted that catalyst 5, which includes a cyclooctyl group on the amine-N, gives an activity of 27200 Kg PE mol cat^{-1} h^{-1}, which compares favorably with Cp_2TiCl_2 or Cp_2ZrCl_2 combined with MAO. Additionally, in each of catalysts presented in Table 1, sterically more encumbered substituents on the amine-Ns generally afford higher-molecular-weight PEs (compare entries 1, 7, 12, 17 and 35-36). The steric congestion provided by the substituent, which probably reduces the rate of chain transfer more significantly than that of chain propagation, is responsible for the enhancement of the product molecular weight [30].

4. Effects of polymerization conditions

Ethylene polymerization behavior using catalysts 3, 4 and 5 were studied under different reaction times (Table 1, entries 20-34) and temperatures (entries 5-19). The highest activity of catalyst 5 was obtained at 10 °C (31200 Kg PE mol cat^{-1} h^{-1}). However, the highest activity of catalysts 4 and 3 occurred at about 35 °C (24500 and 18200 Kg PE mol cat^{-1} h^{-1}, respectively). The reduction in catalyst activity in the polymerization performed at the lower and upper temperature than the optimum value could be attributed to a low propagation rate and catalyst irreversible deactivation respectively. Ethylene polymerization carried out at different reaction times showed catalyst 5 had a shorter lifetime, but catalysts 3 and 4 showed more stable activities during the polymerization.

Additionally, the activity of the catalysts 1 and 8 and M_w values of resultant polymers increased with increasing the monomer pressure (entries 1-3 and 37-39). This behavior is mainly due to the high concentration of the monomer close to the catalyst active centers.

As could be predicted, as molar ratio of [Al]/[Ti] increased, activity of catalysts increased, but M_w values of resultant polymers decreased [19-21].

The large amount of hydrogen concentration could slightly increase the activity of the catalyst 1, 8 and 9 (Figure 3) [20]. A reasonable explanation for this effect might be increase of homogeneity of polymerization system and return of catalytically less reactive species, such as those resulting from 2,1-insertions, to the catalytic cycle through their fast hydrogenation [20].

Moreover, as it can be seen in Figure 4 [20], polydispersity and M_n of the polymer obtained using the catalyst 1 increases with time.

5. DFT studies of catalyst

Since a bis(aminotropone) Ti catalyst contains a pair of non-symmetric bidentate ligands, it potentially displays five isomers (A–E) arising from the coordination modes of the two ligands in an octahedral geometry (Scheme 3) [19]. DFT studies suggested that bis(aminotropone)Ti catalyst 1 assumed isomer A, with a trans-carbonyl-Os, cis-amine-Ns and cis-Cls disposition. Additionally, DFT calculations were performed on a methyl cationic complex (an initial active species generated from bis(aminotropone) Ti catalyst 1 with MAO) in the presence of ethylene, to obtain information about the structure of the catalytically active species (Figure 5) [19] .The calculations revealed that an ethylene-coordinated cationic species assumed an octahedral geometry with a trans-carbonyl-Os, cis-amine-Ns and cis-Me/coordinated ethylene disposition, which fulfills the pivotal requirement for a high efficient catalyst, i.e., a growing polymer chain and a coordinated-ethylene group in the cis-position. An inspection of the calculated structure indicated that the phenyl group on the amine-N was located in close proximity to the active site, suggesting the substituent on the amine-N is the strategic substituent vis-a`-vis catalyst design.

As discussed, unlike catalyst 1, catalysts 8 and 9, with the methyl and t-Butyl groups adjacent to the carbonyl-Os, display very low productivity in the polymerization of ethylene. The calculated structure of methyl cationic complexes originating from 1 and 9 is displayed in Figure 6 [19]. The t-Butyl group of complex 9 seems to provide steric congestion near the polymerization center, which diminishes the rate of chain propagation. This is probably because it obstructs ethylene from gaining access to the active site and subsequent insertion into the Ti-carbon bond. On the other hand, DFT studies show that the active species derived from bis(aminotropone) Ti catalyst 1 possesses higher electrophilicity

Entry	Cat.	Cat. (μ mol)	MAO (mmol)	Time (min)	T (°C)	P (bar)	TOF [a]	Mv (×10³) [b]	Refs.
1	1	1	1.25	15	25	1	8000	260	19-21
2	1	1	1.25	15	25	3	9100	450	20
3	1	1	1.25	15	25	5	9800	630	20
4	2	1	1.25	15	25	1	9650	530	19
5	3	1	1.25	15	0	1	10450	2600	19
6	3	1	1.25	15	10	1	11800	2580	19
7	3	1	1.25	15	25	1	15200	2300	19
8	3	1	1.25	15	35	1	18200	1860	19
9	3	1	1.25	15	45	1	18000	1740	19
10	4	1	1.25	15	0	1	14650	3500	19
11	4	1	1.25	15	10	1	18000	3120	19
12	4	1	1.25	15	25	1	22600	2950	19
13	4	1	1.25	15	35	1	24500	2560	19
14	4	1	1.25	15	45	1	14650	2240	19
15	5	1	1.25	15	0	1	26400	5620	19
16	5	1	1.25	15	10	1	31260	4950	19
17	5	1	1.25	15	25	1	27200	3540	19
18	5	1	1.25	15	35	1	21100	3140	19
19	5	1	1.25	15	45	1	14560	2800	19
20	3	1	1.25	5	25	1	5800	-	19
21	3	1	1.25	10	25	1	10400	-	19
22	3	1	1.25	15	25	1	15200	2300	19
23	3	1	1.25	25	25	1	22500	-	19
24	3	1	1.25	40	25	1	19500	-	19
25	4	1	1.25	5	25	1	7200	-	19
26	4	1	1.25	10	25	1	13600	-	19
27	4	1	1.25	15	25	1	22600	2950	19
28	4	1	1.25	25	25	1	26500	-	19
29	4	1	1.25	40	25	1	17300	-	19
30	5	1	1.25	5	25	1	8200	-	19
31	5	1	1.25	10	25	1	15240	-	19
32	5	1	1.25	15	25	1	27200	3540	19
33	5	1	1.25	25	25	1	21550	-	19
34	5	1	1.25	40	25	1	10500	-	19
35	6	1	1.25	15	25	1	410	285	19
36	7	1	1.25	15	25	1	680	340	19
37	8	1	1.25	15	25	1	3520	271	19,20
38	8	1	1.25	15	25	3	3750	320	20
39	8	1	1.25	15	25	5	4500	390	20
40	9	1	1.25	15	25	1	<100	352	19,20
41	10	1	1.25	15	25	1	3480	368	18
42	Cp_2TiCl_2	1	1.25	15	25	1	16700	1253	18
43	Cp_2ZrCl_2	1	1.25	15	25	1	20000	1000	18

kg PE mol cat^{-1} h^{-1}, [b)] Calculated from intrinsic viscosity

Table 1. Ethylene polymerization results with catalysts 1–9, Ti-FI catalyst 10, Cp_2TiCl_2 and Cp_2ZrCl_2

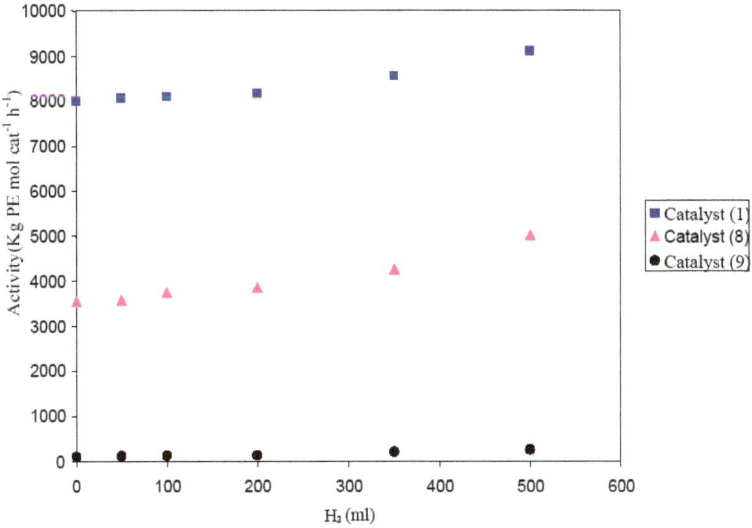

Figure 3. Influence of H_2 concentrations on the catalyst activity. Polymerization conditions: [Al]/[Ti] = 1250, polymerization time = 15 min, temperature = 25 °C, monomer pressure = 1 bar, [Ti] = 1 mmol, toluene = 250 mL [20].

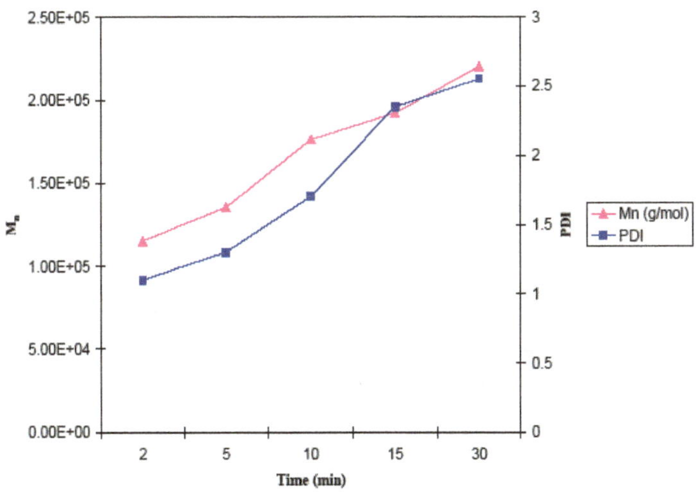

Figure 4. PDI and Mn versus time for the polymer obtained by catalyst 1. Polymerization conditions: [Al]/[Ti] = 1250, temperature = 25 °C, monomer pressure = 1 bar, [Ti] = 1 μmol, toluene =250 mL. M_n and PDI were obtained from GPC [20].

at the Ti center compared with that of Ti-FI catalyst 10 (Mulliken charge of the Ti in atomic unit, catalyst 1=2.212 [19], catalyst 10=2.005 [16]). Considering that the active species derived from a typical metallocene (Cp_2TiCl_2) has a Mulliken charge (Ti) of 1.308 [16], bis(aminotropone) Ti catalyst 1 exhibits particularly high electrophilicity. Therefore, a bis(aminotropone) Ti catalyst generates a catalytically active species that has higher electrophilicity than a Ti-FI catalyst.

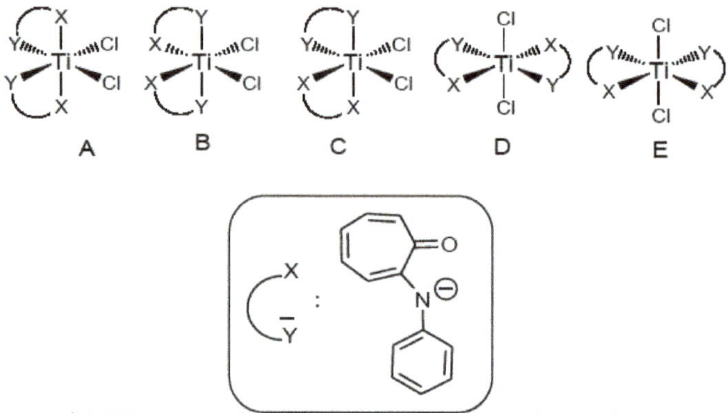

Isomer	A	B	C	D	E
RFE (KJ/Mol)	0	24.9	8.5	53.6	65.3

RFE: Relative Formation Energy

Scheme 3. The relative formation energies of the isomers [19].

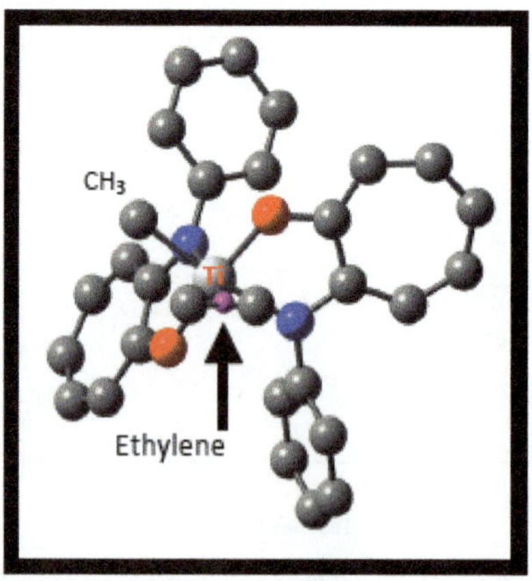

Figure 5. Calculated structure of ethylene-coordinated cationic species derived from bis(aminotropone) Ti catalyst 1. The hydrogen atoms are omitted for clarity [19].

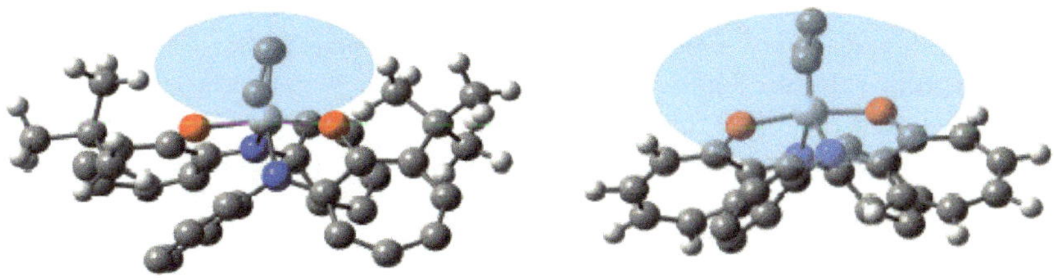

(A) (B)

Figure 6. Calculated structures of cationic active species generated from catalyst 1 (A) and catalyst 9 (B). Some of the hydrogen atoms are omitted for clarity [19].

6. Nickel (II) catalysts based on anilinotropone ligands

6.1. Introduction

Olefin polymerization catalysts based on early-metal metallocene complexes were introduced nearly two decades ago and are seeing increasing commercial utilization [31-33]. Their single-site nature makes them attractive for ligand tailoring, and as a result these catalysts have been modified in innumerable ways to enhance polymerization activity, improve catalyst lifetime, increase the α-olefin/ethylene reactivity ratio in copolymerizations, and control microstructures of polypropylene and other poly-α-olefins [31-33]. A drawback of metallocene and classical Ziegler catalysts is the extreme oxophilicity of the early-metal center. This oxophilicity renders metallocenes inactive toward most functionalized monomers and highly sensitive to polar solvents and impurities. The sensitivity of metallocenes to polar substituents is largely responsible for an increase in interest in late-transition-metal complexes as olefin polymerization catalysts over the past several years [34-37]. Late-metal catalysts complement early-metal catalysts in several ways. (1) Polymers exhibiting quite different microstructures are frequently obtained [36,38-41]. This arises from the ability of the metal to walk along the growing polymer chain via a series of β-elimination and reinsertion reactions which can occur at rates competitive with or faster than olefin insertion. This process results information of branched polymers from ethylene and "chain-straightened" polymers from α-olefins [36,38]. (2) Certain monomers such as trans-2-butene, which cannot be polymerized by early-metal systems, can be successfully polymerized with late-metal systems [42]. (3) Many late-metal catalysts are compatible with protic solvents and nucleophilic impurities. Water compatibility has led to successful emulsion polymerization of ethylene [43-46]. (4) Expanded functional group tolerance has permitted the copolymerization of alkyl acrylates (polar monomers) with ethylene and α-olefins (non polar monomers) [47-50].

Much of efforts has focused on cationic systems of α-diimine ligands (Figure 7) as well as catalysts incorporating closely related neutral ligands [38,40-42,48,49,51-53]. The α-diimine

Ni complexes are very reactive toward ethylene and α-olefins and, as noted above, produce polymers with unique microstructures. The cationic Ni catalysts are electrophilic and sensitive to protic solvents. Functional group compatibility is diminished relative to Pd analogues; however, it should be noted that copolymerization of ethylene and alkyl acrylates has been achieved at high ethylene pressures (> 500 psig) and temperatures above 80 °C [47,54].

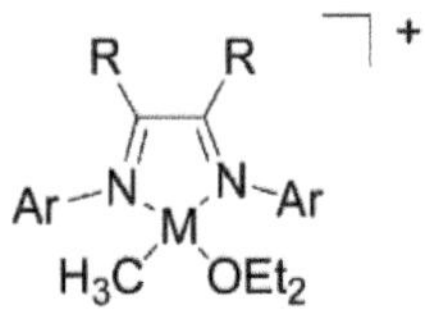

Catalyst 11; M = Ni and Pd

Figure 7. The α-diimine complexes [38,40-42,48,49,51-53].

In view of the high sensitivity of the cationic Ni and Pd catalysts, there has been substantial interest in developing neutral Ni catalysts to overcome these limitations. Several examples of neutral Ni catalysts based on SHOP-type (Shell Higher Olefin Process) ligands have been reported, although productivities and molecular weights are often low [45,50,55,56]. Neutral Ni catalysts modeled after α-diimine complexes, containing bulky ortho-disubstituted arylimine functionality, have been reported. Catalysts based on salicylaldimines (Figure 8), have been described by both the DuPont [57] and the Grubbs groups [58,59]. The most active systems contain either electron-withdrawing nitro substituents in the aromatic ring [57] or bulky substituents at C-3, with a 9-anthracenyl group being most effective [58,59]. In the latter case, activities of 1.3 ×10^5 kg PE mol Ni^{-1} h^{-1} and lifetimes in excess of 6 h have been observed at 45-50 °C [58].

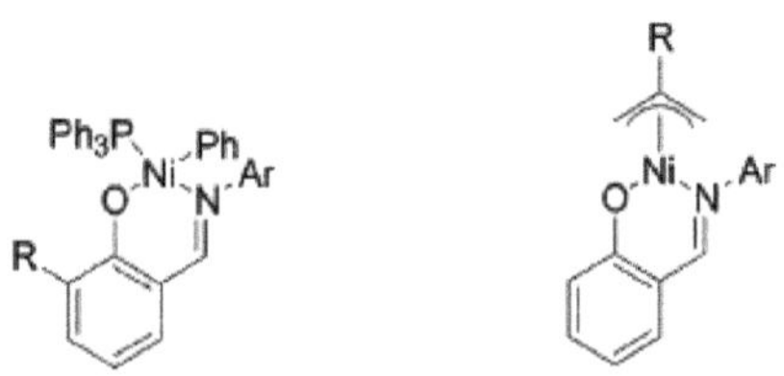

12 a: R=H, 12 c
12 b: R=9-anthracenyl

Figure 8. The salicylaldimines catalysts [57-59].

Brookhart [60,61] designed a ligand which incorporated the key elements of the six-membered chelate salicylaldimine ligand but which would lead instead to a five-member chelate. He chose for this purpose the 2-anilinotropone moiety because it contained the desired anionic N,O chelate, a hindered N-aryl group, and complete conjugation between the N and O. Several variations can be introduced by using hindered or electron-poor N-aryl group and modified tropone skeleton. This family of ligands were synthesized using the palladium catalyzed cross-coupling of aniline with 2-triflatotropone [60-62]. The corresponding highly neutral nickel catalyst is almost high active in ethylene polymerization and it does not require an activator (Figure 9).

Figure 9. Anilinotropone ligands (a) and Ni (II) catalysts based on Anilinotropone (b) [60,61].

To determine the effects of varying the ortho-aryl substituents on ethylene polymerization, in an effort a series of anilinotropone ligands and corresponding nickel complexes was synthesized (Scheme 4). Among these catalysts, they incorporating alkyl substituents at the 2- and 6-positions of the N-aryl ring generated high-molecular weight polyethylenes with monomodal molecular weight distributions of ca. 2 (or less at low temperatures).

Under optimized conditions (Table 2), catalyst 14 bearing isopropyl substituents at the 2- and 6-positions of the N-aryl ring can produce PE, without the addition of a cocatalyst, with a TOF of 8.8 $\times 10^3$ kg of PE mol Ni^{-1} h^{-1} in a 10 min run. Additionally, 14 produced PE with a substantially higher M_n when compared with the salicylaldimino type bearing the anthryl substituent in the *ortho* position (89.6 vs. 54 kg mol^{-1}). Nevertheless 14 showed a short lifetime when compared to 2-salicylaldimino type bearing the anthryl substituent [62].

As can be seen in Table 3, with the exception of the run at 40 °C, the M_n value of the resultant polymers decreases with increasing temperature. The degree of branching steadily increases from 40 to 100 °C (8-67 branches per 1000 carbon atoms). This implies that the lower molecular weights obtained at higher temperatures are the result of an increase in the ratio of the chain transfer rate relative to the chain propagation rate, a normal feature of such polymerizations. The catalyst TOF is at a maximum at 80 °C. Ethylene pressure has also been shown to have dramatic effects on the resultant PE. Increasing the pressure from 1.01 to 41.34 bar at 80 °C caused a decrease in the branching number from 113 to 41 branches per 1,000 C [62].

Unlike salicylaldimine catalysts whose activates dramatically increase [58,63], no significant changes take placed in TOF, M_n, PDI, or branching numbers for catalyst 14 in the presence of added PPh_3. The observation that scavengers do not increase turnover numbers suggests that under the reaction conditions PPh_3 is essentially fully dissociated from the active nickel catalyst. Additionally, addition of polar solvent to the polymerization media decreased catalytic activity, but didn't change branching numbers [62].

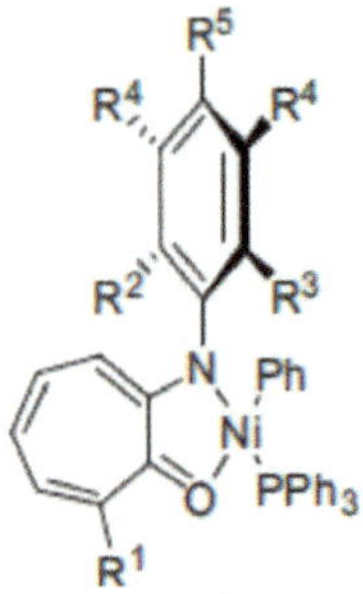

Catalyst	R^1-R^5
13	$R^1 = R^4 = R^5 = H$; $R^2 = R^3 = Me$
14	$R^1 = R^4 = R^5 = H$; $R^2 = R^3 = {}^iPr$
15	$R^1 = R^4 = R^5 = H$; $R^2 = R^3 = {}^tBu$
16	$R^1 = R^4 = R^5 = H$; $R^2 = Me$; $R^3 = {}^tBu$
17	$R^1 = R^4 = R^5 = H$; $R^2 = R^3 = Ph$
18	$R^1 = R^4 = R^5 = H$; $R^2 = R^3 = Cl$
19	$R^1 = R^4 = R^5 = H$; $R^2 = R^3 = Br$
20	$R^1 = H$; $R^2 = R^3 = R^4 = R^5 = F$
21	$R^1 = R^2 = R^3 = R^5 = H$; $R^4 = CF_3$
22	$R^1 = R^3 = R^4 = R^5 = H$; $R^2 = Me$
23	$R^1 = R^4 = R^5 = H$; $R^2 = Me$; $R^3 = CF_3$
24	$R^1 = R^4 = R^5 = H$; $R^2 = R^3 = F$
25	$R^1 = Ph$; $R^2 = R^3 = {}^iPr$; $R^4 = R^5 = H$
26	$R^1 =$ 1-naph; $R^2 = R^3 = {}^iPr$; $R^4 = R^5 = H$

Scheme 4. Ni (II) catalysts based on anilinotropone [62].

Replacement of the standard 2,6-diisopropyl substituents on the aryl group with 2,6-dimethyl (13), 2,6-dichloro (18), 2,6-dibromo (19), and 2-methyl-6-trifluoromethyl(23) substituents had little effect on productivity. A slight increase in TOF was observed for the 2,6-diphenyl-substituted catalyst (17). Significant reduction in TOF was observed for the 2-methyl-6-t-Butyl (16), 2-methyl-(22), 2-t-Butyl-(15), and 2,3,4,5,6-pentafluoro-substituted (20) catalysts. Molecular weights generally increase with increasing steric bulk of the ortho substituents (Table 2) [62].

Substitution of the 2-(2,6-diisopropylanilino)tropone ligand with either phenyl (25) or naphthyl (26) groups resulted in a small increase in productivities and lifetimes at 80 °C and 13.78 bar. However, at 40 °C these catalysts exhibited much longer lifetimes

($t_{1/2}$>1 h) and higher total turnover numbers could be achieved relative to 80 °C polymerizations [62].

Catalyst	P (bar)	T (°C)	TOF [b]	M_w	M_w/M_n	Br/1000 C
13	13.78	80	5424	73.1	1.7	61
14	13.78	80	8800	165.6	1.8	61
16	13.78	80	1014	230.0	2.0	73
17	13.78	80	10038	171.0	1.8	53
18	13.78	80	3924	20.0	2.0	53
19	13.78	80	4152	41.8	1.9	56
20	13.78	80	1152	4.8	3.0	49
21	13.78	80	936	112.8	2.4	57
23	13.78	80	6924	167.2	1.9	59
25	13.78	80	2338	184.8	2.1	63
26	13.78	80	2461	151.2	2.7	62

[a] Polymerization time = 10 min , Catalyst = 5.2 μmol; [b] Kg PE mol cat^{-1} h^{-1}

Table 2. Selected ethylene polymerization data of the Ni (II) catalysts based on anilinotropone [a] [62]

Molecular weights of the polyethylenes increased with pressure, which suggests that chain transfer at least in part occurs through classical β-H elimination rather than chain transfer to monomer. The catalyst decay product is the Ni (II) bis-ligand complex, whose formation must be initiated by reductive elimination of the ligand from a Ni (II) species [62].

Similar to other natural nickel ethylene polymerization catalysts which have already been reported for oligomerization of α-olefins [64,65], oligomerization of 1-hexene at 40 and 60 °C by catalyst 14 led to low yields of oligomers with degrees of polymerization of 21 and 16, respectively [62]. Branching numbers are 147 and 152 branches per 1000 carbon atoms, indicating a very minor amount of chain straightening via a 2,1-insertion and chain walking [39].

Salicylaldimine-based neutral Ni (II) catalyst bearing electron-withdrawing NO_2 group displayed high activity for polymerization of ethylene and resulted in high M_w polymer with fewer branches than the unnitrated catalysts [63]. Accordingly, Brookhart [66] prepared anilinotropone-based neutral Ni (II) catalysts 27-29 (Scheme 5) to examine the effect on catalyst activity and stability of adding strong electron-withdrawing NO_2 groups to the tropone and N-aryl ring.

As can be seen in Table 4, in comparison with unnitrated catalyst 13, the addition of two NO_2 groups dramatically enhances catalytic activity of 27 at 80 °C (entries 6 vs 4) from a TOF of 5423 to 27428 while the M_w of PE were comparable.

entry	Cat. (μmol)	P (psig)	T (°C)	TOF [b]	Mn ($\times10^3$)	Mw/Mn	Br/1000 C
1	14.8	400	60	5554	57	2.8	64
2	7.6	400	40	947	204	2.8	8
3	7.6	400	60	5368	292	2.0	27
4	7.6	400	80	9000	119	1.8	49
5	7.6	400	100	3157	61	1.9	67
6	5.2	1 atm	80	173	6.7	2.0	113
7	5.2	50	80	4615	50	1.7	90
8	5.2	100	80	7629	63	1.9	76
9	5.2	200	80	10615	90	1.8	61
10	5.2	200	80	8769	92	1.8	61
11	5.2	400	80	8192	104	2.0	45
12	5.2	600	80	3000	120	2.0	41
13	5.2	200	80	8653	78	1.9	66
14	5.2	400	80	12576	108	1.9	48
15	5.2	600	80	10153	111	2.0	43

[a] Polymerization time = 10 min, [b] Kg PE mol cat^{-1} h^{-1}

Table 3. Ethylene polymerization with 14 [a] [62]

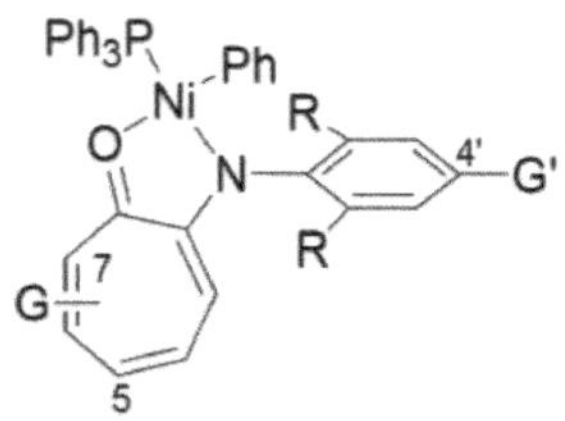

Catalyst	G, G′ and R	R
27	R = Me, G = 5,7-NO_2, G′ = Me	H
28	R = ^{i}Pr, G = 5,7-NO_2, G′ = H	H
29	R = ^{i}Pr, G = 5,7-NO_2, G′ = NO_2	H

Scheme 5. Nitrated catalysts based on anilinotropone 27-29 [66]

Moreover, catalyst 27 exhibited higher thermal stability than salicylaldimine-based Ni (II) catalysts [58] as well as higher lifetime than catalyst 13 (entries 3, 4 vs. 5, 6). As the polymerization temperature was increased from 60 °C to 80 °C, the polymer Mw decreased while catalytic activity and branching increased for catalyst 27.

Similar to 27, the addition of two or three NO_2 groups led to dramatically enhanced catalytic activities of diisopropyl analog 28 and 29, with a TOF of 80666 for 28 (entry 9) and 39600 for 29 (entry 10) compared with unnitrated parent catalysts 13 and 14.

entry	Cat.	Cat. (μmol)	t (min)	T (°C)	ToF [b]	Mw	Mw/Mn	Br/1000 C
1	27	5.5	60	60	800	289000	2.9	10
2	27	6.9	10	60	2260	230000	2.7	8
3	27	2.0	60	80	9650	128000	2.4	18
4	27	2.1	10	80	27428	65000	-	16
5	13	7.6	30	80	1657	-	1.7	-
6	13	5.2	10	80	5423	73000	-	61
7	28	0.9	10	60	4000	-	-	-
8	28	0.45	60	80	6666	-	-	-
9	28	0.9	10	80	80666	-	-	2
10	29	0.5	10	80	39600	-	-	7
11	14	5.2	60	80	1769	162000	1.8	61
12	14	5.2	10	80	8769	165000	1.8	61

[a] Ethylene pressure = 13.78 bar. [b] Kg PE mol cat^{-1} h^{-1}

Table 4. Ethylene polymerizations with 27-29 [a] [66]

All PEs produced by nitro-substituted catalysts 27-29 had monomodal GPC traces (PDIs between 2 and 4) and lower branching densities than those of parent catalysts 13 and 14 (entry 4 vs 6 and entries 9,10 vs. 12).

Catalyst 27 was able to incorporate vinyltrimethoxysilane or 1-hexene into polyethylene, but at a reduced rate compared to ethylene homopolymerization [66]. 1-Hexene or 1-octene oligomerization was also studied with catalysts 27 or 28 [66]. The oligomers were produced with low activities and lower amount of branching than expected (for 1-hexene, the expected number of branches after 2,1-insertion without chain walking is 166 branches/1000C and for 1-octene is 125 branches/1000C), indicating minor chain walking of catalysts during oligomerization.

7. The mechanistic investigation of the polymerization of ethylene

Combined DFT/stochastic studies were undertaken on the mechanism of ethylene polymerization catalyzed by the neutral Ni-anilinotropone catalysts [67]. Chain propagation and isomerization as well as influence of reaction conditions on the branching formation were investigated. Similarly to the case of nickel-salicylaldiminato catalysts the activation barriers for the insertion of ethylene in complexes with the trans alkyl group to the oxygen donor were found higher than those encountered in cis/trans isomerization. Interestingly, stochastic simulation allowed for establishing temperature and pressure dependence of the polymer microstructure. In agreement with experimental evidences the model predicts a decrease with the number of branches with the increase of pressure. Temperature dependence behaves oppositely as a result of an increase in secondary insertions with the temperature.

To get a full mechanistic picture of a typical neutral catalyst system, an elegant and in-depth NMR study of ethylene polymerizations catalyzed by the anilinotropone complexes was also performed [67]. Detailed information concerning the chain propagation process, the barrier to ethylene insertion, the nature, and dynamics of the intermediate Ni alkyl complexes, and the chain transfer and catalyst decay processes were obtained. The rate of insertion of ethylene into the Ni-phenyl bonds of both unsubstituted and aryl-substituted of anilinotropone nickel (II) complexes (14 and 25) was monitored by low temperature NMR spectroscopy. The ethylene insertion, initially, forms the corresponding [Ni(L)(PPh_3)(CH_2CH_2Ph)] complexes determined by monitoring the change in concentrations of starting complexes. First-order rate constants were measured as a function of temperature and ethylene concentration.

Even at high ethylene concentrations, only PPh_3 complexes were observed (no ethylene complexes were detected). As expected, rates accelerate with temperature, increasing about an order of magnitude between -10 and 10 °C. Insertion is dramatically inhibited by phosphine concentration. Thus, it was concluded that for these systems, the catalyst resting state(s) are an equilibrium mixture of phosphine and ethylene complexes (Figure 10). Nevertheless, at high ethylene pressures, the equilibrium is shifted nearly completely to the side of the ethylene complex and TOF becomes independent of ethylene pressure (saturation conditions).

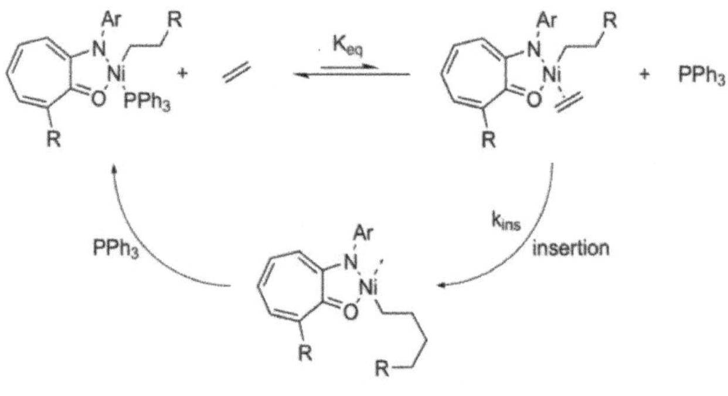

Figure 10. Ethylene polymerization catalyzed by Ni-anilinotropone catalysts [66].

The barriers to migratory insertion in (N–O)Ni(R)-(C_2H_4) complexes, from measurements of TOFs under saturation conditions, were determined to lie in the range of 16–17 kcal mol^{-1}, that is only about 2–3 kcal mol^{-1} greater than those observed for cationic diimine complexes.

The intermediate alkyl complexes have α-agostic interactions with dynamic behavior similar to the cationic alkyl diimine complexes. The barrier to β-H elimination and reinsertion is estimated in ca. 17 kcal mol^{-1}. Free energy barrier to nickel–carbon bond rotation in these complexes occurs with a barrier of 11.1 kcal mol^{-1}. These isomerization processes account for branched polyethylenes generated from these catalysts and resemble those previously observed with diimine catalysts.

Thermolysis of (N,O)Ni(hexyl)(PPh_3) generates the nickel hydride complex and 1-hexene via β-H elimination, through two pathways, one (dominant) independent of phosphine concentration and one involving reversible loss of phosphine, and thus inhibited by PPh_3. This process is a model for chain transfer and clearly explains why polymer molecular weights decrease with increasing phosphine concentration. The rate of propagation is retarded by increasing [PPh_3], while the rate of chain transfer is unchanged, resulting in an increase in R_{ct}/R_{prop}. Since chain transfer to monomer would exhibit a rate inhibition equal to that for R_{prop} with added PPh_3 it was concluded that the major chain transfer route is a simple β-elimination process, and not chain transfer to monomer.

At much slower rates, reductive elimination of the free ligand occurred from the hydride complex, and is inhibited by added phosphine. Catalyst decay under polymerization conditions was shown to occur by a similar process to generate free ligand and a bis-ligand complex formed by reaction of free ligand with an active catalyst species (Figure11).

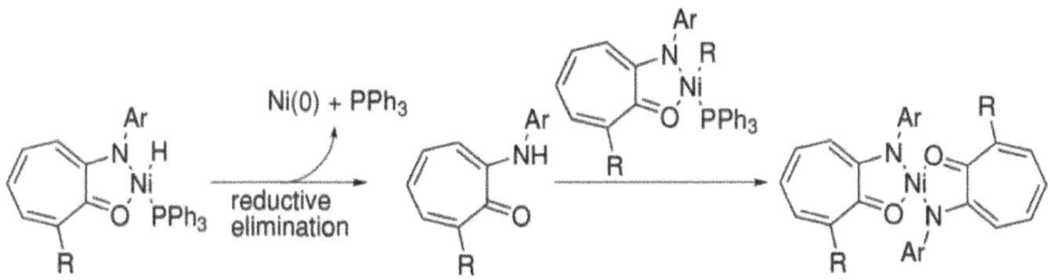

Figure 11. Deactivation of the anilinotropone nickel catalysts [67]

Author details

Reza Sandaroos
Department of Chemistry, Faculty of Science, University of Birjand, Birjand, Iran

Tomás Cuenca
Departamento de Química Inorgánica, Universidad de Alcalá, Campus Universitario, Alcalá de Henares, Spain

Gholam Hossein Zohuri
Department of Chemistry, Faculty of Science, Ferdowsi University of Mashhad, Mashhad, Iran

Saman Damavandi
Department of Chemistry, Sarvestan Branch, Islamic Azad University, Iran

Saeid Ahmadjo
Department of Catalyst, Iran Polymer and Petrochemical Institute, Tehran, Iran

8. References

[1] Ziegler, K, Holzkamp E, Breil H, Martin H (1955) Das Mülheimer Normaldruck-Polythylen-Verfahren. Angew. Chem. 67: 541-547.

[2] Alesso G, Sanz M, Mosquera MEG, Cuenca T (2008) Monocyclopentadienyl Phenoxido–Amino and Phenoxido–Amido Titanium Complexes: Synthesis, Characterisation, and Reactivity of Asymmetric Metal Centre Derivatives. Eur. J. Inorg. Chem. 29 : 4638–4649.

[3] Alesso G, Tabernero V, Mosquera MEG, Cuenca T (2011) Studies on the Active Species in Olefin Polymerisation Generated from Phenoxo-amido titanium "chiral-at-metal" Compounds. J. Organomet. Chem. 696: 2330-2337.

[4] Maupoey MG, Cuenca T, Frutos LM, Castaño O, Herdtweck H, Rieger B (2007) Alkylmono(cyclopentadienyl)titanium Complexes Containing the 2,2-Methylenebis(6-tert-butyl-4-methylphenoxido) Ligand. Studies on the Nature of the Catalytic Species Present in α-Olefin Polymerisation Processes. Eur. J. Inorg. Chem. 1: 147–161.

[5] Bochmann M, (2010) The Chemistry of Catalyst Activation: The Case of Group 4 Polymerization Catalysts. Organometallics, 29:(21) 4711-4740.

[6] Brintzinger HH, Fischer D, Mülhaupt R, Rieger B, Waymouth RM (1995) Stereospecific Olefin Polymerization with Chiral Metallocene Catalysts. Angew. Chem. Int. Ed. Engl. 34: 1143-1170.

[7] Abbo HS, Titinchi SJJ (2011) Bis(Pyrazolyl)Pyridine Late Transition Metal Complexes as Single-Site Catalysts for Ethylene Polymerization to Highly Linear Polyethylene. Catal. Lett. 139: 90–96.

[8] Ittel SD, Johnson LK, Brookhart M (2000) Late-Metal Catalysts for Ethylene Homo-and copolymerization. Chem. Rev. 100: 1169-1203.

[9] Zohuri, G.; Seyedi, S.; Sandaroos, R.; Damavandi, S.; Mohammadi, A. (2010) Novel Late Transition Metal Catalysts Based on Iron: Synthesis, Structures and Ethylene Polymerization. Catal. Lett. 140: 160-166.

[10] Damavandi S, Galland GB, Zohuri GH, Sandaroos R (2011) FI Zr-type Catalysts for Ethylene Polymerization. J. Polym. Res. 18: 1059-1065.

[11] Damavandi S, Zohuri G, Sandaroos R, Ahmadjo S (2012) Novel Functionalized bis(imino)pyridine Cobalt(II) Catalysts for Ethylene Polymerization. J. Polym. Res. 19: 9796-9800.

[12] Zohuri GH, Damavandi S, Sandaroos R, Ahmadjo S (2011) Ethylene Polymerization Using Fluorinated FI Zr-Based Catalyst. Polym. Bull. 66: 1051–1062

[13] Ventura M, Ramırez de Arellano C, Mosquera MEG, Jimenez G, Tomas Cuenca (2011) Synthesis and Structural Characterization of Novel Tetranuclear Organotitanoxane Derivatives. Dalton Trans. 40: 5728-5733.

[14] Ahmadjo S, Zohuri GH, Damavandi S, Sandaroos R (2010) Comparative Ethylene Polymerization Using FI-like Zirconium Based Catalysts, Reac. Kinet. Mech. Cat. 101: 429-442.

[15] Yoshida Y Matsui S, Takagi Y, Mitani M, Nitabaru M, Nakano T, Tanaka H, Fujita T (2000) Post Metallocenes: New Bis(pyrrolyl-2-aldiminato) Titanium Complexes for Ethylene PolymerizationChem. Lett. 1270-1271.

[16] Yoshida Y, Matsui S, Fujita T (2005) Bis(pyrrolideimine) Ti Complexes with MAO: a New Family of High Performance Catalysts for Olefin Polymerization. J. Organometal. Chem. 690: 4382-4397.

[17] Stephan DW Guérin F, Spence, REVH, Koch L, Gao X, Brown SJ, Swabey JW, Wang Q, Xu W, Zoricak P, Harrison DG (1999) A Remarkably Active Non-Metallocene Ethylene Polymerization Catalyst. Organometallics, 18: 2046-2048.

[18] Murtuza S, Casagrande OLJr, Jordan RF (2001) Olefin Polymerization by Tris(pyrazolyl)borate Titanium Catalysts. Polym. Mater. Sci. Eng. 2001, 84: 109-110.

[19] Sandaroos R, Damavandi S, Farhadipour A, (2010) A New Family of High-Performance Ti Catalysts for Olefin Polymerization. Macromol. Chem. Phys. 211: 2339-2346.

[20] Goldani M, Sandaroos R, Mohmmadi A, Goharjoo M, (2012) A comparative study of ethylene polymerization by bis(aminotropone) Ti catalysts. Polym. Bull. 68: 755-773.

[21] Sandaroos R, Damavandi S, Nazif A, Goharjoo M, Mohammadi A (2011) Highly Efficient Bis(aminotropone) Ti Catalyst for Ethylene Polymerization. Chinese Chemical Letters, 22: 213–216.

[22] Ishii SI, Saito J, Mitani M, Mohri JI, Matsukawa N, Tohi Y, Matsui S, Kashiwa N, Fujita T (2002) Highly Active Ethylene Polymerization Catalysts Based on Titanium Complexes Having Two Phenoxyimine Chelate Ligand. J. Macromol. Catal. 179: 11-16.

[23] Mitani M, Saito J, Ishii S, Nakayama Y, Makio H, Matsukawa N, Matsui S, Mohri J, Furuyama R, Terao H, Bando H, Tanaka H, Fujita T (2004) FI Catalysts: New Olefin Polymerization Catalysts for the Creation of Value-Added Polymers. Chem. Rec. 4: 137–158.

[24] M Mitani, T Nakano, T Fujita (2003) Unprecedented Living Olefin Polymerization Derived from an Attractive Interaction Between a Ligand and a Growing Polymer Chain. Chem. Eur. J. 9:(11) 2396-2403.

[25] Makio H, Fujita T (2005) Propene Polymerization with Bis(phenoxy–imine) Group 4 Transition Metal Complexes. Bull. Chem. Soc. Jpn. 78: 52-66.

[26] Nakayama Y, Bando H, Sonobe Y, Fujita T (2004) Development of Single-site New Olefin Polymerization Catalyst Ssystems Using MgCl2-Based Activators: MAO-Free $MgCl_2$-Supported FI Catalyst Systems. Chem. Soc. Japan, 77: 617-625.

[27] Yoshida Y, Saito J, Mitani M, Takagi Y, Matsui S, Ishii SI, Nakano T, Kashiwa N, Fujita T (2002) Living ethylene/norbornene copolymerisation catalyzed by titanium complexes having two pyrrolide-imine chelate ligands. Chem. Commun. 1298-1299.

[28] Yoshida, Y, Mohri J, Ishii S, Mitani M, Saito J, Matsui S, Makio H, Nakano T, Tanaka H, Onda M, Yamamoto Y, Mizuno A, Fujita T (2004) Living Copolymerization of Ethylene with Norbornene Catalyzed by Bis(Pyrrolide–Imine) Titanium Complexes with MAO. J. Am. Chem. Soc. 126: 12023-12032.

[29] Yoshida Y, Matsui S, Takagi Y, Mitani M, Nakano T, Tanaka H, Kashiwa N, Fujita T (2001) New Titanium Complexes Having Two Pyrrolide–Imine Chelate Ligands: Syntheses, Structures, and Ethylene Polymerization Behavior. Organometallics. 20:(33) 4793-4799.

[30] Matsugi T, Matsui S, Kojoh S, Takagi Y, Inoue Y, Nakano T, Fujita T, Kashiwa N (2002) New Titanium Complexes Bearing Two Indolide–Imine Chelate Ligands for the Polymerization of Ethylene. Macromolecules, 35: 4880-4487.

[31] Alt HG, Köppl A, (2000) Effect of the Nature of Metallocene Complexes of Group IV Metals on Their Performance in Catalytic Ethylene and Propylene Polymerization. Chem. Rev. 100: 1205-1221.

[32] Coates GW (2000) Precise Control of Polyolefin Stereochemistry Using Single-Site Metal Catalysts. Chem. Rev. 100: 1223-1252.

[33] Resconi L, Cavallo L, Fait A, Piemontesi F, (2000) Selectivity in Propene Polymerization with Metallocene Catalysts. Chem. Rev. 100: 1253-1345.

[34] Boffa LS, Novak BM (2000) Copolymerization of Polar Monomers with Olefins Using Transition-Metal Complexes. Chem. Rev. 100:1479-1493.

[35] Vatankhah-Varnoosfaderani M, Pourmahdian S, Afshar-Taromi F (2011) High Performance Bulky α-Diimine Nickel(II) Catalysts for Ethylene Polymerization. 20:(11), 897-912.

[36] Liu HR, Gomes PT, Costa SI, Duarte MT, Branquinho R, Fernandes AC, Chien JCW, Singh RP, Marques MM (2005) Highly active new α-diimine nickel catalyst for the polymerization of α-olefins, J. Organomet. Chem. 690: 1314-1323.

[37] Mecking S (2000) Water as a Reaction Medium for Nickel(II)-catalyzed Ethylene Polymerization. Coord. Chem. Rev. 203: 325-351.

[38] Gates DP, Svejda SA, Onate E, Killian CM, Johnson LK, White PS, Brookhart M (2000) Synthesis of Branched Polyethylene Using (α-Diimine)nickel(II) Catalysts: Influence of Temperature, Ethylene Pressure, and Ligand Structure on Polymer Properties Macromolecules 33: 2320-2334.

[39] Guan Z, Cotts PM, McCord EF, McLain SJ (1999) Chain Walking: A New Strategy to Control Polymer Topology. Science, 283: 2059-2062.

[40] Johnson LK, Killian CM, Brookhart M (1995) New Pd(II)- and Ni(II)-Based Catalysts for Polymerization of Ethylene and α-Olefins. J. Am. Chem. Soc. 117: 6414-6415.

[41] Killian CM, Tempel DJ, Johnson LK, Brookhart M (1996) Living Polymerization of α-Olefins Using Ni II α-Diimine Catalysts. Synthesis of New Block Polymers Based on α-Olefins. J. Am. Chem. Soc. 118: 11664-11665.

[42] Leatherman MD, Brookhart M (2001) Ni(II)-Catalyzed Polymerization of trans-2-Butene. Macromolecules, 34: 2748-2750.

[43] Bauers FM, Mecking S (2001) Aqueous Homo- and Copolymerization of Ethylene by Neutral Nickel(II) Complexes. Macromolecules, 34: 1165-1171.

[44] Mecking S, Held A, Bauers FM (2002) Aqueous Catalytic Polymerization of Olefins. Angew. Chem., Int. Ed. 41: 544-561.

[45] Soula R, Broyer JP, Llauro MF, Tomov A, Spitz R, Claverie J, Drujon X, Malinge J, Saudemont T (2001) Very Active Neutral P,O-Chelated Nickel Catalysts for Ethylene Polymerization. Macromolecules, 34: 2438-2442.

[46] Soula R, Saillard B, Spitz R, Claverie J, Llaurro MF, Monnet C (2002) Catalytic Copolymerization of Ethylene and Polar and Nonpolar α-Olefins in Emulsion. Macromolecules, 35: 1513-1523.

[47] Johnson L Bennett A Dobbs K, Hauptman E, Ionkin A, Ittel S, McCord E, McLain S, Radzewich C, Yin Z, Wang L, Wang Y, Brookhart M (2002) Polym. Mater. Sci. Eng. 86: 319-322.

[48] Johnson LK, Mecking S, Brookhart M (1996) Copolymerization of Ethylene and Propylene with Functionalized Vinyl Monomers by Palladium(II) Catalysts. J. Am. Chem. Soc. 118: 267-268.
[49] Mecking S, Johnson LK, Wang L, Brookhart M (1998) Mechanistic Studies of the Palladium-Catalyzed Copolymerization of Ethylene and α-Olefins with Methyl acrylate. Journal of the American Chemical Society, 120(5): 888–899.
[50] Ostoja Starzewski KA, Witte J (1985) Highly Active Ylide-Nickel Catalysts for the Polymerization of Ethylene. Angew. Chem., Int. Ed. Engl. 1985, 24, 599-601.
[51] Tamburlin-Thumin I, Crozet MP, Barriere JC (1999) A Convenient Method for the Alkylation of Tropolone Derivatives and Related Alpha-keto Hydroxy Compounds. Synthesis, 7: 1149-1154.
[52] Small B. L, Brookhart M, Bennett AMA Highly Active Iron and Cobalt Catalysts for the Polymerization of Ethylene. (1998) J. Am. Chem. Soc. 120: 4049-4050.
[53] Small B. L, Brookhart M (1998) Iron-Based Catalysts with Exceptionally High Activities and Selectivities for Oligomerization of Ethylene to Linear -Olefins. J. Am. Chem. Soc. 1210: 7143-7144.
[54] McLain SJ, Sweetman KJ, Johnson LK, McCord E (2002) Polym. Mater. Sci. Eng. 86: 320-321.
[55] Hirose, K, Keim W (1992) Olefin Oligomerization with Nickel PO Chelate Complexes. J. Mol. Catal. 73: 271-276.
[56] Peuckert M, Keim W (1983) A New Nickel Complex for the Oligomerization of Ethylene Organometallics, 2: 594-597.
[57] Johnson LK, Bennett AMA, Ittel SD, Wang L, Parthasarathy A, Hauptman E, Simpson RD, Feldman J, Coughlin EB (1998) Polymerization of Olefins. WO 98/30609.
[58] Younkin TR, Connor EF, Henderson JI, Friedrich SK, Grubbs RH, Bansleben DA (2000) Neutral, Single Component Nickel (II) Polyolefin Catalysts that Tolerate Heteroatoms. Science 287: 460-462.
[59] Yang BH, Buchwald SLJ (1999) Palladium-catalyzed Amination of Aryl halides and Sulfonates. Organomet. Chem. 576: 125-146.
[60] Hicks FA, Brookhart M (2000) Synthesis of 2-Anilinotropones via Palladium-Catalyzed Amination of 2-Triflatotropone. Org. Lett. 2: 219-221.
[61] Hicks FA, Brookhart M (2001) A Highly Active Anilinotropone-Based Neutral Nickel(II) Catalyst for Ethylene Polymerization. Organometallics, 20: 3217-3219.
[62] Hicks FA, Jenkins JC, Brookhart M, (2003) Synthesis and Ethylene Polymerization Activity of a Series of 2-Anilinotropone-Based Neutral Nickel(II) Catalysts. Organometallics, 22: 3533-3545.
[63] Wang C, Friedrich S, Younkin TR, Li RT, Grubbs RH, Bansleben DA, Day MW (1998) Neutral Nickel(II)-Based Catalysts for Ethylene Polymerization. Organometallics, 17: 3149-3159.
[64] Killian CM, Tempel DJ, Johnson LK, Brookhart M, (1996) Living Polymerization of α-Olefins Using NiII−α-Diimine Catalysts. Synthesis of New Block Polymers Based on α-Olefins. J. Am. Chem. Soc. 118: 11664-11665.

[65] Carlini C, De Luise V, Fernandes EG, Martinelli M, Galletti AMR, Sbrana G (2005) Effect of Free Trimethylaluminum Content in Methylaluminoxane on Performances of Bis(salicylaldiminate)nickel(II)-Based Catalysts for Ethylene Polymerization. Macromol. Rapid Comm. 26, 808-812.
[66] Brookhart, MS, Zhang L (2006) Catalyst for ethylene homopolymerization and copolymerization thesis (UMI number 3219473).
[67] Jenkins JC, Brookhart M (2004) A Mechanistic Investigation of the Polymerization of Ethylene Catalyzed by Neutral Ni(II) Complexes Derived from Bulky Anilinotropone Ligands. J. Am. Chem. Soc. 126, 5827-5842.

Chapter 13

Block and Graft Copolymerization by Controlled/Living Radical Polymerization Methods

Hülya Arslan

Additional information is available at the end of the chapter

1. Introduction

Copolymerization is a versatile synthetic tool for controlling the functionality and, therefore, tailoring of the physical and mechanical properties of macromolecules. Block copolymers are one of the most important polymeric materials in technological applications and theoretical research because of their exceptional properties due to the microphase separation [1]. Many basic characteristics of block copolymers and graft copolymers are similar. Graft copolymers also exhibit good phase separation and are used for a variety of applications, such as impact-resistant plastics, thermoplastic elastomers, compatibilizers and polymeric emulsifers. Because of their branched structure they generally have also lower melt viscosities, which is advantageous for processing. They have great potential to realize new properties because of their structural variables (composition, backbone length, branch length, branch spacing, etc.) [2,3]. Various polymerization methods have been used to synthesis block and graft copolymers, such as free radical, anionic, cationic, coordination, group transfer, coupling, changing polymerization mechanism, step growth, macroinitiators, metathesis and many books [4,5], chapters of encyclopaedia [6-10], papers [11,12] offer excellent reviews of research on synthesis block and graft copolymers by these methods.

According to literature, the most common and widely used approach is free-radical polymerization method. Free-radical polymerization is particularly interesting because of its adaptability to a wide range of functinal monomers under less stringent synthesis conditions. Unfortunately, uncontrolled nature of conventinal free-radical polymerization leads to homopolymerzation during the polymerization resulting in a product that is a mixture of homopolymer and copolymer. A major drawback of conventional radical polymerization is that almost no control over the molecular weight, molecular weight distribution and architecture of the polymer is possible, making its macroscopic properties very difficult to be tailored. Living ionic polymerizations can provide a methodology for preparation of copolymers with well-defined structures, such as controlled molecular weights and narrow molecular weight distribution, defined copolymer compositions, branching and end-group functionalities. There are various drawbacks that often retard the practical application of living ionic systems to prepare copolymers. These drawbacks are limitation of the method to certain monomers and their sensitivity to moisture, carbon dioxide, and numerous other acidic or basic compounds.

Polymer chemists have long tried to combine the advantages of both radical and ionic polymerizations in a single polymerizations process. These attempts have been principally successful by controlled/living radical polymerization (CRP) techniques. Controlled/living radical polymerization in brief is based on an understanding and integration of chemistry developed over 60 years in the fields of organic chemistry, conventional radical polymerization and living ionic polymerizations. Recently, extensive research has been carried out in the field of controlled/living radical polymerization (CRP), because it can be employed for the polymerization of numerous vinyl monomers under mild reaction conditions and these methods enable the synthesis of well-defined polymer structures with controlled molecular weight and narrow molecular weight distribution. CRP includes atom transfer radical polymerization (ATRP), nitroxide-mediated polymerization (NMP), and reversible addition-fragmentation chain transfer (RAFT) polymerization. Each of these methods relies on establishment of a rapid dynamic equilibrium between a low concentration of active propagating chains and a predominant amount of dormant chains that are unable to propagate or terminate, and is more tolerant towards functional groups and impurities. Recently, many studies have been reported in the literature about the synthesis of macromolecules with various compositions (homopolymers, random, periodic, block, graft and gradient copolymers) and novel topologies (linear, star, comb, branched, hyper branched, networks, brushes etc.) using CRP techniques [13-30]. It is observed that the significant improvement in block and graft copolymer synthesis has become with development of controlled/living radical polymerization (CRP) techniques. This chapter focuses on the preparation of block and graft copolymers *via* controlled/living radical polymerization (CRP) techniques which are new and efficient synthetic strategies.

2. Controlled/living radical polymerization (CRP)

CRP can be described as "A radical polymerization that can be stopped and re-initiated under external control" and shown schematically in Figure 1.

General CRP mechanism can be summarized as follows;

$$\text{Dormant species} \underset{\text{deactivation}}{\overset{\text{activation}}{\rightleftharpoons}} \underset{\underset{\text{monomer}}{\circlearrowright k_p}}{\text{Active species}} + \text{Capping agent (trap)} \qquad (1)$$

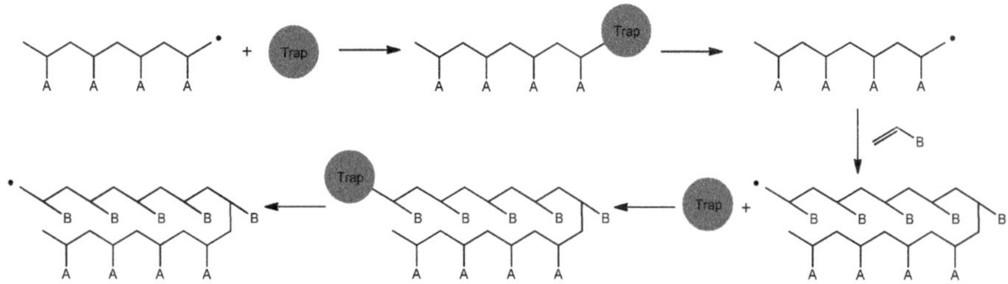

Figure 1. Schematic representation of CRP

Equilibrium is established between dormant, end-capped chains, unable to propagate or terminate, and the active species. The equilibrium should be shifted towards the dormant species and exchange between active and dormant species should be fast relative to propagation. CRP distinguishes itself from conventional radical polymerization by involving a reversible activation process.

2.1. Historical evolution of CRP

Table 1 illustrate development of fundamental concepts in organic radical processes, some concepts in living ionic polymerization and advances in radical polymerizations which all have contributed to controlled/ "living" radical polymerization. The success of controlled/ "living" radical polymerization is an integration of advances in synthetic organic chemistry, living ionic polymerization and conventional radical polymerization (Figure 2) [31,32].

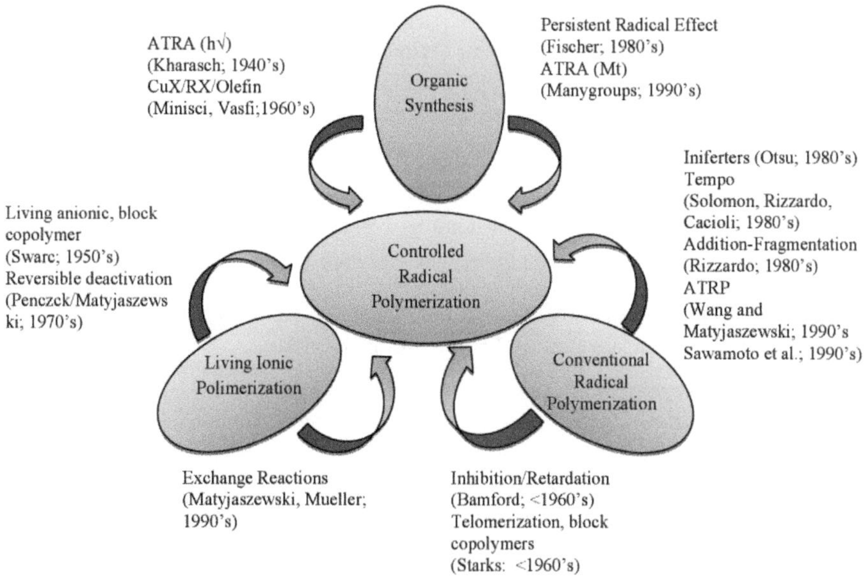

Figure 2. Development of CRP by integration of advances in several fields of chemistry

Organic synthesis	**Living ionic polymerization**	**Conventional radical polymerization**	**Controlled/ "living" radical polymerization**
1940s Kharash: First ATRA (hv) 1960s Minisci, Vosfi: CuX/RX/Olefin Rosantsev: Nitroxide Kochi: Free radicals&Mt 1980s Fischer: Persistent Radical Effect (PRE) 1990s Many groups: ATRA (Mt)	1950s Szwarc: Living anionic polym., block copolymers 1970s Penczek/Matyjaszewski: Reversible deactivation in Cationic Ring Opening Polymerization (CROP) 1980s Kennedy: Inifers, Kennedy, Higashimura, Sawamoto, Sigwalt, Matyjaszewski: Carbocationic 1990s Matyjaszewski, Mueller: Exchange reactions	<1960s Bambord: Inhibition/retardation Starks: Telomerization, block copolymers 1970s Minoura: Cr(II)acetate/MMA 1980s Otsu: "1st" Living radical polym., Iniferters Solomon, Rizzardo, Cacioli: Stable Free Radical Polymerization (SFRP)-Synthesis of oligomers with nitroxide (TEMPO) Rizzardo: Addition-fragmentation (RAFT) 1990s Georges: TEMPO mediated styrene polym. (NMP) Wang and Matyjaszewski: ATRP Sawamoto: ATRP	1980s Otsu: **Iniferters** Solomon, Rizzardo, Cacioli: **(SFRP)-(TEMPO)** Rizzardo: **(RAFT)** 1990s Georges: TEMPO/styrene **(NMP)** Wang and Matyjaszewski: **ATRP** Sawamoto: **ATRP**

Table 1. Evolution of controlled/ "living" Radical Polymerization [31,32]

Radical addition reactions have for a long time been considered to be very difficult to control due to termination reactions which not be avoided. Probably the first example of a successful addition of halogenated compounds to alkenes via radical intermediates (atom Transfer Radical Addition, ATRA) was provided by Kharash under photochemical conditions. This atom transfer radical process was subsequently converted to metal catalyzed reactions by Minisci, Vosfi and others during the 1960s.

In 1982 Otsu et al., for the first time, used the term *living* radical polymerization to describe a free radical polymerization in the presence of dithiocarbamates. In analogy to the inifer used in carbocationic systems, they proposed that dithiocarbamates acts as iniferters, i.e. agents which initiate, transfer and terminate. Unfortunately, as in the previously discussed systems, polydispersities were always relatively high, molecular weights did not evolve linearly with conversion and initiation efficiency was low. A new system for controlling radical polymerization based on nitroxides as stable free radicals is developed by Solomon, Rizzardo and Cacioli in 1986. They synthesized methyl acrylate oligomers *via* reversible capping of growing radical chain by using 2,2,6,6-tetramethyl-1-piperidinyloxy (TEMPO) as a stable free radical. The reversible capping of growing chains defined the first mechanism of three general routes to controlled/ "living" radical polymerization. This route was named as "Stable Free Radical Polymerization (SFRP)" by its discovers. Three main types of techniques have been developed over the years: 1) stable free radical polymerization (SFRP) with nitroxide-mediated radical polymerization (NMP), 2) atom transfer radical polymerization (ATRP) and 3) reversible addition-fragmentation transfer polymerization (RAFT).

2.2. Equilibrium (reversible activation/deactivation) for the three major CRP processes

The reversible activation/deactivation reactions (dynamic equilibrium) in the most effective CRP process may be classified mechanistically into three types, which are (a) dissociation-combination (DC) (coupling), (b) atom transfer (AT) and (c) degenerative chain transfer (DT).

(a) Reversible deactivation by dissociation-combination (coupling). e.g. Nitroxide-mediated polymerization (NMP)

$$\text{Pn-T} \underset{\text{kdeact}}{\overset{\text{kact}}{\rightleftharpoons}} \underset{\circlearrowright\,\text{kp, monomer}}{\text{Pn}^*} + \text{T}^* \qquad (2)$$

(b) Reversible deactivation by atom transfer. e.g. ATRP

$$\text{Pn-X} + \text{Mt}^n/\text{L} \underset{\text{kdeact}}{\overset{\text{kact}}{\rightleftharpoons}} \underset{\circlearrowright\,\text{kp, monomer}}{\text{Pn}^*} + \text{X-Mt}^{n+1}/\text{L} \qquad (3)$$

(c) Degenerative chain transfer. e.g. RAFT

$$\underset{\circlearrowright\,\text{kp, monomer}}{\text{Pn}^*} + \text{Pm-X} \underset{k_{-tr}}{\overset{k_{tr}}{\rightleftharpoons}} \text{Pn-X} + \underset{\circlearrowright\,\text{kp, monomer}}{\text{Pm}^*} \qquad (4)$$

Each of these methods relies on establishment of a dynamic equilibrium between a low concentration of active propagating chains and a predominant amount of dormant chains that are unable to propagate or terminate as a means of extending the lifetime of the propagating chains.

2.3. General features of CRP

It is widely accepted that a controlled polymerization process should display the following features:

1. First-order Kinetic Behavior
2. Pre-determinable Degree of Polymerization
3. Designed (Usually) Narrow Molecular Weight Distribution
4. Long-lived Polymer Chain with Preserved End Functionalities

2.3.1. First-order kinetic behavior:

The polymerization rate (R_p) with respect to the log of the monomer concentration ([M]) is a linear function of time. This is due to the negligible contribution of non-reversible termination, so that the concentration of the active propagating species ([P*]) is constant.

$$R_p = \frac{-d[M]}{dt} = k_p[P^*][M] \qquad (5)$$

$$\ln\frac{[M]_0}{[M]} = k_p[P^*]\,t = k_p^{app}\,t \;\; (\text{if } [P^*] \text{ is constant}) \qquad (6)$$

k_p is the propagation constant. The consequence of equation (6) and the effect of changes in P* are illustrated in below:

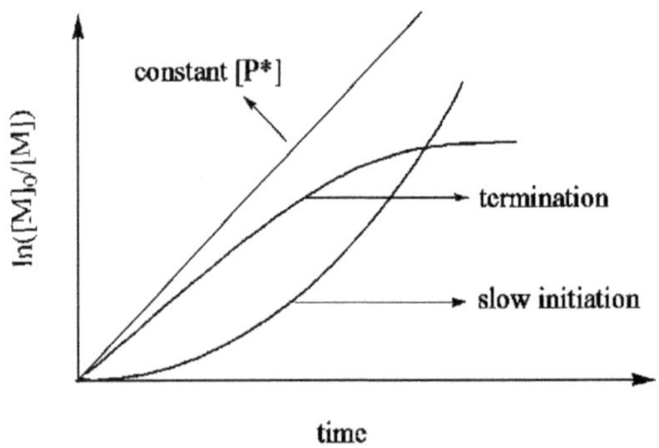

Figure 3. Illustration of the dependence of ln([M]$_0$/[M]) on time

This semi logarithmic plot is very sensitive to any change of the concentration of the active propagating species. A constant [P*] is revealed by a straight line. A steady [P*] in a CRP is established by balancing the rates of activation and deactivation and *not* by balancing the rates of initiation and termination as in a conventional radical polymerization. An upward curvature in the kinetic plot indicates an increase in [P*], which occurs in case of slow initiation. On the other hand, a downward curvature suggests a decrease in [P*], which may result from termination reactions increasing the concentration of the persistent radical, or some other side reactions such as the catalytic system being poisoned or redox processes on the radical.

It should also be noted that the semi logarithmic plot is not sensitive to chain transfer processes or slow exchange between different active species, since they do not affect the number of the active propagating species.

2.3.2. Pre-determinable degree of polymerization:

Degree of polymerization (DP_n), i.e. the number average molecular weight (M_n) is a linear function of monomer conversion.

$$DP_n = \frac{M_n}{M_0} = \frac{\Delta[M]}{[I]_0} = \frac{[M]_0}{[I]_0}(\text{ conversion}) \qquad (7)$$

This result comes from maintaining a constant number of chains throughout the polymerization, which requires the following two conditions:

1. initiation should be sufficiently fast so that essentially all chains are propagating before the reaction is stopped; and
2. an absence of chain transfer reactions that increases the total number of chains.

Figure 4 illustrates feature 2 and shows the ideal growth of molecular weight with conversion, as well as the effects of slow initiation and chain transfer on the molecular weight evolution.

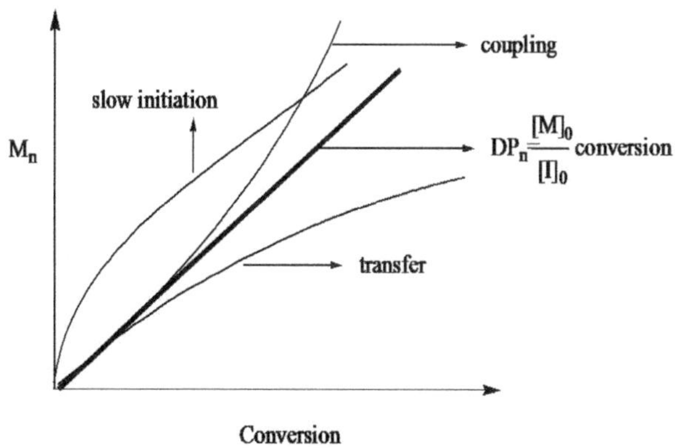

Figure 4. Illustration of the dependency of molecular weight on conversion

It is important to recognize that the evolution of molecular weight is not very sensitive to chain termination, since the number of chains remains unchanged. The effect of termination is only observable on the plot when coupling reactions, forming higher molecular weight polymers, start to play a significant role.

2.3.3. *Designed, usually narrow, molecular weight distribution:*

Although this feature is often desirable, it is not necessarily the result of a controlled polymerization, which only requires the absence of chain transfer and termination, but ignores the effect of rate of initiation, exchange and depropagation.

In order to obtain a polymer with a narrow molecular weight distribution, each of the following requirements should be fulfilled.

i. **The rate of initiation is fast in comparison to the rate of propagation**. This condition allows simultaneous growth of all the polymer chains.
ii. **The exchange between species of different reactivity is fast in comparison with the rate of propagation.** This condition ensures that all the active chain termini are equally susceptible to reaction with monomer allowing uniform chain growth.
iii. **There must be negligible chain transfer or termination**.
iv. **The rate of depropagation is substantially lower than propagation.** This guarantees that polymerization is essentially irreversible.

This should yield a Poisson distribution, as quantified in equation (8).

$$\frac{X_w}{X_n} = \frac{M_w}{M_n} = 1 + \frac{X_n}{(X_n+1)^2} \cong 1 + \frac{1}{X_n} \tag{8}$$

According to equation (8), polydispersity (PDI = M_w/M_n) decreases with increasing molecular weight.

2.3.4. *Long-lived polymer chains with preserved end-functionalities:*

This is a consequence of negligible irreversible chain transfer and termination. Hence, all the chains retain their active centers after the full consumption of the monomer. Propagation resumes upon introduction of additional monomer. This unique feature enables the preparation of block copolymers by sequential monomer addition. The significance of controlled polymerization as a synthetic tool is widely recognized and polymers having uniform predictable chain length are readily available. Controlled polymerization provides the best opportunity to control the bulk properties of target materials through control of the multitude of possible variations in composition, functionality and topology now attainable at a molecular level. Through appropriate selection of the functional (macro) initiator, copolymers formed in a controlled/"living" polymerization process can have essentially any desired topology. Further, the nature of the mechanism enables specific end functionalization and addition of a second monomer to make a block copolymer.

2.4. Potential applications of CRP processes

The development of CRP has allowed the preparation of many new materials with vastly differing properties simply by varying the topology of the polymer (i.e., comb, star, dendritic, etc.), the composition of the polymer (i.e., random, periodic, graft, etc.), or the functional groups at various sites on the polymer (i.e., end, center, side, etc.) (Figure 5) [33].

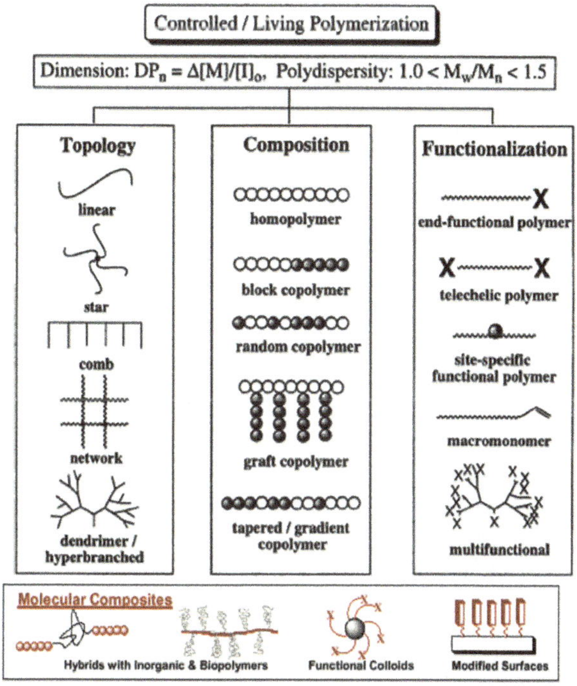

Figure 5. A schematic representation of how new polymers and materials can be prepared using controlled/living radical polymerizations

3. Blok copolymerization

Polymers which are composed of two different monomers are usually referred to as copolymers. The sequential arrangement of these different monomers determines the type of copolymer that is formed. In the case of a random distribution of the two monomer over the polymer chain, the product is called a random copolymer. In a linear block copolymer, the monomeric residues are arranged in such a way that one block consists of monomeric residue A and another block of monomeric residue B. Block copolymers are defined as having a linear arrangement of blocks of varying monomer composition. That is, a block copolymer is a combination of two or more polymers joined end-to-end. Moreover, one can synthesize more exotic structures such as four- or six-arm starblock copolymers and comb-like block copolymers. These have branched structures.

The increasing importance and interest in block copolymers arises mainly from their unique properties in solution and in the solid state which are a consequence of their molecular structure. In particular, sequences of different chemical composition are usually incompatible and therefore have a tendency to segregate in space. Amphiphilic properties in solution and microdomain formation in solid state are directly related to this specific molecular architecture, which can be designed by using existing monomers or polymers. Block copolymers are generally applied as adhesives (e.g. surfactants and viscosity improvers), compatibilizers, thermoplastic elastomers, etc.

The concept of block copolymer synthesis started in 1956 when Szwarc discovered living anionic polymerization. A living process implies that all polymer chains start growing simultaneously; while during chain growth no termination or chain transfer takes place. Consequently, all chains grow for a similar period, and a narrow molecular weight distribution is obtained. When all monomer has been consumed, the active center persists, and upon addition of a new batch of monomer, polymerization continues to form a block copolymer. Anionic polymerization however requires very stringent reaction conditions, as the carbon-centered anion is very sensitive to traces of impurities. Industrially, anionic polymerization is therefore not frequently used. Later, block copolymer synthesis was also achieved by other living polymerization techniques, such as ring opening polymerization and cationic polymerization or by pseudo-living techniques like Ziegler-Natta catalysis. These polymerization techniques, however, suffer from a number of disadvantages: most of them do not tolerate even extremely low levels of impurities (like e.g. moisture or oxygen), and are compatible only with a limited number of monomers. Also these processes are very expensive, due to the special equipment and the reaction conditions that are needed to perform these reactions. This obviously leads to an expensive polymer product. An entirely different synthetic concept towards (multi)block copolymers makes use of the polycondensation principle. Since the early 1980s, there were several early attempts to synthesize block copolymers via regulated free-radical polymerizations. These methods utilized so-called iniferters, i.e. compounds which could serve as INItiators, transFER agents and TERminating agents. However, while useful, these techniques did not offer the desired level of control over macromolecular structure, due to poor molecular weight control, high polydispersities, and low blocking efficiency. This was attributed to slow initiation, slow exchange, direct reaction of counter radicals with monomers, and thermal decomposition of the iniferter. In the last decade, some polymerization techniques that combine the versatility of free radical polymerization with the control of anionic polymerization emerged. These techniques are referred to as controlled radical polymerizations, such as Nitroxide-Mediated Polymerization (NMP), Atom Transfer Radical Polymerization (ATRP) and Reversible Addition-Fragmentation chain Transfer (RAFT) and are based on two principles: reversible termination and reversible transfer. These techniques have enabled researchers to synthesize block copolymers under less stringent conditions than those necessary for living ionic polymerizations: it is possible to work under similar conditions as used in free-radical polymerizations. The synthesis of block copolymers by general polymerization methods such as free radical, anionic, cationic, coordination, coupling, step growth, ring opening, changing polymerization mechanism (active centers transformation reactions), have summarized several times in the literature [1,4,6-10, 34]. In this chapter we will focuses on the preparation of block and graft copolymers *via* controlled/living radical polymerization (CRP) techniques which are new and efficient synthetic strategies.

Block copolymers can be prepared by controlled polymerization routes in two strategies which are the sequential monomer addition strategy and the use of a difunctional initiator. In the sequential monomer addition strategy, certain experimental conditions governing the growth of the first block should be handled carefully to increase the blocking efficiency. It is essentially important to stop the polymerization of the first monomer before it is used up completely for the fact that end functionalities might be lost, resulting in the formation of dead blocks. Another feature to be considered, as in all sequential monomer additions, is the order of introducing each monomer [35]. In brief, the very active monomer should be polymerized first, which is valid for either of the controlled polymerization mechanisms. Different from sequential monomer addition technique utilized in anionic living polymerization, controlled routes allow for the polymerization of monomers A and B to be performed separately in two distinct steps, allowing chain functionalities to be preserved. The first block functions as a macromolecular initiator, a so-called macroinitiator, in the preparation of the second block.

By this way AB type diblock copolymers can be prepared. Further, the nature of the mechanism of CRP enables specific end functionalization and addition of a another monomer (C) to make ABC type triblock copolymer:

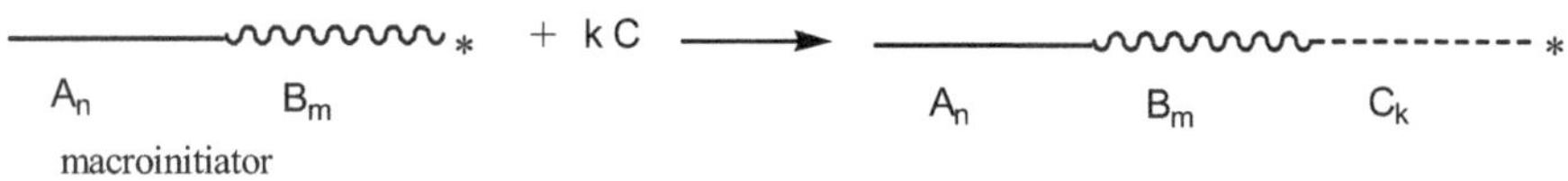

Another route involves the use of a difunctional initiator and has been employed in the preparation of ABA type symmetric triblock copolymers. In this methodology, a compound possessing two initiating sites is utilized in the formation of the middle block first, followed by the polymerization of the second monomer to synthesize the first and the third blocks. This allows the preparation of the ABA type triblock copolymer in two steps, instead of three steps:

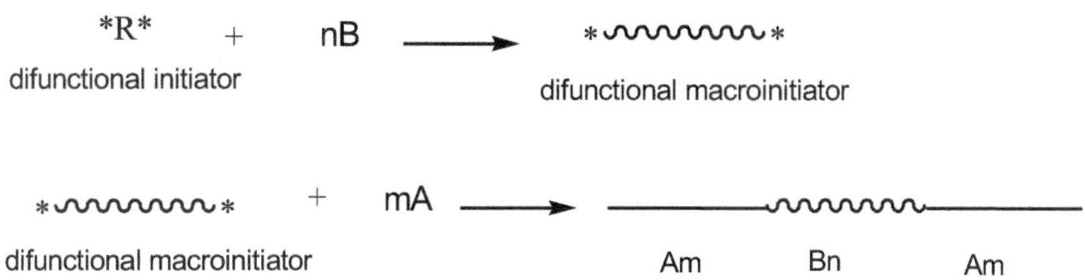

3.1. Block copolymerization *via* Nitroxide Mediated Radical Polymerization (NMP)

Two basic strategies have been applied to initiate NMP. The first method involves use of a conventional initiator (e.g. AIBN, BPO) in the presence of a nitroxide and this system is called as "bimolecular NMP" (Scheme 1).

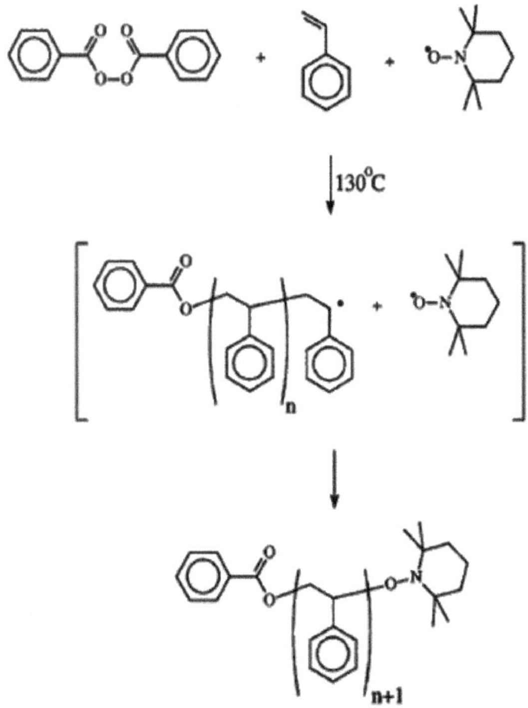

Scheme 1. Bimolecular initiation in NMP

The second method is called as "unimolecular NMP" (Scheme 2) and it is possible when the initiator is a low molecular weight alkoxyamine. In this system the free radical initiator is not required and the C-O bond of the small molecule alkoxyamine derivative is therefore expected to be thermolytically unstable and decompose on heating to give an initiating radical [27].

Bosman et al. reported the preparation of poly(ethylene glycol)-based block copolymers by initial reaction of mono-hydroxy terminated poly(ethylene glycol) with sodium hydride followed by the chloromethyl-substituted alkoxyamine. The PEG-based macroinitiator is used in the nitroxide mediated radical polymerization of styrene to give amphiphilic block copolymer (Scheme 3) [36].

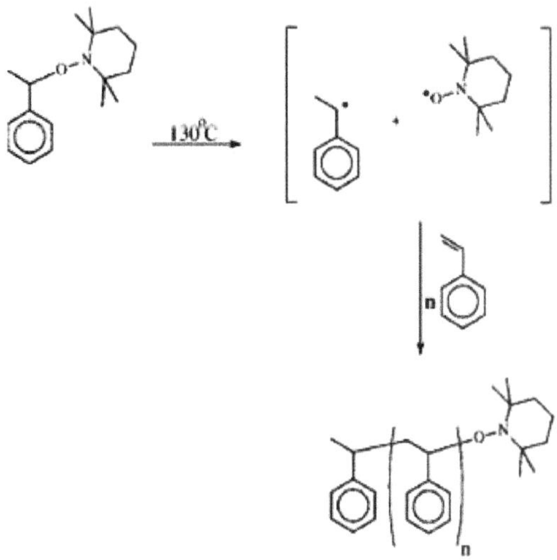

Scheme 2. Unimolecular initiation in NMP

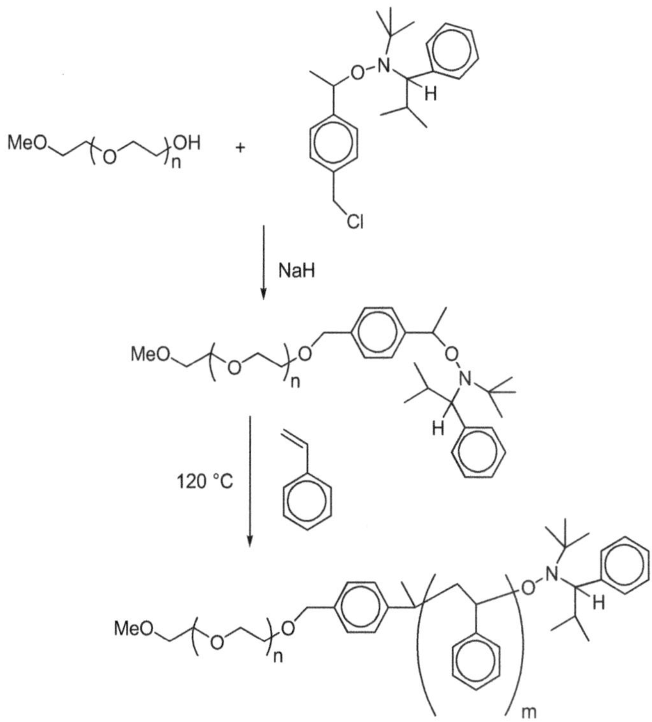

Scheme 3. The synthesis of PEG-*b*-PS by NMP

Choi et al. reported the preparation of functional segmented block copolymers containing localized pyrene groups through convenient NMP processes for the versatile non-destructive functionalization of carbon nanotubes (CTAs). Polystyrene was employed as an organic tail to improve the solubility and dispersibility of CTA in organic mediums and polystyrene compatible polymer matrices (Scheme 4) [37].

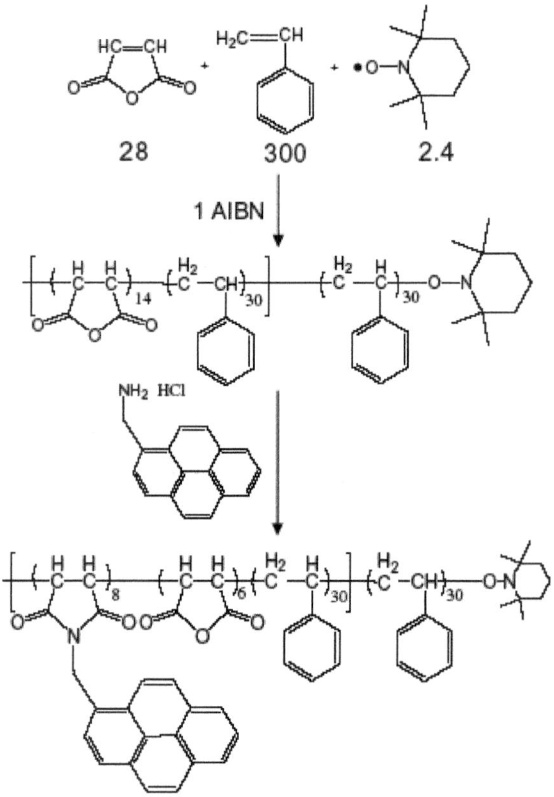

Scheme 4. Schematic representation for the preparation of pyrene functionalized poly(styrene-*co*-maleic anhydride)-*b*-polystyrene (PMAS)

Britze et al. reported the synthesis of poly(*para*-phenylene)-*b*-polystyrene (PPP-*b*-PS) block copolymers using a combination of Suzuki-polycondensation (SPC) and NMP. Alkoxyamine-functionalized poly(*para*-phenylene)s were initially synthesized by SPC and then its block copolymer with styrene by NMP (Scheme 5) [38].

Groison et al. reported the synthesis of well-defined P(MAA_{41}-*co*-SS_{10})-*b*-P(MMA-*co*-St)-SG1 amphiphilic block copolymer nanoobjects *via* nitroxide-mediated emulsion polymerization by using water-soluble macroalkoxyamine (P(MAA_{41}-*co*-SS_{10})-SD1) as macroinitiator (Scheme 6) [39].

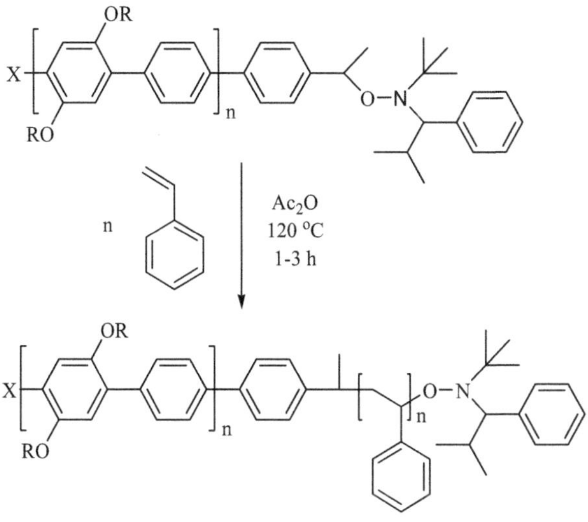

Scheme 5. Synthesis of PPP-*b*-PS under NMP-conditions

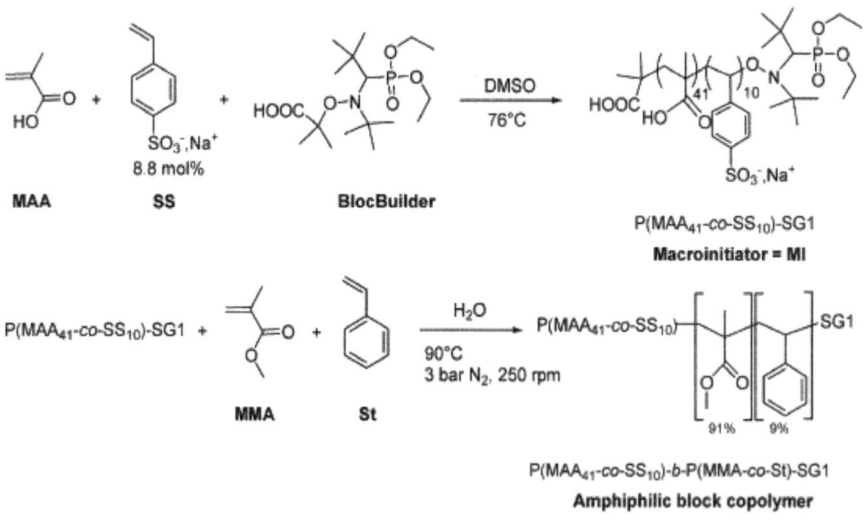

Scheme 6. Synthesis of P(MAA41-*co*-SS10)-*b*-P(MMA-*co*-St)-SG1 amphiphilic block copolymers by using water-soluble macroalkoxyamine (P(MAA41-co-SS10)-SD1) *via* nitroxide-mediated emulsion polymerization

Ruehl et al. reported the synthesis of bidirectional alkoxyamine initiator (Scheme 7) and the preparation of a variety of symmetrical ABA type triblock copolymers based on styrene, n-butyl acrylate, t-butyl acrylate, isoprene and dimethylacrylamide by using bidirectional alkoxamine as bifunctional initiator for NMP [40].

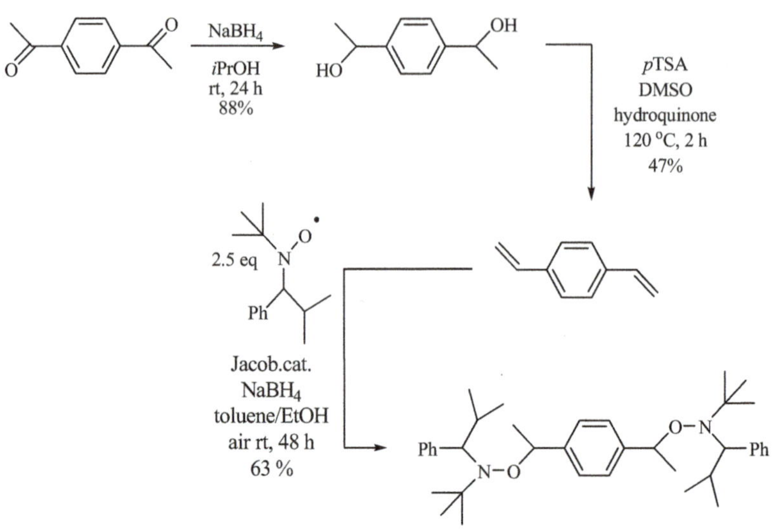

Scheme 7. Sythesis of bidirectional alkoxamine initiator for the preparation of ABA type triblock copolymer *via* NMP

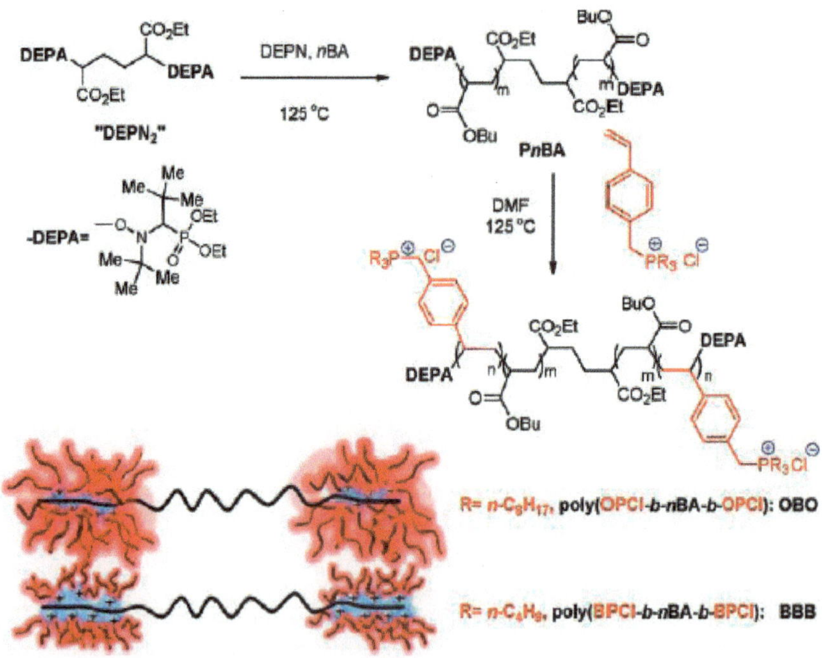

Scheme 8. Synthesis of Phosphonium-Containing Triblock Copolymers and Schematic Representation of Polymer Structures

Phosphonium ion-containing acrylate triblock (ABA) copolymers were synthesized by Long and coworkers using NMP **[41]**. The polymerization of styrenic phosphonium-containing ionic liquid monomers using a difunctional alkoxyamine initiator, $DEPN_2$, afforded an ABA triblock copolymer with an n-butyl acrylate soft center block and symmetric phosphonium-containing external reinforcing blocks (Scheme 8).

The synthesis of amphiphilic triblock copolymer of polystyrene and poly(4-vinylbenzyl glucoside) via TEMPO-mediated living radical polymerization was reported by Kakuchi and coworkers **[42]**. 4-Vinyl benzyl glucoside peracetate **1** was polymerized with difunctional NMP initiator **2** and thus poly(4-vinylbenzyl glucoside peracetate) having TEMPO moieties on both sides of chain ends **3** was obtained. Then the polymerization of styrene was carried out using **3** as a difunctional nmp macroinitiator in order to obtain polystyrene-*b*-poly(4-vinylbenzyl glucoside peracetate)-b-polystyrene triblock copolymers (Scheme 9).

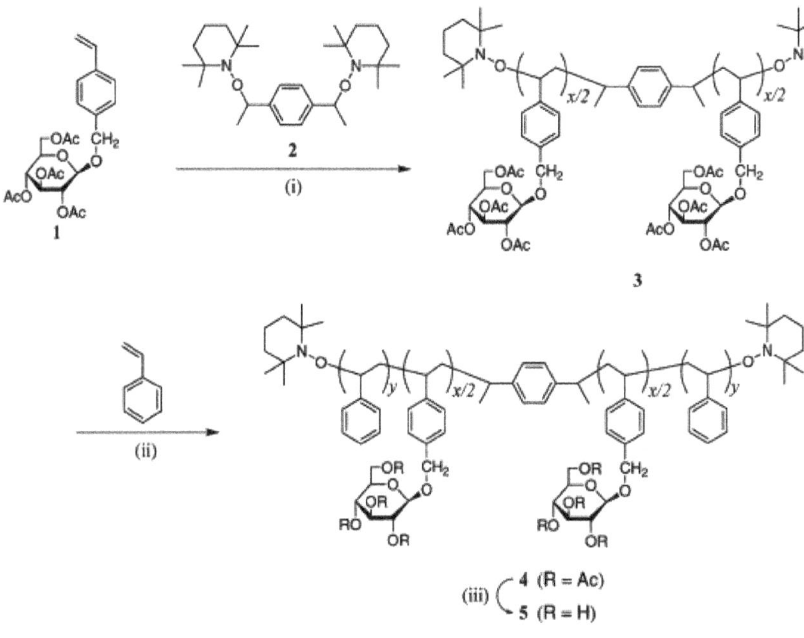

Scheme 9. Synthesis of glycoconjugated triblock copolymer *via* NMP

Some of block copolymers derived using NMP is given in Table 2.

3.2. Block copolymerization *via* Reversible Addition Fragmentation Chain Transfer polymerization (RAFT)

The key component in controlled RAFT polymerization is the chain transfer agent (CTA) (Scheme 10). The CTAs used are thiocarbonylthio compounds and have the general structure

RSC(=S)Z. Examples of RAFT agents (CTA) span all thiocarbonylthio families including dithioesters, xanthates, dithiocarbamates and trithiocarbonates. Z and R groups in the thiocarbonylthio compounds represent the activating group and homolytically leaving group, respectively. These determine the rates of addition and fragmentation. In fact, the choice of RAFT agent for a specified monomer is rather significant and affects the degree of control.

1st Segment	2nd Segment	3rd Segment	Type of block	Ref.
PS	PMS	-	AB	[43]
PS	PBA	-	AB	[43]
PVAc	PAN	PS	ABC	[44]
PVAc	PAN	PBA	ABC	[44]
PVAc	PAN	PVP	ABC	[44]
P(MMA-*ran*-S)	P(MMA-*ran*-AS)		AB	[45]
PEO	PS		AB & ABA	[46]
PS	PSSA		AB	[47]

Polystyrene (PS), Poly(*p*-methylstyrene) (PMS), Poly(*n*-butyl acrylate) (PBA), Poly(vinyl acetate) (PVAc), Poly(acrylonitrile) (PAN), Poly(4-vinylpyridine) (PVP), Poly(methyl methacrylate-*ran*-styrene) (P(MMA-*ran*-S)), Poly(methylmethacrylate-*ran*-4-aminostyrene) (P(MMA-*ran*-AS)), Poly(ethyleneoxide) (PEO), Polystyrenesulfonic acid (PSSA).

Table 2. Some of block copolymers prepared using NMP

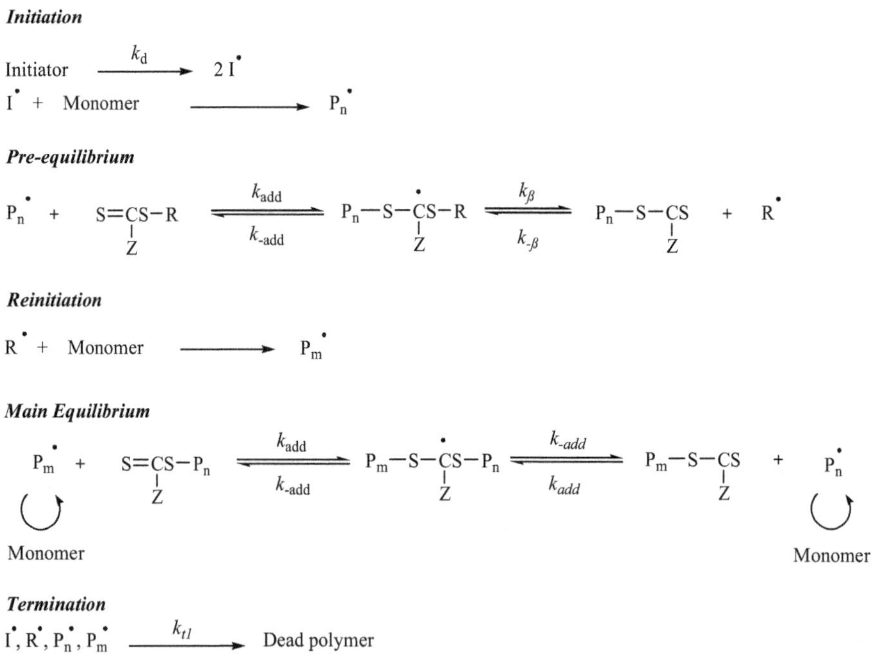

Scheme 10. General scheme for the RAFT process

The first and most thoroughly studied route towards block copolymers via the RAFT process, employs monofunctional RAFT agents (Scheme 11). The block copolymers that have been synthesized in this way all carry the RAFT functionality at the end of the chain. In the first polymerization step, monomer is inserted between the -(C=S)S moiety and the leaving group R. After polymerization is complete, the second monomer is added, polymerization is started by addition of a fresh amount of initiator, and monomer is inserted between the -(C=S)S- moiety and the end of the first block, with the RAFT functionality remaining at the chain end. Repeating this cycle of monomer addition and polymerization, it is possible to produce block copolymers. However, with this method the blocks have to be introduced in the chain step-by-step, which obviously means that to produce an n-block copolymer, a series of n polymerization steps is needed. To avoid inefficient blocking, this procedure should be performed starting with the monomers having the highest transfer rate to the RAFT agent. For the production of multiblock copolymers, this obviously is not the preferred route.

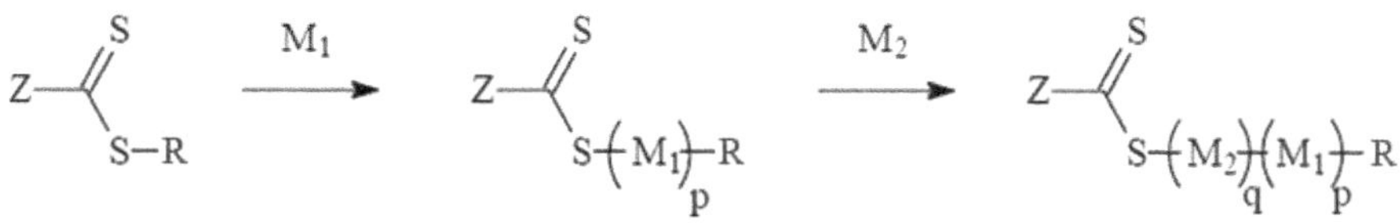

Scheme 11. Block copolymer synthesis *via* a monofunctional RAFT agent

A second route towards block copolymers *via* RAFT comprises the use of symmetrical, bifunctional RAFT agents as well as symmetrical trithiocarbonates (Scheme 12 (a)). The symmetry in these compounds ensures that both arms are of approximately the same length and composition. With these RAFT agents, the number of polymerization steps required for (multi)block copolymer synthesis can be reduced. Scheme 1 clearly demonstrates that for the synthesis of n-block copolymers, (n+1)/2 polymerization steps are required. Note that only block copolymers with an odd number of blocks can be synthesized with these compounds. The major difference between a trithiocarbonate and a bifunctional RAFT agent is the way the monomer is inserted into the RAFT agent. Using a bifunctional RAFT agent, monomer is inserted into the RAFT agent at both sides of leaving group R, while both dithiocarboxylate moieties will remain at the chain ends. In block copolymerization, the second block is thus built at the chain ends of the first block, see Scheme 12 (a). Using trithiocarbonate, monomer is inserted between both A groups and the trithiocarbonate moiety, and the trithiocarbonate moiety remains in the middle of the chain. When chain extending the first block, the second block is built in the middle of the chain, and the first block will shift to the outside of the chain as is depicted in Scheme 12 (b). In the case of trithiocarbonates, the R-S- moiety functionsas the activating group Z. [34]

The well-defined block copolymer of poly[ethylene oxide)-*b*-poly(*N*-isopropylacrlamide) (PEO-*b*-PNIPAM) was synthesized by Liang and coworkers *via* RAFT polymerisation using monofunctional PEO macro-RAFT agent [48] (Scheme 13).

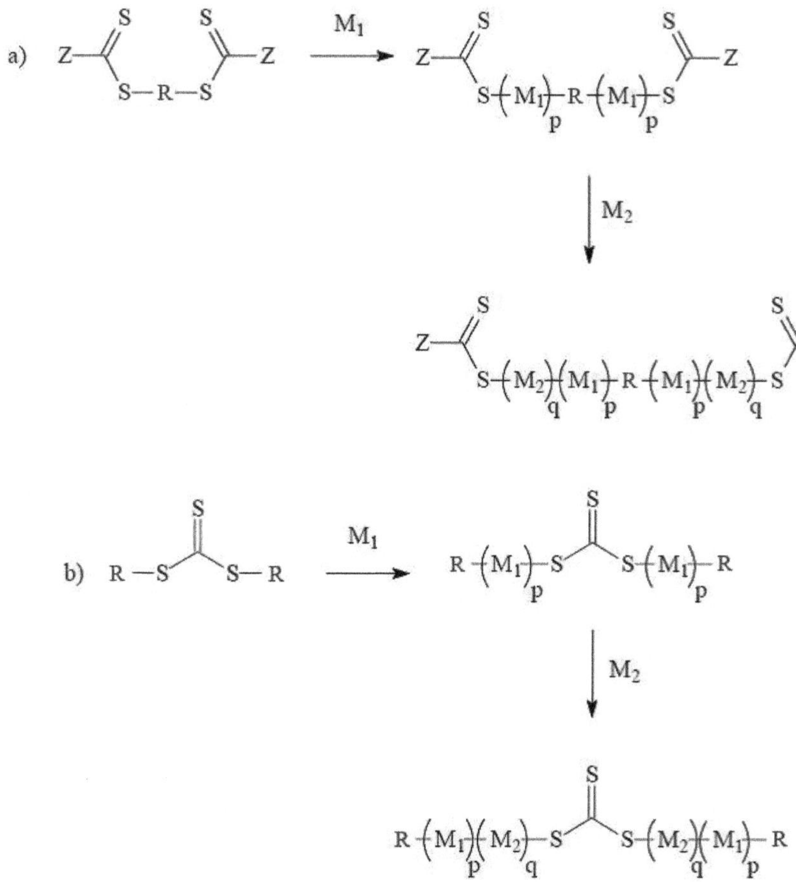

Scheme 12. Block copolymer synthesis *via* a bifunctional RAFT agent

Stenzel and coworkers reported the synthesis of amphiphilic block copolymers based on 2-methacrylamido glucopyranose (MAG) and 5′-*O*-methacryloyl uridine (MAU) via RAFT copolymerization using monofunctional macro-RAFT agent (Scheme 14) [49].

Block copolymers of methyl methacrylate (MMA) and *tert*-butyldimethylsilyl methacrylate (TBDMSMA) have been synthesized by Bressy and coworkers *via* RAFT polymerization technique (Scheme 15) [50].

New block copolymers with narrow molecular weight distribution (MWD=1.04-1.14) based on (2,3-epithiopropylmethacrylate) (ETMA), methyl methacrylate (MMA) and *n*-butylmethacrylate (*n*BMA) have been synthesized by Petzhold and coworkers *via* RAFT polymerization (Scheme 16) [51]

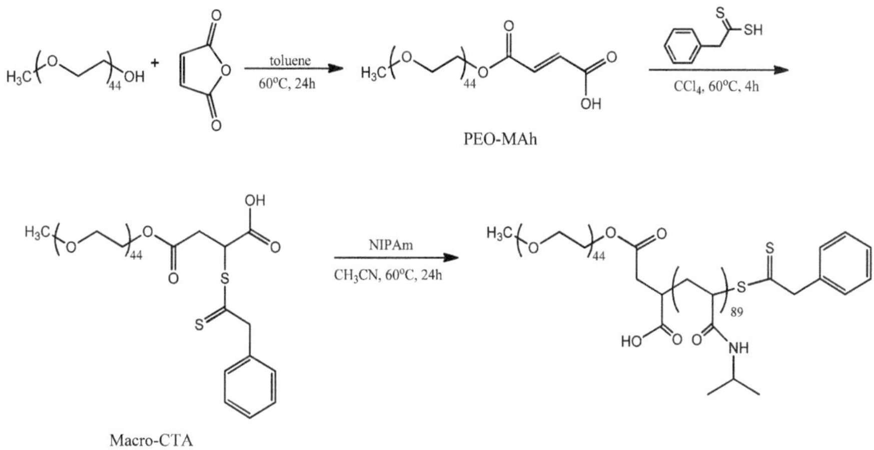

Scheme 13. Synthetic procedure for PEO-*b*-PNIPAM by RAFT

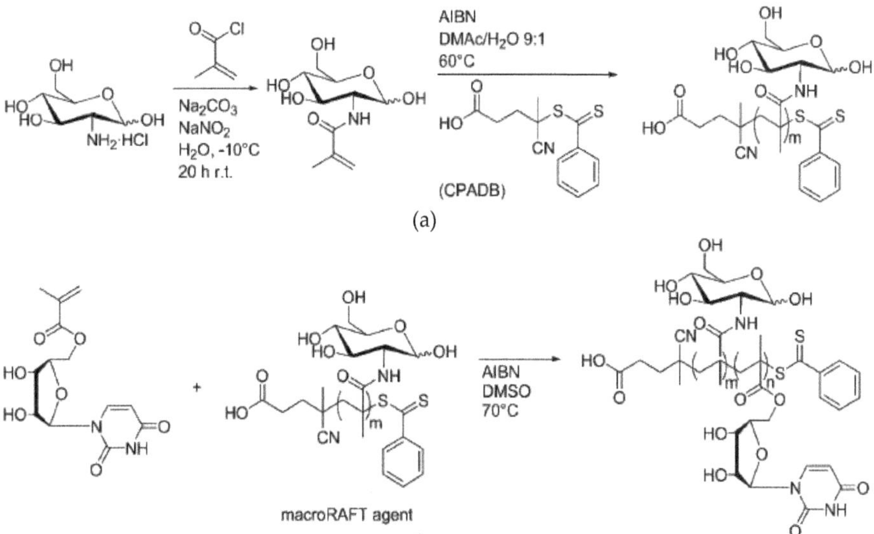

Scheme 14. (a) Synthesis of poly(2-methacrylamido glucopyranose) as a macroRAFT agent (b) synthesis of poly(2-methacrylamido glucopyranose)-*b*-poly(5′-*O*-methacryloyl uridine) via RAFT using monofunctional macro-RAFT agent

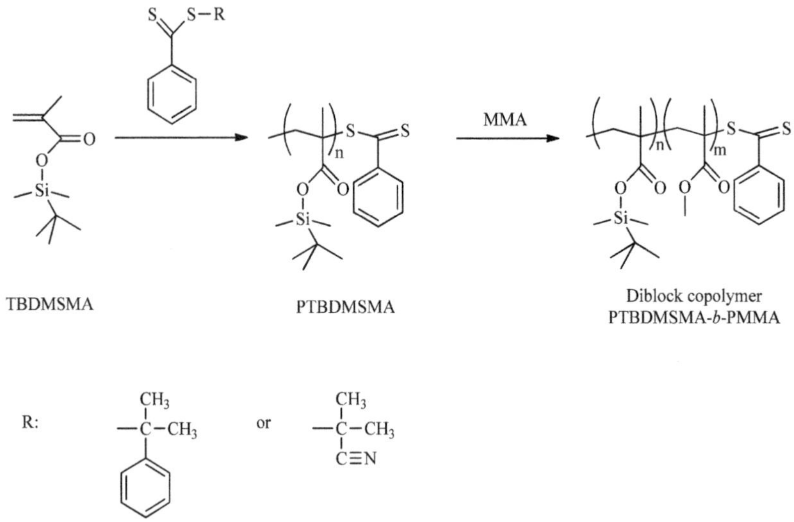

Scheme 15. Block copolymerization of TBDMSMA and MMA by the RAFT process

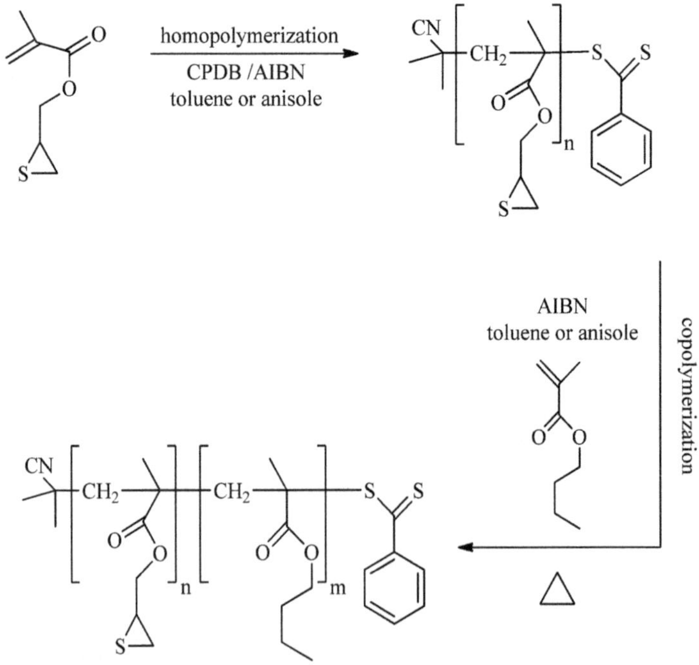

Scheme 16. Schematic representation of the synthesis of the diblock copolymer *via* RAFT polymerization

Nanogel particles and double hydrophilic block copolymers were prepared by Hawker and co-workers using hydrophilic as well as amphiphilic poly(dimethylacrylamide) (PDMA) chain transfer agents (CTA) in RAFT precipitation polymerization (Scheme 17) [52].

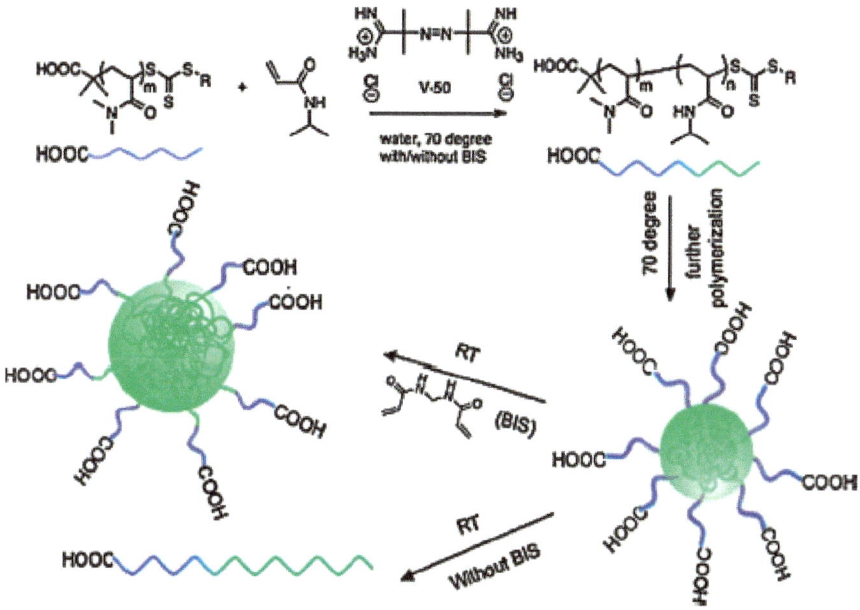

Scheme 17. Precipitation RAFT polymerization of NIPAM using a PDMA macroCTA followed by cross-linking with BIS

Guan et al. reported the synthesis of poly(dimethylsiloxane)-b-poly(2,2,3,3,4,4,4-heptafluorobutyl methacrylate)-*b*-poly(styrene) (PDMS-*b*-PHFBMA-*b*-PS) triblock copolymers by two-step RAFT polymerization [53]. For this purpose, firstly xanthate-capped PDMS was prepared as the PDMS-macro-RAFT agent. Then triblock copolymers were prepared by two step RAFT polymerization using the PDMS-macro-RAFT agent (Scheme 18).

Zhang and co-workers reported the synthesis of polystyrene-*b*-poly(acrylic acid)-*b*-polystyrene (PS-*b*-PAA-*b*-PS) tri block copolymers by RAFT polymerization using S,S′-Bis(α,α'-dimethyl-α''-acetic acid)-trithiocarbonate (BDATC) as chain transfer agent (Scheme 19) [54].

Qupicky and co-workers reported the synthesis of dually responsive (temperature- and redox-responsive) multiblock copolymers of poly(*N*-isopropylacrlamide) (PNIPAAM) and poly(dimethylaminoethylmethacrylate) (PDMAEMA) by RAFT polymerization **[55].** Well-defined bis(dithioester)-functionalized PNIPAM **3** and PDMAEMA **4** were prepared using 1,4-bis(thiobenzoylthiomethyl)benzene **1** and 1,4-bis(2-(thiobenzoylthio)prop-2-yl)benzene **2** as RAFT agents, respectively. Multiblock copolymers were synthesized in a single aminolysis/oxidation step from the bis(dithioester)-terminated PNIPAAM and PDMAEMA (Scheme 20).

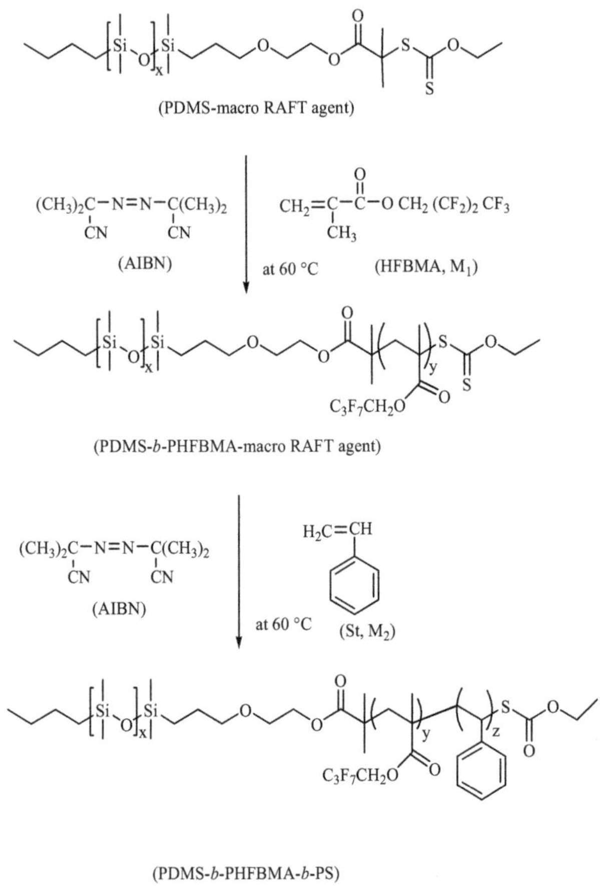

Scheme 18. The synthesis of PDMS-*b*-PHFBMA-*b*-PS triblock copolymers by two-step RAFT polymerization

Baum and Brittain reported the synthesis of surface immobilized diblock copolymer brush (Si/SiO_2//PS-*b*-PDMA) using RAFT [56]. Styrene, methyl methacrylate, and *N,N*-dimethylacrylamide brushes were prepared under RAFT conditions using silicate surfaces that were modified with surface-immobilized azo initiators. Films with controlled thicknesses were produced. RAFT was also used to synthesize PS-*b*-PDMA and PDMA-*b*-PMMA block copolymer brushes that displayed reversible surface properties upon treatment with block-selective solvents. (Scheme 21).

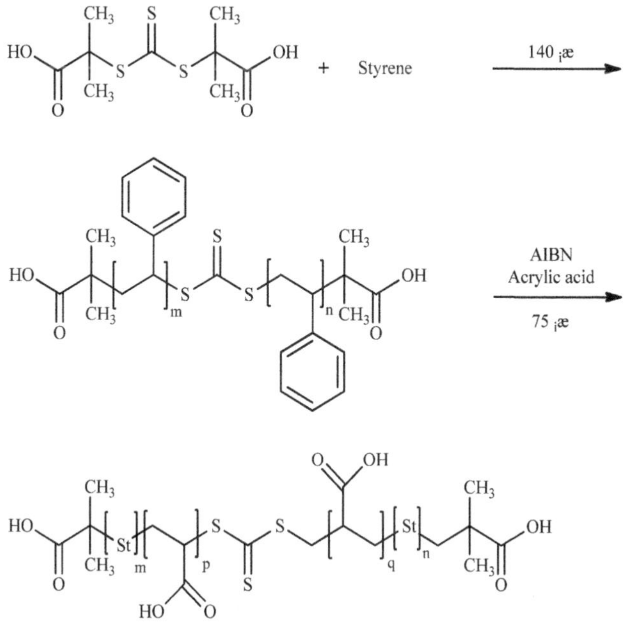

Scheme 19. The synthesis of PS-*b*-PAA-*b*-PS tri block copolymers by RAFT polymerization

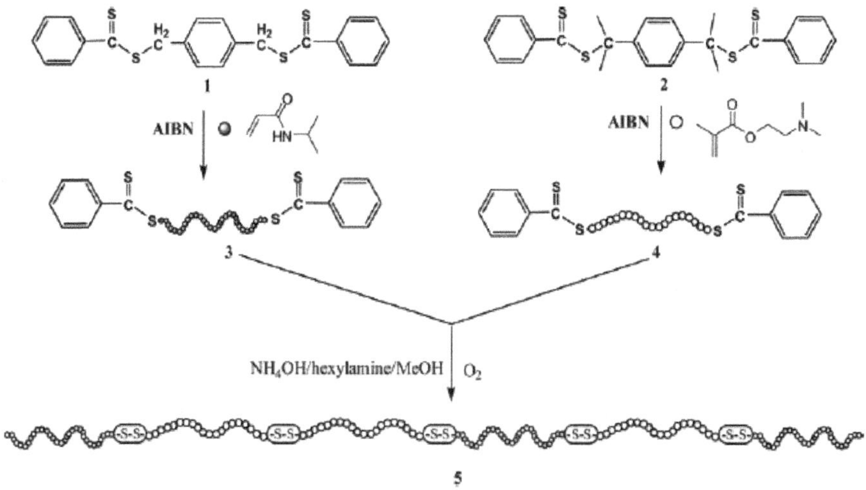

Scheme 20. The synthesis of dually responsive multiblock copolymers of PNIPAM and PDMAEMA by RAFT polymerization

Some of block copolymers derived using RAFT polymerization is given in Table 3.

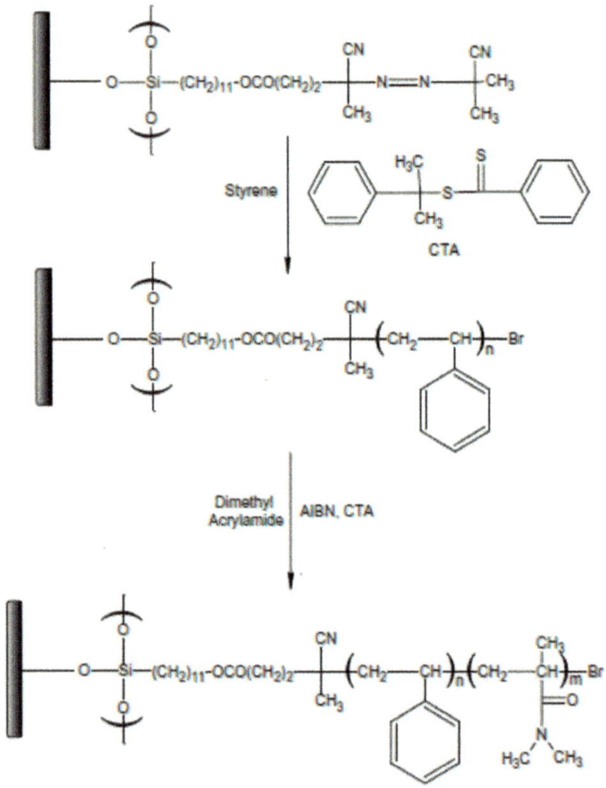

Scheme 21. Synthesis of surface immobilized diblock copolymer brush (Si/SiO_2//PS-*b*-PDMA) using RAFT.

1st Segment	2nd Segment	3rd Segment	Type of block	Ref.
PMAOEGlc	P(6-*O*-MAMMan)	-	AB	[57]
PMAOEGlc	P(6-*O*-MAMGlc)	-	AB	[57]
PNIPAAM	$PMAGlcC_5$	-	AB	[58]
PNIPAAM	PDEAEMA	-	AB	[59,60]
PNIPAAM	PAGA	-	AB	[61
PVAc	PVPv	-	AB	[62]
PVAc	PVBz	-	AB	[62]
P(6-*O*-MAMGlc)	PHEMA	-	AB	[63]

Poly(2-methacryloxyethyl glucoside) (PMAOEGlc), poly(methyl 6-*O*-methacryloyl-D-mannoside) P(6-*O*-MAMMan), poly(methyl 6-*O*-methacryloyl-D-glucoside) P(6-*O*-MAMGlc), Poly(*N*-isopropylacrylamide) (PNIPAAM), Poly(D-glucofuranosyl)-6-methacrylamido hexanoate ($PMAGlcC_5$), Poly(2-(diethylamino)ethyl methacrylate) (PDEAEMA), poly(acryloyl glucosamine) (PAGA), poly(vinyl acetate) (PVAc), poly(vinyl pivalate) (PVPv), poly(vinyl benzoate) (PVBz), poly(2-hydroxyethyl methacrylate) (PHEMA).

Table 3. Some of block copolymers prepared using RAFT polymerization

3.3. Block copolymerization *via* Atom Transfer Radical Polymerization (ATRP)

Of the several controlled/"living" free radical polymerization techniques, ATRP seems to be the most versatile, being able to polymerize a variety of monomers, method to prepare block and graft copolymers. ATRP enables the synthesis of a wide range of (co)polymers with controlled molecular weight, narrow molecular weight distribution, and range of architectures and functionalities.

The synthesis of well-defined AB type diblock copolymers from styrene (S) and methyl acrylate (MA) *via* ATRP was reported by Schubert and co-workers [64]. Both synthetic routes starting with styrene as first block or methyl acrylate as first block are shown in Scheme 22.

1st Segment	2nd Segment	3rd Segment	Type of block	Ref.
PMMA	P(*n*-BA)	PMMA	ABA	[67]
PMMA	PMA	PMMA	ABA	[67]
PDMS	PS	-	AB	[68]
PS	PMAIpGlc	-	AB	[69]
PS	P(*n*-BA)	-	AB	[70]
PS	P(*n*-BA)	PS	ABA	[70]
PS	PE	PS	ABA	[71]
PS	PB	PS	ABA	[71]
MPEG	PDMAEMA	PBMA	ABC	[72]
MPEG	PBMA	PDMAEMA	ABC	[72]
PS	PTMDS	PS	ABA	[73]
PMPC	PDMAEMA	-	AB	[74]
PMPC	PDPAEMA	-	AB	[74]
PBMA	PDMAEMA	-	AB (brushes)	[75]
PAMA	PS	PAMA	ABA	[76]
PAMA	PMMA	PAMA	ABA	[76]
PAMA	PBA	PAMA	ABA	[76]
PSSNa	PMMA	-	AB	[77]
PSSNa	PMMA	PSSNa	ABA	[77]
PCL	P(MAPEO-*co*-DIGaMA)	-	AB	[78]
PGEMA	PDEGMA	-	AB	[79]

Polymethylmethacrylate (PMMA), Poly(n-butylacrylate) (P(*n*-BA)), polymethylacrylate (PMA), poly(dimethylsiloxane) (PDMS), polystyrene (PS), poly(3-*O*-methacryloyl-1,2:5,6-di-*O*-isopropylidene-D-glucofuranose) (PMAIpGlc), polyester and/or polyether (PE), polybutadiene (PB), methoxy-poly(ethyleneglycol) MPEG, poly(2-(dimethylamino)ethylmethacrylate) (PDMAEMA), Poly(n-butylmethacrylate) (PBMA), poly(tetramethyldisiloxane) (PTMDS), poly(2-(methacryloyloxy)-ethylphosphorylcholine) (PMPC), poly(2-(diisopropylamino)ethylmethacrylate) (PDPAEMA), poly(allyl methacrylate) (PAMA), poly(sodium styrene sulfonate) (PSSNa), polycaprolactone (PCL), ω-methacrylate poly(ethylene oxide) (MAPEO), 6-*O*-methacryloyl-1,2:3,4-di-*O*-isopropylidene-D-galactopyranose (DIGaMA), poly(2-glucosyloxyethyl methacrylate) (PEGMA), poly(diethyleneglycolmethacrylate) (PDEGMA)

Table 4. Block copolymers prepared using ATRP

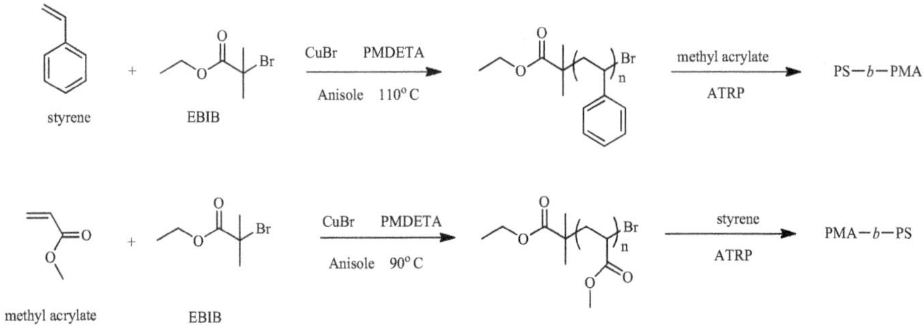

Scheme 22. Schematic representation of synthetic strategy to block copolymers from styrene and methyl acrylate

The synthesis of ABA type triblock copolymers of poly(2-(dimethylamino)ethylmethacrylate)-*b*-polycaprolactone-*b*-poly(2-(dimethylamino) ethylmethacrylate) (PDMAEMA-*b*-PCL-*b*-PDMAEMA) *via* ATRP of DMAEMA using difunctional polycaprolactone (Br-PCL-Br) as macroinitator was reported by Jhurry and Motala-Timol (Scheme 23) [65].

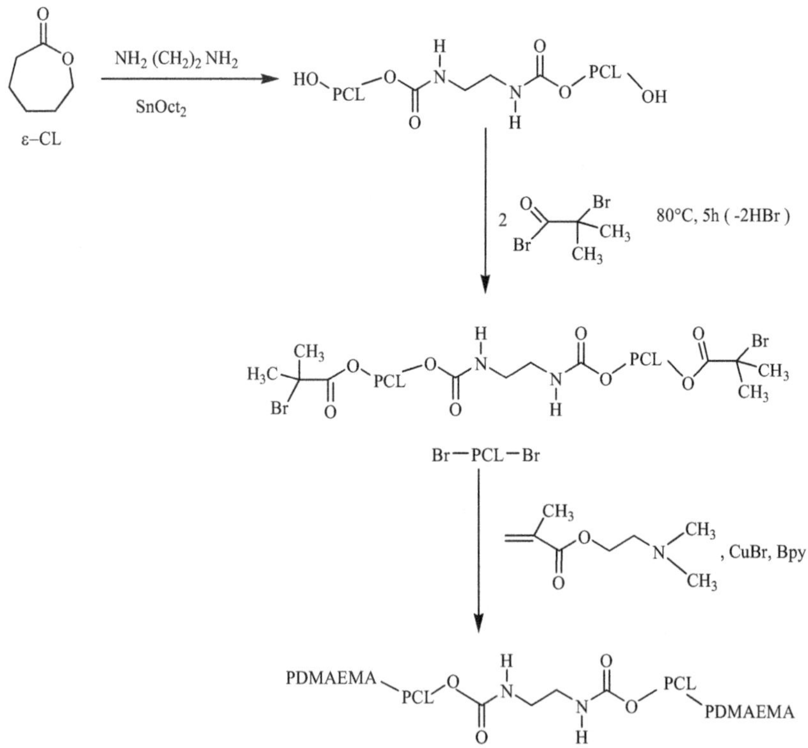

Scheme 23. Synthesis of triblock PDMAEMA-*b*-PCL-*b*-PDMAEMA copolymer

Huang et al. reported the preparation of amphiphilic triblock PMMA-*b*-PDMAEMA-*b*-PMMA and poly(methyl acrylate)-*b*-poly(methyl methacrylate)-*b*-poly(hydroxyethyl methacrylate) (PMA-*b*-PMMA-*b*-PHEMA) copolymer brushes grown from a surface *via* sequential low-temperature ATRP (Scheme 24) [66]

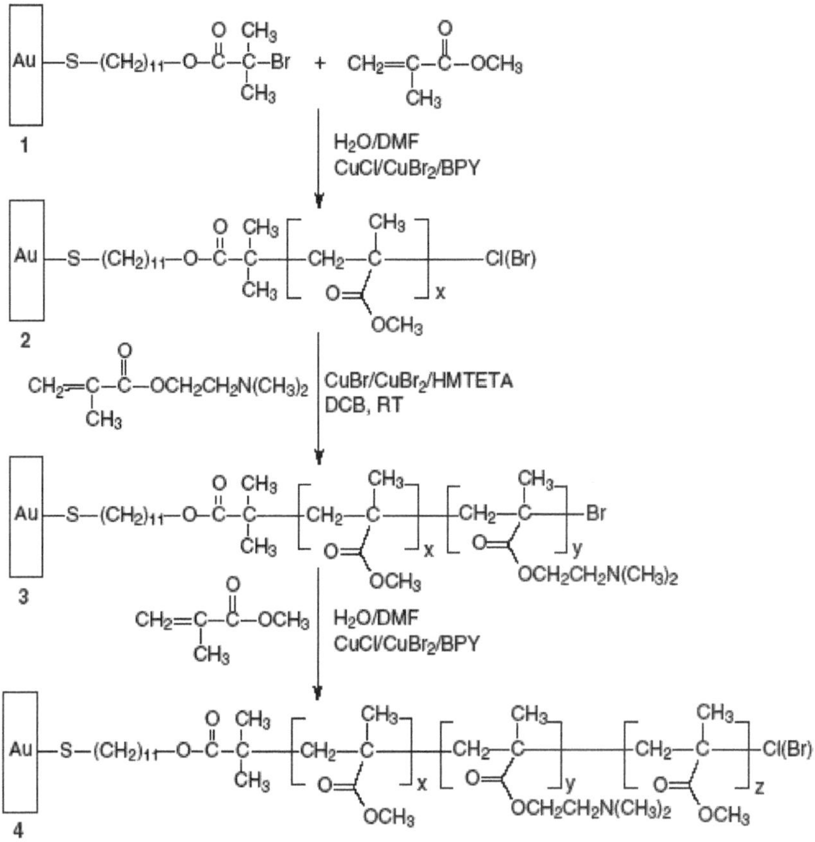

Scheme 24. Synthesis of triblock copolymer brushes from gold surfaces by sequential ATRP.

Recently many studies on the synthesis of block copolymers *via* ATRP in the literature have been reported due to its advantages. Some of them are listed in Table 4.

4. Graft copolymerization

Graft copolymers are another class of segmented copolymers. As informed previously, many basic characteristics of block copolymers and graft copolymers are similar and no major difference exists between block copolymer synthesis and graft copolymer synthesis. The location of sites and functions are at the chain ends or on the chains, respectively:

"Grafting" is a method wherein monomers are covalently bonded (modified) onto the polymer chain. Graft copolymers can be obtained with three general approachs [80]:

1. Grafting-onto: A preformed polymer with a reactive end-group is used a precursor, and then attached to the backbone. This method involve reaction of functional groups (Y) located at chain ends of one kind of polymer with another functional groups (X) which distributed randomly on the main chain of the other polymer (backbone).

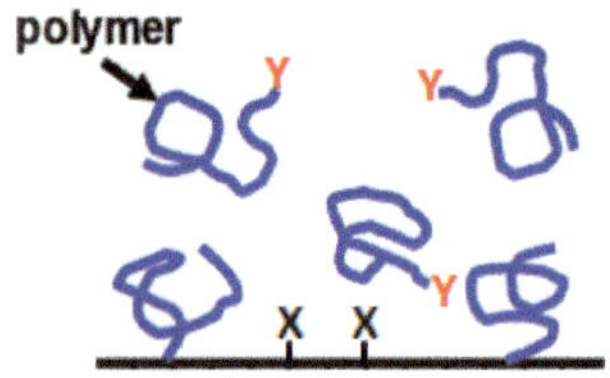

Figure 6. Schematic presentation of Grafting-onto

2. Grafting-from: The monomer is grafted from the backbone. "Grafting-from" is significantly more versatile than "grafting-onto".

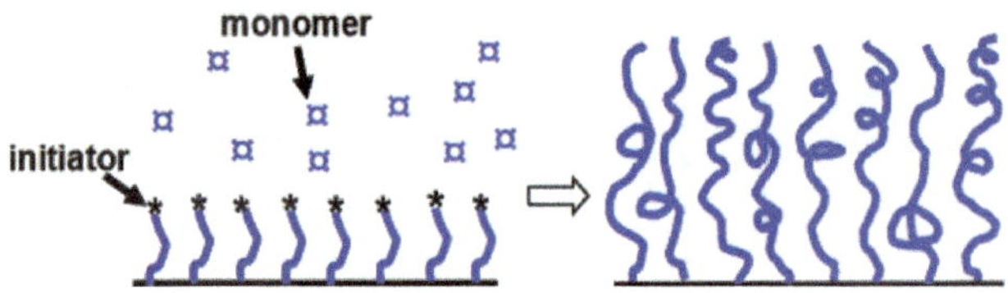

Figure 7. Schematic presentation of Grafting-from

3. Grafting-through in other words grafting *via* surface-attached monomers (macromonomers): In this approach, polymerizations are carried out at the presence of polymers (backbone) onto which functionalized monomers have been attached. The surface-attached monomers are incorporated into growing polymer chains in the same way as the monomers in solution (the monomers in solution are usually different from the surface-attached ones).

4.1. Graft copolymerization *via* NMP

Hua et al. prepared polystyrene grafted chitosan [81] (Scheme 25) and poly(sodium 4-styrenesulfonate) grafted chitosan [82] (Scheme 26) *via* NMP using chitosan-TEMPO macroinitiator.

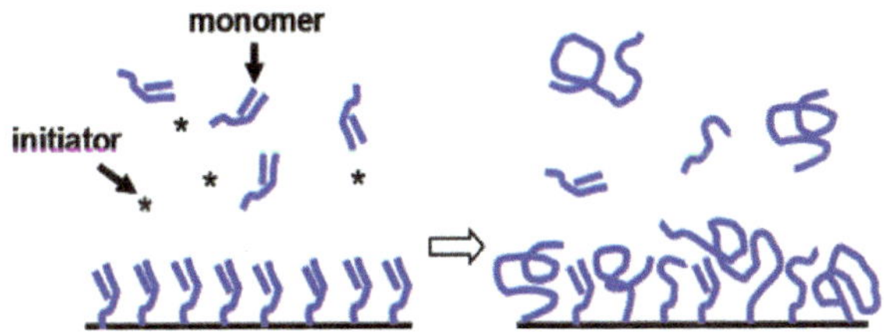

Figure 8. Schematic presentation of Grafting-through

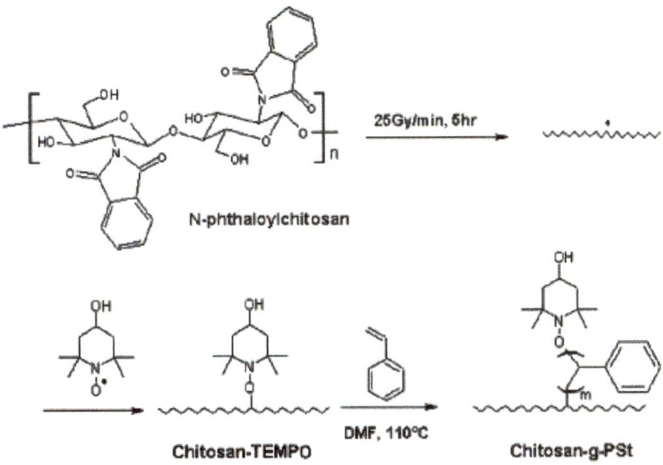

Scheme 25. Synthesis of Chitosan-*g*-PSt by NMP

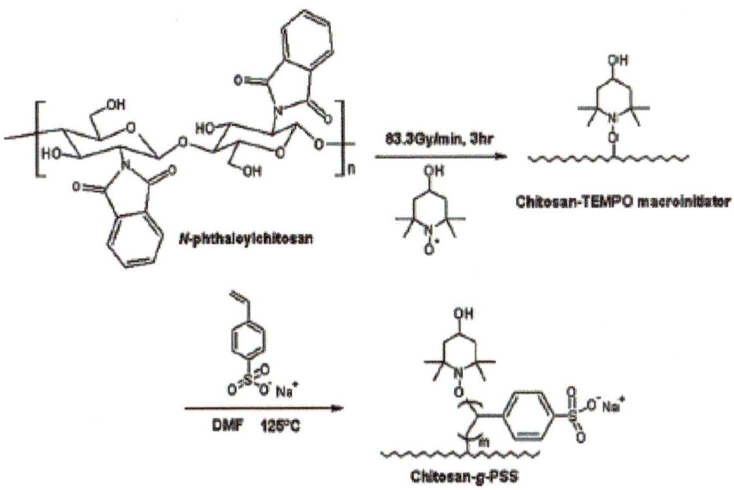

Scheme 26. Synthesis of Chitosan-*g*-PSS by NMP

Jaymand reported the synthesis and characterization of novel type poly(4-chloromethyl styrene-*graft*- 4-vinylpyridine)/TiO_2 nanocomposite *via* NMP [83]. Firstly, poly(4-chloromethyl styrene)/TiO_2 nanocomposite carrying TEMPO groups (PCMS-TEMPO/TiO_2) was synthesized (Scheme 27) and then the controlled graft copolymerization of 4-vinylpyridine was initiated by PCMS-TEMPO/TiO_2 as a macroinitiator (Scheme 28).

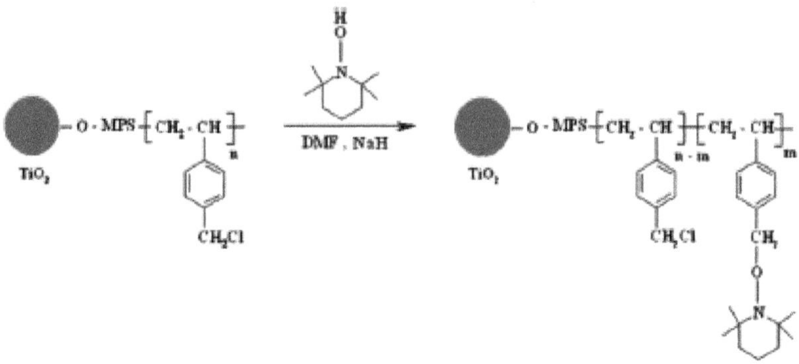

Scheme 27. Synthesis of PCMS-TEMPO/TiO_2 macroinitiator

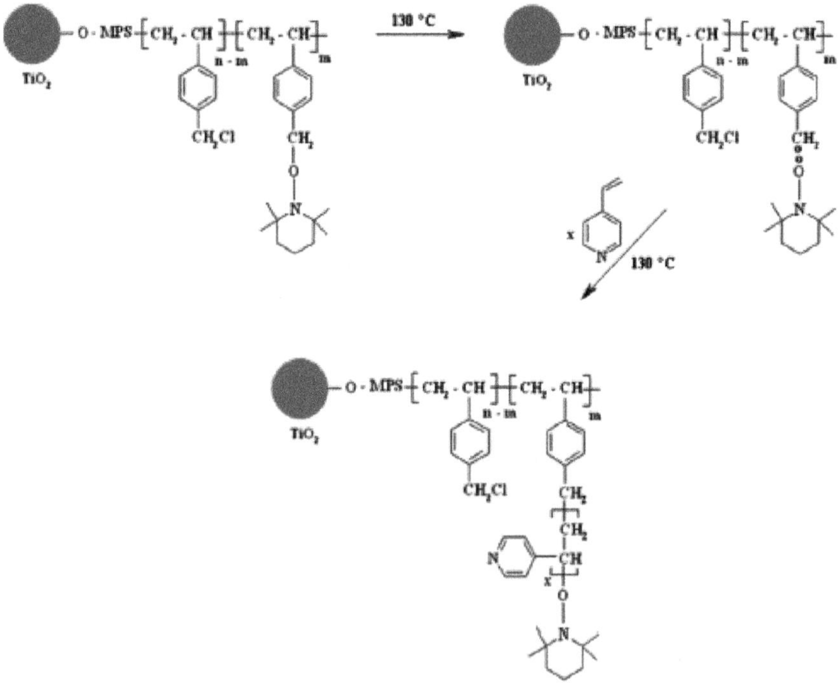

Scheme 28. Nitroxide-mediated living radical polymerization of 4-vinylpyridine onto PCMS-TEMPO/TiO_2 macroinitiator

4.2. Graft copolymerization *via* RAFT polymerization

Neoh and co-workers reported the preparation of graft copolymers of poly(ethylene glycol)methyl ether methacrylate (PEGMA) with poly(vinylidene fluoride) (PVDF) [84] and poly(acrylic acid) (PAAc) with poly(vinylidene fluoride) (PVDF) [85] *via* RAFT polymerization. The peroxides generated on the ozone-pre-treated PVDF facilitated the thermally initiated graft copolymerization of PEGMA and PAAc in the RAFT-mediated process respectively (Scheme 29).

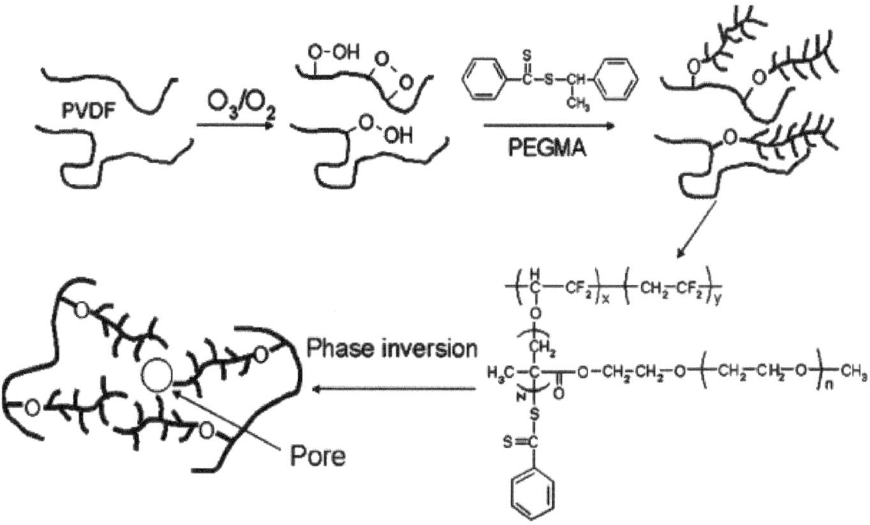

Scheme 29. Schematic illustration of the processes of PEGMA graft copolymerization with the ozone-preactivated PVDF backbone by RAFT-mediated polymerization and the preparation of the PVDF-*g*-PEGMA MF membrane by phase inversion.

Perrier and co-workers performed grafting of styrene from cellulose substrate *via* RAFT polymerization [86]. The hydroxyl groups of cellulose fiber were converted into thiocarbonyl-thio chain transfer agent, and were further used to mediate the RAFT polymerization of styrene (Scheme 30).

Stenzel and co-workers prepared stimuli-responsive glycopolymer brushes composed of *N*-acryloyl glucosamine (AGA) and NIPAAM using RAFT polymerization [87]. The RAFT agent was immobilized on the surface of a treated silicon waver *via* covalent attachment using the Z-group (Scheme 31)

Hua et al. reported the controlled grafting modification of chitosan via RAFT polymerization using chitosan-RAFT agent. The chitosan was first modified with *S*-1-dodecyl-*S'*-(α,α'-dimethyl-α''-acetic acid)trithiocarbonate (DDACT) to serve as RAFT agent, and then the controlled grafting polymerization of acrylic acid were performed [88]. Morimoto et al. reported the grafting of NIPAAM on polysaccarides by RAFT polymerization [89].

Mespouille et al. also reported the grafting of NIPAAM on aliphatic polycarbonates by RAFT polymerization [90]. Synthesis of poly(*tert*-butl methacrylate) *graft*-poly(dimethylsiloxane) graft copolymers *via* RAFT polymerization was reported by Hou and co-workers [91].

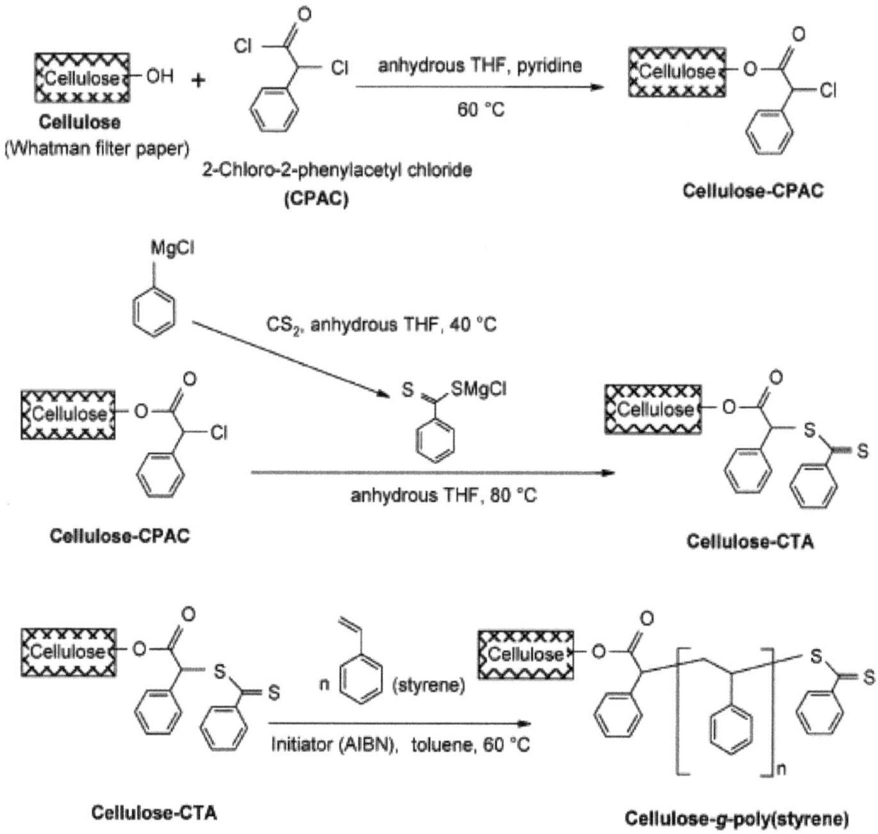

Scheme 30. Synthesis of cellulose chain transfer agent for RAFT polymerization and their use to mediate styrene polymerization

4.3. Graft copolymerization *via* ATRP

In recent year's chemical modification and particularly surface modification of natural polymers such as collagen, cellulose, chitosan etc. has become a popular method for providing a material with desirable properties for practical applications. Particularly ATRP has been proven to be versatile for the grafting of vinyl monomers on to natural polymers.

El Tahlawy & Hudson prepared chitosan macroinitiators by acetylation of chitosan with 2-bromo-isobutyryl bromide (EBİB) in the presence of pyridine as a base. Then chitosan macroinitiator was used to polymerize a methoxy-poly(ethylene glycol)methacrylate

(MeO(PEG)MA) monomer [92]. Using a similar approach, Li et al. proposed the controlled synthesis of chitosan beads grafted with polyacrylamide *via* surface-initiated ATRP [93]. The bromide-end groups were immobilized on the surfaces of chitosan beads through reaction of $-NH_2$ or –OH groups with EBİB (Scheme 32).

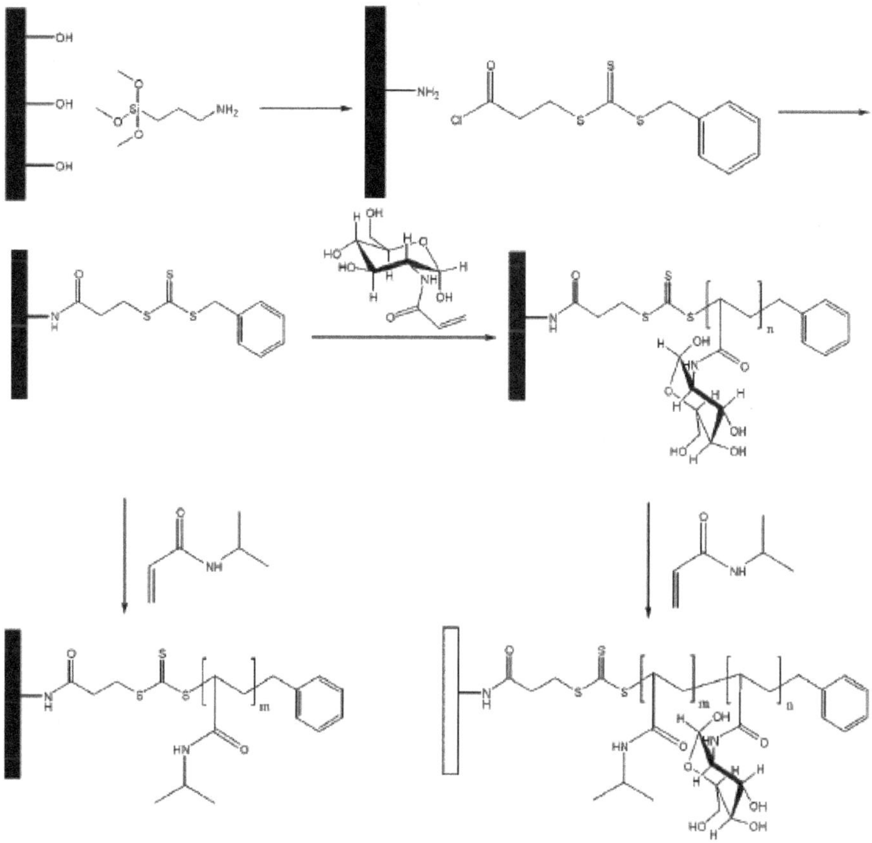

Scheme 31. Synthesis of stimuli-responsive glycopolymer brushes using RAFT polymerization *via* Z-group approach

Arslan et al. reported the synthesis of brush type graft copolymers of poly(3-hydroxybutyrate) (PHB) and poly(3-hydroxyoctanoate) (PHO) with methylmethacrylate, (MMA), styrene, (S), and n-butylmethacrylate, (n-BuMA) by using ATRP via "grafting from" technique [94,95]. Firstly PHB and PHO were chlorinated by passing chlorine gas through their solution in $CHCl_3$/ CCl_4 (75/25 v/v) mixture and CCl_4, respectively, in order to prepare chlorinated PHB, PHB-Cl, and chlorinated PHO, PHO-Cl, with different chlorine contents. Then ATRP of vinyl monomers was initiated by using PHA-Cl as macro initiators in order to obtain brushes containing PHAs (Scheme 33).

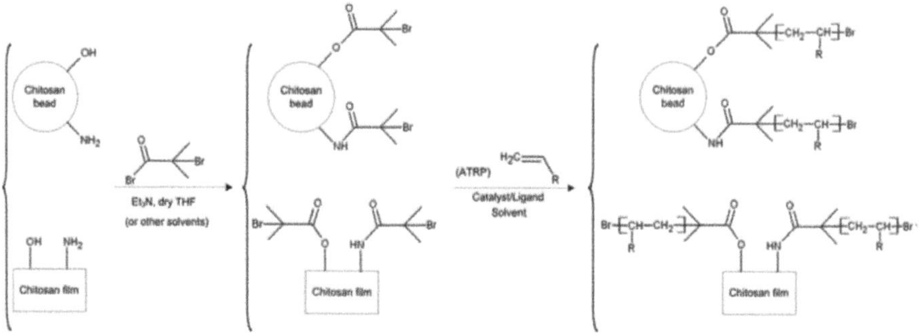

Scheme 32. Schematic presentation of the immobilization of surface initiator and polymerization of vinyl monomers *via* ATRP

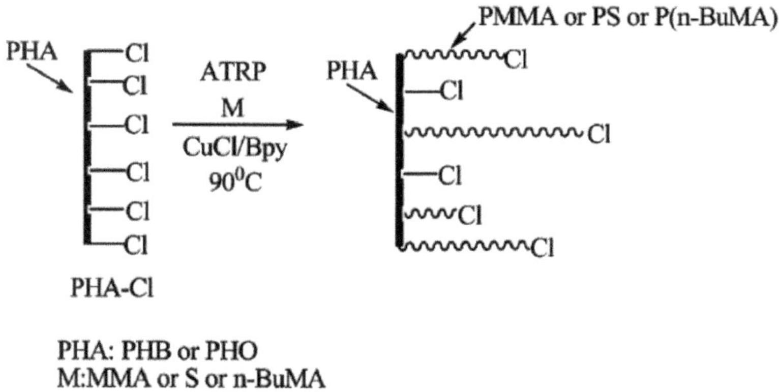

Scheme 33. The synthesis of PHA-*g*-PMMA, PHA-*g*-PS, PHA-*g*-PBMA brush type graft copolymers *via* ATRP

Recently many studies on the synthesis of graft copolymers *via* ATRP in the literature have been also reported. Some of them are listed in Table 5.

5. Block and graft copolymerzation by combined techniques

In some cases, different controlled polymerization systems are combined to produce block and graft copolymers (see Table 6).

Lu et al reported the synthesis of block and graft copolymers of β-pinene and styrene by combination of living cationic polymerization and atom transfer radical polymerization methods (Scheme 34) [105].

Block copolymer synthesis *via* a combination of ATRP and RAFT using click chemistry was reported by Webster and co-workers (Scheme 35) [106].

1st Segment	2nd Segment	Ref.
Chitosan	P(OEGMA)	[96]
Cellulose	PMMA	[97]
Cellulose	PS	[97]
Si/PGMA	PS	[98]
PBIEM-I	PMAIpGlc	[99]
MWNTs	PMAIpGlc	[100]
PVDF	PMAGlc	[101]
Si	PMMA	[102]
Si	PHEMA	[103]
Si	PDMAEMA	[103]
AuNPs	PHEMA	[104]

Poly(oligoethylene glycol methacrylate) (POEGMA), poly(glycidyl methacrylate) deposited on silicon wafer (Si/PGMA), poly(3-O-methacryloyl-1,2:5,6-di-O-isopropylidene-D-glucofuranose) (PMAIpGlc), poly(2-(2-bromoisobutyryloxy)ethyl methacrylate) (PBIEM-I), poly(vinylidene fluoride) (PVDF), poly(3-O-methacryloyl-D-glucofuranose) (PMAIpGlc), silicon substrate (Si), Polymethylmethacrylate (PMMA), polystyrene (PS), poly(2-(dimethylamino)ethylmethacrylate) (PDMAEMA), poly (2-hydroxyethyl methacrylate) (PHEMA), multiwalled carbon nanotubes (MWNTs), golg nanoparticules (AuNPs).

Table 5. Graft copolymers prepared using ATRP

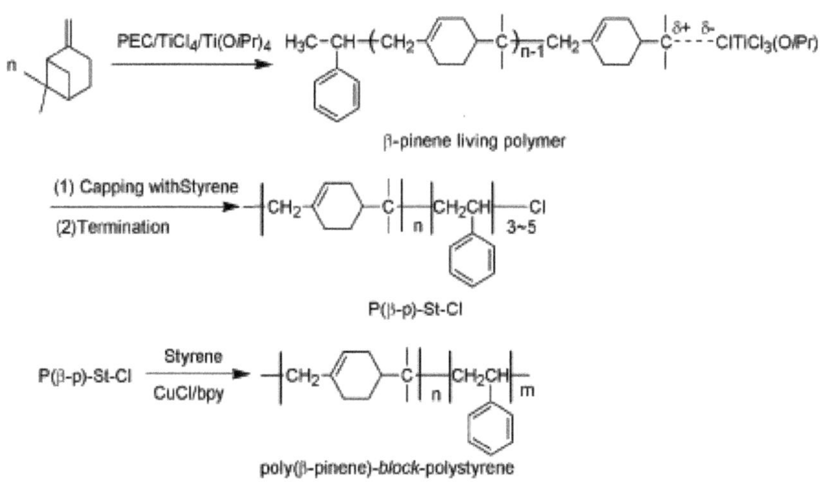

Scheme 34. Synthesis of poly(β-pinene)-*b*-polystyrene by combination of living cationic polymerization and atom transfer radical polymerization methods

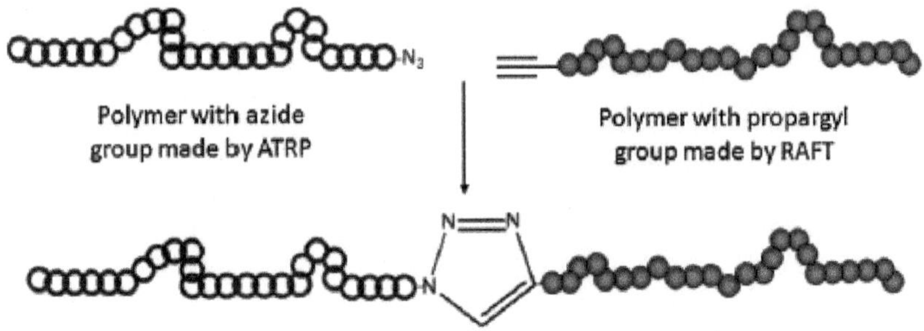

Scheme 35. Concept of combination of ATRP and RAFT using click chemistry.

Combined polymerization techniques	Obtained composition	References
Ring opening-ATRP	Dendrimer-like star block	[107]
ATRP-cationic polym.	AB type diblock	[108]
ATRP-macromonomer method	Graft	[109]
NMP-ATRP	Block-graft	[110]
NMP-RAFT	Graft	[111]
RAFT/MADIX-click Chemistry	Graft-block	[112]
NMP-*iniferter* technique	Graft	[113]
RAFT-click Chemistry	Block	[114]
Ring opening-NMP	Block	[115]
RAFT-ATRP-NMP	Bottle brush	[116]

Table 6. Some examples of polymers obtained by combination of different controlled polymerization mechanisms

6. Conclusion

The synthesis of block and graft copolymers can be achieved *via* different techniques such as free radical, anionic, cationic, coordination, coupling, step growth, ring opening, changing polymerization mechanism (active centers transformation reactions). CRP enables the synthesis of a wide range of (co)polymers with controlled molecular weight, narrow molecular weight distribution, and wide range of architectures (i.e., comb, star, dendritic, etc.), composition (i.e., random, periodic, graft, etc.) and functionalities at moderate experimental conditions.

Author details

Hülya Arslan
Bülent Ecevit University, Department of Chemistry, Zonguldak, Turkey

Acknowledgement

It is a pleasure to acknowledge for their contributions to my graduate students, at the Bülent Ecevit University.

7. References

[1] Lodge T.P (2003) Macromol. chem. phys. 204: 265-273.
[2] Ueda J, Kamigaito M, Sawamoto M (1998) Macromolecules 31(20): 6762-6768.
[3] Shinoda H, Miller P. J, Matyjaszewski K (2001) Macromolecules 34(10): 3186-3194.
[4] Noshay A and McGrath J. E (1977) Block Copolymers: Overview and Critical Survey, Academic Pres, New York .
[5] Ceresa R.J (ed.) (1976) Block and Graft Copolymerization, Vols. 1 and 2, Wiley, London.
[6] Hazer B (1989) Handbook of Polymer Science and Technology, Vol.1: Synthesis and Properties, Edited by Nicholas P. Cheremisinoff, by Marcel Dekker Inc. pp.133-176.
[7] Hazer B (1997) Handbook of Engineering Polymeric Materials, Edited by Nicholas P. Cheremisinoff, by Marcel Dekker Inc. pp: 725-734.
[8] Yağcı Y and Mishra M.K (1996) Polymeric Materials Encyclopaedia, Vol. 1, Salamone J.(ed.), by CRC Press Inc. USA, pp. 789-796.
[9] Fradet A (1996) Polymeric Materials Encyclopaedia, Vol. 1, Salamone J.(ed.), by CRC Press Inc. USA, pp. 797-807.
[10] Mishra M. K and Yağcı Y (1996) Polymeric Materials Encyclopaedia, Vol. 1, Salamone J.(ed.), by CRC Press Inc. USA, pp. 808-814.
[11] Ueda M (1999) Prog. polym. sci. 24: 699-730.
[12] Bhattacharya A, Misra B. N (2004) Prog. polym. sci. 29: 767-814.
[13] Arslan H, Yesilyurt N, Hazer B (2007) Journal of applied polymer science 106: 1742–1750.
[14] Arslan H, Yesilyurt N, Hazer B (2008) Macromol. symp. 269: 23–33.
[15] Cai Y, Hartenstein M and Müller A.H.E (2004) Macromolecules 37: 7484-7490.
[16] Zhang M, Breiner T, Mori H, Müller A.H.E (2003) Polymer 44: 1449-1458.
[17] Sarbu T, Lin K, Spanswick J, Gil R R, Siegwart D.J and Matyjaszewski K (2004) Macromolecules 37: 9694-9700.
[18] Chung I.S and Matyjaszewski K (2003) Macromolecules 36: 2995-2998.
[19] Shen Y, Zhu S, Zeng F. and Pelton R (2000) Macromolecules 33: 5399-5404.
[20] Coessens V, Pintauer T, Matyjaszewski K (2001) Progress in polymer science 26: 337-377.
[21] Yuan X, Lu J, Xu Q, Wang L (2005) Polymer 46: 9186-9191.
[22] Muthukrishnan S, Jutz G, Andre X, Mori H and Müller A.H.E (2005) Macromolecules 38: 9-18.
[23] Muthukrishnan S, Mori H and Müller A.H.E (2005) Macromolecules 38: 3108-3119.
[24] Ydens I, Degee P, Dubois P, Libiszowski J, Duda A, Penczek S (2003) Macromol. chem. phys. 204: 171-179.
[25] Beers K.L and Matyjaszewski K (2001) J. macromol. sci.-pure appl. chem. A38(7): 731-739.
[26] Zhang Z, Ying S, Zhang Q, Xu X (2001) J. polym. sci.: part A: polym. chem. 39: 2670-2676.
[27] Hawker C. J, Bosman A. W.and Harth E (2001) Chem. rev.101: 3661-3688.
[28] Nicolas J, Charleux B, Guernet O and Magnet S (2005) Macromolecules 38: 9963-9973.
[29] Boyes S.G, Granville A. M, Baum M, Akgün B, Mirous B.K, Brittain W. J (2004) Surface science 570: 1-12.
[30] Lipscomb C. E and Mahanthappa M. K (2009) Macromolecules 42: 4571-4579.

[31] Matyjaszewski K (2002) General Concepts and History of Living Radical Polymerization, In Handbook of Radical Polymerization; Matyjaszewski, K., Davis, T.P., eds. John Wiley and Sons, Inc.:Hoboken, pp 383-386.

[32] Matyjaszewski K, Spanswick J (2005) Controlled/living radical polymerization, Materials Today 8(3): 26-33.

[33] Patten T.E and Matyjaszewski K (1998) Adv. Mater. 10: 901-915.

[34] Bussels R (2004) (Multi)block copolymer synthesis via controlled radical polymerization in aqueous dispersions / by Raf Bussels, Eindhoven:Technische Universiteit Eindhoven, Proefschrift. – ISBN 90-386-2825-0,CIP-DATA LIBRARY TECHNISCHE UNIVERSITEIT EINDHOVEN.

[35] Müftüoğlu A. E, Taşdelen M. A, Yağcı Y and Mishra M.K (2009) Handbook of Vinyl Polymers: Radical Polymerization, Process and Technology CRC Press Taylor &Francis Group, New York, Ch11: pp. 307-344.

[36] Bomsan A. W, Frechet J. M. J, Hawker C. J (2001) Polym. mater. sci. eng., 84: 376-??

[37] Choi H, Park M, Lee S, Hong S.C (2008) European polymer journal 44: 3087-3095.

[38] Britze A, Jacob J, Choudhary V, Moellmann V, Grundmeier G, Luftmann H, Kuckling D (2010) Polymer 51: 5294-5303.

[39] Groison E, Brusseau S, D'Agosto F, Magnet S, Inoubli R, Couvreur L and Charleux B (2012) ACS Macroletters 1: 47-51.

[40] Ruehl J, Nilsen A, Born S, Thoniyot P, Xu L, Chen S, Braslau R (2007) Polymer 48: 2564-2571.

[41] Cheng S, Beyer F.L, Mather B.D, Moore R.B and Long T.E (2011) Macromolecules 44: 6509-6517.

[42] Narumi A, Matsuda T, Kaga H, Satoh T, Kakuchi T (2002) Polymer 43: 4835-4840.

[43] Thakur S, Cohen N. A, Tillman E.S (2008) Polymer 49: 1483-4189.

[44] Detrembleur C, Debuigne A, Jerome C, Phan T. N. T, Bertin D and Gigmes D (2009) Macromolecules 42: 8604-8607.

[45] Maric M, Lessard B.H, Consolante V, Ling E. J. Y (2011) Reactive&Functional Polymers 71: 1137-1147.

[46] Beaudoin E, Dufils P. E, Gigmes D, Marque S, Petit C, Tordo P, Bertin D (2006) Polymer 47: 98-106.

[47] Okamura H, Takatori Y, Tsunooka M, Shirai M (2002) Polymer 43: 3155-3162.

[48] Yan J, Ji W, Chen E, Li Z, Liang D (2008) Macromolecules 41: 4908-4913.

[49] Pearson S, Allen N, Stenzel M. H (2009) *J.of polym. sci.:part A:polymer chem.*, 47:1706-1723.

[50] Nguyen M. N, Bressy C, Margaillan A (2009) Polymer 50: 3086-3094.

[51] Sordi M. L. T, Riegel I. C, Ceschi M. A, Müller A. H. E, Petzhold C. L (2010) Eur. polym. j. 46: 336-344.

[52] An Z. S, Shi Q. H, Tang W, Tsung C. K, Hawker C. J, Stucky G. D (2007) J. am. chem. soc. 129: 14493-14499.

[53] Guan C, Luo Z, Qiu J, Tang P (2010) Eur. polym. j. 46: 1582-1593.

[54] Zheng L, Chai Y, Liu Y, Zhang P (2011) *e-Polymers* 39: 1-8.

[55] You Y, Zhou Q, Manickam D. S, Wan L, Mao G and Qupicky D (2007) Macromolecules 40: 8617-8624.

[56] Baum M and Brittain W J (2002) Macromolecules 35: 610-615.
[57] Albertin L, Stenzel M. H, Barner-Kowallik C, Foster L. J. R and Davis T. P (2005) Macromolecules 38: 9075-9084.
[58] Öztürk Z, Komber H, Gram S, Schmaljohann D, Müler A. H. E, Voit B (2007) Macromolecular chemistry and physics 208:1035-1049.
[59] Liu L, Wu C, Zhang J, Zhang M, Liu Y, Wang X, Fu G (2008) J. polym. sci.: part A: polymer chemistry 46: 3294-3305.
[60] Smith A. E, Xu X, Kirkland-York S. E, Savin D. A and McCormick C. L, Macromolecules (2010) 43: 1210-1217.
[61] Zhang L, Bernard J, Davis T. P, Barner-Kowallik C, Stenzel M. H (2008) Macromolecular Rapid Communications 29: 123-129.
[62] Lipscomb C. E and Mahanthappa M. K (2009) Macromolecules 42: 4571-4579.
[63] Albertin L, Stenzel M, Barner-Kowallik C, Foster L. J. R and Davis T. P (2004) Macromolecules 37:7530-7537.
[64] Can A, Altuntas E, Hoogenboom R, Schubert U. S (2010) Eur. polym. j. 46: 1932-1939.
[65] Motala-Timol S, Jhurry D (2007) Eur. polym. j. 43: 3042-3049.
[66] Huang W, Kim J, Baker G. L and Bruening M. L (2003) Nanotechnology 14: 1075-1080.
[67] Shipp D. A, Wang J and Matyjaszewski K (1998) Macromolecules 31: 8005-8008.
[68] Nakagawa Y, Miller P. J and Matyjaszewski K (1998) Polymer 39(21): 5163-5170.
[69] Ohno K, Tsujii Y, Fukuda T (1998) J. of polym. sci.: part A: polym. chem. 36: 2473-2481.
[70] Cassebras M, Pascual S, Polton A, Tardi M, Vairon J (1999) Macromol. rapid commun. 20: 261-264.
[71] Wang X, Luo N, Ying S, Liu Q (2000) Eur. polym. j. 36: 149-156.
[72] Lee S. B, Russell A. J and Matyjaszewski K (2003) Biomacromolecules 4: 1386-1393.
[73] Strissel C, Matyjaszewski K, Nuyken O (2003) Macromol. chem. phys. 204: 1169-1177.
[74] Licciardi M, Tang Y, Billingham N. C and Armes S. P (2005) Biomacromolecules 6: 1085-1096.
[75] Xu C, Wu T, Batteas J. D, Drain C. M, Beers K. L, Fasolka M. J (2006) Applied surface science 252: 2529-2534.
[76] Paris R, Luis de la Fuente J (2008) Eur. polym. j. 44: 1403-1413.
[77] Oikonomou E. K, Pefkianakis E. K, Bokias G, Kallitsis J. K (2008) Eur. polym. j. 44: 1857-1864.
[78] Suriano F, Coulembier O, Degee P, Dubois P (2008) J. of polym. sci.: part A: polym. chem. 46: 3662-3672.
[79] Pasparakis G and Alexander C (2008) Angew. chem. int. ed. 47: 4847-4850.
[80] Advincula R. C, Brittain W. J, Caster K. C, Ruhe J (2004) Polymer Brushes: Synthesis, Characterization, Applications (John Wiley &Sons, New York).
[81] Hua D, Deng W, Tang J, Cheng J, Zhu X (2008) International journal of biological macromolecules, 43: 43-47.
[82] Jiang J, Hua D, Jiang J, Tang J, Zhu X (2010) Carbohydrate polymers 81: 358-364.
[83] Jaymand M (2011) Polymer 52: 4760-4769.
[84] Chen Y, Ying L, Yu W, Kang E.T and Neoh K.G (2003) Macromolecules 36: 9451-9457.
[85] Ying L, Yu W.H, Kang E.T and Neoh K.G (2004) Langmuir 20: 6032-6040.

[86] Roy D, Guthrie J.T and Perrier S (2005) Macromolecules 38: 10363-10372.
[87] Stenzel M.H, Zhang L, Huck W.T.S (2006) Macromol. rapid commun. 27: 1121-1126.
[88] Hua D, Tang J, Cheng J, Deng W, Zhu X (2008) Carbohydrate polymers 73: 98-104.
[89] Morimoto N, Qiu X, Winnik F.M and Akiyoshi K (2008) Macromolecules 41: 5985-5987.
[90] Mespouille L, Nederberg F, Hedrick J.L and Dubois P (2009) Macromolecules 42: 6319-6321.
[91] Li J, Yi L, Lin H, Hou R (2011) Journal of polymer science part A: polymer chemistry 49: 1483-1493.
[92] Tahlawy K and Hudson S. M (2003) Journal of applied polymer science 89: 901-912.
[93] Li N, Bai R and Liu C (2005) Langmuir 21: 11780-11787.
[94] Arslan H, Yeşilyurt N, Hazer B (2007) J. of app. polym. sci. 106: 1742-1750.
[95] Arslan H, Yeşilyurt N, Hazer B (2008) Macromol. symp. 269: 23-33.
[96] Munro N. H, Hanton L. R, Moratti S. C, Robinson B. H (2009) Carbohydrate polymers 77: 496-505.
[97] Meng T, Gao X, Zhang J, Yuan J, Zhang Y, He J (2009) Polymer 50: 447-454.
[98] Liu Y, Klep V, Zdyrko B and Luzinov I (2004) Langmuir 20: 6710-6718.
[99] Muthukrishnan S, Zhang M, Burkhardt M, Drechsler M, Mori H and Müler A. H. E (2005) Macromolecules 38: 7926-7934.
[100] Gao C, Muthukrishnan S, Li W, Yuan J, Xu Y and Müler A. H. E (2007) Macromolecules 40: 1803-1815.
[101] Wang J, Xu Y, Xu H, Zhang F, Qian Y, Zhu B (2008) J. of app. polym. sci. 109: 2914-2923.
[102] Wang Y, Pei X, He X, Lei Z (2005) Eur. polym. j. 41: 737-741
[103] Xu F. J, Cai Q. J, Kang E. T and Neoh K. G (2005) Langmuir 21: 3221-3225.
[104] Bach L. G, Islam Md. R, Jeong Y. T, Gal Y. S, Lim K. T (2012) Applied surface science 258: 2816-2822.
[105] Lu J, Liang H, Li A, Cheng Q (2004) European polymer journal 40: 397-402.
[106] Nasrullah M. J, Vora A, Webster D. C (2011) Macromol. chem. phys. 212: 539-549.
[107] Hedrick J. L, Trollsas M, Hawker C. J, Atthoff B, Claesson H, Heise A and Miller R. D (1998) Macromolecules 31: 8691-8705.
[108] Düz A. B, Yağcı Y (1999) European polymer journal 35: 2031-2038.
[109] Shinoda H, Miller P. J and Matyjaszewski K (2001) Macromolecules 34: 3186-3194.
[110] Tasdelen M. A, Yagci Y, Demirel A. L, Biedron T, Kubisa P (2007) Polymer bulletin 58: 653-663.
[111] Wang S, Cheng Z, Zhu J, Zhang Z, Zhu X (2007) Journal of polymer science: part A: polymer chemistry 45: 5318-5328.
[112] Quemener D, Le Hellaye M, Bissett C, Davis T. P, Barner-Kowallik C, Stenzel M. H (2008) Journal of polymer science: part A: polymer chemistry 46: 155-173.
[113] Gromadzki D, Makuska R, Netopilik M, Holler P, Lokaj J, Janata M, Stepanek P (2008) European polymer journal 44: 59-71.
[114] Xue X, Zhu J, Zhang Z, Cheng Z, Tu Y, Zhu X (2010) Polymer 51: 3083-3090.
[115] Habraken G. J. M, Peeters M, Thornton P. D, Koning C. E and Heise A (2011) Biomacromolecules 12: 3761-3769.
[116] Zehm D, Laschewsky A, Liang H and Rabe J. P (2011) Macromolecules 44: 9635-9641.

Chapter 14

PolyADP-Ribosylation in Postfertilization and Genome Reprogramming: Implications for Carcinogenesis

Tomoharu Osada and Mitsuko Masutani

Additional information is available at the end of the chapter

1. Introduction

Pottranslational modification of proteins (PTM) is involved in molecular targeting or signal transduction as a basis of a variety of biological processes of the cells. Approximately 300 PTMs are existed in the cells, some of which are supposed to have important roles in cell physiology or embryogenesis. In addition, some modifications are reciprocally interacted, which regulates gene expression or protein dynamics or processing. We focus here polyADP-ribosylation (PARylation), which is found to contribute importantly at fertilization and postfertilization development. Moreover, we discuss the possible drug discovery of ADP-ribosylation inhibitors in carcinogenesis or fertility control.

Upon fertilization, two kinds of genetic materials meet together to generate a new organism [1]. Meanwhile, the organism develops without major expression of genes from genomes. Although maternal proteins are supposed essential for the period, the details are ill understood. Transition of zygotic gene expression is also essential for further development. Some PTMs on transcriptional factors may be key regulation to express the genes which supports early development. Parental DNAs are differently regulated in the postfertilized eggs, although the biological significance is controversial. Theses implications suggest importance of PTMs on postfertilization and transition of zygotic development. Recently we found that PARylation, a PTM of protein is important for postfertilization development, which seems similar with those of carcinogenesis.

The PARylation reaction is a PTM of proteins, which is synthesized with poly(ADP-ribose) polymerase (PARP) and metabolized with poly(ADP-ribose) glyceraldehyde (PARG) [2,3]. The reciprocal regulation is speculated as a key mechanism that underlies reversible regulation of gene expression. The poly(ADP-ribose) is generated from NAD, an energy reservoir of the cells by PARP. The PARP synthesizes poly(ADP-ribose) using NAD. PARP polymerizes ADP-ribose residues to generate poly(ADP-ribose) chains onto proteins. PARylation reaction itself is regulated by auto-PARylation of Parp1. Since the ADP-ribose residue is negatively charged, the acceptor proteins become negatively charged by the addition of ADP-ribose residues. The electrical charge-shift of the protein induced by PARylation may decrease an easy access of proteins into the DNA structure or may induce structural instability of protein-DNA interactions, because the negative charge of poly(ADP-ribose) supports its association with DNA-binding proteins, which are positively charged. The recognition of DNA sequences by protein structure may also be affected by PARylation. Upon postfertilization development, zygotic gene expression is essentially activated prior to the 4-cell cleavage stage in mice and human. Therefore, behavior of maternal molecules plays important roles at postfertilization development before zygotic gene activation. PTM of protein is supposed to contribute to protein dynamics at postfertilization development. Pharmacological blockage of PARylation revealed defects of postfertilization development in mice [4]. The data raise a question regarding the roles of PARylation at fertilization development. A wealth of study revealed that NAD is rich in eggs and rapidly degraded upon fertilization in Xenopus larvae. Parp activation at fertilization may contribute to consumption of NAD, which may bring a plausible explanation towards the uncovered subject. Postfertilization development is specific biological window to highlight the significance of PTM, because transcription is minor mechanisms for the organisms at the period. In this chapter, we discuss the roles of PARylation in eggs and the function of postfertilization development. Further we discuss the possible implications to drug discovery focusing on PARylation regulation. We focus on polyADP-ribosylation because we previously showed that no fertilized eggs were obtained by pharmacological blockage of PARylation by a PARP inhibitor, PJ-34 [4]. Other inhibitors of PARylation showed similar results. Therefore, a hypothesis was raised that posttranslational regulation is key regulation of maternal genetic materials at fertilization because no robust transcription is occurred at the beginning of life.

2. PARylation in postfertilization development

Recent research showed that pharmacological blockage of PARylation leads to defects in pronuclear fusion during postfertilization in mice. Based on the observations, several approaches were achieved to investigate the roles of PARylation in vivo and mechanisms of reprogramming applicable for regenerative medicines or elucidation of human diseases including carcinogenesis.

Based on our observations regarding the disorganized microtubule assembly in oocytes by PARylation inhibitors, polyADP-ribosylated proteins of oocytes and postfertilized eggs were searched. We found that the tubulins (α1c, β2c) [4] and glutathione *S*-transferees μ5 (GSTμ5) (Osada et al., unpublished data) was highly polyADP-ribosylated after fertilization.

Because PARylation is known as a regulatory machinery of DNA surveillance system, this data provided us with a novel sight of view on the roles of PARylation regarding cell signaling and metabolisms. As a major upstream signaling cascade after sperm entry, MAP-kinase (MAPK) signaling has been postulated as the key regulatory mechanism for pronucleus formation, and Parp1 is reported to be involved in the regulation of MAPK signaling [5]. Our data suggest that Parp1 is a novel nuclear component for PPN (Pseudo pronuclei) formation, which may be mediated by MAPK signaling. Phosphorylation level of Erk was decreased in the MII phase of Parp-1 null oocyte. Because the downregulation of Erk phosphorylation is known to disturb microtubule organization, these results raise a possibility that the transient administration of a Parp inhibitor may block the early phase of post-fertilization process.

Second, we focused on the roles of polyADP-ribosylation on epigenetic regulation. Accumulating study suggests that PARylation has a fundamental role in transcription as well as DNA surveillance, which mediates topological changes of the DNA structure [6,7]. We first examined the effects of PARylation on chromatin modification during postfertilization development. An immunofluorescence study using histone acetylation and methylation antibodies were initially carried out. Upregulation of dimethyl-H3k4 in the PARylation-inhibited embryos at 6 hpf, and downregulation of dimethyl-H3K9 and dimethyl-H3K27 at 15 hpf were observed (unpublished data, Osada et al.). Recently, functional links between histone methylation and DNA methylation have been elucidated. Based on our observations of distinct effects of PARylation inhibition on histone modification regulation, we next sought for the roles of PARylation in DNA methylation. Global methylation of the maternal genome is maintained and sperm DNA is quickly demethylated during the postfertilization period [8]. The low immunoreactivity to the MetC antibody in the PARylation inhibited embryos was observed and it persisted at least until 15 hpf (unpublished data, Osada et al.). The data suggest that epigenetic regulation is asymmetrically activated during postfertilization development, although gene expression is not activated. Therefore, PARylation affects the integrity of epigenetic regulation of oocytes and postfertilized eggs (Figure 1).

An additional implication of this finding is that PARylation regulates the epigenetic nonequivalence of the pronuclei in mice. The molecular mechanisms regulating the asymmetric DNA methylation of one-cell embryos are largely unknown, and we showed that PARylation is involved in the epigenetic regulation of the mouse early development. PGC7/Stella-deficient eggs show similar defects in the protection of female pronuclei from DNA demethylation [9]. The nucleo-cytoplasmic transport of the responsible proteins appears to be important for the regulation of DNA methylation. The role of PARylation in DNA methyltransferase (DMNT) regulation is yet to be fully elucidated, although the interaction of DNMT1 with PARP1 and the indirect repression of DNMT1 activity by interacting with PAR have been suggested [10]. Based on our data showing an association between PARylation or PARP and histone modification, it is speculated that PARylation may regulate the accessibility of DNMT or demethylase to DNA, in a manner mediated by chromatin remodeling. Our study provides a novel avenue for better understanding of establishment of chromatin organization through the histone codes and DNA methylation that may underlie at the beginning of zygotic development.

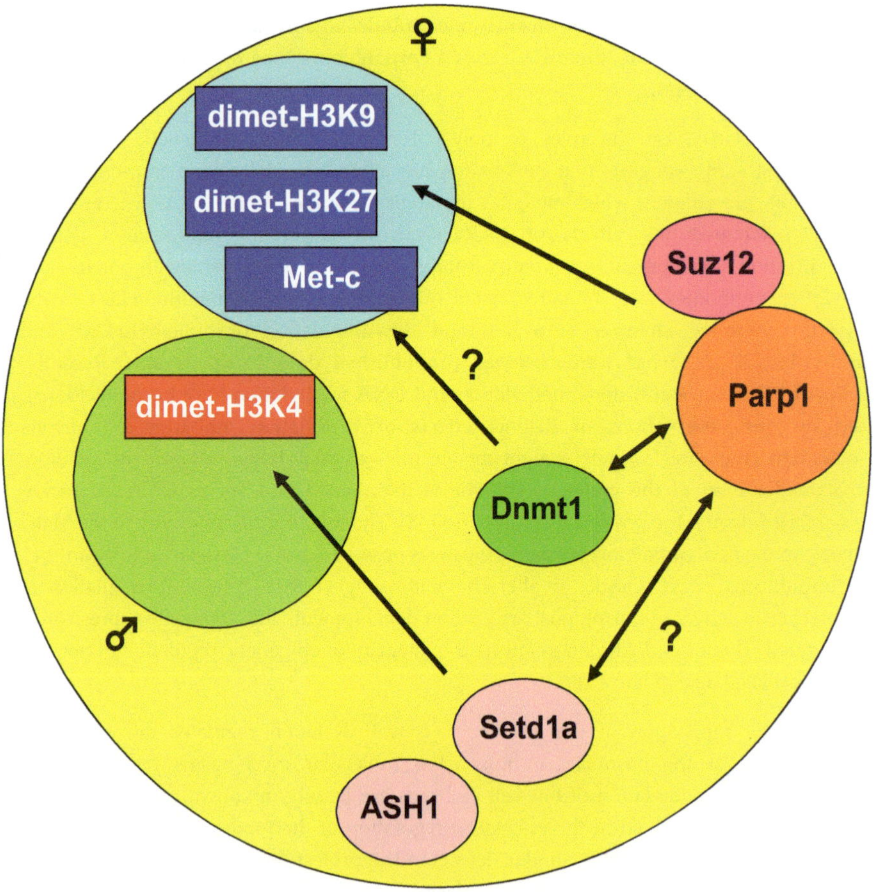

Figure 1. Scheme of putative roles of PARylation in the chromatin dynamics of one-cell embryos. In the one-cell embryos (yellow circle), PARP1 may interact with histone modifying enzymes including Suz12, that might affect the asymmetric regulation of histone modifications in the pronuclei (male pronucleus in blue circle and female pronucleus in green).

To examine whether there is a direct or indirect linkage of epigenetic regulation with PARylation, we screened molecules that interacts with Parp1. To elucidate the molecular basis of the effect of PARylation on histone modification, we performed yeast two-hybrid screening using the bait vector carrying N-terminal and automodification domain. The Suz12, a histone methyltransferase, was identified as a candidate molecule interacting with Parp1 (unpublished data, Osada et al.). Suz12 is a component of PRC2/3/4 (Polycomb Repressive Complex 2/3/4), which regulates the repressive status of transcription by the methylation of histone H3K9 and H3K27. Our findings support the idea that Parp1 acts as a regulatory scaffold for the access of PRC2/3 to target DNA, because none of the components of the PRC2/3/4 complexes are DNA binding proteins. OCT4, bidirectional transcriptional regulator, is suggested to play a role of transcriptional regulation by recruiting PRC2/3/4 to the target genes, which are indispensable for the embryonic development. Modification of PARylation triggers loosening of chromatin structure, which may enable the access of transcriptional factors to DNA duplex structure.

Third, as shown in Figure 2, defects in pronuclear formation by inhibition of PARylation raise a possibility that PARylation is involved in laminar formation of pronuclei [4]. In fact, HP1, an anchor protein of lamina and heterochromatin is polyADP-ribosylated [11]. Our findings showed prolonged presence of lamin-A/C, a core protein of lamina by PARP inhibitor PJ-34 treated eggs. Interestingly, lamin-A/C is predominantly existed in undifferentiated cells. Lamin-A/C is known as a marker of carcinogenesis [4]. Our observation provided a new insight into regulatory machinery of laminar formation in pluripotent cells by PARylation. Our initial data suggest that cell-cycle is not disturbed when embryos were subjected to treatment of PARP inhibitor.

During the first cell cycle of mouse embryos, a few genes are transcribed mainly from paternal genome. Inhibition of transcription during one-cell embryos by RNA polymerase inhibitors showed dispensable roles of transcription in normal development. Zygotic gene activation is required for progression from 2-cell to 4-cell embryos. These indicate that posttranslational regulation of protein should act as a stem mechanism of the development of one-cell embryos. Upon fertilization, highly compacted chromatin of gametes was acutely decondensed to form pronuclei (PN) within a few hours. Protamines of sperm chromatin are replaced by maternal histone H1 during this process, which may be associated with global hypomethylation of sperm-derived PN. In contrast, maternal chromatin arrested at metaphase II progresses rapidly into G1 phase and subsequently forms the female PN. DNA synthesis from paternal genome is preceded to that from maternal DNA. A minor transcription is activated solely from male PN. This evidence suggests that the requirement for the posttranslational regulations of parental genomes before mingling of both gamete DNA to begin the proper zygotic development. Of posttranslational modification of proteins, we examined here the effects of PARylation during the first cell cycle of mouse embryos. Metabolism of NAD, which is the substrate of PARPs, is acutely activated upon fertilization. Further analysis will elucidate the biological functions of PARylation upon fertilization and downstream target molecules of them.

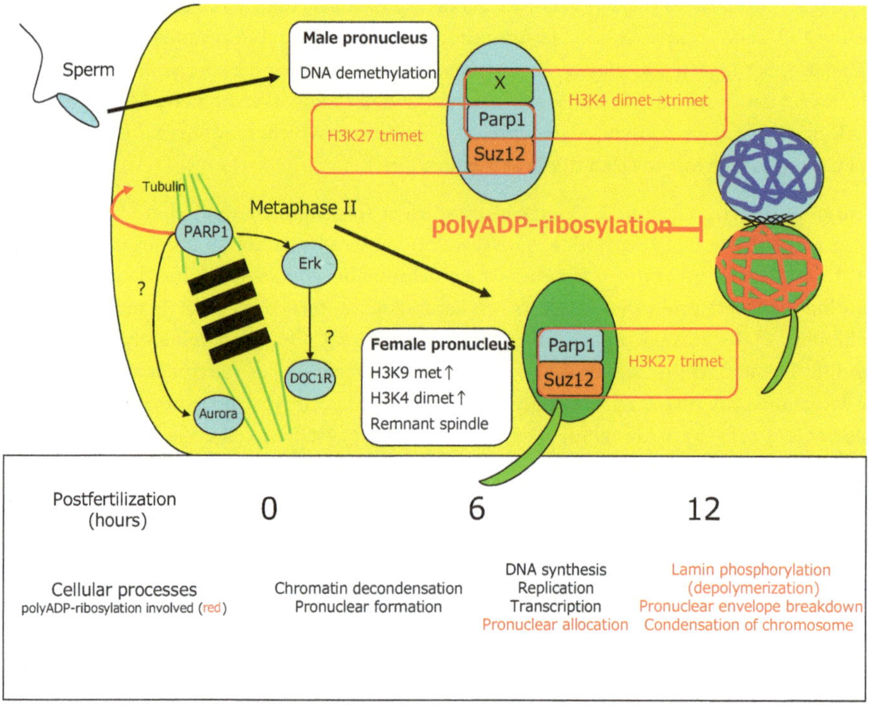

Figure 2. Model for regulation of pre- and postfertilization process by PARylation. Biological processes occurred during pre- and postfertilization were shown. Putative correlation with polyADP-ribosylation were indicated in red.

3. Implication towards reprogramming

In 1997, the first cloned animals were generated [12]. Cytogenetic analyses have shown that karyotypic aberrations occur in cloned embryos during the first mitotic cleavage. Epigenetic errors in cloned animals are also argued to be the major reason for the limited success rates of cloned animal births. However, whether genome-wide chromatin remodeling during nuclear reprogramming causes DNA damage, or whether defects in DNA repair cause the inefficiency of cloned animal births have not been investigated. Recently, it was reported that double strand break (DSB)–mediated chromatin remodeling regulates transcription, and that Parp1 is critically involved in this process. Various roles for Parp1 have been described, including in the chromatin remodeling involved in transcriptional regulation [2,3]. The roles of Parp1 in NT-embryo development in the contexts of DNA repair and chromatin remodeling was examined. To do this, we used *Parp1*-null mutant cells as recipient oocytes and as a source of donor nuclei.

We observed that the activation process in NT eggs was enhanced in *Parp1*–null NT embryos, although some genomic instability was observed (Osada T., unpublished data). Dynamic changes in histone acetylation and methylation were induced under *Parp-1* deficiency in NT embryos (Figure 3). The lack of Parp1 may thus facilitate chromatin condensation or transcriptional silencing. Parp1 is involved in both the regulation of DNA strand break repair and the epigenetic control of gene expression. Our findings suggest that Parp1 is important for chromatin remodeling possibly through histone modification during the nuclear reprogramming of NT embryos. The phosphorylation of histone H2AX at DSBs is believed to be crucial for the recognition and repair of DNA damage. The foci-like presence of γH2AX in the NT embryos implied that DNA repair was taking place in the PNN. H2AX is a direct target of ATM kinase, which is expressed at the spindle of MII oocytes, and Parp1 interacts with ATM. The prolongation of the γH2AX foci in the *Parp1*$^{-/-}$ NT embryos could therefore be related to a delay in repairing DNA strand breaks, as is the case with *Parp1*$^{-/-}$ MEF and ES cells, which show prolonged γH2AX foci after DNA damage induced by neocartinostatin. In *Parp1*$^{-/-}$ NT embryos, mitotic arrest of embryogenesis at the 2-4-cell stage was frequently observed, accompanied by polynucleated blastomeres. Since Parp1 is located in centrosomes as well as in nuclei, the lack of Parp1 may have disturbed the normal cell division cycle and the synchronous pattern of cell division among the blastomeres. Transient Parp-1 functional inhibition may be therefore useful to improve the efficiency of NT by modulating the dynamic organization of chromatin without causing genomic instability.

4. Implication towards human diseases

PARylation has been known to possess important roles in carcinogenesis. PARP mainly affects DNA surveillance system in the cells, which deficit increases the risk of carcinogenesis. In addition, recent study showed a functional relationship between PARP and MAPK signaling, known as a major intracellular signaling of transformation into cancer cells. Although morphological change of nuclear envelope has been known in cancer cells, the molecular mechanism of the processes is not well understood. Recent studies revealed molecular signaling of laminar formation is altered during carcinogenesis. Our study revealed that PARylation is involved in these molecular events. Oocytes are capable of being isolated without other cell types by adding hyarulonidase, which is far easier compared with the difficulty in homogenous isolation of purified cancer cells. Our data suggest oocytes provide with a unique biological window for elucidating the mechanism of PARylation in carcinogenesis. Further comparative analysis between cancer cells and oocytes may highlight the uncovered mechanisms underlying the carcinogenesis processes.

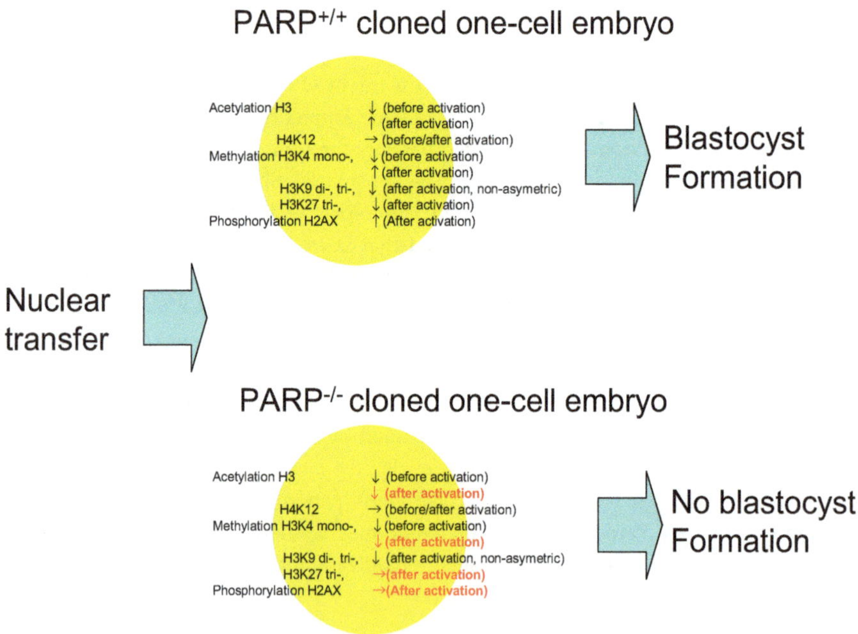

Figure 3. The scheme of histone modification and the effects of *Parp1* deficiency during the first cleavage of NT embryogenesis. Histone modifications of the one-cell reconstructed oocytes were indicated in *Parp*$^{+/+}$, and those which *Parp1* deficiency influences were indicated in *Parp*$^{-/-}$ in red.

We also described here that PARylation is a novel target of anti-conception. Inhibition of PARylation may easy to be handled, because exposure of an inhibitor to oocytes is effective enough to stop fertilization. Further extensive analysis should be carried out regarding the optimized dose of inhibitors and non-toxic dose on other tissues. Biological phenomena, in which PARylation is involved, are variable and its inhibitors could serve as possible pharmaceutical targets including inflammation or brain injury as well as carcinogenesis and reproduction as discussed in this chapter. Further basic investigation of the roles of PARylation in the cells will broaden the view for the understanding the embryogenesis and proof of mechanism of human diseases for drug discovery.

Author details

Tomoharu Osada and Mitsuko Masutani
Mitsubishi Chemical Medience Corporation, Kashima Laboratory, Japan
Division of Genome Stability Research, National Cancer Center Research Institute, Japan

Acknowledgement

We thank Dr. Ryuzo Yanagimachi and Dr. Takashi Sugimura for giving informative insight throughout the study. This study is partly supported by a Grand-in Aid for the Third Comprehensive 10-Year Strategy for Cancer Control from the Ministry of Health, Labor and Welfare of Japan.

5. References

[1] Yanagimachi, R (1994) Mammalian fertilization. In: Knobil E, Neil JD, editors. Physiology of Reproduction. 2nd edition. New York: Raven Press; pp. 189-317.
[2] Miwa, M. and Masutani (2007) M. PolyADP-ribosylation and cancer. Cancer Sci. 98:1528-1535.
[3] Schreiber V, Dantzer F, Ame JC, de Murcia G (2006) Poly(ADP-ribose): novel functions for an old molecule. *Nat Rev* Mol Cell Biol. 7: 517-528.
[4] Osada T, Ogino H, Hino T, Ichinose S, Nakamura K, Omori A, et al. (2010) PolyADP-ribosylation is required for pronuclear fusion during postfertilization in mice. PLoS One 5:e12526. doi:10.1371/journal.pone.0012526
[5] Cohen-Armon M, Visochek L, Rozensal D, Kalal A, Geistrikh I (2007) DNA-independent PARP-1 activation by phosphorylated ERK2 increases Elk1 activity: a link to histone acetylation. Mol. Cell. 25:297-308.
[6] Tulin A, Spradling A (2003) Chromatin loosening by poly(ADP)-ribose polymerase (PARP) at Drosophila puff loci. Science 299: 560-562.
[7] Caiafa P, Guastafierro T, Zampieri M (2009) Epigenetics: poly(ADP-ribosyl)ation of PARP-1 regulates genomic methylation patterns. FASEB J 23: 672-678.
[8] Mayer W, Niveleau A, Walter J, Fundele R, Haaf T (2000) Demethylation of the zygotic paternal genome. Nature 402: 501-502.

[9] Nakamura T, Arai Y, Umehara H, Masuhara M, Kimura T, et al (2007) PGC/Stella protects against demethylation in early embryogenesis. Nat. Cell Biol. 9: 64-71.

[10] Reale A, Matteis GD, Galleazzi G, Zampieri M, Caiafa P, 2005 Modulation of DNMT1 activity by ADP-ribosylation. Oncogene 24: 13-19.

[11] Quenet D, Gasser V, Fouillen L, Cammas F, Sanglier-Cianferani S, et al (2008) The histone subcode: poly(ADP-ribose) polymerase-1 (Parp-1) and Parp-2 control cell differentiation by regulating the transfcriptional intermediary factor TIF beta and heterochromatin protein HP1alpha. FASEB J. 22: 3853-3865.

[12] Wilmut I, Schnieke AE, McWhir J, King AJ. Campbell KH (1997) Viable offspring derived from fetal and adult mammalian cells. Nature 385:810-813.

Section 4

Polymer Physics

Chapter 15

Rheology – Theory and Application to Biomaterials

Hiroshi Murata

Additional information is available at the end of the chapter

1. Introduction

Rheology is science which treats the deformation and flow of materials. The science of rheology is applied to the physics, chemistry, engineering, medicine, dentistry, pharmacy, biology and so on. Classical theory of elasticity treats the elastic solid which is accordant to Hooke's law [1]. The stress is directly proportional to strain in deformations and is not dependent of the rate of strain. On the other hand, that of hydrodynamics treats the viscous liquid which is accordant to Newton's law [1]. The stress is directly proportional to rate of strain and is not dependent of the strain. However, most of materials have both of elasticity and viscosity, that is, viscoelastic properties. Mainly viscoelastic properties of the polymers are analyzed in rheology. Rheology would have two purposes scientifically. One is to determine relationship among deformation, stress and time, and the other is to determine structure of molecule and relationship between viscoelastic properties and structure of the materials.

2. Theory of rheology in polymeric materials

A strong dependence on time and temperature of the properties of polymers exists compared with those of other materials such as metals and ceramics. This strong dependence is due to the viscoelastic nature of polymers. Generally, polymers behave in a more elastic fashion in response to a rapidly applied force and in a more viscous fashion in response to a slowly applied force. Viscoelasticity means behavior similar to both purely elastic solids in which the deformation is proportional to the applied force and to viscous liquids in which the rate of deformation is proportional to the applied force. The behavior of polymers is very complex due to this dual nature.

Generally a mechanical model is used to explain viscoelastic behavior of materials which consist of springs and dashpots. The spring complies with Hooke's law as a mechanical model with an elastic element. When a spring is fixed at one end and a load is applied at the other, it becomes instantaneously extended. Then, after the load is removed, it immediately recovers its original length [2] (Fig 1). On the other hand, the dashpot complies with Newton's law of viscosity as a mechanical model with a viscous elememt. When a load is applied to a dashpot, it opens gradually, strain being a function of time. Then, after the load is removed, the dashpot remains open and recovery does not occurs [2] (Fig 2). The mechanical model which consists of a Hookean spring and a Newtonian dashpot in parallel is a Voigt element (Fig 3), and that in series is a Maxwell element (Fig 4) [3]. These are simple models and the behavior of polymers can be explained by combining these models.

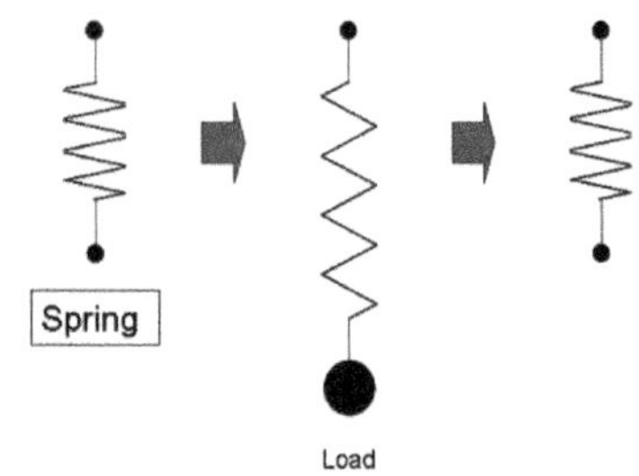

Figure 1. A spring which represents an elastic material.

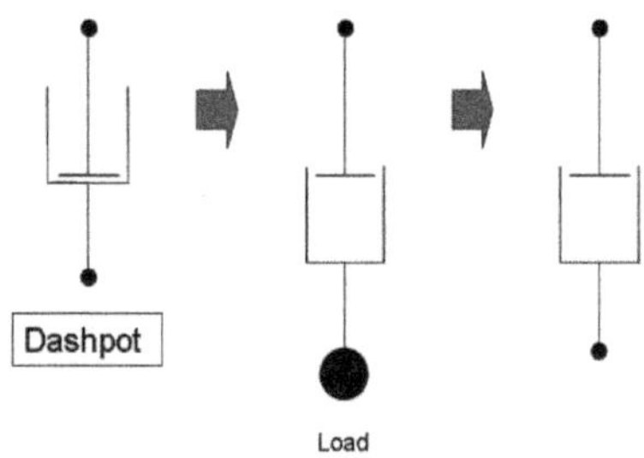

Figure 2. A dashpot which represents a viscous material.

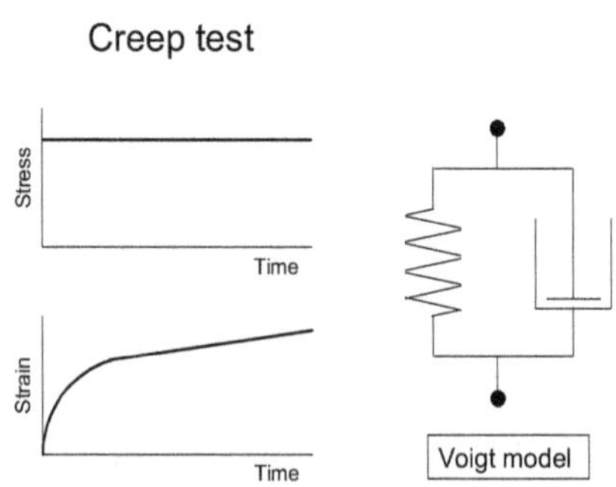

Figure 3. Creep test and a Voigt model.

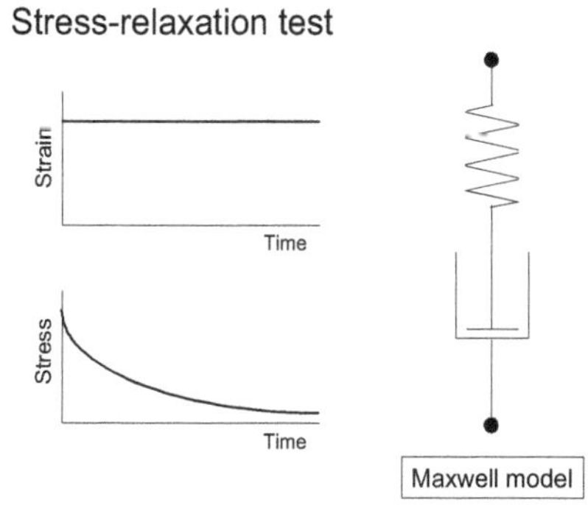

Figure 4. Stress relaxation test and a Maxwell model.

Measurement of viscoelastic properties of polymeric materials is achieved by several tests. These include static measurements such as creep test and stress relaxation test and dynamic measurement.

2.1. Creep

Creep test measures strains caused by a specific load, i.e. a definite force, in a certain range of time. This phenomenon is analyzed using a Voigt model (Fig 3).

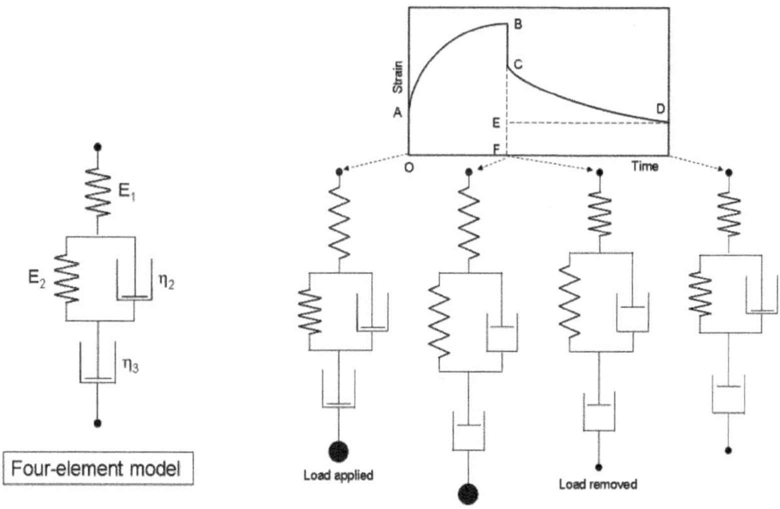

(a) (b)

Figure 5. (a) Four-element model for creep and creep recovery; (b) creep and creep recovery.
OA: Instantaneous strain; AB: Time dependent opening; BC: Instantaneous recovery (=OA); CD: Slow recovery; EF: Permanent deformation.

Furthermore, creep and creep recovery of a material can be analyzed by a four-element model (Fig 5). When a constant stress is applied at $t = 0$, the material shows an instantaneous increase in strain (the spring E_1), followed by a gradual increase in strain (the spring E_2 / dashpot η_2 and dashpot η_3). On removal of the stress, the spring E_1 recovers instantaneously followed by gradual recovery of the spring E_2 / dashpot η_2. Some permanent deformation remains as a consequence of the dashpot η_3 [2].

2.2. Stress relaxation

Stress relaxation test measures the stress required to maintain the level of strain as a function of time after rapidly deforming a specific volume of the material. The Stress relaxation behavior of cross-linked amorphous polymer is different from that of uncross-linked amorphous polymer.

The cross-linked amorphous polymer does not flow. Thus stress relaxation curves of the cross-linked polymers are evaluated by the analogies of a three-element model in which one Maxwell element and one spring are connected in parallel (Fig 6) [4]. The equilibrium modulus to be E_e, then the relaxation modulus $E_r(t)$ for the three-element model is defined as:

$$\begin{aligned} E_r(t) &= \sigma(t) / \gamma_0 \\ &= \frac{\sigma_1(0)\exp(-t/\tau_1) + \sigma_e}{\gamma_0} \\ &= E_1(0)\exp(-t/\tau_1) + E_e \end{aligned}$$

where total stress is $\sigma(t)$, strain is γ_0, stress on Maxwell element and degenerated element only with the spring are σ_1 and σ_e, elastic modulus E_1, coefficient of viscosity η_1, and relaxation time τ_1.

The instantaneous moduli E_0 (= $E_r(0)$) and η_1, respectively, were represented by:

$$\begin{aligned} E_0 &= E_1 + E_e = E_r(0) \\ \eta_1 &= E_1\tau_1 \end{aligned}$$

On the other hand, uncross-linked amorphous polymer flows. Thus stress relaxation curves of the uncross-linked polymers are evaluated by the analogies of a four-element model in which two Maxwell elements are connected in parallel (Fig 7) [5]. $E_r(t)$ for the four-element model is defined as:

$$\begin{aligned} E_r(t) &= \sigma(t) / \gamma_0 \\ &= \frac{\sigma_1(0)\exp(-t/\tau_1) + \sigma_2(0)\exp(-t/\tau_2)}{\gamma_0} \\ &= E_1(0)\exp(-t/\tau_1) + E_2(0)\exp(-t/\tau_2) \end{aligned}$$

where the stress on Maxwell elements is σ_1 and σ_2, elastic moduli are E_1 and E_2, the coefficient of viscosity η_1 and η_2, and the relaxation time τ_1 and τ_2 ($\tau_1 > \tau_2$).

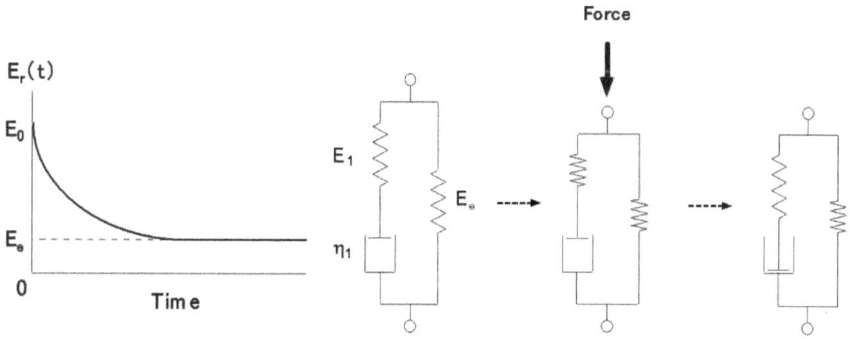

(a) (b)

Figure 6. (a) Stress relaxation curve of cross-linked polymers; (b) Three-element model in which one Maxwell element and one spring are connected in parallel (Reference [4]).

E_0, the steady-flow viscosity η_0, and η_1, η_2 of each Maxwell element, respectively, are represented as follows:

$$E_0 = E_1 + E_2 = E_r(0)$$
$$\eta_1 = E_1\tau_1$$
$$\eta_2 = E_1\tau_2$$
$$\eta_0 = \eta_1 + \eta_2$$

To make comparison among the materials, the above rheological parameters are used.

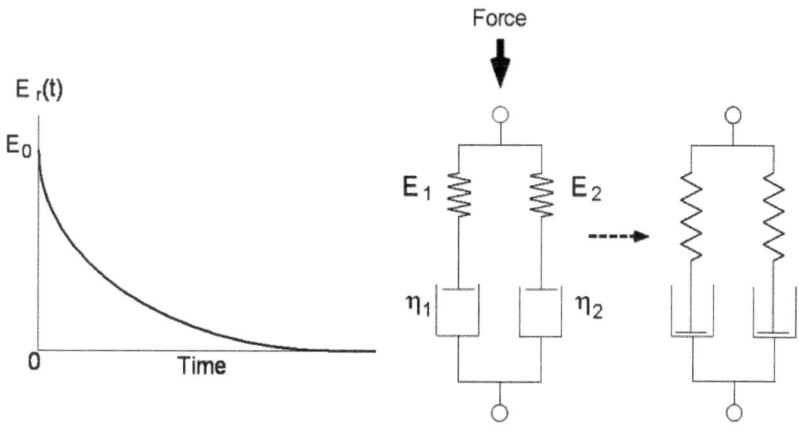

(a) (b)

Figure 7. (a) Stress relaxation curve of uncross-linked polymers; (b) Four-element model in which two Maxwell elements are connected in parallel (Reference [5]).

2.3. Dynamic viscoelastic properties

Viscoelastic properties of polymeric materials have been characterized by means of creep test and stress relaxation test. These static measurements have provided valuable information on the materials. However, these methods are not suitable for rigorous evaluation of the rheological characteristics of the materials. Dynamic mechanical test, which measures the response of the material to sinusoidal or other cyclic stresses, is suitable for rigorous evaluation of the rheological characteristics of the materials.

A schematic representation of the relationship between stress and strain, with a sinusoidally varying stress, for a perfectly elastic material, a viscoelastic material and perfectly viscous liquid is shown in Figure 8 to explain analysis of dynamic viscoelasticity [6, 7]. Using a dynamic viscoelastometer, a sinusoidal tensile strain is added to one end of a specimen and the stress response is detected at the other end. For a perfectly elastic solid, the strain is exactly in phase with the stress. For a perfectly viscous liquid, the strain is 90° out of phase. In the case of a viscoelastic material, the strain is somewhere in between ($\pi/2 > \delta > 0$). Amplitude attenuation of the sinusoidal strain and delay of the strain wave are known to be dependent on the viscoelastic properties of the material [1, 6, 7]. The complex modulus E^*, which is determined experimental by applying a sinusoidal stress, is resolved into two components, i.e. storage modulus E' and loss modulus E'' (Fig 8). E' is the ratio of the stress in phase with the strain to the strain, whereas E'' is the ratio of the stress 90° out of phase with the strain to the strain. E' represents the elastic component of material behavior and it directly proportional to the energy storage in a cycle of deformation. E'' represents the viscous component of material behavior and it isdirectly proportional to the average dissipation or loss of energy as heat in a cycle of deformation. The tangent of the phase angle (δ) between stress and strain, the loss tangent ($\tan \delta$), is a useful parameter and a measure of the ratio of energy lost to energy stored during cyclic deformation [6].

The complex modulus E^* of each material is calculated as follows [6]:

$$|E^*| = \Delta F / S \times Lt / \Delta L$$

where ΔF is the dynamic load,S the area of specimen, Lt the length of specimen and ΔL the dynamic displacement.

The storage modulus E' and loss modulus E'', are defined as:

$$E^* = E' + iE''$$
$$E' = |E^*| \cos\delta$$
$$E'' = |E^*| \sin\delta$$

where $i = \sqrt{-1}$. The loss tangent $\tan \delta$ is given by

$$\tan\delta = E'' / E'$$

The rheological parameters of E', E'' and tan δ are often used to evaluate the temperature or frequency - dependence of the materials.

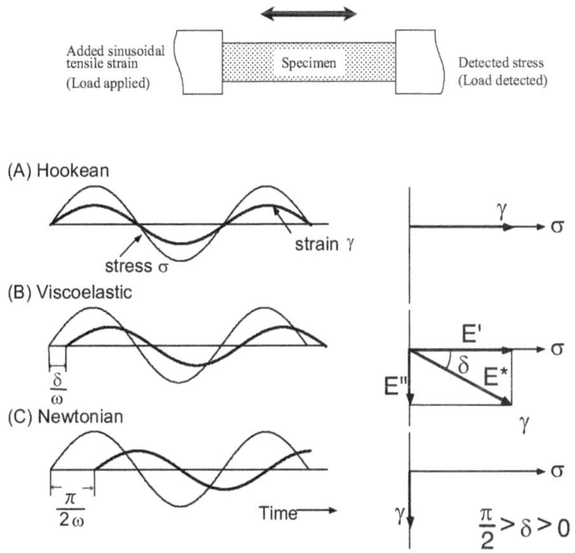

Figure 8. Schematic representation of dynamic mechanical test instrument and relationship between stress and strain of perfectly elastic solid (Hookean body) (A), viscoelastic material (B) and perfectly viscous liquid (Newtonian fluid) (C) with sinusoidally varying stress (Reference [6, 7]).

3. Experimental methods

Experimental methods depend on the condition of materials (liquid, solid, or sol-gel transition).

3.1. Viscoelastic liquid and gelation

There are many methods of measurements for viscoelastic liquid and gelation. In this section, the method by means of rheometer is introduced. An apparatus and schematic representation of dynamic viscoelasticity during the sol – gel transition are shown in Figure 9 and 10, respectively. The rheometer with a parallel plate or cone and plate test configuration is often used in oscillatory mode. The paste was placed on the plate of the rheometer. Torque is monitored at constant oscillating frequency and angular displacement.

In the parallel plate system, the shear strain (γ) and shear stress (τ) are determined experimentally as follows:

$$\gamma = F\gamma\omega$$
$$\tau = F\tau T$$

where $F\gamma$ (=R/d) is the shear strain factor; $F\tau$ (=$2/\pi R^3$) the shear stress factor; ω the angular displacement; T the torsional force; R the radius of the plate and d the shear gap.

The complex dynamic shear modulus (G^*), shear storage modulus (G'), shear loss modulus (G'') and loss tangent (tan δ) are defined as follows [8]:

$$|G^*| = \tau / \gamma$$
$$G^* = G' + iG''$$
$$G' = |G^*| \cos\delta$$
$$G'' = |G^*| \sin\delta$$
$$\tan\delta = G''/G'$$

where $i = \sqrt{-1}$, and δ is the phase angle between stress and strain.

G' represents the elastic component of material behavior, whereas G'' represents the viscous component of material behavior.

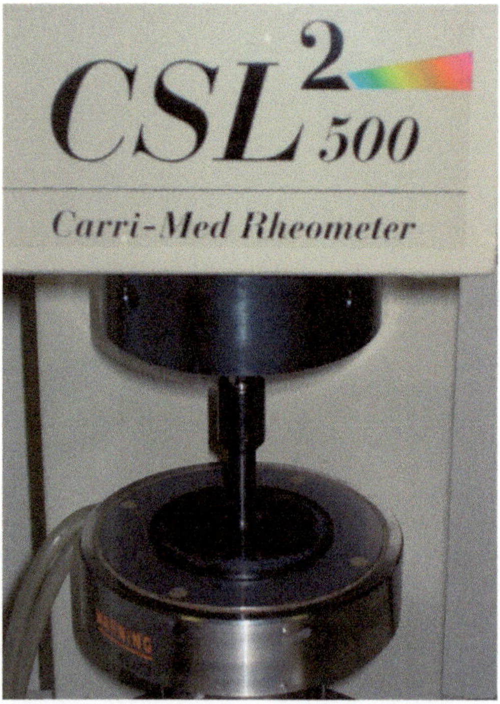

Figure 9. Controlled-stress rheometer.

Polymers undergo phase transition from liquid to solid at a critical extent of reaction during gelation. The instant of this transition is called the gel point [3]. The polymer is said to be at the gel point if its steady shear viscosity is infinite and its equilibrium modulus is zero, and at least one chain of infinite molecular weight has been created. The following scaling law at the gel point is suggested [9];

$$G'(\omega) = G''(\omega) \sim \omega^{1/2}$$

At the gel point, G' and G'' are congruent, *i.e.* tan δ =1, and this behavior is valid in the entire range 0 < frequency ω < ∞. Thus the gelation time is defined as the time to reach the gel point at which tan δ =1 (G'=G'') [3, 8, 9] (Fig 10).

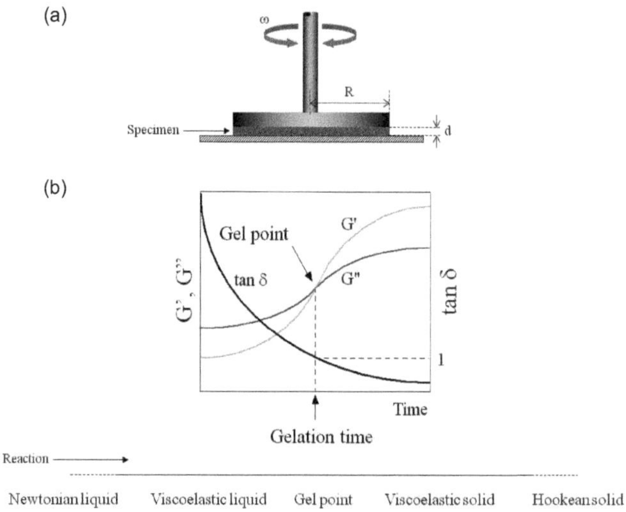

Figure 10. (a) Schematic diagram of a controlled-stress rheometer in a parallel plate configuration; (b) Schematic representation of dynamic viscoelastic properties of gelation system as a function of reaction time (Reference [8]).

3.2. Viscoelastic solids

The dynamic viscoelastic properties of polymeric solids are measured mainly by one of three methods; namely, the free tortional vibration method, resonance forced vibration method and non-resonance forced vibration method. The free damped oscillation of a torsion pendulum is used in the free torsional vibration method. However, this method is limited at the low end of the frequency range (less than 0.1 Hz) due to the effect of air resistance. The resonance forced vibration method also operates most effectively at high frequencies (greater than 10 Hz) and requires large specimens for accurate measurements. The non-resonance forced vibration method overcomes the deficiencies of the other two methods. The method enables the frequency-dependent properties of the polymeric solids to be determined over a wide range of frequencies, such as 0.01 - 100 Hz, and thus allows predictions of behavior under various conditions [6, 10].

One example of dynamic viscoelastometer based on a principle of non-resonance forced vibration is shown in Figure 11. This device consists of a measurement operation block, high/low constant temperature chamber, main unit, power unit, data processing device and testing jigs. The main unit performs high speed analogue-to-digital, digital-to-analogue conversion and transmits input-output signals to the measurement operating block, high/low temperature chamber and data processor. The measurement operating block consists of a magnetic exciter, tension control motor, amplitude detecting sensor and amplifier, load detecting load cell and chucks. During testing a suitable tension is applied to the loaded specimen and the dynamic displacement, i.e. sine wave vibration (sinusoidal stress), is added as a forced power. At the other end of the specimen, the dynamic load is detected and this is converted to familiar rheological parameters such as dynamic strain and dynamic stress, complex dynamic tensile modulus (E^*), tensile storage modulus (E'), tensile loss modulus (E'') and loss tangent ($\tan \delta$) [6].

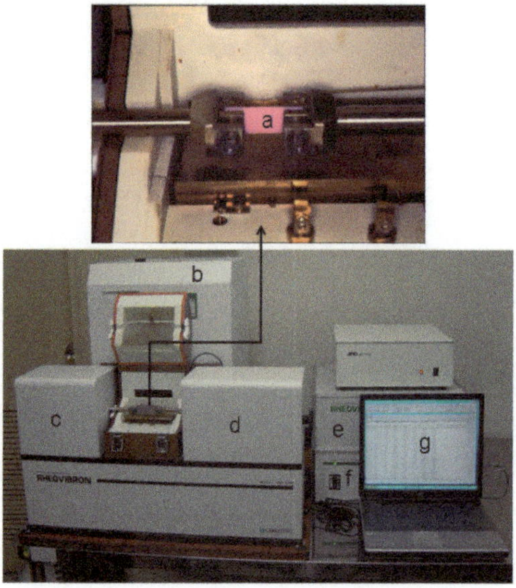

Figure 11. Dynamic viscoelastometer. (a): specimen, (b): high/low constant temperature chamber, (c): driver and displacement detector, (d): load detector, (e): main unit, (f): power amplifier, and (g): personal computer.

4. Rheology in biomaterials

Many biomaterials are available in medicine and dentistry. In this section, rheological properties of dental materials for dentures such as denture base and denture liners are referred (Fig 12). Denture is a removable artificial replacement for one or more teeth carried on a removable plate or flame. Denture liners are of two types, hard denture reline resins and soft denture liners [11].

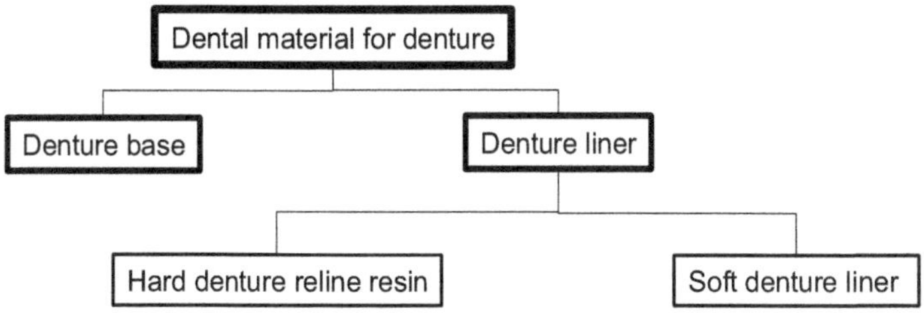

Figure 12. Dental materials for denture.

4.1. Denture base resin and hard denture reline resin

Denture base materials generally consist of heat-polymerized acrylic resins. Hard denture reline resins are used to improve the stability and retention of ill-fitting complete and removable partial dentures. Several types of these resins are available, including autopolymerized and visible light-polymerized materials (Fig 13). Visible light-polymerized, hard denture reline resins can be divided into 2 groups according to the supplied form: a powder-liquid type and a paste-type. Some visible light-polymerized reline resins are dual-polymerized. The powder form of the visible light-polymerized and autopolymerized, powder-liquid type materials consists of poly (ethyl methacrylate), poly (methyl methacrylate), or poly (methyl methacrylate/ethyl methacrylate), along with a peroxide initiator, and pigment. The liquid components are mixtures of methacrylate-based monomers: methyl methacrylate, n-butyl methacrylate, or cyclohexyl methacrylate, along with a cross-linking agent (ethoxylated bisphenol A dimethacrylate or 1,6-hexanediol dimethacrylate). In addition, the liquid forms of the visible light-polymerized materials and the autopolymerized materials include a photoinitiator, such as camphorquinone, and a chemically activated accelerator, such as N,N-dimethyl p-toluidine, respectively. The dual-polymerized materials contain both compounds. The visible light-polymerized, paste-type materials are a mixture of methacrylate copolymers, a polyfunctional monomer (such as urethane dimethacrylate), and a silica filler, which acts as a thickening agent. The differences in composition and structure will influence the mechanical properties of the resulting polymer. Increased cross-linking and microfine silica may result in higher flexural modulus values. The degree of cross-linking of heat-polymerized denture base resins is higher than that of direct denture reline resins. The paste-type materials include the silica filler [2, 7].

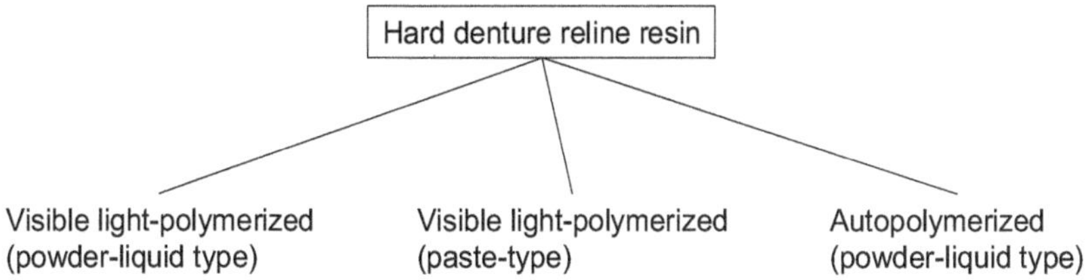

Figure 13. Classification of hard denture reline resins.

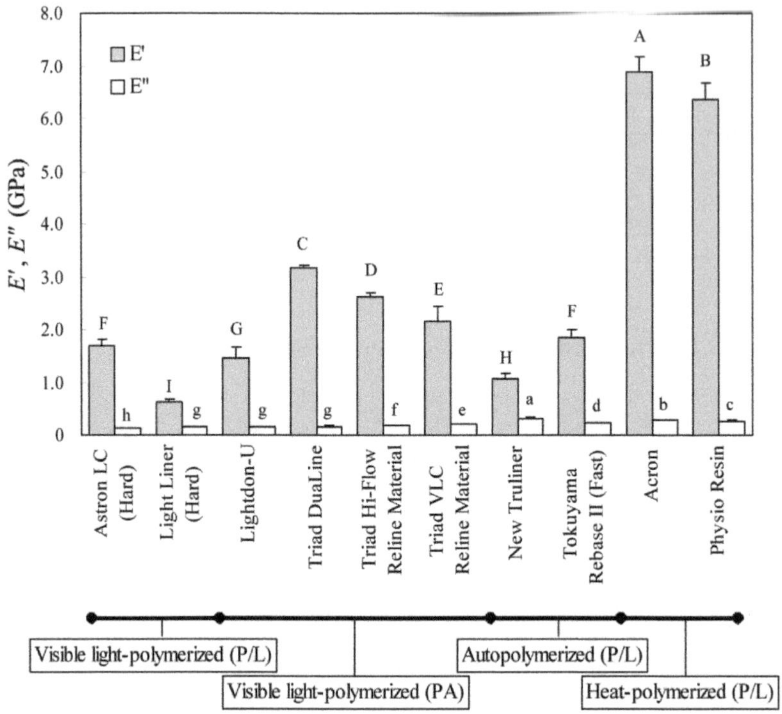

Figure 14. Storage (E') and loss moduli (E'') at 1 Hz of hard denture reline resins and heat-polymerized denture base resins. Within test parameter, values having similar uppercase letters (E') or lowercase letters (E'') are not statistically different. Within each material type, $E' > E''$ (Reference [7]).

The previous study evaluated the dynamic mechanical properties of 8 hard denture reline resins and 2 denture base resins using a dynamic viscoelastometer based on a principle of non-resonance forced vibration [7]. The storage modulus (E') values of 3 visible light-polymerized, paste-type reline resins were significantly higher than those of the other 5 reline resins. However, the E' values of all reline resins were significantly lower than those of the 2 heat-polymerized denture base resins. Except for 1 autopolymerized reliner, all reline materials had significantly lower loss modulus (E'') than the heat-polymerized denture base resins (Fig 14). The loss tangent (tan δ) values of all visible light (except for 1) and autopolymerized reliners were significantly higher than those of the heat-polymerized denture base materials (Fig 15). It has been found that the visible light-polymerized, paste-type reline resins tend to show greater stiffness than the visible light- or autopolymerized, powder-liquid type reline resins. This finding may be attributed to the presence of silica fillers in these products, which act to reinforce the resin. All of the hard denture reline resins have been also found to exhibit greater flexibility compared to the heat-polymerized denture base resins. Differences in dynamic mechanical properties are most likely due to differences in material composition and character of the resulting polymerized network. For example, the type, molecular weight and particle size of the polymer powder, the type and content of the monomer and cross-linking agent, and the powder/liquid ratio may all have an influence on mechanical properties of the polymerized material. In the case of the paste-type materials, type, concentration, and particle size of the filler may also have a large influence on the properties. A lower degree of cross-linking and a smaller quantity of silica filler in the reline resins may result in increased flexibility characteristics [7].

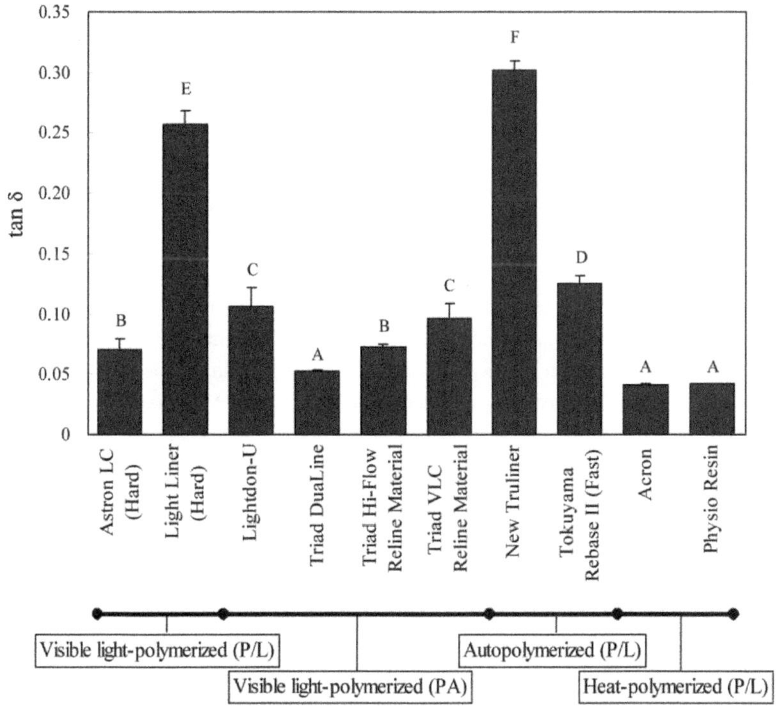

Figure 15. Loss tangent (tan δ) at 1 Hz for hard denture reline resins and heat-polymerized denture base resins. Within test parameter, values having similar letters are not statistically different. Increase in tan δ values indicates material is more able to absorb more energy without elastically returning it, acting more as shock absorber (Reference [7]).

4.2. Soft denture liner

4.2.1. Dynamic viscoelastic properties

Soft denture liners are widely used for denture wearers who complain of masticatory pain. These patients have a thin and non-resilient oral mucosa and/or severe alveolar resorption. These materials are applied to the denture intaglio surface to distribute and absorb masticatory forces by means of the cushioning effect. The clinical effect of the materials is considered to be influenced by viscoelastic properties and durability [11].

Soft denture liners are divided into permanent soft liners and acrylic temporary soft liners (Fig 16). The permanent soft liners can be classified into mainly: (1) autopolymerized silicone, (2) heat-polymerized silicone, (3) autopolymerized acrylic resin, and (4) heat-polymerized acrylic resin. The acrylic temporary soft liner is classified as tissue conditioner [11].

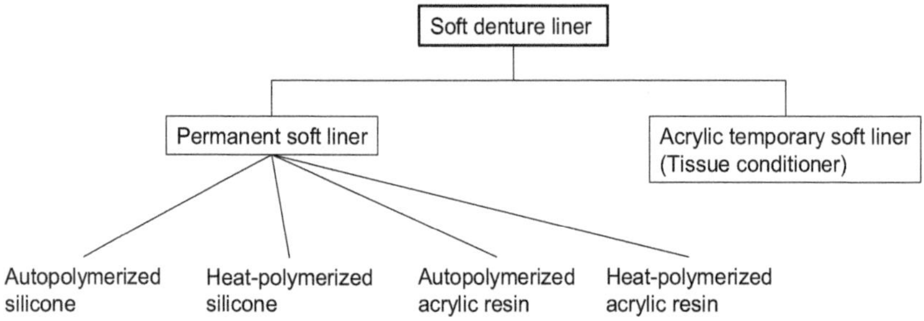

Figure 16. Classification of soft denture liners.

The autopolymerized silicone permanent soft liners are supplied in the form of a two-paste cartridge. These materials are based on a polyvinylsiloxane system. They do not involve the production of by-product such as ethyl alcohol after setting because this setting reaction is the self-curing addition. The heat-polymerized silicone permanent soft liners are supplied as a one-paste system with a free radical initiator. This product consists of a polydimethylsiloxane polymer with pendant or terminal vinyl group that cross-linking occurs [2, 11].

Both the autopolymerized and heat-polymerized acrylic permanent soft liners are supplied as a powder and liquid. The powder generally consists of poly (ethyl methacrylate) or poly (butyl methacrylate) along with some peroxide initiator. The liquid of autopolymerized acrylic material consists of 2-ethylhexyl methacrylate, tertiary amine and plasticizer. Those of heat-polymerized material are a mixture of methyl methacrylate and plasticizer [11].

The tissue conditioner also supplied as powder and liquid components. The main component of the polymer powder of most of the materials is poly (ethyl methacrylate) or a related copolymer. The liquid is a mixture of a plasticizer, such as butyl phthalyl butyl glycolate, dibutyl phthalate and dibutyl sebacate, and ethyl alcohol. Some material based on poly (butyl methacrylate) or a related copolymer contains no ethyl alcohol in the liquids [11].

Tissue conditioners contain no monomers in the liquid and no initiator in the powder. Thus the materials are uncross-linked amorphous polymers. On the other hand, both the silicone and acrylic permanent soft liners are cross-linked amorphous polymers. The differences in structure would influence the clinical behavior of the relined denture [11, 12].

The previous study evaluated the dynamic viscoelastic properties of soft denture liner over a wide range of frequency by a dynamic viscoelastometer (Fig 17) [6, 13]. The dynamic viscoelastic behavior of the acrylic soft liners showed sensitivity to changes in frequency, while that of the silicone was not markedly frequency dependant. The acrylic soft liners had higher tan δ than the silicone soft liners. That is, the acrylic materials demonstrated viscoelastic properties, and the silicones were found to be elastic with tan δ approaching zero. Furthermore, the acrylic permanent materials differed from the temporary materials through the nature of the dependence of tan δ on frequency. Tan δ of the acrylic permanent materials increased as the frequency increased from 0.01 to 10 Hz, then decreased again at higher frequencies. Conversely, tan δ of the acrylic temporary materials decreased as the frequency increased from 0.01 to 10 Hz, then increased again at higher frequencies. Large differences in dynamic viscoelasticity among the materials are most likely due to the differences in the composition and structure. In high-damping materials and especially at the damping peak, much of the energy to deform a material is dissipated directly into heat [3]. Therefore, tan δ is considered to reflect the cushioning effect of the soft denture liners required in the clinical situation [13].

This parameter is also a sensitive indicator of cross-linking [14]. The acrylic temporary materials are formed by polymer chain entanglements of uncross-linked polymers. These entanglements act as temporary and relatively weak cross-links. At temperatures well above the glass transition temperature, damping decreases with an increasing degree of cross-linking. The ability of the acrylic materials, having higher values of tan δ, to absorb energy and relieve stress will be greater than that of the silicones with lower tan δ values [13].

4.2.2. Durability

Changes in dynamic viscoelasticity of soft denture liners over a 3 years period have been determined [6]. The changes in water storage varied markedly among the soft denture liners (Fig 18). The acrylic soft liners exhibited a greater increase in the storage modulus, loss tangent and especially loss modulus with time than the silicone permanent soft liners. The low molecular weight plasticizer contained in the acrylic soft liners is leached out into the water, and, at the same time, water is absorbed into the materials [15]. This will lead to loss of initial viscoelasticity and deterioration in the material conditions. The silicone materials remained stable over time. This would be due to the low water absorption and low solubility of components [16]. From the standpoint of durability, the silicone soft liners are better than the acrylic materials clinically.

4.2.3. Clinical consideration

Relationship between dynamic viscoelasticity and masticatory function has been determined [11,13]. The masticatory function of 10 complete-denture wearers was evaluated by means of maximum bite forces, chewing times and frequencies for test food samples (ham, pickled radish). The patients' subjective assessments of satisfaction with the relined denture were also conducted by means of visual analogue scales (VAS). One acrylic temporary soft liner (tissue conditioner), 1 silicone permanent soft liner and 1 acrylic permanent soft liner were applied to the mandibular dentures at a thickness of 2 mm. Hard resin-based dentures were also evaluated.

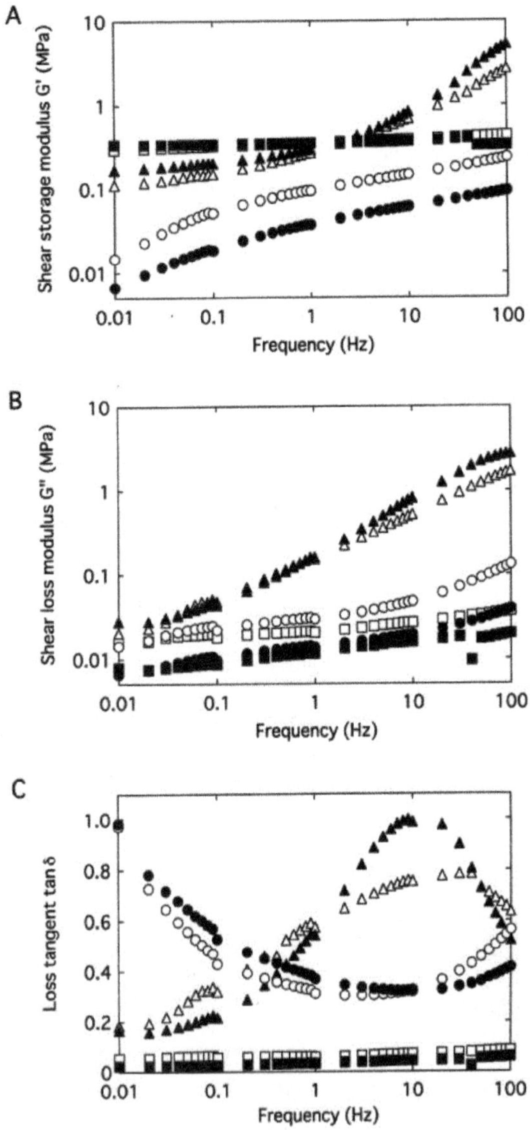

Figure 17. Dependence of storage modulus G' (A), loss modulus G'' (B) and loss tangent tan δ (C) on frequency for 6 soft denture liners. Acrylic temporary soft liners A (●) and B (○), silicone permanent soft liners C (■) and D (□), and acrylic permanent soft liners E (▲) and F (Δ) (Reference [13]).

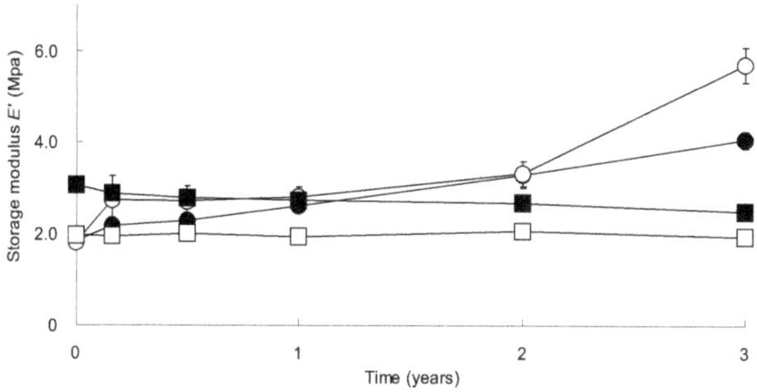

Figure 18. Variation of storage modulus E' with time of storage of 4 soft denture liners at 1 Hz. Acrylic permanent soft liners A (•) and B (○), and silicone permanent soft liners C (■) and D (□) (Reference [6], a part is altered).

Higher maximum bite force and VAS values mean the higher masticatory function. The viscoelasticity of soft denture liners was found to have a great influence on the masticatory function of the complete denture wearers (Fig 19). The greater improvement in masticatory function was observed in dentures lined with the acrylic permanent soft liners, which have higher loss tangent and storage modulus, than in those lined with the silicone permanent soft liners, which have lower loss tangent and higher storage modulus at 1 Hz (Fig 17). The improvement by dentures lined with the acrylic temporary soft liners, which have higher loss tangent and lower storage modulus, was smaller than that by the other soft denture liners.

A higher value of loss tangent (*i.e.*, viscoelastic properties) is likely to exhibit a degree of stress relief under masticatory forces. Theoretically the physical properties of basal seat mucosa should be equal to those of soft denture liners. The Young's moduli E of the oral mucosa was reported to range from approximately 0.4 to 4.4 MPa [17]. E of acrylic temporary soft liners is lower than that of oral mucosa, while E of both acrylic and silicone permanent soft liners is with in the range of E for mucosa. Higher storage modulus may produce more ability of instantaneous crush of food by the soft-lined dentures. It was found that the application of a soft denture liner having a relatively high value of both loss tangent (*i.e.*, viscoelastic properties) and storage modulus to the mandibular complete denture leads to the most marked improvement in masticatory function [11, 13].

4.2.4. Sol-gel transition

The influence of composition and structure on dynamic viscoelasticity of concentrated polymer solutions based on poly (ethyl methacrylate) (PEMA) used as acrylic temporary soft liners (tissue conditioners) through the sol-gel transition was evaluated [8]. The acrylic temporary soft liners are concentrated polymer solutions based on PEMA which are applied to the denture intaglio surface to condition abused tissues underlying ill-fitting dentures. The liquid consists of ester plasticizer and ethanol. They are uncross-linked (formed by polymer chain entanglement but not cross-linked), amorphous polymers, formed in situ from the mixture of a polymer powder and a liquid plasticizer. The polymer powder generally consists of PEMA of molecular weights ranging between 1.79 X 10^5 and 3.25 X 10^5 and contains no initiator. The liquid consists of an ester-based plasticizer and 4 - 50 wt% ethanol (EtOH), and contains no monomer. Mixing of the powder and liquid results in polymer chain entanglement and formation of a coherent gel characterized by viscoelastic behavior. The PEMA particles are slowly penetrated by the large molecules of the ester-based plasticizer. The alcohol accelerates plasticizer penetration into the polymer to produce a clinically acceptable gelation time [18].

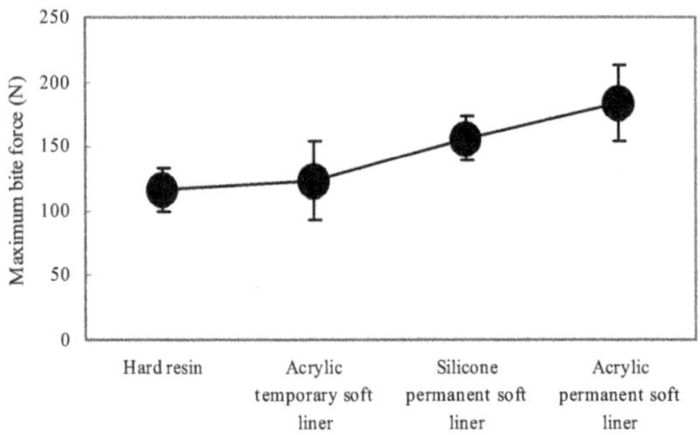

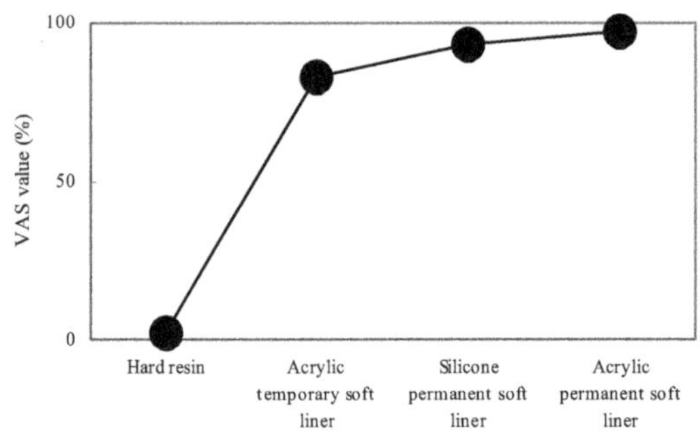

Figure 19. Masticatory function for a subject (Reference [11]).

Four PEMA polymer powders, with weight average molecular weight (M_w) from 1.08X10^5 to 5.30X10^5 were used. The average particle size was approximately 20 μm and polydispersity coefficients were approximately 2. Three aromatic esters butyl phthalyl butyl glycolate (BPBG), dibutyl phthalate (DBP) and benzyl benzoate (BB), and an aliphatic ester dibutyl sebacate (DBS) containing ethanol at 2, 5, 10 and 20wt% were used as liquid components. The powder/liquid (P/L) ratio was varied from 0.8 to 1.4 by weight. The various combinations of the 16 different materials were tested to determine the contribution ratio (ϱ) of each factor to the gelation times by means of an orthogonal array $L_{16}(4^5)$ (Table 1). Secondly, 26 different materials were used for the determination of the effect of each factor by a one-way layout method [8].

Dynamic viscoelastic properties of the materials during gelation were determined at 37°C. A controlled-stress rheometer was used with a parallel plate test configuration (diameter=20 mm, gap=1000 μm) in oscillatory mode (Fig 9). Torque was monitored every 30 sec at constant oscillating frequency (1Hz) and angular displacement (3 mrad). The complex dynamic shear modulus (G^*), shear storage modulus (G'), shear loss modulus (G'') and loss tangent (tan δ) were determined. The gelation time was defined as the time at which tan δ =1 ($G'=G''$) was reached, that is the gel point [8].

Viscoelasticity of the acrylic temporary soft liners during gelation can conveniently be discussed by dividing the curves of tan δ into three regions; region 1, where tan δ decreases with time rapidly; region 2, where tan δ =1 (the gel point); region 3, where tan δ continues to decrease. At the initial stage in region 1, the PEMA polymers dissolve in the plasticizer, and polymer chain entanglements are still scarce. All the solutions are essentially viscous Newtonian fluids. Entanglements then begin to occur and the viscosity gradually increases with time until the solutions can no longer be described as viscous fluids. Before the gel point, although G' is always smaller than G'', G' increases much more rapidly than G'' and thus tan δ decreases rapidly. In the region 2, an infinitely long molecule is produced and the system reaches the gel point (liquid-solid transition), which occurs near where tan δ =1. In the region 3, pseudo cross-links consisting of chain entanglements are forming. The materials behave more elastically with increasing reaction time. G' increases rapidly and becomes larger than G'' and thus tan δ continues to decrease [8].

The compatibility of different polymers and solvents can be predicted by means of the solubility parameter (SP), which is the square root of the cohesive energy density or vaporization energy (ΔE_v) divided by the molar volume (V):

$$SP = (\Delta E_v / V)^{1/2}$$

In general, polymers and solvents should be mutually soluble when their solubility parameters are equal. It has been reported that a method, which considers solvents as poorly, moderately, or strongly hydrogen bonded, could give relatively accurate predictions of the solubility of a solvent in a polymer (Tables 2 and 3) [12, 19].

The gelation time decreased exponentially with an increase in ethanol content. The higher-molecular-weight polymer powders and a higher P/L ratio produced the shorter gelation times. Among the plasticizers, the shortest gelation time was with benzyl benzoate, and the longest was observed with dibutyl sebacate (Figs 20 and 21). Furthermore, results from the orthogonal array $L_{16}(4^5)$ indicate that the ethanol (strong polar bonding) content (ϱ=53.8%) had more influence on gelation time than the molecular weight of polymer powders (ϱ=26.7%), which in turn had more influence than the type of plasticizer (molar volume) (ϱ=9.0%) and the P/L ratio (concentration of polymers) (ϱ=4.5%). It was found that the gelation, *i.e.*, entanglement speed, of concentrated polymer solutions based on PEMA can be controlled over a wide range by varying the polymer molecular weight and especially ethanol content [8].

No.	Factor / Col.	A / 1	B / 2	3	C / 4	D / 5
1		1	1	1	1	1
2		1	2	2	2	2
3		1	3	3	3	3
4		1	4	4	4	4
5		2	1	2	3	4
6		2	2	1	4	3
7		2	3	4	1	2
8		2	4	3	2	1
10		3	2	4	3	1
11		3	3	1	2	4
12		3	4	2	1	3
13		4	1	4	2	3
14		4	2	3	1	4
15		4	3	2	4	1
16		4	4	1	3	2

Factor	Level			
A: Plasticizer	A_1 = Butyl phthalyl butyl glycolate		A_2 = Dibutyl phthalate	
	A_3 = Benzyl benzoate		A_4 = Dibutyl sebacate	
B: Ethanol (wt%)	B_1 = 2	B_2 = 5	B_3 = 10	B_4 = 20
C: M_w of polymer powders	$C_1 = 1.08 \times 10^5$	$C_2 = 2.39 \times 10^5$	$C_3 = 3.75 \times 10^5$	$C_4 = 5.30 \times 10^5$
D: P/L ratio	D_1 = 0.8	D_2 = 1.0	D3 = 1.2	D_4 = 1.4

Table 1. Assignment by Orthogonal Array $L_{16}(4^5)$. Reference [8]

	Molecular Weight	Viscosity (cP) at 20 °C	Molar Volume (cm^3/mol)	Solubility Parameter ($(J/cm^3)^{1/2}$)[a]
Butyl phthalyl butyl glycolate	336.38	51.3	305.0	20.65
Dibutyl phthalate	278.35	20.3	265.6	19.00
Benzyl benzoate	212.25	8.5	190.5	21.98
Dibutyl sebacate	314.46	10.3	337.1	18.50
Ethanol	46.07	1.2	58.5	26.00

[a] The solubility parameter values are those quoted by Reference [12]

Table 2. Characteristics and Solubility Parameters of Plasticizers and Ethanol. Reference [8]

	Solubility Parameter $((J/cm^3)^{1/2})^a$		
	Poorly hydrogen bonded solvent	Moderately hydrogen bonded solvent	Strongly hydrogen bonded solvent
Poly (ethyl methacrylate)	17.4 - 22.7	16.0 - 27.2	19.4 - 23.3
Poly (methyl methacrylate)	18.2 - 26.0	17.4 - 27.2	0

[a] The solubility parameter values are those quoted by Reference [12]

Table 3. Solubility Parameters of Polymers for Solution. Reference [8]

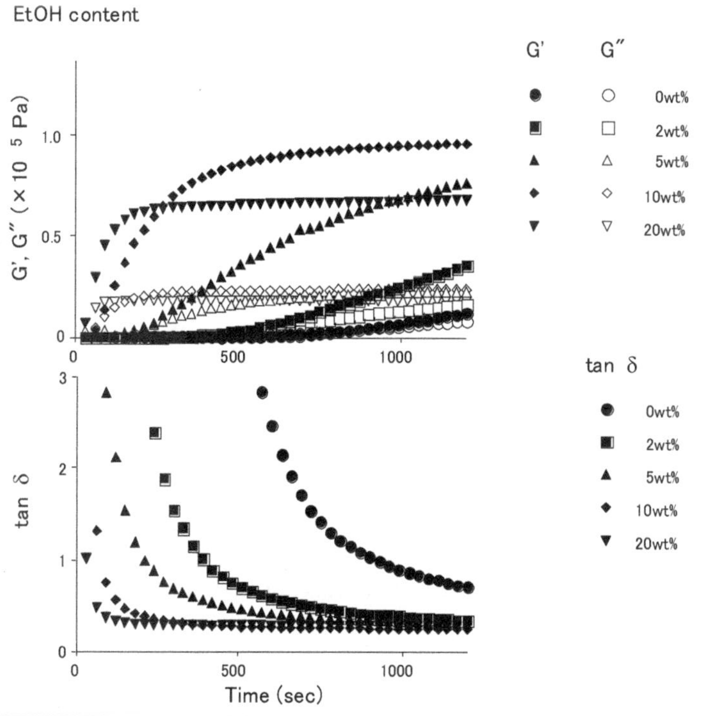

Figure 20. Relationship between variations of G', G'' and tan δ of concentrated polymer solutions based on PEMA with time and ethanol content in liquids (Reference [8]).

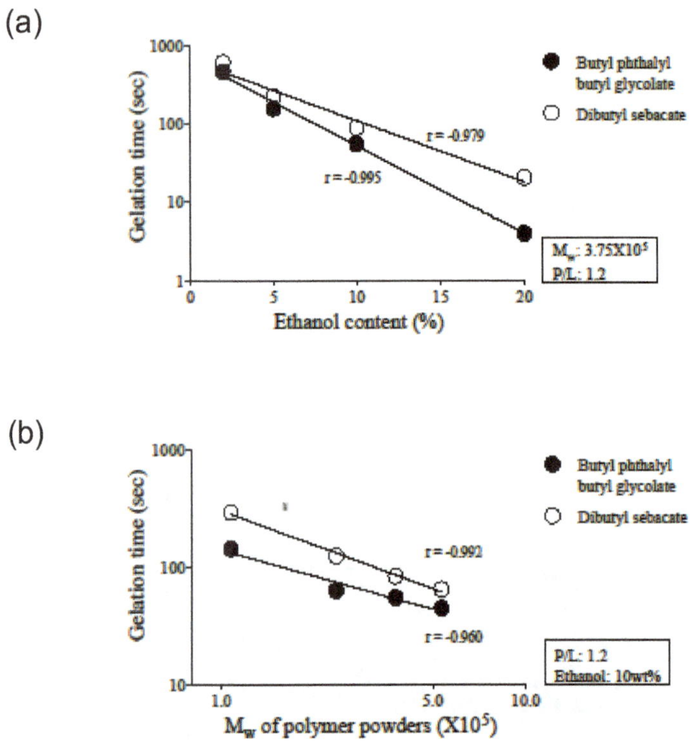

Figure 21. Relationship between gelation times of concentrated polymer solutions based on PEMA, and (a) ethanol content in liquids and (b) molecular weight (Mw) of polymer powders (Reference [8]).

5. Conclusion

In this chapter, basis and theoretical considerarion of rheology, and application in biomaterials are described. Rheology will contribute to the science of biomaterials, medicine, dentistry and biology in addition to the polymer science much more.

Author details

Hiroshi Murata
Department of Prosthetic Dentistry, Graduate School of Biomedical Sciences, Nagasaki University, Nagasaki, Japan

6. References

[1] Ferry JD. (1980) Viscoelastic Properties of Polymers. 4th edition. New York: John Wiley & Sons Inc. pp. 1-33.

[2] McCabe JF. (2008) Applied Dental Materials. 9th edition. Oxford: Blackwell Publishing Ltd. pp. 6-21.

[3] Nielsen LE, Landel RF. (1994) Mechanical Properties of Polymers and Composites. 2nd edition. New York: Marcel Dekker Inc. pp. 63-232.

[4] Murata H, Haberham RC, Hamada T, Taguchi N. (1998) Setting and stress relaxation behavior of resilient denture liners. J. Prosthet. Dent. 80: 714-722.

[5] Murata H, Hamada T, Djulaeha E, Nikawa H. (1998) Rheology of tissue conditioners. J. Prosthet. Dent. 79: 188-199.

[6] Murata H, Taguchi N, Hamada T, McCabe JF. (2000) Dynamic viscoelastic properties and the age changes of long-term soft denture liners. Biomaterials. 21: 1421-1427.

[7] Murata H, Seo RS, Hamada T, Polyzois GL, Frangou MJ. (2007) Dynamic mechanical properties of hard, direct denture reline resins. J. Prosthet. Dent. 98: 319-326.

[8] Murata H, Chimori H, Hamada T, McCabe JF. (2005) Viscoelasticity of dental tissue conditioners during the sol-gel transition. J. Dent. Res. 84: 376-381.

[9] Winter HH, Chambon F. (1986) Analysis of linear viscoelasticity of a crosslinking polymer at the gel point. J. Rheology 30: 367-382.

[10] Clarke RL. (1989) Dynamic mechanical thermal analysis of dental polymers. 1. Heat-cured poly (methyl methacrylate)-based materials. Biomaterials 10: 494-498.

[11] Murata H, Hamada T, Sadamori S. (2008) Relationship between viscoelastic properties of soft denture liners and clinical efficacy. JDSR. 44: 128-132.

[12] Parker S, Braden M. (1990) Formulation of tissue conditioners. Biomaterials. 11: 579-584.

[13] Murata H, Taguchi N, Hamada T, Kawamura M, McCabe JF. (2002) Dynamic viscoelasticity of soft liners and masticatory function. J. Dent. Res. 81: 123-128.

[14] Murayama T, Bell JP. (1970) Relationship between the network structure and dynamic mechanical properties of a typical amine-cured epoxy polymer. J. Polymer Sci. A2 8: 437-445.

[15] Kazanji MNM, Watkinson A. (1988) Soft lining materials: their absorption of, and solubility in, artificial saliva. Br. Dent. J. 165: 91-94.

[16] Kalachandra S, Minton RJ, Taylor DF, Takamata T. (1995) Characterization of some proprietary soft lining materials. J. Mater. Sci. Med. 6: 647-652.

[17] Inoue K, Arikawa H, Fujii K, Shinohara N, Kawahata N. (1985) Viscoelastic properties of oral soft tissue. Dent. Mater. J. 4: 47-53.

[18] Murata H, Toki K, Hong G, Hamada T. (2002) Effect of tissue conditioners on the dynamic viscoelastic properties of a heat-polymerized denture base. J. Prosthet. Dent. 88: 409-414.

[19] Bellenger V, Kaltenecker-Commercon J, Verdu J, Tordjeman P. (1997) Interactions of solvents with poly (methyl methacrylate). Polymer 38: 4175-4184.

Chapter 16

Two-Photon Polymerization Fabrication of Doped Microstructures

Daniel S. Correa, Leonardo De Boni,
Adriano J. G. Otuka, Vinicius Tribuzi and Cleber R. Mendonça

Additional information is available at the end of the chapter

1. Introduction

Scientists and engineers have sought to design more compact and efficient devices by means of microfabrication techniques. Examples of microfabrication processes include lithography, chemical vapor deposition, sol-gel, dry etching, among others. Although every technique has its own specificities and advantages, most of them are multi-step processes and demand long time of fabrication. Because of such limitations, and also because of the small range of materials that can be used in these techniques, there has been extensive search for alternative microdevice fabrication methods. A new microfabrication technique, called two-photon polymerization (2PP), emerged in 1997 opening a wider range of material possibilities with the advantage of allowing tridimensional fabrication. In this technique, Maruo et al (Maruo, et al., 1997) employed a laser-based-apparatus to fabricate three-dimensional polymeric microstructures with no topological constrains, high penetration depth without surface modifications and resolution bellow the diffraction limit. This technique usually employs femtosecond laser pulses to promote two-photon absorption (2PA) of a photosensitive molecule dissolved into the bulk of an unpolymerized resin, which creates a radical and triggers polymerization in a confined spatial region.

In a 2PA transition, an atom or molecule is taken to an excited state by simultaneously absorbing two photons in a single quantum event. Considering that 2PA is difficult to be attained by using conventional (low intensity) light sources, normally pulsed laser light is employed. Owing to the nonlinear nature of the 2PA, it displays a quadratic dependence to the light intensity, providing high spatial resolution and low light scattering. As an outcome for a polymerization relied on 2PA, chemical reaction takes place in a tiny focal volume, allowing the fabrication of tridimensional polymeric structures with resolution below the diffraction limit. For instance, two-photon polymerization has already been applied to fabricate micro and nanodevices for a wide variety of applications (Kawata, et al., 2001, LaFratta, et al., 2007, Lee, et al., 2008, Marder, et al., 2007, Sun and Kawata, 2004).

To further enhance the technological potentiality of two-photon polymerization (2PP) to fabricate microstructures, one can dope the basic resin with dyes, nanoparticles, special polymers, and other materials. Such methodology leads to the fabrication of active micro-structures with chemical, biological or optical properties, with potential applications in, for instance, optical data storage, photonic crystals, fluorescent and bioactive microstructures (Correa, et al.,2012, Correa, et al.,2009, Cumpston, et al.,1999, Drakakis, et al.,2006, Farsari, et al.,2008, Liska, et al.,2007, Mendonca, et al.,2007, Ovsianikov, et al.,2007, Serbin, et al.,2004, Sun, et al.,2001, Takeyasu, et al.,2008, Tayalia, et al.,2008). In this chapter, we will discuss theoretical and experimental aspects of the two-photon polymerization process, and will review the last advances in the fabrication of doped microstructures. Such microstructures present unique optical, electrical or biological properties (according to the dopant characteristics), desired for technological applications.

2. Two-photon absorption: Fundamentals

Two-photon absorption is a process related to the change in the electronic state of an atom or molecule, which is excited from a lower energy level $|n\rangle$ (ground state, in the most of cases) to a higher electronic state $|n+1\rangle$, absorbing simultaneously two-photon of same (called "degenerate process") or different energies (called "nondegenerate process"). An intermediate virtual state is created from the interaction between the electromagnetic field and matter. Two-photon absorption was theoretically predicted by Maria Göppert-Mayer in 1931 during her doctoral thesis (Göppert-Mayer,1931). Only after thirty years, the experimental observation of the phenomenon became possible, thanks to the invention of the LASER. The first experimental verification of 2PA was confirmed by Kaiser et al, who detected the fluorescence signal emitted by a europium-doped crystal (Kaiser and Garrett,1961), whose electrons were excited via a two-photon transition. Furthermore, this effect was experimentally proved by Isaac Abella (Abella,1962) in 1962, using Cesium vapor as the two-photon absorbing material. Nowadays, two-photon absorption processes have been extensively studied because of the enormous number of technological applications, such as, multiphoton optical limiters, multiphoton fluorescence spectroscopy, as well as multiphoton polymerization. The importance of multi-photon absorption studies is not only restricted to technological applications, but also lies in basic research, which enables to characterize and gather information about different materials. In this field, the quantification of such effect can provide insights on the molecular geometry and electronic charge distribution of an excited molecule. Additionally, two-photon absorption measurements provide properties (symmetry and energy) of some electronic states that can not be seen using traditional spectroscopic experiments.

To promote an excitation by absorption of two photons, the sum of the energies of each isolated photon need to match the energy of the electronic transition, as illustrated in Fig. 1, for degenerated and nondegenerated cases. Such absorption process is classified as a simultaneous event. According to the uncertainty principle of Heisenberg, however, photons that participate in the transition can have a time interval given by h/ΔE, where ΔE is the difference of energy between the material's virtual state and the closest real state, and h is the Planck constant. In other words, the absorption of two photons does not need to be totally instantaneous, once it occurs in an interval given by the Heisenberg principle. Therefore the difference between the virtual and closest real state depends on the energy of the first photon that arrives in the molecule or atom. It means that if the photon presents energy close to the one between the states involved in the transition, according to the uncertainty principle, the 2PA probability will be enhanced.

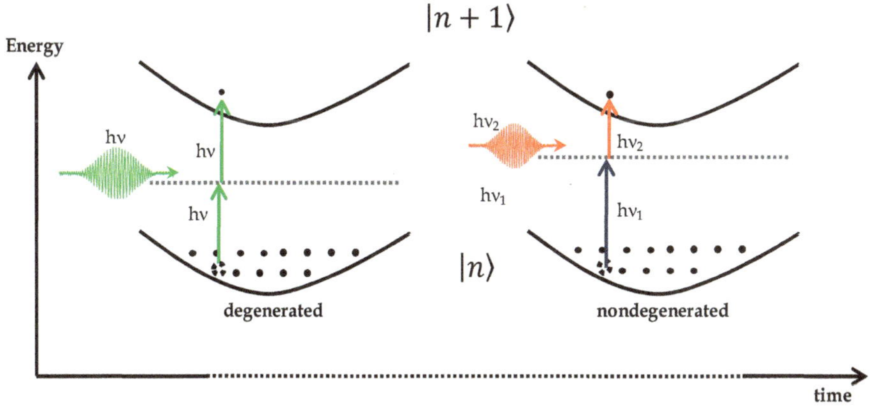

Figure 1. Representation of a degenerated and nondegenarated two-photon transition between $|n\rangle$ and $|n+1\rangle$ electronic states of an atom or molecule. The dotted lines represent a virtual state which intermediate the two-photon absorption process.

Due to the low probability of the 2PA to occur (compared to conventional one-photon absorption (1PA)), a high number of photons per unit of time and per unit of area is required to reach the atoms or molecules. As the effect exhibits a quadratic dependence on the number of photons, 2PA can be achieved by using ultrashort laser pulses. Once the strength of the two-photon absorption effect is several orders of magnitude smaller than the one-photon one, it is very important that the material presents no linear absorption at the laser wavelength employed.

In order to better understand how the two-photon effect appears in the light-matter interaction, one can imagine that when the electromagnetic field, which interacts with the matter, has a magnitude in the order of the interatomic field, it is able to modify the charge distribution in the material in a anharmonic way. As a consequence, the polarization of the material will be described by nonlinear terms, which are dependent on the incident electromagnetic field. Depending on the phase between the generated polarization and the applied electric field, the results can be either the Kerr effect (Abella,1962) or two-photon absorption. When two-photon absorption is analyzed from the nonlinear optics point of view, one can see that it is related to the imaginary part of the third-order nonlinear susceptibility, $\chi^{(3)}$, (Boyd,1992, Shen,1984), as shown by Eq. 1 (Sheik-Bahae,1990):

$$\mathrm{Im}\,\chi^{(3)} = \chi_I^{(3)} = \frac{n_0^2 \varepsilon_0 c^2}{\omega}\beta \tag{1}$$

in which c is the speed of light, ε_0 is the vacuum permittivity, ω is the photon frequency, n_0 is the refractive index of the material, and β is the two-photon absorption coefficient. Consequently, if the two-photon absorption process is significant, the total light absorption (α) of the material is described by:

$$\alpha(I) = \alpha_0 + \beta I_0 \tag{2}$$

where, α_0 is the linear absorption, and I_0 is the light intensity reaching the sample. Equation 2 shows that the two-photon effect becomes more pronounced when the light intensity is considerable high. By using Eq. 2, the attenuation of light intensity as it propagates in the material (represented in Fig. 2), is described by:

$$\frac{dI}{dz} = -\left(\alpha_0 I_0 + \beta I_0^2\right) \tag{3}$$

with I being the light intensity and z the optical path.

Considering that the material presents negligible linear absorption, the first term on the right side of Eq. 3 can be eliminated. Consequentially, the attenuation of the light will only depend on the two-photon absorption coefficient.

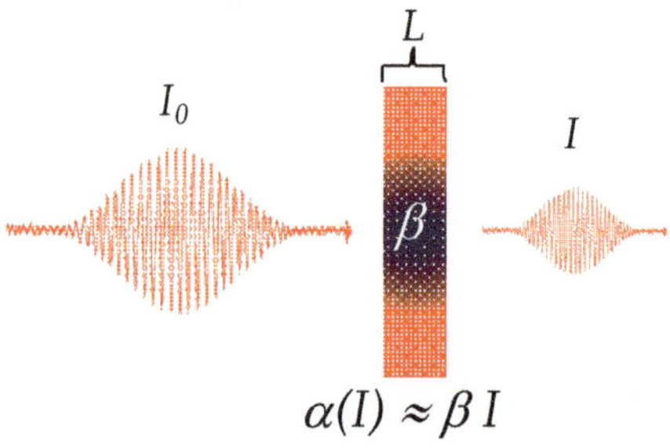

Figure 2. Representation of light attenuation when a materials of thickness L presents a two-photon absorption process.

After some mathematic manipulation, the solution for the attenuation of the light (I) due to the two-photon absorption process can be written as:

$$I = \frac{I_0}{1 + \beta L I_0} \tag{4}$$

By Eq. 4, one can see a decrease of the light intensity as the two-photon absorption coefficient increases. This coefficient carries information about the probability of a molecule or atom to absorb two photons simultaneously. In other words, the two-photon absorption coefficient is directly associated with the two-photon absorption cross-section (δ), and is described by Eq. 5:

$$\beta = \frac{\delta N}{h\nu} \tag{5}$$

where, h is the Planck constant, and N is the number of absorbers per unit of volume. The molecular two-photon absorption cross-section, δ, is usually given in units of Göppert-Mayer (GM), in which, 1 GM is equivalent to 1×10^{-50} cm^4.s.$photon^{-1}$. It is important to mention that similar to the one-photon absorption spectrum, a molecule or atom can present a two-photon absorption spectrum as well. Nowadays, in order to quantify and understand the two-photon absorption cross-sections spectra, monochromatic light sources have been employed. This is normally achieved when optical parametric amplifiers devices are used. In addition, supercontinuum light sources have also been employed to observe the magnitude of this effect (Balu, et al.,2004, De Boni, et al.,2004).

In terms of quantum mechanics analysis, two-photon absorption presents selection rules distinct from one-photon absorption ones. For example, if the molecule is centrosymmetric, one and two-photon transitions for a certain energy level are not allowed at the same time. In other words, one-photon allowed state is not a two-photon allowed one and vice-versa. This arises as a consequence from the angular momentum, which needs to be conserved after the electronic transition. Since photons present spin ±1, one-photon absorption requires that the states involved in the transition differ in its orbital wavefunctions by an angular momentum of ±1. In the same way, for a two-photon transition, the energy levels involved require a change of ±2. This particular characteristic in the quantum mechanics selection rules turns the linear spectroscopy unable to indentify two-photon allowed states. In the opposite way, nonlinear spectroscopy, such as two-photon absorption measurements are no able to excite one-photon transitions. On the other hand, when a molecule is not centro-symmetric, the selection rules can be relaxed; enabling the existence of final states reached both by one- or two-photon absorption.

3. Two-photon absorption polymerization: Principles and applications

Photopolymerization can be defined as a chemical reaction that turns molecules of low molecular weight (monomers) into macromolecules consisting of repeating units, by using light as the reaction trigger. For one-photon absorption polymerization, the radical species are formed by using a conventional light source (e.g. UV lamp). However, in a two-photon absorption polymerization, electronic transition occurs by simultaneous absorption of two photons, so that radical formation and subsequent polymerization occurs only in the vicinity of the focused light source (laser beam). This results in a small solidified volume (voxel) around the focal spot. By either scanning the laser beam through the resin volume or by moving the sample in three perpendicular directions (X, Y and Z), arbitrary tridimensional microstructures can be created.

Also, by changing the monomer or the mixture of monomers used in the resin, one can change some characteristics of the final polymer. In this way, it is possible to tailor some properties of the produced microstructure. For example, polymer hardness and the amount of shrinkage the material undergoes due to polymerization can be controlled. Even more, by doping the base resin with different compounds, it is possible to change physical, chemical, biological and optical properties of the final microstructure.

two-photon polymerization microfabrication has a set of unique advantages over conventional microfabrication processes. Because several pulsed laser systems operate in the near infra-red region, where most curable monomers and polymers are transparent, and because two-photon absorption occurs only in defined spatial regions (where light intensity is high enough), it is possible to reach higher light penetration depths. In this manner, this process is able to polymerize, or solidify only a small region in the bulk of the liquid resin, without making any changes to its surface or the surrounding region.

The spatial confinement of the excitation, based on a two-photon process, is illustrated in Fig. 3, which depicts a comparison between one and two-photon induced fluorescence (Marder, et al.,2007) in a liquid sample contained in a silica fused cuvette. Note the fluorescence localization of the two-photon induced process, while one-photon excitation promotes fluorescence along the whole optical path.

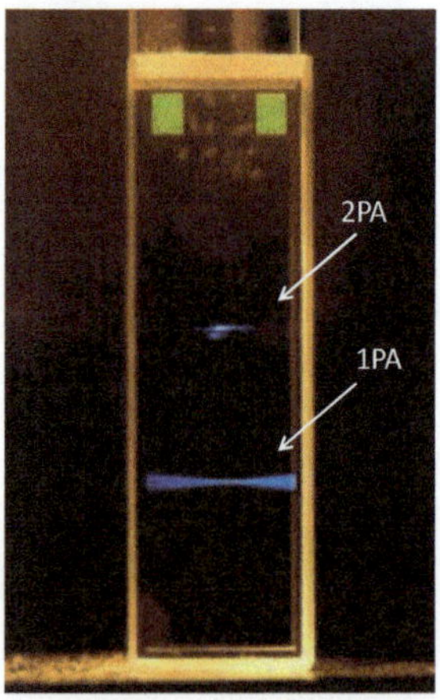

Figure 3. Image of induced fluorescence for one and two-photon absorption process (1PA and 2PA). Note the fluorescence localization of two-photon induced processes. Reprinted with permission from (Marder, et al.,2007).Copyright [2007], Materials Research Society.

The resolution in microdevice fabrication technique is of prime importance. The nonlinear nature of two-photon polymerization allows for localized excitation and gives a fine resolution to this fabrication method. In fact, the two-photon absorption process pushes the fabrication resolution bellow the diffraction limit. All optical systems resolutions are limited by its optical instruments diffraction. The diffraction limit is the fundamental maximum resolution that any optical system can achieve. The minimum spot size that can be achieved is given by Abbe's expression (Fowles,1989):

$$r = \frac{0.61\lambda}{NA} \tag{6}$$

where, r is the focal spot radius, λ is the light wavelength, and NA is the numerical aperture of the system. The numerical aperture is a number defined by Abbe as:

$$NA = n\sin(\theta) \tag{7}$$

where, n is the index of refraction in the focusing medium, and θ is the convergence angle of the beam. Figure 4 shows a schematic diagram for objectives with different NA.

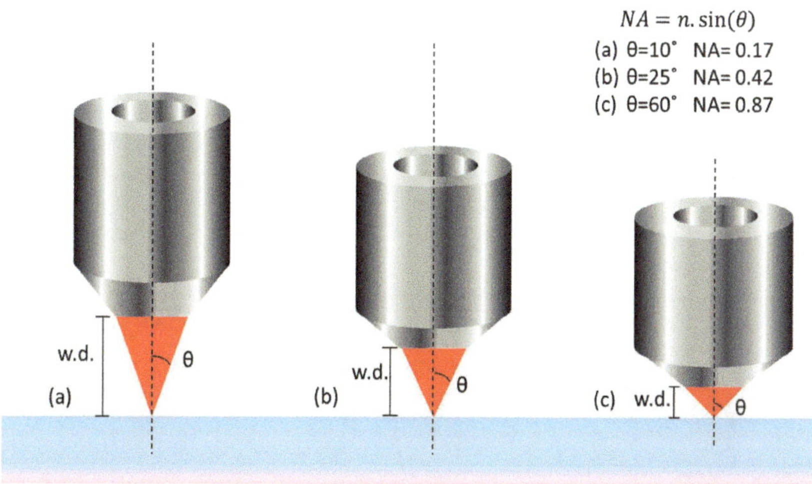

Figure 4. Light focusing through microscope objective lens with different NA. W.D. stands for working distance.

The greater the numerical aperture, the smaller the working distance, i.e., the distance between the focal plane and the surface of the objective, according to Fig. 4. However, Abbe's equation works for plane waves only. Since most pulsed lasers have an approximately Gaussian beam profile, the minimum beam waist is given by:

$$w_0 = \frac{\lambda}{\pi NA} \tag{8}$$

where ω_0 is the beam waist at the focal plane. The beam waist is defined as the distance from the center of the Gaussian beam up to the point where the amplitude of the electric field falls down to $1/e$ of its maximum. However, Eq. 8 is still an approximated value; it works well for small NA optics (NA<<1). Generally, however, two-photon polymerization setups use high numerical aperture optics, such as microscope objective lenses. For high numerical aperture systems, such as immersion microscope objectives, a more accurate expression would be:

$$w_0 = \frac{\lambda}{\pi NA}\sqrt{n^2 - NA^2} \tag{9}$$

Equation 9 shows that the smallest feature that one system can produce is limited by the microscope objective lens numerical aperture and by the wavelength. Nevertheless, once two-photon absorption rate is proportional to the square of the incident beam intensity and, therefore, depends on the squared Gaussian intensity profile, the beam presents a narrower waist than the one defined by Eq. 9.

Figure 5 shows the difference between a Gaussian and a squared Gaussian beam profile. Therefore, due to the nonlinear nature of this process, it is possible to achieve resolution below the diffraction limited spot size. This phenomenon reduces the beam waist by a factor of $\sqrt{2}$. In addition to the square dependence, there is still another phenomenon that can yield even better resolution to two-photon polymerization process, which is the polymerization threshold. The polymerization threshold imposes a minimum power below which, even if the sample is irradiated, no polymerization takes place. This phenomenon usually occurs due to the presence of oxygen in the resin. Molecular oxygen inhibits the action of the photoinitiator radicals, preventing the polymerization reaction from occurring. Although this might be a problem for polymer film curing, it can be used as an advantage in the two-photon polymerization fabrication, in the sense that it can decrease even more the polymerized voxel size.

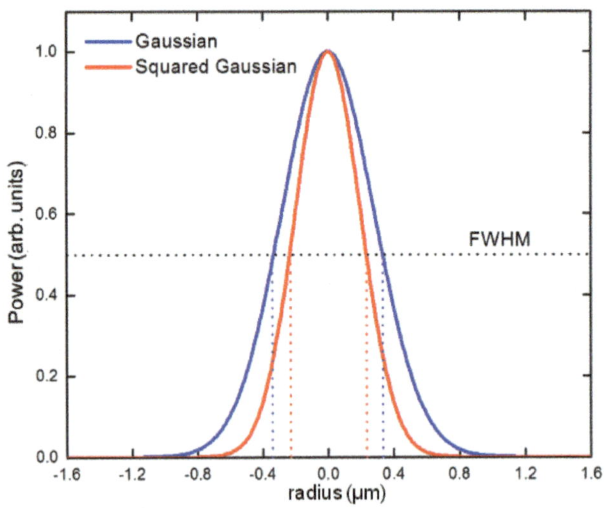

Figure 5. Gaussian and squared gaussian beam intensity profile. For comparison, it is also displayed (dotted line) the laser beam full width at half maximum (FWHM).

For a Gaussian profile, the central part of the beam is more intense than its borders. Therefore, even though the outer regions of the beam might not have enough power to start the reaction, the central part of the beam can overcome the threshold. It is important to realize that polymerization will also be contained in the Z direction. Equation 10 shows the intensity envelope of a Gaussian beam in cylindrical coordinates:

$$I(r,z) = I_0 \frac{w_0^2}{w(z)^2} e^{-2\left(\frac{r}{w(z)}\right)^2} \tag{10}$$

If we consider the laser beam waist at the focal plane and its intensity profile, only the central part of it will effectively solidify the resin. The beam waist (radius) at the focal plane is ω_0 and it grows with Z (distance from the focal plane) as:

$$w(z) = w_0 \sqrt{1 + \left(\frac{2\lambda}{\pi n w_0^2}\right)} \equiv \sqrt{\frac{\lambda}{\pi n}\left(z_R + \frac{z^2}{z_R^2}\right)} \tag{11}$$

where λ is the wavelength, n is the index of refraction and z_R is the Rayleigh length, defined as the distance from the focal plane in which the beam waist increases by a factor of $\sqrt{2}$. The Rayleigh length can be written as:

$$z_R = \frac{\pi n w_0}{\lambda} \tag{12}$$

If we set $I_{(r,z)} = I_{th}$ in Eq. 5, (*th* stands for threshold) we can obtain the diameter (*D*) and the length (*L*) of the focal region where the intensity exceeds the threshold value (Juodkazis, et al.,2005), according to:

$$D(r) = w_0 \sqrt{2 \ln\left(\frac{I(r)}{I_{th}}\right)} \tag{13}$$

$$L(z) = 2 z_R \sqrt{\left(\frac{I(z)}{I_{th}}\right) - 1} \tag{14}$$

As the two-photon absorption process depends on the square of the intensity and therefore on the focal volume in which the polymerization reaction in fact occurs, the diameter (*D*) and length (*L*) will be smaller. The dimensions of the polymerization voxel for two-photon polymerization should be written as (Juodkazis, et al, 2005):

$$D(r) = w_0 \sqrt{\ln\left(\frac{I(r)}{I_{th}}\right)} \tag{15}$$

$$L(z) = 2z_R\sqrt{\left(\frac{I(z)}{I_{th}}\right)^{1/2} - 1} \tag{16}$$

By controlling laser power in order to have it near the polymerization threshold value, it is possible to create structures with resolution well below the diffraction limit. Figure 6 illustrates how polymerization threshold can yield further improvements in fabrication resolution regardless of the optical system. These are some of the characteristics that made two-photon polymerization microfabrication technique so appealing for many research groups in the last decade.

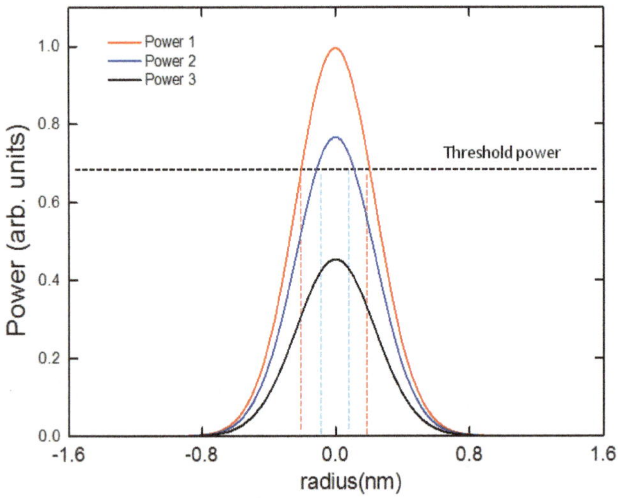

Figure 6. Threshold power relative to various gaussian profiles. The more the peak power gets closer to the threshold power the smaller is the voxel size. Polymerization is not observed if peak power is below the threshold power.

The first two-photon polymerized microstructures were reported by Maruo et al (Maruo, et al.,1997) and date back to 1997. In that work, their intention was to demonstrate the capabilities of this new fabrication technique. No other process offers both, complex tridimensional fabrication and resolution below the diffraction limit. In order to illustrate such features, Kawata et al (Kawata, et al.,2001) used a computer model and created a micro sculpture of a bull with subdiffraction-limit resolution. In addition, to show that the microstructures retained good structural and physical properties, they fabricated a micro spring attached to one cube and demonstrated its oscillation. Figure 7 (a) and (b) shows both structures.

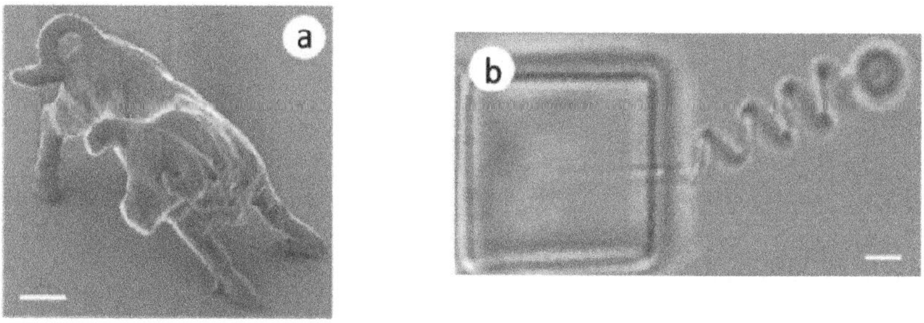

Figure 7. Micro sculpture of (a) a bull and (b) oscillating spring system. Scale bar is 2μm. Reprinted with permission from (Kawata, et al.,2001). Copyright [2001], Nature Publishing Group.

4. Materials employed and experimental setup

The polymeric resins employed for two-photon absorption polymerization vary widely. For instance, Maruo et al reported the use of urethane acrylate monomer (Maruo, et al.,1997), Baldacchini et al (Baldacchini, et al.,2004) and Haske et al (Haske, et al.,2007) used a mixture of two-acrylic resin, while Chichkov's group employed Inorganic-organic hybrid polymers (Schlie, et al.,2007, Serbin, et al.,2004). Our group has used the same resin formulation proposed by Baldacchini et al (Baldacchini, et al.,2004), whose formulation consists of a mixture of two tri-acrylate monomers (see Fig. 8): (a) tris(2-hydroxyethyl)isocyanurate triacrylate and (b) ethoxylated(6) trimethylolpropane triacrylate. The first one increases the microstructure hardness, while the second one reduces shrinkage upon polymerization. Ethyl-2,4,6-trimethylbenzoylphenylphosphinate, commercially known as Lucirin TPO-L, has been used as the photoinitiator, due to its favorable two-photon absorption properties (Baldacchini, et al.,2004, Mendonca, et al.,2008). The chemical structures of the two acrylic resins and the photoinitiator are displayed in Fig. 8 (a), (b) and (c) respectively.

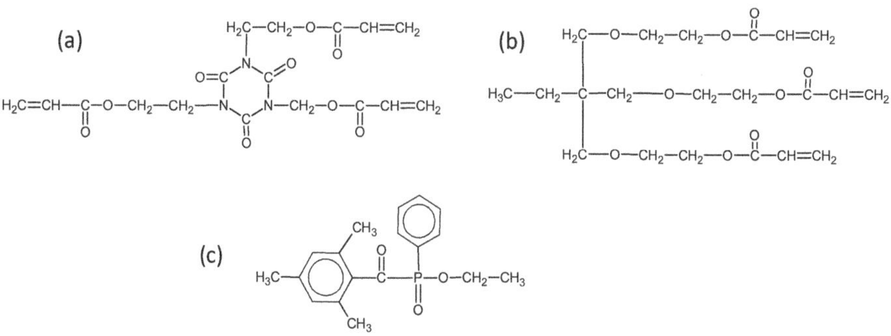

Figure 8. Chemical structures of materials used in the two-photon polymerization. (a) tris(2-hydroxyethyl)isocyanurate triacrylate and (b) ethoxylated(6) trimethylolpropane triacrylate and (c) Ethyl-2,4,6-trimethylbenzoylphenylphosphinate (photoinitiator).

The experimental setup used in two-photon polymerization typically employs femtosecond lasers and pulse energies of nanojoules. Although the average energy is low, the pulse peak intensity is high enough to promote two-photon absorption polymerization. In our specific case, we use a Ti:sapphire laser oscillator system, which delivers pulses of 100 fs centered at 800 nm. The laser beam is focused onto the resin to be polymerized by using a 0.65-NA microscope objective. The average power used depends on the nature of the materials employed (such as type of polymeric resin, photoinitiator, dopants etc), once it must be high enough to overcome polymerization threshold, but low enough to avoid material degradation. We typically employ pulse energies in the order of 0.1 nJ, measured before the objective. The sample preparation consists in placing a drop of the resin on a substrate (glass slide) previously cleaned. The resin is retained between spacers and enclosed by a cover slip, as illustrated in Fig. 9.

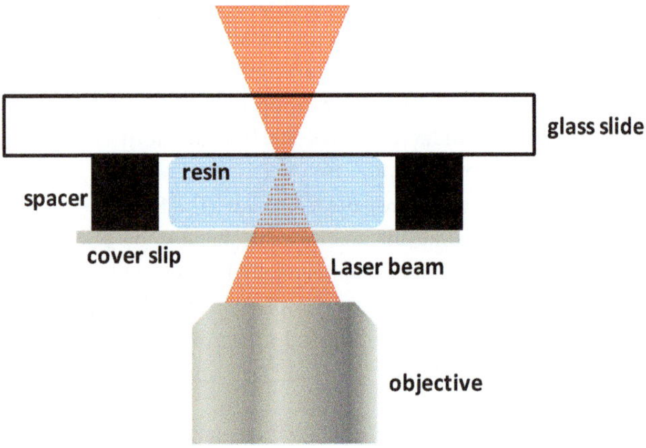

Figure 9. Scheme of sample preparation, where the resin to be polymerized by the laser beam is placed between a glass slide and a cover slip, separated by spacers.

The sample is then positioned and moved in the axial *z* direction using a motorized stage, while the laser beam scans the sample across the *xy* plane, by using a pair of galvano mirrors. In addition, a CCD camera is coupled to the experimental setup to monitor two-photon absorption polymerization in real time. An illustration of the two-photon polymerization experimental setup is displayed in Fig. 10.

After the desired microstructure is fabricated, the sample is immersed in ethanol to wash away the unsolidified resin, leaving behind solely the desired microstructure on the glass substrate, according to the scheme sequence depicted in Fig. 11.

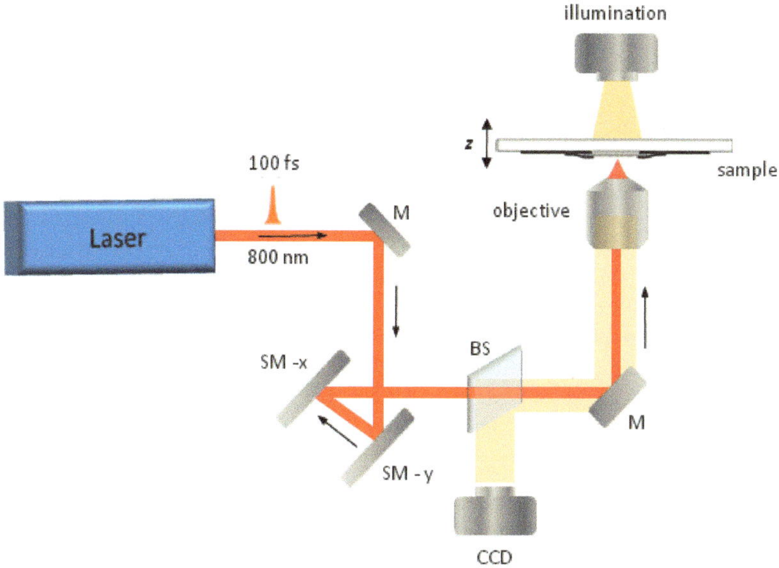

Figure 10. Scheme of the experimental setup employed for the two-photon absorption polymerization. M, BS, SM-x and SM-y stands for mirror, beam splitter, scanning mirror in x and y direction, respectively.

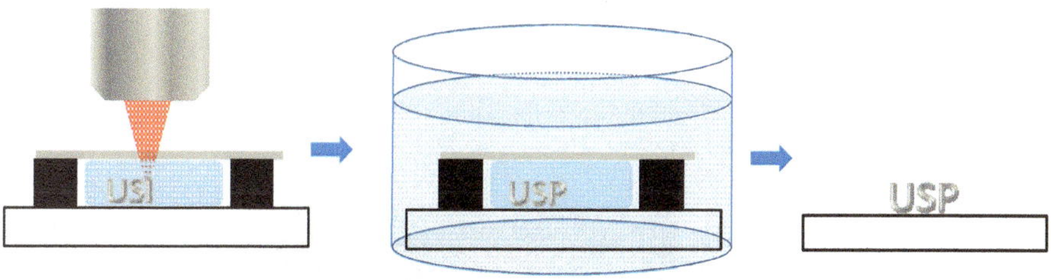

Figure 11. Scheme of the sample washing with ethanol after microfabrication to remove the unpolymerized resin.

5. Doped microstructures and applications

Most of the initial studies reporting microfabrication of structures via two-photon absorption polymerization dealt with undoped microstructures. Although in this case the structures still present high definition and functionality, the final structures are of limited application, once the properties cannot be probed or altered by external sources. In the last few years, however, several research groups have attempted to dope the resin formulations with a large variety of advanced materials, aiming at giving new functionalities to microstructures, such as chemical, biological or optical properties. Next we will present some results that diverse research groups have obtained by using methodologies to incorporate distinct dopants in the resin formulations, aiming at obtaining two-photon polymerized microstructures with unique properties for technological applications.

- *Incorporation of conjugated polymers*

Conjugated polymers are materials of great interest to be used in technological devices, once the electron delocalization along the polymer backbone can render them conductivity and nonlinear optical properties. For instance, in a methodology proposed by our group (Mendonca, et al.,2009), poly (2-methoxy-5-2-ethylhexyloxy-1,4-phenylenevinylene) (MEH-PPV), widely known for its conductivity (Yu, et al.,1995), electroluminescence (Nguyen, et al.,2000, Parker,1994), and nonlinear optical properties (Chung, et al.,2002, Correa, et al.,2007, De Boni, et al.,2004, Oliveira, et al.,2006) has been incorporated into the basic resin in a guest-host strategy. In the proposed methodology, the host consists of the two triacrylate monomers and the photoinitiator displayed in Fig. 8. To this solution, it is added up to 1.5% of MEH-PPV, which was then mixed up for 2 hours. Ethanol was eliminated by evaporation at room temperature for 24 h, yielding a viscous liquid. From this liquid, we extracted some drops and place on the glass substrate, as illustrated in Fig. 9.

Two-photon absorption polymerization was employed to fabricate waveguides containing MEH-PPV, using as substrates mesoporous silica films, which have a low refractive index (*n* = 1.185) throughout the visible spectrum and, therefore, minimize waveguiding losses (Konjhodzic, et al.,2005). Figure 12 (a) shows a fluorescence microscopy top view image of 100 μm-long waveguide fabricated on a mesoporous silica substrate, illuminated in its central region by a CW laser beam at 532 nm. The MEH-PPV fluorescence is guided through the microstructure and scatters at both microstructure ends. When the experiment was carried out in waveguides fabricated on top of a conventional glass slide, no waveguiding was observed due to light coupling to the substrate. Figure 12 (b) displays a side view scheme of the microstructure excitation and the corresponding waveguiding. Such waveguide is promising to be used in photonics applications such as microLEDs.

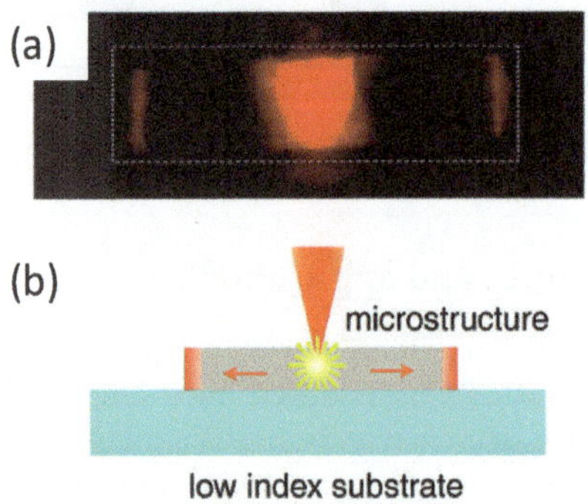

Figure 12. (a) Fluorescence optical microscopy (top view image) of a waveguide containing MEH-PPV fabricated by two-photon polymerization on mesoporous silica substrate. Light is guided through the microstructure ends when it is excited at its center. In (b) is displayed a schematic side view of the structure. Reprinted with permission from (Mendonca, et al.,2009).Copyright [2009], American Institute of Physics.

- *Incorporation of azoaromatic dyes*

Farsari et al (Farsari, et al.,2008) reported the fabrication of photonic crystals via two-photon absorption polymerization, by incorporating into the basic resin a nonlinear optically active chromophore, based on reacting Disperse Red 1 (DR1) with (3-isocyanopropyl) triethoxysilane. The structures reported by them had a bright red color, typical of Disperse Red 1 and presented stop-gaps at specific frequencies, close to the near infrared region, typical of photonic crystals.

Birefringence in microstructures have been explored by Mendonça et al (Mendonca, et al.,2007), where they have fabricated, via two-photon absorption polymerization, microstructures incorporating the azodye DR13, as the one displayed in the SEM image in Fig. 13 (a). Birefringence could be optically induced and erased in these microstructures due to a reversible trans-cis photoisomerization and a subsequent molecular orientation of the azo group. After exposure to laser, the microstructure exhibited residual birefringence because the chromophore molecules become oriented in the direction perpendicular to the laser polarization. Figure 13 (b) and (c) show the corresponding transmission microscopy images, where the microstructure is visible when the angle between the sample axis, defined by the Ar+ ion laser exposure and the polarizers, is an odd multiple of 45°. This birefringence can be completely erased with circularly polarized light or by heating the sample close to the polymer glass transition temperature.

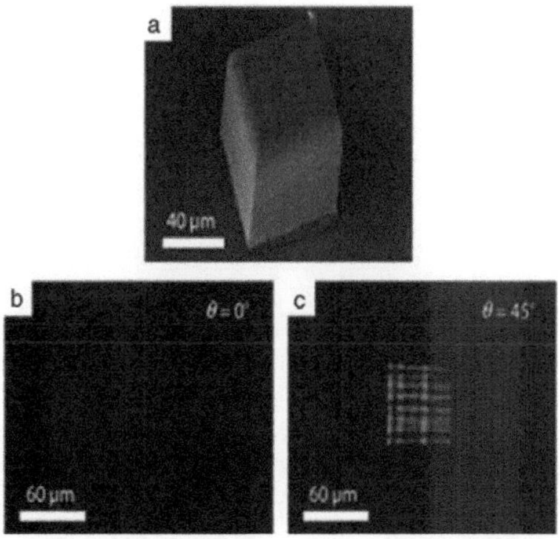

Figure 13. (a) Scanning electron micrograph of a cubic microstructure containing DR13. Polarization microscope images of a microstructure (like the one shown in (a)) at two angles between the polarizer and sample axis, displaying the birefringence behavior owing to DR13 incorporation. Reprinted with permission from (Mendonca, et al.,2007). Copyright [2007], American Institute of Physics.

- *Incorporation of magnetic nanoparticles*

Desinging and fabricating microdevices by two-photon absorption polymerization that would be easily manipulated are import to understand the mechanics on the micro/nanomachine motion. Wang et al, (Wang, et al.,2009) proposed an interesting methodology where the basic resin was doped with surface-modified Fe_3O_4 nanoparticles, and yielded structures with magnetic properties, whose movement could be externally controlled by magnets. Figure 14 displays a magnetic microspring fabricated by two-photon absorption polymerization, where one end of the spring remains attached to an anchor, which in turn, is attached to the substrate, and the other end is attached to a freely moving sphere. By applying a magnetic field, the objected was moved in the direction of higher magnetic field strength. Therefore, by changing the ferromagnet position, distinct motions could be induced, as displayed in Fig. 14 (a) and (b), which shows the microspring in its non-alongated state (a) and in its alongated state (b).

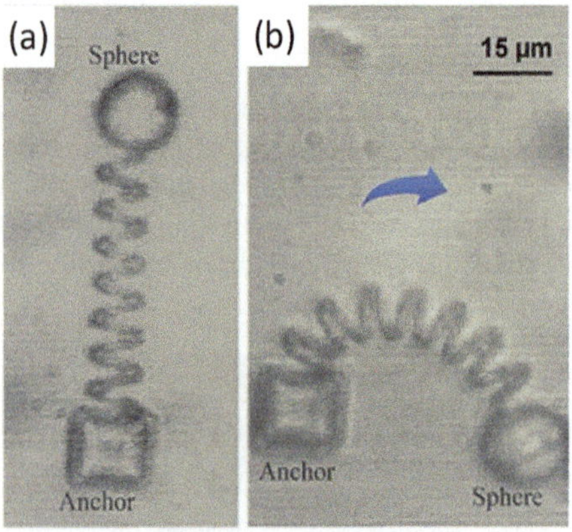

Figure 14. Photograph of a magnetic micro-spring fabricated via two-photon absorption polymerization, by using a resin doped with magnetic nanoparticles. (a) Microspring in its natural non-alongated state and (b) in its alongated state. Reprinted with permission from (Wang, et al.,2009). Copyright [2009], Optical Society of America.

- *Incorporation of bio-related materials*

Two-photon polymerized structures incorporating biomaterials have also been subject of investigation by a few research groups. For instance, chitosan-containg microstructures fabricated by two-photon polymerization have been fabricated by Correa et al (Correa, et al.,2009). Chitosan is a biodegradable and biocompatible polymer, with applications in blood coagulation, soft tissue and bone regeneration, which could render to the microstructures appealing properties for medical applications. In that work, it was demonstrated, by Raman microscopy, that chitosan does not react chemically with the matrix resin, retaining its characteristics after the fabrication process, which is suitable for biomedical applications.

A methodology employing two-photon absorportion polymerization to fabricate antibactericide microneddles was proposed by Gittard et al (Gittard, et al.,2010). Nowadays, the use of microneedles for drug delivery is still limited because of the risk of infection associated with contamination of the micronnedles. Such problem may be overcome by imparting these devices with antimicrobial properties. Gittard et al (Gittard, et al.,2010) used a photopolymerization-micromolding technique to fabricate microneedle arrays containing Polyethylene Glycol-Gentamicin Sulfate, which inhibited growth of *Staphylococcus aureus* bacteria.

Claeyssens et al (Claeyssens, et al.,2009) proposed using two-photon polymerization to fabricate microstructures by using the biodegradable triblock copolymer poly(ε-caprolactone-co-trimethylenecarbonate)-b-poly(ethyleneglycol)-b-poly(ε-caprolactone-co-trimethylenecarbonate). Samples of the structures fabricated are displayed in the SEM images in Fig. 15, which present good resolution. Besides, the initial cytotoxicity tests employed showed that the material does not affect cell proliferation, which demonstrates the capability of using two-photon polymerization to fabricate biodegradable scaffolds for tissue engineering.

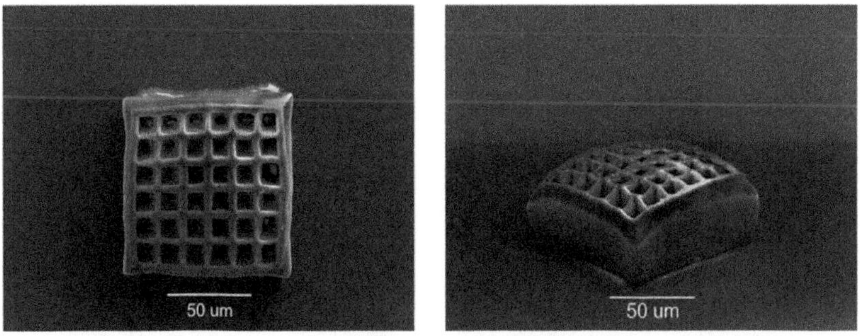

Figure 15. Scanning electron micrographs of structures fabricated with biodegradable triblock copolymer poly(ε-caprolactone-co-trimethylenecarbonate)-b-poly(ethyleneglycol)-b-poly(ε-caprolactone-co-trimethylenecarbonate). Reprinted with permission from (Claeyssens, et al.,2009). Copyright [2009], American Chemical Society.

Farsari et al (Drakakis, et al.,2006, Farsari, et al.,2010) have functionalized the surface of tridimensional structures made with a biocompatible organic-inorganic hybrid matrix, and subsequently immobilized photosensitive biotin on it. The integrity of biotin was confirmed by binding and detecting the presence and distribution of avidin via fluorescence microscopy. Experiments with peptides have also been carried out by the same group. Such methodology seems promising for applications in scaffolds for cell growth and tissue engineering.

- *Incorporation of other organic and inorganic materials*

Research groups have used a considerable variety of doping materials and methodologies for incorporating such materials into the base resins, in order to render, for instance, optical properties to the two-photon polymerized microstructures. Sun et al (Sun, et al.,2007) proposed a methodology to produce microstructures containing embedded CdS nanoparticles. The materials were prepared by using Cd acrylate derivatives as the base resin, and after the microfabrication process, the structures were treated with H_2S gas for about 48 h to form CdS-polymer nanocomposites. By using this methodology, they were able to fabricate photonic crystals and structures presenting a high fluorescence signal.

Jia et al (Jia, et al.,2010) fabricated microstructures using a organic–inorganic hybrid photosensitive polymer containing dispersed PbS quantum dots. Such material exhibits high third-order nonlinearity, and was used to fabricate woodpile-shape photonic crystals with stop gaps in the telecommunication wavelength region. The authors also prospected these nanocomposite materials for applications in active micro/nanodevices, such as ultrafast switching, signal regeneration, and high speed demultiplexing systems.

Shukla et al (Shukla, et al.,2010) demonstrated a new method for producing subwavelength plasmonic and conductive patterned structures within a polymeric host. The method involves both, two-photon-initiated photoreduction of a gold precursor and two-photon polymerization using a negative photoresist. By this methodology, they several types of structures, including optically active planar chiral structures and plasmonic nanostructures, which have potential use for the development of metamaterials. Example of such structures are displayed in Fig. 16 (a) and (b) (Shukla, et al.,2010), which shows confocal microscopy plasmonic donuts fabricated by the mentioned method.

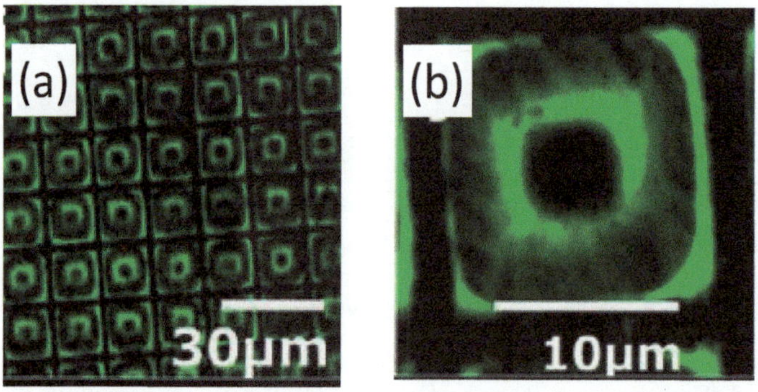

Figure 16. (a,b). Confocal fluorescence image of plasmonic donuts structures fabricated by a combination of two-photon-initiated photoreduction of a gold precursor and two-photon absorption polymerization. Reprinted with permission from (Shukla, et al.,2010). Copyright [2010], American Chemical Society.

Microstructures containing rodhamine have been fabricated by two-photon absorption polymerization by a few research groups (Correa, et al.,2011, Zukauskas, et al.,2010). Rhodamine is a remarkable material to dope resins owing to its chemical stability, strong luminescence and nonlinear optical properties. Correa et al. (Correa, et al.,2011) fabricated two-photon polymerized structures containing rodhamine and excited them with silica wires with diameters of approximately 1 μm, , which are produced through the fiber tapering technique (flame-heated fiber drawing) (Tong, et al.,2003). Figure 17 (a) shows a schematic view of the excitation of a two-photon polymerized microstructure with silica tapered fiber. Figure 17 (b) and (c) show optical micrographs of (a) frontal and (b) lateral excitation at 514 nm of microstructures containing rhodamine by tapered silica fibers (Correa, et al.,2011, Correa, et al.,2012).

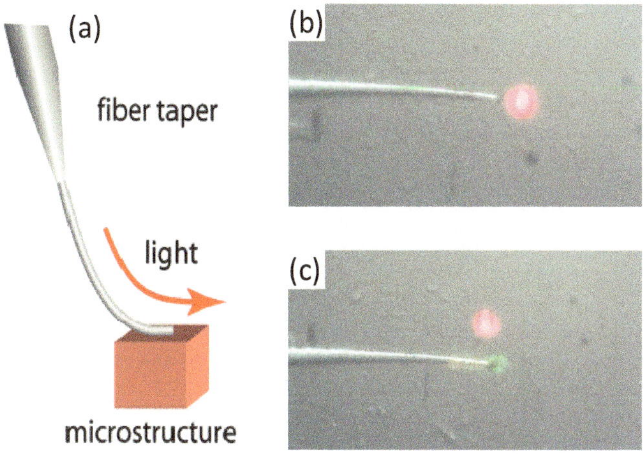

Figure 17. Excitation at 514 nm of a two-photon polymerized microstructure doped with rhodamine using silica tapered fiber. (a) Schematic view, (b) frontal excitation and (c) lateral excitation. (b) and (c) reprinted with permission from (Correa, et al.,2011). Copyright [2011], Springer Science+Business Media B.V.

To further advance the use of the structures' optical properties, integration between them was explored, aiming at exciting or collecting light by using silica tapered fibers. The integration between structures was carried out with silica tapered fibers similar to the one displayed in Fig. 17. Figure 18 shows a set of two-photon polymerized structures connect by such tapered optical fibers. This step represents a potential approach to integrate arrays of optical microstructures, aiming at the fabrication of micro-optical photonics circuitry.

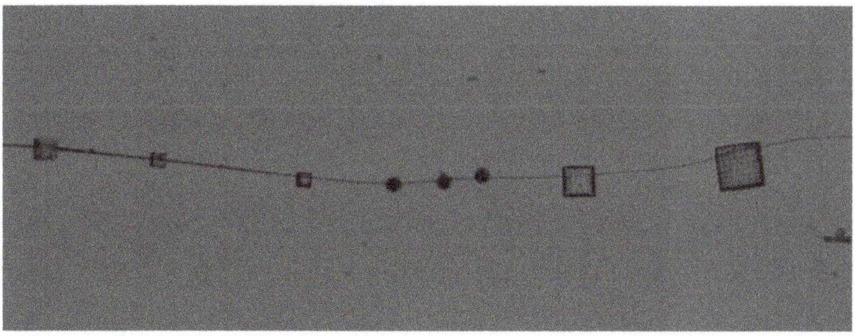

Figure 18. Transmission optical micrograph of a set of two-photon polymerized structures connected by tapered optical fibers.

By incorporating two or more distinct dopants into the base resin, one can obtain two-photon polymerized microstructures that present, for instance, fluorescence at two or more distinct wavelengths. This approach would be useful, for instance, to produce multi-colored pixels presenting a RGB pattern. In Fig. 19, we show microscope images of undoped (gray) and rhodamine doped (purple) microstructures fabricated by two-photon polymerization in the same substrate.

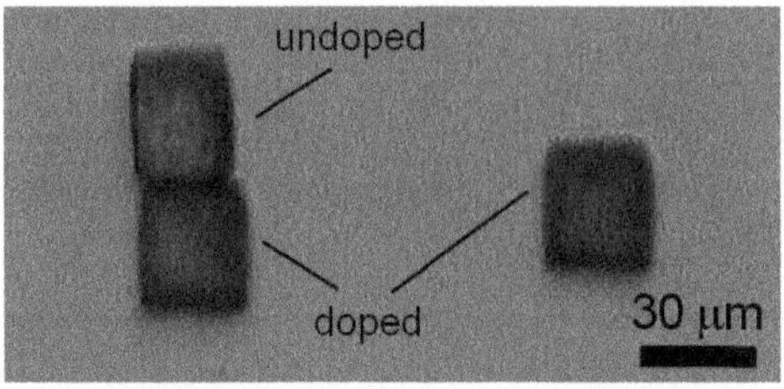

Figure 19. Optical microcopy image of two-photon polymerized structures fabricated on the same substrate. The gray structure was fabricated only with the base resin (undoped), while the purple one (doped) contains rhodamine.

In addition to the based on the methodology of incorporating materials into the bulk of basic resin, another interesting approach is to functionalize the fabricated microstructure's surface. In this case, after functionalization, other materials can be adhered to the surface, giving to the microstructures unique properties. For instance, functionalization has been carried out by several groups to achieve metallization of structures' surface, with gold, silver, and copper (Farrer, et al.,2006, Rill, et al.,2008, Takeyasu, et al.,2008). Such metallic structures can be highly conductive, opening doors for applications in microelectrical circuitry.

6. Conclusions

This chapter aimed at demonstrating fundamental and experimental aspects of fabrication of doped microstructures via two-photon absorption polymerization. Due to the efforts of many research groups in the last years, the use of this technique and its applications has spread out all over the world. By using this technique, tridimensional microstructures of complex shape, high definition and structural integrity can be fabricated to be used in microdevices. Distinct dopants incorporated into the basic resin formulation can yield to the microstructure features such as biocompatibility, luminescence, magnetic properties and waveguiding. Such microstructures with enhanced properties are highly desired for applications in optical devices, photonic crystals, micro-waveguides, microfluidics and scaffolds for tissue engineering, whose commercial applications in the market are expected to happen in the next years.

Author details

Daniel S. Correa
Embrapa Instrumentação, São Carlos, SP, Brazil

Leonardo De Boni, Adriano J. G. Otuka, Vinicius Tribuzi and Cleber R. Mendonça
Instituto de Física de São Carlos, Universidade de São Paulo, São Carlos, SP, Brazil

Acknowledgement

We thank financial support from Fundação de Amparo à Pesquisa do Estado de São Paulo (FAPESP) , Conselho Nacional de Desenvolvimento Científico Tecnológico (CNPq), and Embrapa.

7. References

Abella, I. D. (1962). Optical double-photon absorption in cesium vapor. *Physical Review Letters*, vol. 9, 453-&

Baldacchini, T., LaFratta, C. N., Farrer, R. A., Teich, M. C., Saleh, B. E. A., Naughton, M. J., and Fourkas, J. T. (2004). Acrylic-based resin with favorable properties for three-dimensional two-photon polymerization. *Journal of Applied Physics*, vol. 95, 6072-6076

Balu, M., Hales, J., Hagan, D. J., and Van Stryland, E. W. (2004). White-light continuum Z-scan technique for nonlinear materials characterization. *Optics Express*, vol. 12, 3820-3826

Boyd, R. W., Nonlinear optics. Boston: Academic Press, 1992.

Chung, S.-J., Maciel, G. S., Pudavar, H. E., Lin, T.-C., He, G. S., Swiatkiewicz, J., Prasad, P. N., Lee, D. W., Lin, J.-I., and Menard, M. (2002). Two-photon properties and excitation dynamics of poly(p-phenylenevinylene) derivatives carrying phenylanthracence and branched alkoxy pendents. *J. Phys. Chem. A*, vol. 106, 7512-7520

Claeyssens, F., Hasan, E. A., Gaidukeviciute, A., Achilleos, D. S., Ranella, A., Reinhardt, C., Ovsianikov, A., Xiao, S., Fotakis, C., Vamvakaki, M., Chichkov, B. N., and Farsari, M. (2009). Three-Dimensional Biodegradable Structures Fabricated by Two-Photon Polymerization. *Langmuir*, vol. 25, 3219-3223

Correa, D., S, Tribuzi, V., Cardoso, M. R., Misoguti, L., and Mendonca, C. R. (2011). Selective excitation through tapered silica fibers of fluorescent two-photon polymerized structures. *Applied Physics A-Materials Science & Processing*, vol. 102, 435-439

Correa, D. S., Cardoso, M. R., Tribuzi, V., Misoguti, L., and Mendonca, C. R. (2012). Femtosecond Laser in Polymeric Materials: Microfabrication of Doped Structures and Micromachining. *IEEE Journal of Selected Topics in Quantum Electronics*, vol.18, 176-186

Correa, D. S., De Boni, L., Balogh, D. T., and Mendonca, C. R. (2007). Three- and four-photon excitation of poly(2-methoxy-5-(2'-ethylhexyloxy)-1,4-phenylenevinylene) (MEH-PPV). *Advanced Materials*, vol. 19, 2653-2656

Correa, D. S., Tayalia, P., Cosendey, G., Dos Santos Jr., D. S., Aroca, R. F., Mazur, E., and Mendonca, C. R. (2009). Two-photon polymerization for fabricating structures containing the biopolymer chitosan. *Journal of Nanoscience and Nanotechnology*, vol. 9, 5845-5849

Cumpston, B. H., Ananthavel, S. P., Barlow, S., Dyer, D. L., Ehrlich, J. E., Erskine, L. L., Heikal, A. A., Kuebler, S. M., Lee, I. Y. S., McCord-Maughon, D., Qin, J. Q., Rockel, H., Rumi, M., Wu, X. L., Marder, S. R., and Perry, J. W. (1999). Two-photon polymerization initiators for three-dimensional optical data storage and microfabrication. *Nature*, vol. 398, 51-54

De Boni, L., Andrade, A. A., Correa, D. S., Balogh, D. T., Zilio, S. C., Misoguti, L., and Mendonca, C. R. (2004). Nonlinear absorption spectrum in MEH-PPV/chloroform solution: A competition between two-photon and saturated absorption processes. *Journal of Physical Chemistry B,* vol. 108, 5221-5224

De Boni, L., Andrade, A. A., Misoguti, L., Mendonca, C. R., and Zilio, S. C. (2004). Z-scan measurements using femtosecond continuum generation. *Optics Express,* vol. 12, 3921-3927

Drakakis, T. S., Papadakis, G., Sambani, K., Filippidis, G., Georgiou, S., Gizeli, E., Fotakis, C., and Farsari, M. (2006). Construction of three-dimensional biomolecule structures employing femtosecond lasers. *Applied Physics Letters,* vol. 89, 144108

Farrer, R. A., LaFratta, C. N., Li, L. J., Praino, J., Naughton, M. J., Saleh, B. E. A., Teich, M. C., and Fourkas, J. T. (2006). Selective functionalization of 3-D polymer microstructures. *Journal of the American Chemical Society,* vol. 128, 1796-1797

Farsari, M., Ovsianikov, A., Vamvakaki, M., Sakellari, I., Gray, D., Chichkov, B. N., and Fotakis, C. (2008). Fabrication of three-dimensional photonic crystal structures containing an active nonlinear optical chromophore. *Applied Physics a-Materials Science & Processing,* vol. 93, 11-15

Farsari, M., Vamvakaki, M., and Chichkov, B. N. (2010). Multiphoton polymerization of hybrid materials. *Journal of Optics,* vol. 12, 124001

Fowles, G. R., Introduction to Modern Optics, 2nd ed. New York: Dover Publications, 1989.

Gittard, S. D., Ovsianikov, A., Akar, H., Chichkov, B., Monteiro-Riviere, N. A., Stafslien, S., Chisholm, B., Shin, C. C., Shih, C. M., Lin, S. J., Su, Y. Y., and Narayan, R. J. (2010). Two Photon Polymerization-Micromolding of Polyethylene Glycol-Gentamicin Sulfate Microneedles. *Advanced Engineering Materials,* vol. 12, B77-B82

Göppert-Mayer, M. (1931). Elementary processes with two quantum jumps. *Ann. Phys.,* vol. 9, 273-294

Haske, W., Chen, V. W., Hales, J. M., Dong, W. T., Barlow, S., Marder, S. R., and Perry, J. W. (2007). 65 nm feature sizes using visible wavelength 3-D multiphoton lithography. *Optics Express,* vol. 15, 3426-3436

Jia, B. H., Buso, D., van Embden, J., Li, J. F., and Gu, M. (2010). Highly Non-Linear Quantum Dot Doped Nanocomposites for Functional Three-Dimensional Structures Generated by Two-Photon Polymerization. *Advanced Materials,* vol. 22, 2463-2467

Juodkazis, S., Mizeikis, V., Seet, K. K., Miwa, M., and Misawa, H. (2005). Two-photon lithography of nanorods in SU-8 photoresist. *Nanotechnology,* vol. 16, 846-849

Kaiser, W. and Garrett, C. G. B. (1961). 2-Photon Excitation in CaF2 - Eu2+. *Physical Review Letters,* vol. 7, 229-231

Kawata, S., Sun, H. B., Tanaka, T., and Takada, K. (2001). Finer features for functional microdevices - Micromachines can be created with higher resolution using two-photon absorption. *Nature,* vol. 412, 697-698

Konjhodzic, D., Bretinger, H., Wilczok, U., Dreier, A., Ladenburger, A., Schmidt, M., Eich, M., and Marlow, F. (2005). Low-n mesoporous silica films: structure and properties. *Applied Physics a-Materials Science & Processing,* vol. 81, 425-432

LaFratta, C. N., Fourkas, J. T., Baldacchini, T., and Farrer, R. A. (2007). Multiphoton fabrication. *Angewandte Chemie-International Edition,* vol. 46, 6238-6258

Lee, K. S., Kim, R. H., Yang, D. Y., and Park, S. H. (2008). Advances in 3D nano/microfabrication using two-photon initiated polymerization. *Progress in Polymer Science,* vol. 33, 631-681

Liska, R., Schuster, M., Infuhr, R., Tureeek, C., Fritscher, C., Seidl, B., Schmidt, V., Kuna, L., Haase, A., Varga, F., Lichtenegger, H., and Stampfl, J. (2007). Photopolymers for rapid prototyping. *Journal of Coatings Technology and Research,* vol. 4, 505-510

Marder, S. R., Bredas, J. L., and Perry, J. W. (2007). Materials for multiphoton 3D microfabrication. *MRS Bulletin,* vol. 32, 561-565

Maruo, S., Nakamura, O., and Kawata, S. (1997). Three-dimensional microfabrication with two-photon-absorbed photopolymerization. *Optics Letters,* vol. 22, 132-134

Mendonca, C. R., Baldacchini, T., Tayalia, P., and Mazur, E. (2007). Reversible birefringence in microstructures fabricated by two-photon absorption polymerization. *Journal of Applied Physics,* vol. 102, 013109

Mendonca, C. R., Correa, D. S., Baldacchini, T., Tayalia, P., and Mazur, E. (2008). Two-photon absorption spectrum of the photoinitiator Lucirin TPO-L. *Applied Physics a-Materials Science & Processing,* vol. 90, 633-636

Mendonca, C. R., Correa, D. S., Marlow, F., Voss, T., Tayalia, P., and Mazur, E. (2009). Three-dimensional fabrication of optically-active microstructures containing MEH-PPV. *Applied Physics Letters,* vol. 95, 113309

Nguyen, T. Q., Martini, I. B., Liu, J., and Schwartz, B. J. (2000). Controlling interchain interactions in conjugated polymers: The effects of chain morphology on exciton-exciton annihilation and aggregation in MEH-PPV films. *Journal of Physical Chemistry B,* vol. 104, 237-255

Oliveira, S. L., Correa, D. S., De Boni, L., Misoguti, L., Zilio, S. C., and Mendonca, C. R. (2006). Two-photon absorption cross-section spectrum of a pi-conjugated polymer obtained using the white-light continuum Z-scan technique. *Applied Physics Letters,* vol. 88, 021911

Ovsianikov, A., Chichkov, B., Adunka, O., Pillsbury, H., Doraiswamy, A., and Narayan, R. J. (2007). Rapid prototyping of ossicular replacement prostheses. *Applied Surface Science,* vol. 253, 6603-6607

Parker, I. D. (1994). Carrier tunneling and device characteristics in polymer light-emitting-diodes. *Journal of Applied Physics,* vol. 75, 1656-1666

Rill, M. S. R. M. S., Plet, C., Thiel, M., Staude, I., Von Freymann, G., Linden, S., and Wegener, M. (2008). Photonic metamaterials by direct laser writing and silver chemical vapour deposition. *Nature Materials,* vol. 7, 543-546

Schlie, S., Ngezahayo, A., Ovsianikov, A., Fabian, T., Kolb, H. A., Haferkamp, H., and Chichkov, B. N. (2007). Three-dimensional cell growth on structures fabricated from ORMOCER (R) by two-photon polymerization technique. *Journal of Biomaterials Applications,* vol. 22, 275-287

Serbin, J., Ovsianikov, A., and Chichkov, B. (2004). Fabrication of woodpile structures by two-photon polymerization and investigation of their optical properties. *Optics Express,* vol. 12, 5221-5228

Sheik-Bahae, M., A. A. Said, T. H. Wei, D. J. Hagan, E. W. Van Stryland. (1990). Sensitive Measurement of Optical Nonlinearities Using a Single Beam. *Ieee Journal of Quantum Electronics,* vol. 26, 760-769

Shen, Y. R., The Principles of Nonlinear Optics. New York: John Wiley, 1984.

Shukla, S., Vidal, X., Furlani, E. P., Swihart, M. T., Kim, K. T., Yoon, Y. K., Urbas, A., and Prasad, P. N. (2010). Subwavelength Direct Laser Patterning of Conductive Gold Nanostructures by Simultaneous Photopolymerization and Photoreduction. *Acs Nano,* vol. 5, 1947-1957

Sun, H. B. and Kawata, S., "Two-photon photopolymerization and 3D lithographic microfabrication," in *NMR - 3D Analysis - Photopolymerization.* vol. 170 Berlin: Springer-Verlag Berlim, 2004, pp. 169-273.

Sun, H. B., Tanaka, T., Takada, K., and Kawata, S. (2001). Two-photon photopolymerization and diagnosis of three-dimensional microstructures containing fluorescent dyes. *Applied Physics Letters,* vol. 79, 1411-1413

Sun, Z. B., Dong, X. Z., Nakanishi, S., Chen, W. Q., Duan, X. M., and Kawata, S. (2007). Log-pile photonic crystal of CdS-polymer nanocomposites fabricated by combination of two-photon polymerization and in situ synthesis. *Applied Physics a-Materials Science & Processing,* vol. 86, 427-431

Takeyasu, N., Tanaka, T., and Kawata, S. (2008). Fabrication of 3D metal/polymer microstructures by site-selective metal coating. *Applied Physics a-Materials Science & Processing,* vol. 90, 205-209

Tayalia, P., Mendonca, C. R., Baldacchini, T., Mooney, D. J., and Mazur, E. (2008). 3D cell-migration studies using two-photon engineered polymer scaffolds. *Advanced Materials,* vol. 20, 4494-4498

Tong, L. M., Gattass, R. R., Ashcom, J. B., He, S. L., Lou, J. Y., Shen, M. Y., Maxwell, I., and Mazur, E. (2003). Subwavelength-diameter silica wires for low-loss optical wave guiding. *Nature,* vol. 426, 816-819

Wang, J., Xia, H., Xu, B. B., Niu, L. G., Wu, D., Chen, Q. D., and Sun, H. B. (2009). Remote manipulation of micronanomachines containing magnetic nanoparticles. *Optics Letters,* vol. 34, 581-583

Yu, G., Gao, J., Hummelen, J. C., Wudl, F., and Heeger, A. J. (1995). Polymer photovoltaic cells - enhanced efficiencies via a network of internal donor-acceptor heterojunctions. *Science,* vol. 270, 1789-1791

Zukauskas, A., Malinauskas, M., Kontenis, L., Purlys, V., Paipulas, D., Vengris, M., and Gadonas, R. (2010). Organic dye doped microstructures for optically active functional devices fabricated via two-photon polymerization technique. *Lithuanian Journal of Physics,* vol. 50, 55-61

Chapter 17

A Survey of Equations of State for Polymers

Yuri Guerrieri, Karen Valverde Pontes,
Gloria Meyberg Nunes Costa and Marcelo Embiruçu

Additional information is available at the end of the chapter

1. Introduction

The thermodynamics of polymeric systems play an important role in the polymer industry and are often a key factor in polymer production, processing and material development, especially for the design of advanced polymeric materials. Many polymeric products are produced with a solvent or diluent (or a mixture of them) and often with other low molecular weight compounds (plasticizers, among others). A problem which often arises is how to remove the low molecular weight constituent(s) from the final product (polymer). The solution to this problem involves, among other tasks, solving the vapor-liquid equilibrium (VLE) and/or the vapor-liquid-liquid equilibrium (VLLE) problem. Other applications of polymer thermodynamics directly involve the polymerization processes. For example, several processes such as the production of PET (polyethylene terephthalate) are carried out in two-phase (vapor-liquid) reactors. Phase equilibrium compostions of the reacting components will determine their phase concentrations and thus the outcome of the polymerization reaction. Another example is the case of LDPE (Low Density PolyEthylene) made in autoclave reactors where it may be desirable to perform the polymerization reaction nearby but outside the two-liquid phase region, but close to it, which makes accurate liquid-liquid equilibrium (LLE) information at high pressure essential. During PE (polyethylene) or PP (polypropylene) industrial processing, for example, deposition of the polymer on the reactor surface, heat exchangers and flash drums frequently occurs and this can cause clogging in pipelines. Modeling solid-liquid equilibrium (SLE) is a useful basis from which to gain a better understanding of these industrial polymer problems and thus to avoid their occurrence.

Analogous to the modeling of conventional phase equilibrium, there are two basic approaches available to describe phase equilibrium of polymer-solvent mixtures: activity coefficient models and equations of state (EOS). There are several drawbacks to the activity coefficient approach, for example: it is hard to define standard states, especially for supercritical components; the parameters of the activity coefficient models are very temperature dependent, and critical phenomena are not predicted because different models are used for the vapor and liquid phases. Furthermore, other thermodynamic properties such as densities, enthalpies, entropies, among others, cannot usually be obtained from the same model because the excess Gibbs free energy is rarely known as a function of temperature and pressure.

EOS are powerful tools for investigating thermodynamic properties and phase behavior of pure fluids and their mixtures. There are many well-tested EOS available for fluid mixtures of conventional substances. For mixtures of polymers with solvents, on the other hand, problems arise due to the different characteristics of the components. To address these many polymer-specific EOS have been proposed, which focus on the polymer component(s) of the mixture. Efforts to represent conventional systems with these EOS have not always been very successful; indeed some of these models perform less successfully than traditional cubic EOS in this regard. This may be a handicap when these models are used for the VLE of the polymer-solvent mixtures. In such cases, little or no polymer is present in the vapor phase and the solvent compressibility plays an important role in the phase behavior. Consequently, there is a strong incentive to extend the conventional EOS developed for small molecules to polymers.

There are two basic issues in extending cubic EOS to apply to polymers and their use. The first issue is the description of the pure component EOS parameters for polymers. To obtain these parameters, various techniques have been suggested. The second issue in extending cubic EOS to apply to polymers is the selection of mixing rules (MR) for the EOS parameters. The classical mixing rules of van der Waals (vdW) have already been tested for polymer solvent mixtures, however, it has been observed that, in order to fit the experimental data, some unrealistic values are necessary for the binary interaction parameters (BIP).

The use of equations of state in phase equilibrium modeling instead of activity coefficient models is mainly a result of the recent development of a class of mixing rules that enable the use of liquid activity coefficient models in the EOS formalism. The implication of this change is far-reaching as an EOS offers a unified approach in thermodynamic property modeling. With this approach, the applicability of simple cubic EOS has been extended to complex systems, such as polymeric systems, if coupled with the appropriate activity coefficient model. Therefore, there is much interest in mixture EOS models capable of describing higher degrees of nonideality than that possible with the van der Waals one-fluid model and its modifications.

Future development of EOS for polymer mixtures is unclear and some contradictory statements can be found in the literature. Some authors indicate that cubic equations can be extended to correlate and predict VLE in polymer mixtures accurately. On the other hand, others state that, considering the complexity of this type of mixture, simplicity is not a necessary requirement for an EOS, as the calculation of parameters for the mixture components is more important. There is agreement, however, on the fact that future development of EOS for polymer mixtures must emphasize the study of mixing rules and that EOS input parameters should be related to the commonly measured properties of the polymers.

The most apparent progress toward EOS with the ability to describe phase behavior with polymers has been made by applying statistical mechanics. Some early models derived from statistical thermodynamics assumed molecules to be arranged in a lattice, whereas many of the more recent theories picture molecules to be moving freely in continuous space. In lattice models, the molecules are assumed to have one or more segments, and the partition function of the system can be obtained by counting the possible configuration when these segments are arranged in hypothetical cells which are like the lattices in solid materials. Then the thermodynamic quantities can be calculated from the partition function on the basis of statistical mechanics.

A huge amount of work has been done on the understanding of phase behavior in polymeric mixtures, either from an experimental or theoretical point of view. As well as supplying important data, experiments enable the evaluation of EOS models for the correlation and/or prediction of phase behavior. A model, on the other hand, takes much less experimental effort and can guide the researcher/analyst in the right direction.

A detailed review of the different lines of developing equations of state for the calculation of fluid phase equilibria is given by [1]. Recently [2] presented and discussed in depth both classical and novel thermodynamic models, which have been developed and can potentially be used for industrial applications. A review of the use of some equations of state (EOS) for LDPE process simulation can be found in Orbey *et al.* [3] and Valderrama [4].

Although there have been some analyses on equations of state that can describe the phase equilibria involving polymers, additional assessments are necessary. In general, the available works concern a specific approach, not taking into account others. In addition, these reviews and surveys focus on detailed model theory or theoretical possibilities of model variations, with a few quotes from practical applications. This chapter therefore presents an overview of the progress on EOS models for polymer systems considering the following approaches:

- Cubic EOS (mixing rules incorporationg excess Gibbs free energy models)
- Lattice models [Sanchez-Lacombe (SL) equation of state]
- Perturbation theory (SAFT equation: the original version and its variants)

In EOS applications only works dealing with phase equilibrium are discussed, other types of applications, such as solvent absorption and/or polymer swelling, are not addressed. The timeline diagram in Figure 1 shows some of the key developments and outstanding papers related to the development of equations of state for polymer systems, which are discussed in this chapter. The following notation is used:

- Cubic Equations of State [Huron and Vidal (HV) Mixing Rule (MR); Sako-Wu-Prausnitz (SWP) Equation; Wong and Sandler (WS) MR; vdW Applied to Polymer Solutions (vdW-P) by Kontogeorgis, Harismiadis, Fredeslund and Tassios; Linear Combination of Vidal and Michelsen (LCVM) Mixing Rules by Boukouvalas, Spiliots, Coutsikos, Tzouvaras and Tassios; Zhong and Masuoka (ZM) MR]
- Lattice Models [Flory-Huggins (FH); Sanchez and Lacombe (SL) equation; key Modifications and Applications (MA) of SL by Kleintjens and Koningmeld, Panayiotou and Vera, Kiran and Xiong and Zhuang, Koak and Heidemann, Gauter and Heidemann, Krenz and Heidemann and de Loos]
- Perturbation theory [Wertheim Thermodynamic Perturbation Theory (TPT); Statistical Associating Fluid Theory (SAFT) by Chapman, Gubbins, Jackson, Radosz; Chen and Kreglewski [5] SAFT (CK-SAFT) by Huang and Radosz; Perturbed-Chain SAFT (PC-SAFT) by Gross and Sadowski; Simplified PC-SAFT (sPC-SAFT) by von Solms, Michelsen and Kontogeorgis]

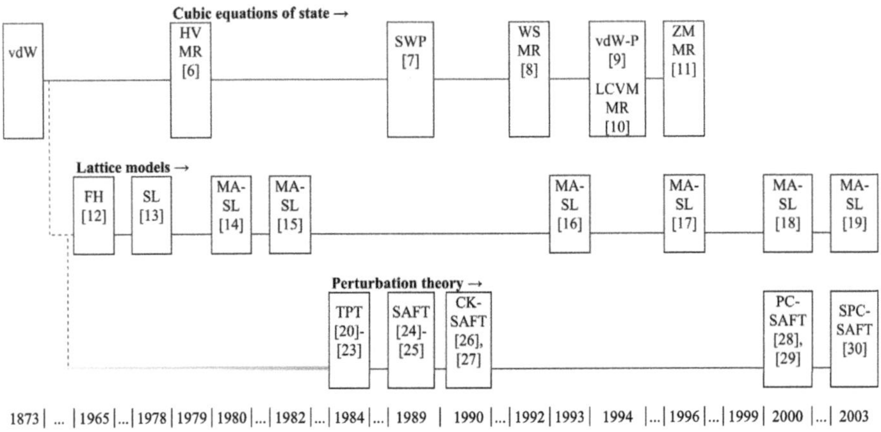

Figure 1. Timeline of some Key Model Developments Addressed in this Chapter from van der Waals (vdW) Equation in 1873.

2. Cubic equations of state and mixing rules

The first group of models to describe the phase behavior (by calculating the equilibrium constant) corresponds to the van der Waals equations of state, known as cubic equations, in either the original version or variants thereof. They are extremely simple and efficient for experimental data correlation. In this group, modifications of the Redlich-Kwong equation stand out, especially the Soave-Redlich-Kwong (SRK) [31] and the Peng-Robinson (PR) [32], which can calculate, often successfully, the vapor-liquid equilibrium for normal fluid and mixtures. However, application of the cubic equations of state for polymeric blends is not immediately obvious as this application does not follow standard procedures. The conventional method for calculating the pure parameters in cubic equations of state requires components' critical properties and vapor pressure, which do not exist for polymers. Therefore, two basic issues should be addressed when extending cubic equations of state for polymers and their mixtures. The first, presented in the following paragraphs is the description of the parameters of pure components, and the second is the choice of the mixing rule, which will be discussed later in sections 2.1-2.3.

There are four conditions to be satisfied when selecting the pure component parameters of a cubic equation of state. First, a polymer is non-volatile and therefore should not exhibit any vapor pressure. If there are oligomers in the mixture though, low vapor pressures might be considered. Therefore, critical properties may be assigned for the oligomers, treating them as conventional components. The second condition is that the equation of state should predict densities of molten polymers. The third condition requires that the parameters reflect the polymers' basic characteristics such as the degree of polymerization. This is important because experimental data demonstrate that these polymer characteristics directly affect the vapor-liquid equilibrium in polymer-solvent mixtures. The fourth point, somewhat connected to the third, requires easily accessible and physical meaning characteristics as input parameters for calculating the parameters of the equation of state. As stated before, the SRK equation of state is expressed by:

$$P = \frac{R \cdot T}{V - b} - \frac{a(T)}{V \cdot (V + b)} \tag{1}$$

and the PR equation is given by:

$$P = \frac{R \cdot T}{V - b} - \frac{a(T)}{V \cdot (V + b) + b \cdot (V - b)} \tag{2}$$

where T is temperature, V is molar volume, P is pressure and R is the universal gas constant. The parameter a is a measure of the attractive forces between molecules, and the parameter b is the co-volume occupied by these molecules.

The first attempt to apply a cubic equation to polymers was made by Sako *et al.* [7] in order to calculate the high pressure vapor-liquid equilibrium for a polyethylene-ethylene system. To overcome the problem of calculating the pure parameters for the polymer, Sako *et al.* [7] calculated the attractive parameter in the SRK equation using the London dispersion formula, and extrapolated the co-volume (b) values from n-alkane data. To take into account external degrees of freedom, they also added a third parameter c, whose values were fitted from density data. The SWP (Sako-Wu-Prausnitz) equation is a member of the cubic family:

$$P = \frac{R \cdot T \cdot (V - b + b \cdot c)}{V \cdot (V - b)} - \frac{a(T)}{V \cdot (V + b)} \tag{3}$$

The SWP equation was used with relative success by Tork *et al.* [33] when calculating the phase equilibria in binary and ternary systems of polyolefins. The calculations focused on the high pressure phase equilibrium for ethylene-polyethylene systems and for liquid-liquid equilibrium in systems containing either high density polyethylene or polyethylene-polypropylene copolymer. The results for the copolymer-solvent system were compared with those provided by the SAFT (Statistical Associating Fluid Theory) equation. The two equations of state can describe the UCST (Upper Critical Solution Temperature) and LCST (Lower Critical Solution Temperature) behavior as well as U-LCST, with similar precision. When using the SAFT equation, the binary interaction parameter is maintained constant, while in the SWP equation this parameter is expressed as a function of temperature. In addition, Sako *et al.* [7] investigated the influence of an inert gas on the LCST for the polyethylene-hexane system. The polydispersity of different polyethylene resins is considered when computing the phase equilibrium using pseudocomponents, chosen using the moments of experimental molecular weight distributions.

Kontogeorgis *et al.* [9] used the van der Waals equation of state to correlate vapor-liquid equilibrium data of polymer solutions. They proposed a method to calculate the interaction parameter a and co-volume b in the equation of state for polymers from two volumetric datasets at low pressure. Both parameters a and b (assumed to be independent of temperature) can be analytically expressed from two experimental molar volumes, each one at a different temperature. The pressure in the van der Waals equation is then considered equal to zero. The parameters a and b are linear functions of molecular weight. When dealing with polymer solutions, these parameters are obtained from van der Waals mixing rules as well as from the classic combination rules. Fitting only one binary parameter, the van der Waals equation of state is able to correlate the equilibrium pressure for various solutions of polyethylene and polyisobutylene accurately. However, large negative values for the binary interaction parameters, very different from typical values, are frequently required, indicating that this procedure, although empirically successful, does not have a significant physical basis. For almost athermic solutions, the Berthelot combining rule [34] is considered, and the binary interaction parameter is predicted by a simple function of the molecular weight of the solvent. Thus, satisfactory results are obtained.

The performance of cubic equations of state is directly related to the efficiency of mixing rules to represent the phase equilibria at high pressures. Basically, the mixing rules can be divided into two classes: van der Waals-type and those that incorporate excess Gibbs energy (G^E).

2.1. van der Waals mixing rules

In order to extend the application of PR and SRK cubic equations of state for polymer-solvent systems, the conventional mixing rules employed are those from van der Waals (vdW) [34], which are expressed as:

$$a = \sum_i \sum_j x_i \cdot x_j \cdot a_{ij} \tag{4}$$

$$b = \sum_i \sum_j x_i \cdot x_j \cdot b_{ij} \tag{5}$$

where x_i and x_j are the mol fractions, a_{ij} is the cross-energy parameter and b_{ij} is the cross co-volume parameter.

It should be stressed that these rules are limited to non-polar fluids and therefore are unable to represent the highly non-ideal behavior of polar or associative fluids.

An empirical approach to overcome the shortcomings of the vdW mixing rule has been to simply add new parameters and composition dependence to the combination rule for parameter *a*, usually keeping the combination rule for parameter *b*. Some examples may be cited: Adachi and Sugie [35]; Panagiotopoulos and Reid [36] and Schwartzentruber *et al.* [37]. These modified rules usually aim to solve specific problems, and the number of binary interaction parameters is quite variable. There are many problems associated with these multiparametric combination rules which limit their use in process design for mixtures containing many components (such as mixtures of isomers). Among them, the dilution effect may be cited: as the number of components in a mixture increases, the molar fraction of any component becomes smaller. This leads to small contributions of the new added parameters and terms that are strongly composition dependent. Consequently, as the number of components increases, the mixing rule is effectively reduced to a quadratic dependence, as in the one-fluid van der Waals fluid theory [38].

2.2. Mixing rules for excess free energy (G^E) models

Like conventional phase equilibrium modeling, there are two basic modeling tools for dealing with polymer-solvent mixtures: excess Gibbs free energy (G^E) models (or activity coefficient models) and equation of state models. There are plenty of models in each category and selecting the best model for a specific project can often be quite difficult. Furthermore, equation of state and activity coefficient models have varying abilities in extrapolating data beyond given ranges of temperature and pressure, which further hampers the choice of the best model. These models also behave differently when predicting vapor-liquid equilibrium from other measured properties, such as the infinite dilution activity coefficient for a polymer in solvent.

Over the last two decades several methods combining activity coefficient models with equations of state have emerged. These methods are useful for correlating/predicting the phase equilibria of conventional mixtures, and are promising for mixtures containing polymers. Moreover, they allow us to investigate an activity coefficient model in two ways: first as a conventional model (i.e. in the approach γ-ϕ); secondly as part of an equation of state. In general, activity coefficient models are considered more flexible to accommodate a highly complex phase behavior. Equation of state models, on the other hand, may take into account the effects of compressibility in a thermodynamically more consistent way, and are most useful at higher pressures.

Huron and Vidal [6], pioneers in this field, incorporated an excess Gibbs free energy model into a mixing rule. Their method is based on three assumptions: (i) the excess Gibbs free energy, calculated from an equation of state at infinite pressure equals the excess Gibbs free energy calculated from an activity coefficient model for the liquid phase (ii) the co-volume parameter *b* is set to the volume at infinite pressure (iii) the excess volume equals zero. Thus the mixing rule is written as:

$$\alpha_V = \frac{1}{q_2} \cdot \frac{G^E}{R \cdot T} + \sum_i x_i \cdot \alpha_i \tag{6}$$

where $\alpha_V = a/(b \cdot R \cdot T)$ and $q_2 = -\ln 2$ for SRK equation. Huron and Vidal [6] showed that their mixing rule gives good results for non-ideal mixtures. Soave [39] showed that the Huron and Vidal rule represents an improvement over the classical quadratic mixing rules and can accurately correlate the vapor-liquid equilibrium for highly nonideal systems. The Huron and Vidal mixing rule has also been applied to several polar and asymmetric systems [40]-[42].

The Huron and Vidal mixing rule has some undesirable characteristics, such as: it does not reproduce the quadratic dependence of the second virial coefficient (QDSVC) with the composition at low pressure; it has no predictive value because the parameters of the activity coefficient model, estimated at low pressure, have to be re-estimated at high pressure; furthermore, these parameters are temperature dependent. Various proposals [43]-[44] have tried to cope with these constraints, however, they fail to succeed as discussed below.

Mollerup [45] suggested an alternative method to that of Huron and Vidal, assuming that the excess volume is zero at low pressure and that the excess Gibbs free energies calculated from an equation of state and from an activity coefficient model can be matched in this condition. Therefore, the activity coefficient parameters do not need be re-estimated if pressure and temperature conditions correspond to those at which they are fitted. However, since this theory cannot be applied to supercritical fluids, as well as the difficulty of computing roots (of the equation of state) for the liquid phase at zero pressure, its application is restricted.

Heidemann and Kokal [46], in accordance with Mollerup [45], also take the reference state at null pressure, attempting, however, to overcome the problem of calculating the root for the liquid phase at zero pressure. The major contribution of this method is to propose an extrapolation procedure from the system pressure, enabling calculation at temperatures near and above the critical point. Important to mention is that this method requires the solution of a transcendental equation when calculating the mixing rule. Comparative studies demonstrate a better performance of the Heidemann and Kokal rule when compared to the Huron and Vidal rule [6].

A method very similar to Heidemann and Kokal's [46] was proposed by Michelsen [47]. The main difference between them lies in the extrapolation method used for supercritical components. The mixing rule in the Michelsen approach also requires the solution of a transcendental equation. The SRK-Wilson model (using SRK equation of state and Wilson activity coefficient model), obtained using this method, was tested to obtain the phase envelope, including the critical points, and to calculate the phase diagrams at high pressure, without re-estimating the parameters of the Wilson model. Good results were achieved.

Michelsen [48] modified his own method, considering an explicit mixing rule, i.e. avoiding the solution of the transcendental equation. The only drawback of this modification lies in the impossibility of ensuring the accurate reproduction of the G^E model at low temperatures. Thus, imposing a linear mixing rule to the parameter a, an expression similar to the Huron and Vidal rule [6] was obtained, and therefore it is called the first-order Modified Huron and Vidal (MHV1) mixing rule.

Dahl and Michelsen [49] found out that replacing the linear approach (MHV1) by a quadratic approximation considerably improves the reproductibility of the G^E model. The resulting mixture rule, a second-order modification of the Huron and Vidal rule, became known as the MHV2 (second-order Modified Huron and Vidal mixing rule), where the linear rule for the co-volume b was maintained. Like the Huron and Vidal mixing rule, the MHV2 does not satisfactorily describe the excess molar volume. Additionally, it is theoretically incorrect at the lower bound pressure (when the pressure goes to zero), where it does not show the QDSVC with the mole fraction.

Attempting to straighten out the theoretical inconsistency of the aforementioned mixing rules, Wong and Sandler [8] proposed a new method in which the rules fulfill the QDSVC with the composition at low pressure condition. The basic idea was to consider the excess Helmholtz free energy as much less dependent on pressure than the excess Gibbs free energy. In this way, the excess Helmholtz free energy at high pressure might be equal to the excess Gibbs free energy at low pressure. Therefore, this mixing rule is given by:

$$a = b \cdot \left(\sum_i \frac{x_i \cdot a_i}{b_i} + \frac{G^E}{c} \right) \tag{7}$$

where:

$$b = \frac{\sum_i \sum_j x_i \cdot x_j \cdot (b - a/R \cdot T)_{ij}}{1 - \sum_i x_i \cdot (a_i / b_i \cdot R \cdot T) - (G^E / c \cdot R \cdot T)} \tag{8}$$

$$(b - a/R \cdot T)_{ij} = \frac{1}{2} \cdot \left[(b - a/R \cdot T)_i + (b - a/R \cdot T)_j \right] \cdot (1 - K_{ij}) \tag{9}$$

and c is a constant, equals –ln(2) for the SRK equation. The mixing rule gives the correct low density limit (the mixture second virial coefficient has a quadratic dependence on mole fraction). The high density limit, in turn, is consistent with experimental data: the equation of state provides the same molar excess Helmholtz free energy at infinite pressure as a function of composition, as well as that obtained from the selected activity coefficient model. The Wong and Sandler [8] mixing rule is not density dependent. It should be highlighted that, unlike the methods proposed above, the Wong-Sandler mixing rule introduced an additional binary interaction parameter besides those predicted by the activity coefficient model, which is the second virial coefficient binary interaction parameter itself. According to the authors, this parameter as well as those from the activity coefficient model, estimated at low pressure, may be interchangeably used (i.e. without restriction) at high pressure.

Boukouvalas *et al.* [10] proposed a new mixing rule for the parameter a in the attractive term of cubic equations of state. The idea was to make a linear combination between the Huron and Vidal and the MHV1 mixing rules, producing the name LCVM (Linear Combination of Vidal and Michelsen mixing rules), which may be expressed as:

$$\alpha = \lambda \cdot \alpha_V + (1-\lambda) \cdot \alpha_M \tag{10}$$

where α_V and α_M are given by Vidal and Michelsen rules, respectively. The contributions related to α are weighted by a factor λ, which is proposed by Boulouvalas *et al.* [10] to be 0.36. For the parameter b, a classical linear rule was considered. The performance of this model was compared to MHV2 and MHV1 models, using the Soave equation and the UNIFAC [UNIQUAC (UNIversal QUAsiChemical) Functional-group Activity Coefficient] for nonpolar and polar systems, symmetrical and asymmetrical, low and high pressures. The results indicate an equivalent performance of LCVM compared to the other two models when investigating systems containing molecules of similar sizes. In mixtures composed of molecules with quite different sizes, in particular gas systems with alkanes, the LCVM rule showed superior results.

Zhong and Masuoka [50], based on experimental data, evaluated the MHV1 mixing rule with SRK equation of state and the original UNIFAC model for G^E. They found out that: 1) SRK equation with MHV1 can not reproduce the G^E of the G^E model used in the mixing rule for asymmetric systems, even at low pressure; 2) the original UNIFAC is not accurate for asymmetric systems with large alkanes. The first point reflects the deficiency of MHV1 for asymmetric systems, while the second is caused by the low predictive ability of UNIFAC for systems containing large alkanes. Moreover, it is evident that, although MHV1 is able to reproduce the G^E model exactly, it may not be as accurate for gas-large alkane systems as UNIFAC can not describe these systems properly. As a result, it is pointless to pursue exact reproduction of the G^E model when attempting to improve the predictive capability for these systems. However, it is interesting to observe that the SRK equation with MHV1 can satisfactorily reproduce G^E experimental data if a correction factor is added to G^E in the original UNIFAC model used in MHV1. With this observation, a new mixing rule was proposed: the MR1 (mixing rule 1), obtained by Zhong and

Masuoka [50]. The MR1 rule is very accurate for systems composed og gas and large alkanes, when the correction parameter is obtained as a simple correlation function of the carbon number for a given gas.

A new mixing rule for cubic equations of state, particularly suitable for highly symmetric systems, was proposed by Zhong and Masuoka [51]. It was validated by two cubic equations of state: a modification of Peng-Robinson equation proposed by Stryjek and Vera (PRSV) [52] and the SRK equation. As there is no critical point for polymers, parameters a and b in the equation of state cannot be calculated from critical properties and because polymers are almost non-volatile, their vapor pressures are very low. Therefore, it is possible to use zero-pressure experimental densities to determine them. Alternatively, the approach adopted by Orbey and Sandler [38] may be employed, i.e. to use densities at low pressure with a hypothetical and very low vapor pressure, for example 10^{-7} MPa. Zhong and Masuoka [51], in the proposal of this new mixing rule, considered a null Helmholtz free energy in the limit when pressure tends to infinity. This is the only difference between this mixing rule and the one proposed by Wong and Sandler [8]. As a result, only one parameter is necessary in the new mixing rule, which is much simpler than the method of Wong and Sandler [8], and is as simple as Kontogeorgis *et al.* method [9]. When using the new mixing rule for ten polymer solutions in a wide temperature range, the results show that it allows cubic equations of state to correlate the vapor-liquid equilibrium of polymer solutions precisely even if just one temperature independent parameter is used. These results verify that the assumption, namely the excess Helmholtz free energy is null at infinite pressure, is feasible, or at least acceptable, for polymer solutions. The authors demonstrate that accurate correlations for polymer solutions are insensitive to parameters a and b in the equation of state.

In recent years, many studies have focused on improving hybrid models, i.e. equations of state which embody G^E models into mixing rules. They attempt to expand their applicability to more complex systems, such as those containing highly polar components or molecules with significantly different sizes (e.g., polymer-solvent), without loosing versatility and simplicity. Recently, Ahlers and Gmehling [53] proposed the VTPR model (Volume Translated Peng-Robinson) which brings together the UNIFAC and Peng-Robinson equation with translated volume. In the VTPR model, the two Flory-Huggins (FH) type combinatorial terms [12], which come from the equation of state and from the UNIFAC model, as well as the Staverman-Guggenheim [54] contribution of the UNIFAC combinatorial term, were eliminated. Moreover, an empirical approach in VTPR incorporates different exponents in the combination rule for the crossed co-volume of the equation of state, depending on the system studied: for those without polymer, it was set to 0.75, whereas for solvent-polymer systems, it was set to 0.5. This empirical approach, however, introduces some uncertainties. For example, it is not possible to set a single parameter for mixtures containing two solvents and a polymer. Furthermore, it is difficulty to choose two exponents for some systems containing a molecule similar to a polymer, for example, propane/hexacontane.

Voutsas *et al.* [55] proposed a new mixing rule, UMR (Universal Mixing Rule), for cubic equations of state applicable to symmetric and asymmetric systems. For the cohesion parameter of the cubic equation, this mixing rule includes the Stavermann-Guggenheim combinatorial term and the residual term of the original UNIFAC model. For the co-volume parameter in the cubic equation, a quadratic mixing rule in the composition is used. This rule has been applied to the *t*-PR (Translated Peng-Robinson) [56] equation [also known as the *t*-mPR (Translated and Modified Peng-Robinson) equation], which is a modification of the PR equation. Very satisfactory results were obtained using the original interaction parameters from the UNIFAC model inpredicting vapor-liquid and liquid-liquid equilibria at low and high pressures for several asymmetric systems including polymer mixtures.

2.3. Modeling polymeric systems with equations of state embodying Gibbs free energy (G^E) models

Orbey and Sandler [57] applied the PRSV cubic EOS, along with the mixing rules proposed by Wong and Sandler [8], to correlate vapor-liquid equilibrium data for some polymer solutions. For pure solvents, they used the conventional method to determine the parameters of the equation of state from the critical properties and the acentric factor. For polymers, however, in order to determine these parameters, they chose an arbitrary value for the vapor pressure, 10^{-7} MPa, and used experimental data of molten polymer densities. As expected, the parameters *a* and *b* are at least slightly dependent on the molecular weight. Orbey and Sandler [57] used the Flory-Huggins [12] expression to calculate the activity coefficient.

Orbey *et al.* [57] used the SRK cubic equation [31], combined with the Flory-Huggins G^E model in the Huron and Vidal [6] mixing rule, to correlate the vapor-liquid equilibrium of polymer-solvent mixtures. To extend the SRK equation for pure polymers, suitable critical constants were selected based on available information about long-chain hydrocarbons. For applications in mixtures, the single binary interaction parameter from the Flory-Huggins [12] model was obtained from activity coefficient data at infinite dilution, without using any experimental data for vapor-liquid equilibrium. The results showed that this approach, i.e. an equation of state coupled with mixing rules which incorporate G^E, may represent the vapor-liquid equilibrium of the polymer-solvent with good accuracy. It was also observed that the binary interaction parameter from Flory-Huggins [12] is much less dependent on temperature and composition when the Flory-Huggins model is coupled with the SRK then when it is used directly in the activity coefficient model.

An equation of state based on ASOG (Analytical Solution Of Groups), called PRASOG (Peng-Robinson-ASOG), was developed by Tochigi [58] to predict the vapor-liquid equilibrium of non-polymeric and polymeric solutions. It makes use of the zero-pressure G^E mixing rule, hence is consistent with the second virial coefficient dependence, in order to compute the mixture parameters of the Peng-Robinson equation of state and it predicts G^E by the ASOG method. To apply PRASOG to polymer solutions, the PRASOG-FV (PRASOG Free Volume) has been proposed calculating G^E from ASOG-FV, and then the vapor-liquid equilibrium in polyisobutylene solutions is predicted.

Tochigi *et al.* [59] extended the application of PRASOG, presented by Tochigi [58] for other polymer solutions. Nine binary systems were investigated in a temperature range from 298.15 K to 361.25 K, with six solvents (benzene, toluene, acetone, methyl ethyl ketone, ethyl acetate, propyl acetate) and four polymers (polystyrene, polyethylene oxide, polypropylene oxide, polyvinyl acetate). When using PRASOG-FV, the accuracy achieved was comparable to those of ASOG-PV and UNIFAC-FV.

Kang *et al.* [60] performed a comparative study for polymers and associating systems using the Peng-Robinson equation with the Wong-Sandler mixing rule (PR-WS), SAFT equation and NLF-HB (Non-Random Lattice Fluid Theory with Hydrogen Bonding) equation. The comparison was based on the prediction accuracy of the bubble point pressure, the molar fraction in vapor phase and the activity of the component in the liquid phase. Several factors were considered for comparison: model evaluation through their modeling errors, characteristics of the estimated parameters and computational issues. In general, when using appropriate parameters, all models provided good results when far away from critical regions, except in the case of non-polar polymers dissolved in non-associating polar solvents.

An evaluation of vapor-liquid equilibrium in polymer-solvent systems with cubic equation of state was performed by Louli and Tassios [61]. In this study the parameters a and b of PR equation were fitted from PVT (Pressure-Temperature-Volume) data of pure polymers, assuming that the ratio parameters/(molecular weight) are independent from the molecular weight. Several polymer-solvent systems were evaluated using three different mixing rules, all requiring only one adjustable parameter: vdW [34], ZM (Zhong and Masuoka) [11] and MHV1 [48]. The ZM rule gave the best results and the same performance was achieved when extrapolating predictions regarding temperature and molecular weight.

Using the PRSV cubic equation of state, Haghtalab and Espanani [62] studied the vapor-liquid equilibrium in polymer binary solutions with different molecular weights and temperatures. The parameters of the cubic equation of state were calculated using the Wong-Sandler mixing rule [8] incorporating the FH-NRTL-NRF (Flory-Huggins Non-Random Two Liquid Non-Random-Factor) excess Gibbs free energy model. The total vapor pressure of the polymer solutions was correlated using two adjustable energy parameters as functions of temperature with six constants for the entire temperature range. The modeling results showed very good agreement with the experimental data of several binary polymer solutions.

Voutsas *et al.* [63] showed that the UMR rule with the binary interaction parameters of the original UNIFAC model, independent of temperature, leads to poor predictions of vapor-liquid equilibrium at high temperatures and poor predictions for the heat of mixing. For this reason, Voutsas *et al.* [63] used the model proposed by Hansen *et al.* [64], which consider the binary interaction parameters temperature dependent, overcoming the drawbacks mentioned before. The performance of the new model was evaluated for the prediction of heat of mixing and also vapor-liquid, liquid-liquid and solid-gas equilibria in binary and multicomponent systems with different degrees of non-ideality and asymmetry, including polymer-solvent systems, showing good results.

The SRK and the Sanchez and Lacombe (SL) equations of state were applied by Costa *et al.* [65]-[66] to the flash simulation in a low-pressure separator (LPS) and also in a high-presure separator (HPS) in an industrial polyethylene facility (specifically, 8 low-density polyethylene resins and 25 linear low-density polyethylene resins were investigated). Three mixing rules were used in the SRK equation: van der Waals (vdW) one-fluid, Wong-Sandler and LCVM. The latter two mixing rules incorporate the Bogdanic and Vidal activity coefficient model [67]. All these models involve two adjustable parameters. The results for the LPS separator indicate that SL and SRK-vdW are the best models. The results for the HPS indicate that the SL is the best model. The SRK-LCVM and SRK-WS are unable to describe the HPS overhead composition.

Costa *et al.* [68] modeled the SLE in polyethylene and polypropylene solutions using SRK and PC-SAFT (Perturbed-Chain SAFT) equations of state. Two mixing rules were coupled with SRK: the Wong-Sandler rule and the LCVM rule, both considering the activity coefficient model from Bogdanic and Vidal [67]. The models were evaluated using SLE data at atmospheric and high pressure, obtained from literature. The binary interaction parameters of SRK and PC-SAFT equations of state were estimated to describe the experimental behavior of 20 different polymer-solvent systems at atmospheric pressure and 31 other polymer-solvent systems at high pressure better. The SRK-LCVM model showed the best performance with the SLE atmospheric data, although when evaluating equations predictive ability, PC-SAFT showed advantages as it is not easy to generate a good correlation of the G^E (SRK-LCVM) parameter with temperature, whereas the PC-SAFT parameter correlated very well with temperature for all the systems analyzed. In high pressure conditions, interaction parameter correlations as a function of molecular weight and polymer concentration were developed for PC-SAFT and SRK-LCVM (SRK-WS model was not appropriate for the high pressure calculations carried out). PC-SAFT provided the best performance with excellent results, showing suitable interpolating and extrapolating (predictive ability) features.

3. Lattice models

In the second group of models for calculating the equilibrium constant, it is assumed that the molecules have one or more segments, and that the partition function of the system can be obtained by counting the number of possible configurations when these segments are arranged in hypothetical cells that resemble the crystal lattice of a solid. The thermodynamic functions can be calculated using the formalism of statistical mechanics. These crystal lattices can be considered compressible or incompressible. Incompressible lattices are generally used to model liquids at low pressures, a condition in which the concept of activity coefficient is used. The most widely used activity coefficient models are based on this formalism, e.g. [12], [69]-[71]. For compressible lattices, equations of state based on lattice models result. An example of such models is the lattice fluid theory [13], [72].

The lattice model, originally developed to describe the liquid phase, considers the liquid in a quasi-crystalline state, in which the molecules do not translate fully chaotically as in a gas, but each one tends to stay in a small region, a more or less fixed position in space, around which it vibrates back and forth. The quasi-crystalline picture of the liquid state supposes that the molecules are regularly arranged in space as in a lattice, and therefore models for liquid and liquid mixtures are called lattice models. Molecular considerations suggest that deviations from ideal behavior in liquid solutions are mainly due to the following effects: first, the attraction forces between unlike molecules are quantitatively different from those between alike molecules, giving rise to a nonzero enthalpy of mixing; second, if the molecules differ significantly in size or shape, the molecular arrangement in the mixture can be appreciably different from that for pure liquids, resulting in a non-ideal entropy of mixing; finally, in binary mixtures, if the attraction forces in one among the three possible interaction pairs are much stronger (or much weaker) than the other two pairs, there will be some preferred orientation of the molecules in the mixture what, in extreme cases, can lead to instability or incomplete miscibility [34].

The most simple lattice model considers a mixture of two liquids whose molecules are small, symmetrically spherical and similar in size (the ratio of their sizes is close to one). This model assumes that the molecules of each pure liquid are regularly arranged and equidistant from each other in the lattice. The molecular movement is limited to vibrations around equilibrium positions and is not affected by the mixing process. This model also assumes that for a fixed temperature the lattice spacing in both pure liquids and in the mixture are the same, regardless of composition (excess volume is null). The first step is to obtain an expression for the potential energy of a pure liquid or a mixture, assuming that the potential energy is pair-to-pair additive for every pair of molecules and that only the nearest neighbors are considered in this sum. This means that the potential energy of a large number of molecules in the lattice is given by the sum of the potential energy of all pairs of molecules situated immediately next to each other. Therefore, considering the excess volume and the excess entropy as null, the excess Gibbs free energy for the two-suffix Margules model can be obtained from the total potential energy in the lattice [34].

This lattice model is particularly useful for describing polymeric solutions in liquid solvents. Flory and Huggins [34] independently developed a theory for polymeric solutions which have formed the foundation of most subsequent developments in the last fifty years. In the Flory-Huggins [12] model the system polymer-solvent is modeled as a lattice structure, where each site is occupied by a molecule of solvent or a polymer segment. The combinatorial contributions to the thermodynamic mixing functions are calculated from the number of possible arrangements of the polymer molecules and solvent in the lattice. These combinatorial contributions correspond to the entropy of mixing. The combinatorial contributions of Flory-Huggins [12] model implicitly state that the mixing volume and the enthalpy of mixing are zero. The number of possible molecular arrangments leads to the well-known Flory-Huggins expression for the entropy of mixing [34]. The Flory-Huggins theory and its variations have been successful in correlating and/or predicting the UCST behaviour and loop phase behavior. Variations of this theory include making the interaction parameter of the enthalpy of mixing dependent on composition and/or temperature. In this context, the works of Cheluget *et al.* [73] and Bae *et al.* [74] may be cited. The UCST behavior, i.e. transition from two-phase to one phase, takes place as a result of energetic effects. The loop behavior normally occurs when specific interactions such as hydrogen bonds take place. Compressibility of a polymeric solution is not the key issue therefore it can be modeled by an incompressible lattice theory. As mentioned above, however, polymeric solutions also exhibit LCST behavior which occurs when polymer and solvent molecules experience different volumetric expansions. For these systemsa theory that takes into account the effects of compressibility is required. Significant work has been done to extend the Flory-Huggins theory for such systems by the inclusion of vacant sites (holes) in the lattice, which may vary to enable compressible lattice representation. Within this approach the models of Kleintjens and Koningveld [14] and Panayiotou and Vera [15] may be cited but the Sanchez and Lacombe model [13] should be highlighted given its wide application in polymeric systems.

3.1. Sanchez and Lacombe (SL) equation

The lattice fluid theory for liquid and gaseous mixtures developed by Sanchez and Lacombe [13], [75] is formally similar to the Flory-Huggins theory. However, the essential and important difference is that the Sanchez and Lacombe theory introduces holes to account for variations in compressibility and density, i.e. the mixture density may vary by increasing the fraction of holes in the lattice. The Sanchez and Lacombe equation uses a random mixing expression for the attractive energy term. Random mixture means that the composition everywhere in the solution equals the total composition, i.e. there are no effects of local composition. The energy of the lattice depends only on nearest neighbors interactions. For a pure component the only non-zero interaction energy corresponds to mer-mer pair interaction. The interaction energies of types mer-hole and hole-hole are zero. The Sanchez and Lacombe equation assumes a random mixture of holes and mers. Therefore, the number of mer-mer nearest neighbors is proportional to the probability of finding two neighboring mers in the system. The Sanchez and Lacombe EOS [13] is given by:

$$\tilde{\rho}^2 + \tilde{P} + \tilde{T} \cdot \left[\ln\left(1 - \tilde{\rho}\right) + \left(1 - \frac{1}{r}\right) \cdot \tilde{\rho} \right] = 0 \tag{11}$$

where:

$$\tilde{T} = \frac{T}{T^*} \quad \tilde{P} = \frac{P}{P^*} \quad \tilde{\rho} = \frac{\rho}{\rho^*} \tag{12}$$

$$T^* = \frac{\varepsilon^*}{k} \quad P^* = \frac{\varepsilon^*}{v^*} \quad \rho^* = \frac{MW}{r \cdot v^*} \tag{13}$$

and T is the absolute temperature, P is the pressure, ρ is the density, MW is the molecular weight, k is the Boltzmann constant, and r, ε^*, and v^* are pure component parameters related to the corresponding scale factors T^*, P^* and ρ^*, respectively. These scale factors are independent of the molecular size of the polymer. For mixtures, the model parameters become composition dependent through the following mixing rules:

$$v_{mix}^* = \sum_i \sum_j \phi_i \cdot \phi_j \cdot v_{ij}^* \tag{14}$$

$$\varepsilon_{mix}^* = \frac{1}{v_{mix}^*} \cdot \sum_i \sum_j \phi_i \cdot \phi_j \cdot \varepsilon_{ij}^* \cdot v_{ij}^* \tag{15}$$

$$\frac{1}{r_{mix}} = \sum_j \frac{\phi_j}{r_j} \tag{16}$$

where the segment fraction of component i, ϕ_i, is calculated as a function of the weight fraction w_i, given by:

$$\phi_i = \frac{\dfrac{w_i}{\rho_i^* \cdot v_i^*}}{\sum_j \left(\dfrac{w_j}{\rho_j^* \cdot v_j^*} \right)} \tag{17}$$

The cross parameters are:

$$v_{ij}^* = \frac{1}{2} \cdot \left[v_{ii}^* + v_{jj}^* \right] \cdot \left(1 - l_{ij}\right) \tag{18}$$

$$\varepsilon_{ij}^* = \sqrt{\varepsilon_{ii}^* \cdot \varepsilon_{jj}^*} \cdot \left(1 - k_{ij}\right) \tag{19}$$

where l_{ij} and k_{ij} are binary interaction parameters.

Thus, the Sanchez and Lacombe equation obtains the PVT properties of pure component assuming that it is broken into parts or mers which are placed on a lattice and can interact with intermolecular potential. In order to calculate the density of the system correctly, an appropriate number of holes are also placed at specific sites in the lattice. In principle, this equation of state is appropriate to describe the thermodynamic properties of fluids in a wide range of conditions, from normal liquid or gaseous state to supercritical fluid at high temperatures and pressures. A real fluid is characterized by three molecular parameters or by three equation of state parameters, which must be known if the equation of state is to be used. In fact these parameters can be determined through any configurational thermodynamic property obtained experimentally. Vapor pressure data though, are particularly useful for solvents because they are readily available for a wide variety of fluids. For polymers, these characteristic parameters can be estimated by experimental data of the liquid density over a wide range of pressures and temperatures, using for example a numerical procedure based on non-linear least squares. When few PVT data are available, the parameters can be estimated from experimental values of density, thermal expansion coefficient and compressibility factor at ambient temperature and pressure.

Gauter and Heidemann [18] proposed a procedure to obtain parameters of the pure solvent from the critical temperature, critical pressure and acentric factor, as usually done with cubic equations of state. The polymer parameters were determined through PVT data regression.

Gauter and Heidemann [76] suggested that the polymer's parameters can be adjusted to simultaneously reproduce cloud-point data of polymer-solvent equilibrium and PVT data. They managed to obtain parameters for the Sanchez and Lacombe equation for polyethylene that could be applied for different samples, regardless of molecular weight and molecular weight distribution. The degree of branching and/or the presence of comonomers may also influence the parameters of the polymer.

3.2. Modeling polymeric systems using the Sanchez and Lacombe equation

Although there are few references in the literature for vapor-liquid equilibrium (e.g. [3], [77]) and one using industrial plant data [65], a large number of SL EOS evaluations have been reported in the literature regarding liquid-liquid equilibrium.

Kiran *et al.* [16] evaluated the efficiency of the Sanchez and Lacombe equation in predicting the high pressure phase behavior of varying molecular weight (16400, 108000 and 420000) polyethylene solutions in n-pentane and in binary solvents [(n-pentane)-(carbon dioxide)]. It was shown that concentration and pressure variations are correctly predicted if the characteristic temperature of the polymer is suitable adjusted using data from a sample with a specific molecular weight. The model also correctly predicts the behavior shift (solvent dependent) from LCST to UCST as the amount of carbon dioxide in solvent [(n-pentane)-(carbon dioxide)] increases.

Xiong and Kiran [78] modeled ternary systems of [polyethylene-(n-pentane)-(carbon dioxide)] using the Sanchez and Lacombe equation. Phase diagrams were generated for pressures up to 300 MPa and temperatures up to 460 K. The results show that the system can exhibit two or three phases depending on the pressure. At a given temperature, the three phase region disappears with increasing pressure. Depending on the pressure, the calculations also predict the displacements observed experimentally from LSCT to UCST, which are illustrated in ternary diagrams as displacements of the phase boundaries with temperature. It was shown that for polymer samples with high molecular weight, ternary calculations can be simplified by assuming that the polymer-poor phase is essentially free of polymer. Xiong and Kiran [79]-[80] investigated polyethylene binary systems with n-butane, n-pentane and CO_2.

Koak and Heidemann [17] studied the phase behavior of polymer-solvent systems under conditions close to the vapor pressure curve of the solvent where the vapor-liquid-liquid equilibrium can occur. Experimental data of High Density Polyethylene (HDPE) in n-hexane were modeled using the following equations: Sanchez and Lacombe, Kleintjens and Koningsfeld [14] and the Perturbed Hard-Sphere-Chain (PHSC) [81]. The phenomena of interest include the LCST behavior and the liquid solvent, vapor solvent and polymer three-phase equilibrium. All the three models examined provided a reasonable representation of the cloud-point for the system HDPE and n-hexane along the three-phase line in the conditions investigated. Phoenix and Heidemann [82] used the SL EOS to develop an algorithm to determine the cloud and shadow point curves of polydisperse polymer/solvent systems using continuous thermodynamics to represent the polymer.

Wang *et al.* [83] compared the performance of the Group Contribution Lattice Fluid equation of state (GCLF) with the Sanchez and Lacombe equation for liquid-liquid equilibrium data in polymer-solvent systems. The authors showed that both equations of state are able to predict the UCST and LCST behaviors, simultaneous or otherwise, as well as the hourglass shape behavior in which there is no LCST or UCST. The systems studied were: (acetic acid)-dodecane, polyisobutylene-(n-pentane), polyethylene-(n-hexane), polystyrene-(n-hexane), polyisobutylene-(n-pentane) and polystyrene-acetone. In all cases the GCLF equation performed better than the SL equation. The GCLF equation showed good sensitivity for the polymer molecular weight, but failed to correctly describe the sensitivity regarding the pressure. The best performance of the GCLF equation was attributed to the simultaneous use of the saturated steam and liquid properties in the regression of group parameters for the equation.

The applicability of equations of state for the modeling and simulation of phase equilibria in polymer production processes is investigated by Orbey *et al.* [57]. A two-stage flash separation of unreacted ethylene from polyethylene, which mimics the separation process in the production of LDPE, is used as a prototype for the simulation, where three equations of state (SAFT, SL and SRK-MHV1-FH) are compared when correlating volumetric, calorimetric and vapor-liquid phase equilibrium properties for ethylene and LDPE. Each equation of state has some unique characteristics that influence the modeling results of the pure components as well as the mixtures. When extended to binary mixtures of ethylene with polyethylene, the results show that the three equations can satisfactorily fit the data, although the best results are obtained with the SRK equation. As expected, all models exhibit less satisfactory results when no binary parameter is fitted to the data. For SRK and SAFT equations, only one binary parameter significantly affected the model performance. On the other hand, the authors observed that in the SL equation the second binary interaction parameter can also make a significant difference.

Koak *et al.* [84] studied the high pressure phase behavior of some industrially important polymer systems: polyethylene-ethylene and polybutene-(1-butene). New experimental data were presented for the system polybutene-(1-butene) in the pressure range from (9 to 17) MPa and in the temperature range from (405 to 447) K. The range of polymer concentration, expressed as polymer weight percent, ranged from 0.31 to 16.65. The system showed LCST behavior. Data from polybutene-(1-butene) and polyethylene-ethylene mixtures, presented by de Loos *et al.* [85], were modeled using the SAFT and the SL equations. The phase equilibrium calculations were carried out for two scenarios: i) the polymer is considered monodisperse ii) the polymer polydispersity is taken into account, characterizing the polymer through pseudo-components. The polymer polydispersity has a significant effect on the phase behavior of the system under investigation. The results show that, although the model is suitable for polyethylene-ethylene, interaction parameters are required, which depend on the system and on the temperature. Additionally, polymer polydispersity has a significant effect on the phase behavior of the mixture, even for reasonably monodisperse samples. The modeling effort for the system polybutene-(1-butene) showed that, if the models are used in their standard forms, alternative strategies are needed to estimate the polymer parameters so that correlation and/or reasonable prediction of the phase behavior of the polymer solution can be obtained.

In order to verify if a single set of parameters can be used to obtain useful correlations for different polyethylene resins with different solvents, Gauter and Heidemann [76] used the Sanchez and Lacombe equation to model the cloud point isotherms for two systems of ethylene and polyethylene and a system of polyethylene in n-hexane. The three polyethylene samples examined differ considerably in average molecular weight and polydispersity. The polymer parameters were obtained by adjusting volumetric data of pure polyethylene, using an additional volume displacement coefficient. The results showed that the cloud point behavior of the polymer-solvent equilibrium for a variety of polymers and solvents can be correlated with the same set of polymer parameters. The required interaction parameters are relatively small in magnitude. Unfortunately, the calculated results are extremely sensitive to these numbers, even to the third decimal place. In addition, the temperature dependence, although slight, is essential to obtain a reasonable data fit.

Trumpi *et al.* [86] measured cloud point data for a binary system of monodisperse LDPE and ethylene. The cloud points were measured between (395 and 440) K and pressures up to 175 MPa. The experimental data were modeled with the SL equation. The LDPE parameters were obtained from a sequence of non-linear regression analysis based on experimental data for both cloud point and PVT data for polyethylene melt. The results show that the SL equation fits the experimental data well for a wide range of temperatures, pressures and compositions. For diluted mass fractions, on the other hand, the data fit is less accurate. The experimental uncertainty in this region is higher than for polymer-rich mixtures, however, it is smaller than the deviations between model prediction and experimental data. The difference in cloud point pressures between calculated and experimental data increases at the lowest polymer mass fractions.

Krenz *et al.* [19] used the technique described by Trumpi *et al.* [86] to adjust the SL parameters to fit both polyethylene-solvent cloud points and polyethylene density data. The molar mass distribution of the various polyethylene samples were represented by a number of pseudocomponents ranging from 7 to 16. When correlating the cloud points of polyethylene in a variety of solvents, it seemed that there was a unique set of polyethylene parameters that would work for all mixtures. The polyethylene energy parameters, ε_i, could also be adjusted to fit the critical point of a polyethylene-solvent mixture and still provide an accurate representation of the cloud points. The polyethylene parameters, derived from fitting the critical point of the mixture, were more consistent than those found from the cloud points alone [87].

The effect of using different molar mass distributions to represent the same polymer on the HDPE-ethylene cloud points, was examined by Krenz *et al.* [88]. The HDPE parameters were taken from Krenz *et al.* [19]. Log-normal and gamma distributions approximated by nine pseudocomponents were used to match the reported average molar masses. The amount of branching was not known for these HDPE samples but it was believed to be reflected in the different HDPE-ethylene binary interaction parameters [87].

Cloud points for three hydrogenated PolyButaDiene (hPBD)-(n-hexane) mixtures were calculated using the SL equation by Schnell *et al.* [89] and compared to experimental measurements. The SL parameters were directly regressed from pure component PVT data and the hPBD samples were assumed to be monodisperse. The cloud point calculations are predictive because no BIP was used.

Correlation and prediction of miscibility involving binary blends of a variety of homopolymers [polypropylene, polybutadiene, polyisoprene, poly(methyl methacrylate), polystyrene, among others] were investigated by Voutsas *et al.* [90] considering three models: EFV-UNIFAC [91], PR and SL. The performances were evaluated in terms of i) liquid-liquid equilibrium correlation in polymer blends ii) prediction of the effect of polymer molecular weight by using interaction parameters obtained from a pair of molecular weights iii) prediction of the effect of the system pressure on miscibility using interaction parameters obtained from miscibility data at low pressures. All the experimental data used correspond to those of monodisperse polymers. Satisfactory correlation results were obtained with all models but their quality depended on whether the interaction parameters were temperature dependent or not. A satisfactory prediction of the effect of polymer molecular weight on the blend was obtained only with the EFV-UNIFAC model and the SL equation. The SL model showed the best performance and also successfully predicted the effect of pressure on the solution critical temperature, albeit with a poorer prediction of the composition at this temperature.

Chen *et al.* [92] set out to measure important phase equilibria for the industrial production of LLDPE (Linear LDPE) using metallocene catalyst technology. The phase equilibria for (n-hexane)-polyethylene and ethylene-(n-hexane)-polyethylene mixtures were measured from (373.2 to 473.2) K at pressures up to 20 MPa. Approximate monodisperse polymers and their mixtures were used to investigate the effect of polymer molecular weight on phase behavior. All the systems exhibit liquid-liquid equilibrium with UCST. The SL equation was used to correlate the phase behavior of these systems, and the effect of adding supercritical ethylene provided quantitative agreement with experimental equilibrium data. The Hosemann-Schramek function [33] provided a suitable characterization of the molecular weight distribution used in some calculations.

Nagy *et al.* [93] measured cloud point, bubble point and liquid-liquid-vapor bubble point data for binary LLDPE-(n-hexane) and ternary LLDPE-(n-hexane)-ethylene systems. Experimental data were collected in the temperature range of (400 to 500) K at pressures up to 14 MPa. Experimental data of LLDPE-ethylene and LLDPE-hexane were modeled with a modified SL equation [94]. The LLDPE parameters were obtained by performing a sequence of non-linear regression analyses from PVT data of molten polyethylene and experimental cloud point data of the LLDPE-(n-hexane) and LLDPE-(ethylene) systems. From this information and from the adjustment of the SL equation for (n-hexane)-ethylene data, the phase behavior of the ternary system LLDPE-(n-hexane)-ethylene could be predicted. Using this procedure, the effect of the ethylene concentration on the cloud point pressure is slightly overestimated. Therefore, the BIP of the pair LLDPE-ethylene was fitted to the cloud point data of the LLDPE-hexane-ethylene triplet. The SL equation provided a good description of the cloud point curve and an almost quantitative prediction of the ternary bubble point and phase-boundary curves of the vapor-liquid-liquid equilibrium.

One or more polyethylene samples with varying molecular configurations can be mixed to produce a blend with different physical characteristics. Krenz and Heidemann [95] used the MSL (Modified Sanchez Lacombe) equation [96] to calculate the cloud points of a blend of two polydisperse LLDPE resins in a hydrocarbon solvent. The MSL equation is a lattice equation that can be used to calculate polydisperse polymer solutions. The considered polyethylene resins were hPBD type. The cloud points were compared with experimental data available for the systems (hPBD-1)-(hPBD-2)-(n-hexane) and (hPBD-3)-(hPBD-4)-(n-pentane). The four hPBD samples have different molecular weight distributions, although the other properties of the mixture are unknown (degree, type and frequency of branching in the polyethylene molecule). The temperature dependent BIP for LLDPE-hydrocarbon were previously fitted to binary mixture cloud points.

Kanellopoulos *et al.* [97] used the SL equation to calculate the solubility of α-olefins in polyolefins over a wide range of temperatures and pressures. The characteristic parameters of the pure components (T^*, P^*, ρ^*) were estimated using a dynamic molecular procedure: using commercial software, each selected species (i.e. penetrating molecules and the polymer chain) had its molecular architecture firstly built and its geometry optimized by minimizing the system energy. For all the binary systems investigated, just a single BIP between the penetrating molecules and the polymer chains was estimated. The binary parameter value depends on the penetrating molecule, the comonomer, the polymer crystallinity, as well as the selected experimental conditions (temperature and pressure). The calculated theoretical solubility showed excellent agreement with the experimental measurements and demonstrated the ability of the SL equation to predict the solubility of olefins in semicrystalline α-polyolefins.

Nagy *et al.* [98] measured equilibrium data at high pressure for the LLDPE-isohexane system which exhibits LCST behavior. The following measurements were performed with weight fractions of polymers ranging from 0 to 0.25, at temperatures of (380-500) K and a pressure of 12 MPa: cloud point data, bubble points and three-phase liquid-liquid-vapor bubble points. The data were modeled using the Sanchez and Lacombe equation.

4. Perturbation theory

Equations of state based on molecular structures not only provide a useful thermodynamic basis for deriving chemical potentials or fugacities (necessary for phase equilibrium simulation) but they can also help separate and quantify the effects of molecular structure and interactions on global properties and phase behavior. Examples of these effects are the molecular size and shape (e.g. chain length and chain branch), energy of association (e.g. hydrogen bonding), average field energy (e.g. dispersion and induction). Ideally, a single equation of state should incorporate all these effects [26].

Much progress has been made in the development of molecular theories of associative solutions and those containing macromolecules. The essence of this progress is the use of statistical mechanics methods, such as perturbation theory, to correlate the molecular properties with the macroscopic properties of the system under study. In perturbation models, a simple system is initially used as reference, which should characterize the essential aspects of the system and it is usually obtained using a theory with well-defined assumptions. The difference between the actual and ideal system (i.e. the reference system) is then computed using some correction terms, called perturbation terms, which are often based on semi-empirical models. The complexity and magnitude of these perturbations depend on the degree of accuracy with which the reference term, representing the ideal system, can be specified.

With this method, Beret and Prausnitz [99] used the results of Carnahan and Starling [100] for hard spheres, which can be characterized by square well potential to describe the reference state, and proposed the so-called Perturbed Hard-Chain Theory (PHCT). A further refinement, the Perturbed Asinotropic Chain Theory (PACT), was made by Vilmalchand and Donohue [101] and Vilmalchand et al. [102]. The PACT equation of state takes into account the effects of different molecular sizes, shape and intermolecular forces, including anisotropic dipole and quadrupole forces. The calculations from Vimalchand et al. [102] show that the explicit inclusion of multipolar forces can predict the properties of highly non-ideal mixtures with reasonable accuracy, without the use of binary interaction parameters. However, for pure fluids, the prediction behavior of the PACT equation of state is similar to other comparable equations of state. Kim et al. [103] developed a Simplified version of the PHCT equation (SPHCT), replacing the attractive term of the PHCT equation with a simpler theoretical expression. This simpler equation has been used in a large number of applications, including mixtures of molecules which greatly differ in size. Ikonomou and Donohue [104] derived the Associated PACT equation (APACT). This equation takes into account isotropic repulsive and attractive interactions, anisotropic interactions due to the dipole and quadrupole moments of molecules and hydrogen bonding, and it can predict the thermodynamic properties of associative pure components as well as associative multicomponent mixtures.

4.1. SAFT equation

More recently, a new model in the family of perturbation models was developed by Chapman *et al.* [24]-[25] and by Huang and Radoz [26]. This model is known as SAFT (Statistical Associating Fluid Theory) and is based on the TPT (Thermodynamic Perturbation Theory) work from Wertheim who presented a series of papers [20]-[23] in which a coherent statistical mechnical theory of associating fluids was proposed, expanding the Helmholtz free energy in a series of integrals of molecular distribution function and potential association. Here molecules are treated as different species according to the number of bonded associating sites, and separate singlet densities are defined for each possible bonding state of a molecule. Chapman *et al.* [24]-[25] derived the expression for the Helmholtz free energy of this new reference fluid and compared the results to Monte-Carlo based simulations, obtaining satisfactory results. Huang and Radoz applied the SAFT theory to a number of real pure compounds in 1990 (also known as CK-SAFT, because they applied a different dispersion term proposed by Chen and Kreglewski [5]) [26] and proposed an extension to mixtures in 1991 [27], concluding that the equation is suitable for most of the components/systems investigated.

The essence of the SAFT equation is to use a reference system which incorporates the chain length (molecular size and shape) and the molecular association, rather than the reference fluid with hard (rigid) spheres, which is much simpler. It is expected that the effects due to other types of intermolecular forces (dispersion, induction, among others) are weaker, and therefore, considered through a perturbation term. Thus, it is expected that this theory is able to describe most real fluids, including polymers and polar fluids. In the SAFT model, the molecules are interpreted as a mixture of spherical segments of equal size, interacting according to a square-well potential. In addition, two types of bonds between these spheres can occur: covalent bonds to form chains and association bonds for specific interactions [26], [25].

When developing the equation of state, it is assumed that the molecules are formed from segments of rigid spheres, according to the diagram in Figure 2. Initially, the fluid is composed only of rigid spheres of equal size, and only the effect of rigid spheres are considered. The reference fluid consists of rigid spheres forming chains (tetramers) via covalent bonds. Hydrogen bonds between terminal sites of different chains result in oligomer chains. The last step takes into account weak dispersion forces.

The equation is derived in terms of the residual Helmholtz free energy a^{res} :

$$\frac{a^{res}(T,\rho)}{R\cdot T}=\frac{a^{total}(T,\rho)}{R\cdot T}-\frac{a^{ideal}(T,\rho)}{R\cdot T} \tag{20}$$

where T is the temperature and ρ is the density of the system. The residual Helmholtz free energy a^{res} is expressed with regard to the Helmholtz free energy of an ideal gas a^{ideal} at the same T and ρ. If the Helmholtz free energy of a fluid is known, all other properties such as pressure, chemical potential, among others, can be calculated using basic thermodynamic equations.

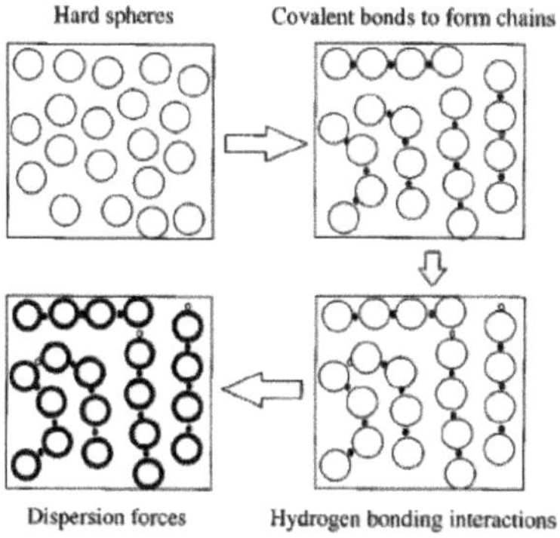

Figure 2. Illustrative Picture of Molecule Formation in the Theory Underlying the SAFT Model [105].

Formation of rigid spheres (a^{hs}) and chains (a^{chain}), as well as association (a^{assoc}) and dispersion (attraction, a^{disp}) interactions, all contribute to the residual Helmholtz free energy:

$$\frac{a^{res}(T,\rho)}{R\cdot T}=\frac{a^{hs}(T,\rho)}{R\cdot T}+\frac{a^{chain}(T,\rho)}{R\cdot T}+\frac{a^{assoc}(T,\rho)}{R\cdot T}+\frac{a^{disp}(T,\rho)}{R\cdot T} \tag{21}$$

For a pure fluid, the formation of one mol of each chain, consisting of m segments, requires m moles of hard sphere. For the hard-sphere term a^{hs}, the Carnahan-Starling [100] expression is used. The parameters of the pure components used to calculate a^{chain} are identical to those used to calculate a^{hs}. No additional parameter is required to take into account connectivity. For the dispersion term a^{disp}, Chapman *et al.* [24]-[25] used the expression originally proposed by Cotterman *et al.* [106] while Huang and Radosz [26] used a polynomial expression based on molecular dynamic simulation with square-well fluid. The contribution due to chain formation a^{chain} is given by Wertheim's association theory [20]-[23] where the association bonds are replaced by chain formation covalent bonds, as well as the association term a^{assoc}. The number of association sites in a single molecule is unlimited, however, it must be specified. Wertheim's contribution terms (chain and association) are essentially unchanged in the several versions of SAFT.

4.2. PC-SAFT equation

Gross and Sadowski [28]-[29] developed a modification to the SAFT equation referred to as PC-SAFT (Perturbed-Chain SAFT). In the structure of the PC-SAFT equation, molecules are assumed to be chains of spherical segments, freely linked and exhibiting attraction forces among them. The repulsive interactions are described by an expression of rigid chain (hard sphere + chain) developed by Chapman *et al.* [107], which is the same as used in the SAFT equation of state. The attraction interactions are in turn divided into dispersion interactions and a contribution due to association. Figure 2 illustrates the formation of a molecule according to the PC-SAFT theory. Earlier versions of SAFT assume that the dispersive interactions of molecule chains are the same as those of spherical molecules. Further investigation, however, demonstrates that equations of state may be improved when the dependence on chain length is considered in the dispersive interactions. A new version, which explicitly takes into account this dependence, has been developed, leading to the PC-SAFT equation. The dispersion term was obtained by extending Barker and Henderson's [108] theory for chain molecules. This theory considers that a chain segment is connected to neighboring segments. It also considers the effect of the nearest neighbor segments in segment interactions.

When the systems under investigation do not contain associative fluids, the term that takes into account such interactions can be ignored.

In the literature, pure component parameters for various substances can be found, either small molecules or macromolecules. The model is already available in commercial software.

4.3. Other modifications of the SAFT equation

Although the PC-SAFT equation provides excellent results when simulating polymeric systems, a brief survey of other modifications involving the original form of the SAFT equation is given in this section. Four comprehensive reviews of the development and application of the various types of SAFT have appeared recently [1], [2], [105], [109].

Instead of using the hard sphere fluid as reference, Blas and Veja [110] used the Lennard-Jones fluid, leading to the soft-SAFT equation. The chain and association terms remained similar to those in the original SAFT formulation. The soft-SAFT equation was successfully applied to pure n-alkanes, 1-alkenes, 1-alcohols and binary and ternary mixtures of n-alkanes including the critical region. In the case of mixtures, two binary parameters should be used even for mixtures of n-alkanes.

Another version of the SAFT equation is the SAFT-VR (SAFT Variable Range) equation [111]-[114]. The differences between SAFT-VR and PC-SAFT arise from the specific treatment of the attractive interactions between segments and the choice of the reference fluid. The SAFT-VR takes as reference the hard-sphere fluid, while PC-SAFT takes the hard-sphere-chain fluid. The SAFT-VR equation describes a fluid of associating molecules with the chain segments interacting through attractive forces of variable range (VR). In SAFT-VR a reference system with interacting monomers is used to build the molecule.

The Simplified PC-SAFT (sPC-SAFT) equation [30], [114], [115] is not in fact a new equation of state, rather it is a simplified version of the original PC-SAFT regarding mixing rules. Therefore, the parameters of the pure components of the original and simplified PC-SAFT are the same. The sPC-SAFT equation assumes that all segments in the mixture have the same average diameter which provides a volume fraction of the mixture which is identical to the actual mixture. This simplified version is simpler to implement and improves computational performance compared to the original PC-SAFT with negligible differences in accuracy.

4.4. Polymeric systems modeling using the SAFT equation and its modifications

This section presents a brief review of studies that used the SAFT equations of state and its modifications for the modeling of polymeric systems. A crucial aspect for the success of modeling with SAFT and PC-SAFT equations is the correct selection of the parameters of the pure components. In general, the estimation of these parameters may not be easy for macromolecular compounds [116]. Moreover, the fit to experimental data through the estimation of binary interaction parameter k_{ij} is another important point for successful modeling. In this sense some works in which the SAFT and PC-SAFT equations are used for polymer solutions will be discussed, highlighting the results achieved.

From the work of Chapman *et al.* [24]-[25] and Huang and Radosz [26] several applications of the SAFT model can be found in the literature. Table 1 presents a summary of some applications of the SAFT model for systems consisting of homopolymers and copolymers. A more detailed review of the application of SAFT model to polymeric systems can be found in [116] and [2].

Chen *et al.* [117]-[118] studied different phase transitions, from liquid to liquid-vapor (L to LV) and liquid to liquid-liquid (L to LL) in binary, ternary and quaternary systems containing the solvents ethylene, propylene, 1 -butene, 1-hexene, n-hexane and methylcyclopentane, and Poly(Ethylene-co-Propylene) (PEP). SAFT modeling was used for PEP of varying molecular weights at low and moderate pressures [(0-500) bar]. The BIP set was defined as an exponential function of the polymer molecular weight and adjusted by three parameters.

Xiong and Kiran [80] compared the performance of SAFT with SL to model cloud point curves in polyethylene systems with n-butane and n-pentane. The pure component parameters were taken from literature [26] and the BIP were assumed to be equal to zero. For all temperature-pressure-composition ranges, the SAFT model was superior to SL.

The approach used by Han *et al.* [119] was to measure the cloud point and the coexistence pressures in propylene and ethylene solutions of alternating PEP of well-controlled polydispersity directly, from monodisperse to broadly polydisperse. These experimental data were modeled using the SAFT equation. More specifically they fited the cloud point pressure for monodisperse PEP and used the model for predicting the cloud point and coexistence pressure of bimodal polydisperse PEP, without any refitting.

Pan and Radosz [120] used the SAFT equation for copolymers to describe the fluid-liquid and solid-liquid transitions in solutions of polyethylene and poly(ethylene-co-olefin-1) in propane as well as the fluid-liquid transition in solutions of polystyrene in n-hexane. The parameters of the pure solutes were estimated based solely on the molecular weight and on the structure. Copolymer SAFT EOS has been also used to model SLE in systems containing polyethylene, m-xylene and amyl acetate [121].

Polymer	Solvent	Reference
Poly(ethylene-glycol) (PEG)	Propane, nitrogen, CO_2	[125]
PolyEthylene (PE)	Ethylene	[33], [84], [126], [127], [128]
	Propane (also SLE)	[120], [129]
	n-butane, n-pentane	[80], [130]
	Toluene	[126]
	Isobutane	[129]
	CO_2	[126], [129]
	1-hexene	[127]
	Cyclohexane	[122]
	Hexane, heptane, octane	[123]
	SLE in amyl acetate and m-xylene	[121]
PolyPropylene (PP)	1-butene, n-butane, propane, propylene	[116]
PolyButylene (PB)	1-butene	[84]
PolyStyrene (PS)	Cyclohexane, metylcyclohexane, ethylbenzene, chlorobenzene, CO_2	[126]
	Propane	[131]
	Cyclohexane	[122]
	Cyclohexane, CO_2	[132], [133]
Polyisobutylene (PIB)	Ethane, ethylene, propane, propylene, dimethyl ether	[134]-[136]
PolyCarbonate (PC)	n-alkane (C_8-C_{12}), alcohol (C_3-C_{10}), benzene, toluene, o, m, p-xylene, ethylbenzene	[137]
Poly(ethylene-co-vinyl-acetate) (EVA)	vinyl acetate, ethylene, alkanes	[127], [138], [139]
Poly(vinyl-acetate) (PVA)	Benzene, Vinyl acetate	[132]
Poly(ethylene-co-olefin)	Ethylene, propilene, propane (also SLE), 1-butene, 1-hexene, n-alkane (C_6-C_8)	[120], [127], [140]
Poly-methyl methacrylate (PMMA)	CO_2–methyl methacrylate	[141]
Poly(Ethylene-co-Propylene) (PEP)	Ethylene, ethane, propylene, 1-butene, and 1-hexene, methylcyclopentane	[33], [117], [118], [119], [142]
Poly(1,1-dihydroperfluorooctylacrylate) (poly(FOA))	CO_2	[143]

Table 1. Polymeric Systems Modeled with SAFT Equation

Dariva [116] applied the SAFT equation of state to model PolyPropylene (PP)-solvent systems at low and moderate pressures. Two non-metallocene polypropylenes (molecular weights: 476745 and 244625 g/mol; and polydispersities: 4.4 and 5.0) and a metallocene polypropylene (molecular weight: 197150 g/mol; and polydispersity: 2.9) were used. As solvents, propylene, n-propane, 1-butene and n-butane were used. The author modeled transitions L to LV, L to LL and LL to LLV using the SAFT model with and without fitting the BIP. A large amount of experimental data for these systems can be found in this work.

Horst *et al.* [122] studied the influence of supercritical gases in the phase behavior of the systems polystyrene-cyclohexane-gas and polyethylene-cyclohexane-gas, modeling the experimental data with the SAFT equation of state. As supercritical gases, the authors used ethane, propane and nitrogen. The experimental data were collected at moderate pressures, and the binary interaction parameters used for the adjustment were defined as quadratic functions of temperature. Good results were obtained in the modeling, although a larger model mismatch in regions of higher polymer concentration can be obseved.

Jog *et al.* [123] used the SAFT equation to model the liquid-liquid equilibrium of LLDPE with hexane, octane and heptanes [124]. The effects of temperature, pressure, polymer concentration and molecular weight on the phase separation were successfully evaluated. The effect of polydispersity on cloud point was also considered. Although the SAFT predictions are sensitive to the binary interaction parameters, a constant value for the binary parameters was considered to model the cloud point in varying conditions (temperature, pressure and polymer concentration) and varying solvents. The SAFT equation showed a good predictive capacity for this system.

Besides the works already mentioned, the SAFT equation of state has also been applied in more recent works as a "reference" model, its performance being compared to PC-SAFT model, as will be shown below. After the work of Gross and Sadowski [28]-[29], some applications of the PC-SAFT model may be found in literature. Table 2 presents a summary of some of the applications of the PC-SAFT model for systems consisting of homopolymers and copolymers. A more detailed review of the application of PC-SAFT model to polymeric systems can be found in [2].

Tumakaka *et al.* [144] used the PC-SAFT equation to model cloud point curves for polymeric systems consisting of polyolefins, using ethane, ethene, propane, propylene, n-butane, 1-butene and CO_2 as solvent. Here, the good results obtained for modeling the systems LDPE-solvents and HDPE-ethylene at high pressures should be highlighted. As well as modeling systems consisting of polyolefins, polyethylene copolymers and PVA (polyvinyl acetate vinyl) systems were also modeled. Good results were obtained when representating a system consisting of polypropylene with moderate polydispersity (MW/MN = 2.2), assuming that the polypropylene was monodisperse. The monodisperse assumption was also considered for the LDPE-solvent system, whereas for the HDPE-ethylene system the polyethylene was modeled using pseudocomponents.

Polymer	Solvent	Reference
Polyethylene (PE)	Ethylene, ethane, propylene, propane, butane, 1-butene, hexane	[73], [152], [153], [154], [155], [159], [161]
	n-heptane	[162]
	Toluene	[126]
	CO_2	[126]
	Cyclohexane, 1-octene	[155]
	SLE in n-alkanes and in m-xylene	[158]
	SLE with a variety of solvents	[68]
Polypropylene (PP)	Propane, n-pentane, CO_2	[126], [151], [163]
	SLE with a variety of solvents	[68]
	Diisopropyl ketone	[147]
Polystyrene (PS)	Cyclohexane, CO_2, metylcyclohexane, ethylbenzene, chlorobenzene	[126], [162]
	Ethylbenzene, butyl acetate	[147]
Polyamide (PA)	Caprolactam, water	[150]
Poly(methyl acrylate) (PMA)	2-octanone	[146], [164]
Poly(ethylene-co-methacrylic acid) (EMA or PE-co-MA)	Ethylene, propylene, butane, 1-butene	[146], [160]
	Propylene, butane	[148]
Poly(ethylene-co-acrylic acid) (EAA or PE-co-AA)	Ethylene	[160]
Poly(ethylene-co-vinyl-acetate) (EVA)	Cyclopentane	[146]
	Ethylene	[148]
Poly(ethylene)-co-olefine	Ethylene, propylene, propane, 1-butene, 1-hexene, n-alkane (C_6-C_8)	[144], [146], [163]
Poly(vinyl-acetate) (PVA)	Methyl ethyl ketone, propyl acetate, 1-propylamine, 2-propylamine, 2-methyl-1-propanol, 2-propanol	[147]
Polyolefins and PVA	Ethane, ethene, propane, propylene, n-butane, 1-butene, CO_2	[144]
PE, PP, PB, PIB, PS	Ethylene, n-butane, 1-butene, n-pentane, cyclohexane	[145]
Biopolymers: poly(d,l-lactide) (PLA), poly(butylene succinate) (PBS) and poly(butylenes succinate-co-adipate) (PBSA)	Cloro difluorometane, CO_2, dimetil ether, difluiorometane, trifluorometane, tetrafluorometane	[165]
Polycarbonate (PC)	CO_2, cyclohexene oxide	[166]
Poly(dimethylsiloxane) (PDMS)	Pentane	[167]
PMMA, poly(butyl methacrylate) (PBMA), PVA, PS, PP, Polybutadiene (BR), PIB, PMA, poly(ethyl acrylate) (PEA), poly(butyl acrylate) (PBA), Polyphthalamide (PPA)	CCl_4, CH_2Cl_2, methyl acetate, methyl ethyl ketone, 1-propanol, 4-heptanone, chlorobutane, octane, cyclohexane, benzene, toluene, ethybenzene, xilene, acetone, diethylketone	[156]
PDMS, PE, PS, PBMA, PIB, PB, poly(alpha-methylstyrene) (P-αMS), PMMA, PVA	Benzene, toluene, alkane (C_5-C_8), cyclohexane, methylcyclohexane, 1-propanol, 2-propanol, 1-butanol, 2-butanol, 2-methyl-1-propanol, methyl acetate, prothyl acetate, methyl ethyl ether, acetone, propylamine, isopropylamine	[157]

Table 2. Polymeric Systems Modeled with PC-SAFT Equation

Gross and Sadowski [145] used the fractionation of LDPE in three pseudocomponents to represent the cloud point curves for the ethylene-LDPE system (MW/MN = 8.56). The pure component parameters for ethylene and polyethylene and one binary interaction parameter k_{ij} (interaction ethylene-LDPE) were estimated from the simultaneous regression of polymer density data and a single cloud point curve, requiring the optimization of four parameters (m, σ, ε/k, k_{ij}). Additionally, the modeling of other polymeric systems in varying equilibrium conditions was carried out for a wide range of temperatures [(75-197) °C], pressures [(0-2000) bar] and weight compositions [(0-100)%]. Polymeric systems consisting of LDPE, HDPE, PP, polybutene, polyisobutene and polystyrene were evaluated. As solvents, ethylene, n-butane, 1-butene, n-pentane and cyclohexane were used. Comparisons of results obtained from SAFT and PC-SAFT models corroborate that PC-SAFT shows the best performance.

Gross and Sadowski [145] proposed changes to the PC-SAFT equation, adding two more parameters concerning the association term. Simulations of liquid-liquid and vapor-liquid equilibrium of systems consisting of simple molecules were compared with the SAFT model. Slightly better results were observed for the PC-SAFT equation. The pure component parameters were obtained from the simultaneous regression of vapor pressure and liquid phase density data. Thus, a total of five parameters were optimized for each component i: segment diameter (σ_i), segment number (m_i), segment energy (ε_i/k), association energy (ε^{AiBi}/k) and effective association volume (k^{AiBi}). The BIP k_{ij} parameter for each system was optimized later.

Cheluget *et al.* [73] applied the PC-SAFT equation of state to model a flash separation system of an industrial LLDPE plant. The system under study consisted of ethylene, 1-butene, cyclohexane and polymer at 267 °C and 33 bar. The PC-SAFT parameters used in this study were estimated from binary system liquid-liquid and vapor-liquid equilibrium data obtained from literature. The authors did not re-estimate the binary interaction parameters to fit the model to the industrial data, to compare the predicted (using interaction parameters from literature) and industrial data. The lack of parameter re-estimation is the most likely cause of the significant deviations between experimental and predicted values.

Gross *et al.* [146] extended the PC-SAFT model [28]-[29] for copolymers. The authors modeled phase equilibrium for ethylene copolymer systems with random alternating chains in a wide range of compositions (including homopolymer) and with molecular weights ranging between 709 and 242000 g/mol. The studied polymers were composed of repeating apolar [poly([ethylene oxide]-co-propylene) and poly([ethylene oxide]-co-[butene-1])] and polar [poly([ethylene oxide]-co-[vinyl acetate]) and poly(ethylene-co-[methyl acrylate])] units. Additionally, the authors reported binary interaction parameters of phase equilibrium for systems consisting of homopolymers, whose repeating units are present in the copolymers, and varying solvents, and some of these interaction parameters also considered the composition of the copolymer. The BIP were estimated from equilibrium data of binary polymer-solvent systems.

Kouskoumvekaki *et al.* [147] implemented a simplified version of the PC-SAFT equation of state, developed by von Solms *et al.* [30] with little repercussion for polymeric systems made up of a variety of solvents, including polar, apolar and associative compounds. Pure component parameters were estimated from vapor pressure and liquid phase density data. The simplified model showed similar results to those obtained by the original PC-SAFT equation, thus presenting some advantages due to its simplicity and lower computational cost.

In the work of Tumakaka and Sadowski [148], the PC-SAFT equation was applied to pure polar compounds as well as to the vapor-liquid and liquid-liquid equilibrium of binary mixtures containing polar compounds, with low molecular weight, and polar copolymers. As the original PC-SAFT is unable to describe polar systems, the authors used an extended version of the equation for polar systems. The dipolar interactions, which contribute significantly to the total intermolecular forces, are explicitly explained in molecular theory [149]. Due to the inclusion of a term of polar interactions in the molecular theory, it was also necessary to include a pure component parameter in the term. When dealing with mixture modeling, the authors defined the binary interaction parameter either as an independent term or as a function of the comonomer molar fraction.

The sPC-SAFT equation (Simplified PC-SAFT equation) was applied by Kouskoumvekaki *et al.* [150] to the vapor-liquid equilibrium of binary and ternary systems of polyamide-6 with several solvents (water, caprolactam, ethyl benzene and toluene). Binary interaction parameters between polyamide-6, caprolactam and water were estimated using experimental data of the binary mixtures. The estimated parameters were used to predict and correlate the ternary mixture of polyamide-6, caprolactam and water. When optimizing the pure parameters of polyamide-6, the corresponding values of caprolactam were considered as initial estimates, and just the segment diameter needed to be adjusted using experimental data of liquid volume. The results showed that the sPC-SAFT equation is a versatile tool for modeling multi-component systems containing polyamide.

Arce and Aznar [151] modeled the systems PP-(n-pentane) and PP-(n-pentano)-CO_2 using the PC-SAFT equation of state. In this work resins of low molecular weight (MW = 50400 and 95400) at moderate pressure (below 350 bar) were considered. The PC-SAFT, Sanchez-Lacombe and Peng-Robinson models were used to predict the cloud point pressures from experimental data on each system. Although all the models were able to describe the system, the PC-SAFT equation showed superior performance. For all the models, the authors used the temperature dependent BIP.

Spyriouni and Economou [152] evaluated the performance of SAFT and PC-SAFT equations of state to describe the phase behavior of mixtures containing polydisperse polymers and copolymers at high pressure. Although there are several studies showing the application of both equations in modeling the phase behavior of polymer systems, the major contribution of this work was to compare the performance of both models for a wide variety of homopolymers and copolymers. The authors concluded that both models show a similar performance in modeling the equilibrium, however, from the data presented, the PC-SAFT model shows superior results for most systems.

Pedrosa *et al.* [153] presented phase equilibrium calculations for polyethylene solutions with varying solvents using two different versions of the SAFT equation: PC-SAFT and the soft-SAFT. The soft-SAFT equation uses the spherical fluid of Lennard-Jones as a reference, including attractive and repulsive interactions, while the reference term in the PC-SAFT equation is the rigid sphere chain. The studies carried out by Pedrosa *et al.* [153], which also dealt with vapor-liquid equilibrium, showed that results with soft-SAFT equation are slightly more accurate than those obtained with PC-SAFT in some cases.

Buchelli *et al.* [154] investigated the performance of the PC-SAFT equation of state for modeling the HPS and LPS units downstream from a low-density polyethylene tubular reactor. Plant data were used to validate the equilibrium stage model prediction for the two gas-liquid flash separators, however, the pure component parameters and BIP of this model were obtained exclusively from experimental data published in the literature. The authors achieved good agreement between the model and LPS plant data, although the predicted solubility was not in agreement with plant-measured values for the HPS.

Guerrieri [155] investigated the behavior of polymeric systems in two industrial polyethylene plants, a LDPE plant and a HDPE/LLDPE plant, using the PC-SAFT equation. The liquid-liquid equilibrium at high pressure, observed in the reactor, and the vapor-liquid equilibrium, observed in the low-pressure separator, were investigated in the LDPE plant. For this study, 8 commercial resins were considered. In the HDPE/LLDPE plant, the vapor-liquid equilibrium in the intermediate pressure separator was investigated. Here, 25 commercial resins were investigated. The experimental data were taken from measurements and mass/energy balances available in both plants, and the modeling of binary and multicomponent systems consisting of ethylene, ethane, propylene, propane, 1-butene, cyclohexane, 1-octene and polymer was carried out.

Tihic *et al.* [156]-[157] developed a group contribution method to be used in PC-SAFT equation to predict their pure parameters. If pure polymer parameters in SAFT-type equations are obtained only from density data, poor predictions of phase equilibrium may result. Therefore, the group contribution method for parameter estimation was developed through the adjustment of vapor pressure and density data based on a database of 400 components of low molecular weight. The data required to calculate the phase equilibrium for polymers using this contribution method are the polymer molecular structure in terms of functional groups and a single interaction parameter for accurate mixture calculations.

Understanding the phase behavior of polymer solutions is of great theoretical and practical importance. Some work has also been done on the development of algorithms for real-time prediction of SLE in solution polymerization of polyethylene based on PC-SAFT EOS and to study the effects of monomer and polymer polydispersity in solution polymerization processes [158]. Costa *et al.* [68] also modeled the SLE in polyolefins (polyethylene and polypropylene) solutions using PC-SAFT EOS for a variety of different polymer-solvent systems at atmospheric and high pressure with very good results. Pressure *versus* temperature (P-T) isopleths can be used to determine the number of phases present at a given T, P, and overall mixture composition. The PC-SAFT EOS was applied by Costa *et al.*

[159] to simulate the curves that describe the borderlines between several distinct regions depicted in *P-T* isopleths for polyethylene solutions. A new strategy was used and the simulation results show good agreement with experimental cloud point isopleths data from the literature. In order to track the operational performance of industrial reaction systems safely, a strategy to calculate the distance between a given operational point (specified through a given pressure and a given temperature) and the corresponding point in the interface, for a fixed molecular weight and a fixed polymer fraction weight, has been developed which could also be extended for real-time prediction applications.

Kleiner *et al.* [160] extended the association term of the PC-SAFT EOS to account for the polydispersity of the copolymer samples. This EOS was used to model cloud-point curves of the systems poly(ethene-co-acrylic acid)-ethene and poly(ethene-co-methacrylic acid)-ethene. Both copolymer composition and molecular weight distribution were varied. To account for polydispersity the concept of pseudocomponents has been applied and they were generated such as to match the molecular weight distribution. An algorithm has been developed for calculating phase equilibria of polydisperse associating copolymer-solvent systems. The PC-SAFT approach turns out to be capable of adequately modeling and even predicting the phase behavior of the polydisperse polymeric systems by using two pseudocomponents for each copolymer, but no additional adjustable parameters.

Phase-dependent BIP were computed by Costa *et al.* [161] with PC-SAFT EOS by correlating the flash results for both high and low pressure separators (HPS and LPS) for the industrially significant mixture ethylene-ethane-propane-propylene-LDPE. HPS and LPS data were correlated for five of eight LDPE resins. A pressure, composition and molar mass dependent binary interaction parameter model was proposed for both the vapour and liquid phase. The resulting model was able to provide a good representation of the experimental data. The polydispersity and branching of the LDPE resins, as well as the temperature, were lumped into the BIP. Clearly, the proposed phase-dependent BIP model provides a good representation of the phase behaviour in two industrial separators for very complex polydisperse mixtures.

5. Conclusions

This chapter has presented a review of equation of state models for polymers, demonstrating their increasing evolution in performance for describing phase equilibria in polymer systems. The development of EOS for polymers remains a very active area of research and it is difficult to recommend a specific EOS [2].

In general the equations of state using G^E models are unable to describe high-pressure phase equilibria with the desired quality [65], thus have a more restricted application. On the other hand, their equations and mixing rules are simple, which facilitates their convergence and the obtaining of terms required for the calculation of other thermodynamic properties, e.g. mixture specific heat.

Regardless of the model used, the calculation of the parameters of the pure components for polymers is a major challenge. When using an equation of state, the need for other applications besides the description of phase equilibria and PVT data, such as the calculation of Joule-Thompson coefficient and general energy balances should be borne in mind. Despite its importance, little research has been done in this respect with polymer systems. Note that the simpler the equation and its mixing rule, the easier it will be to obtain these properties and other important thermodynamic properties.

Although the PC-SAFT equation seems to show certain superiority in performance when compared to other models, no agreement was observed in the reviewed literature on which concept and/or thermodynamic model structure is better. Thus the choice of the model remains dependent on the system and its conditions. Although many studies address polymeric systems with a high concentration of polymers in a wide range of molecular weights, very few studies can be found that model oligomers (low molecular weight polymers), despite being a relevant problem as in many cases the quality of its measurement/prediction is critical for the proper functioning of important process equipment.

Author details

Yuri Guerrieri, Karen Valverde Pontes, Gloria Meyberg Nunes Costa and Marcelo Embiruçu
Programa de Engenharia Industrial (PEI) - Escola Politécnica –
Universidade Federal da Bahia, Salvador-BA, Brazil

Acknowledgement

The authors would like to thank CAPES (Coordenação de Aperfeiçoamento de Pessoal de Nível Superior), CNPq (Conselho Nacional de Desenvolvimento Científico e Tecnológico) and FAPESB (Fundação de Amaro à Pesquisa do Estado da Bahia) for financial support.

6. References

[1] Wei Y S, Sadus R J (2000) Equations of State for the Calculation of Fluid-Phase Equilibria. AIChE J. 6: 169-196.

[2] Kontogeorgis G M, Folas G K (2010) Thermodynamics Models for Industrial Applications. From Classical and Advanced Mixing Rules to Asssociation Theories. Wiley.

[3] Orbey H, Bokis C P, Chen C-C (1998) Equation of State Modeling of Phase Equilibrium in the Low-Density Polyethylene Process: The Sanchez-Lacombe, Statistical Associating Fluid Theory and Polymer-Soave-Redlich-Kwong Equation of State. Ind. Eng. Chem. Res. 37: 4481-4491.

[4] Valderrama J O (2003) The State of the Cubic Equations of State. Ind Eng. Chem. Res. 42: 1603-1618.

[5] Chen S S, Kreglewski A (1977) Applications of the Augmented van der Waals Theory of Fluids. I. Pure Fluids. Bunsen-Ges. Phys. Chem. 81: 1048-1052.
[6] Huron M J, Vidal J (1979) New Mixing Rules in Simple Equation of State for Representing Vapour-Liquid Equilibria of Strongly Nonideal Mixture. Fluid Phase Equilib. 3: 255-271.
[7] Sako T, Wu A H, Prausnitz J M (1989) A Cubic Equation of State for the High-Pressure Phase Equilibria of Mixtures Containing Polymers and Volatile Fluids. J. Appl. Polym. Sci. 38: 1839-1858.
[8] Wong D S H, Sandler S I (1992) A Theoretically Correct Mixing Rule for Cubic Equation of State. AIChE J. 38: 671-680.
[9] Kontogeorgis G M, Harismiadis V I, Fredeslund A, Tassios D P (1994a) Application of the van der Waals Equation of State to Polymers I.Correlation. Fluid Phase Equilib. 96: 65-92.
[10] Boukouvalas C, Spiliots N, Coutsikos P, Tzouvaras N, Tassios D (1994) Prediciton of Vapor-Liquid Equilibrium with the LCVM Model: a Linear Combination of the Vidal and Michelsen Mixing Rules Coupled with the Original UNIFAC and the t-mPR Equation of State. Fluid Phase Equilib. 92: 75-106.
[11] Zhong C, Masuoka H (1996a) A New Mixing Rule for Cubic Equation of State and its Application to Vapor-Liquid Equilibria of Polymer Solutions. Fluid Phase Equilib. 123: 59-69.
[12] Flory P J (1965) Statistical Thermodynamics of Liquid Mixtures. J. Amer. Chem. Soc. 87: 1833-1838.
[13] Sanchez I C, Lacombe R H (1978) Statistical Thermodynamics of Polymer Solutions. Macromolecules 11: 1145-1156.
[14] Kleintjens L A, Koningmeld R (1980) Liquid-Liquid Phase Separation in Multicomponent Polymer Systems XIX. Mean-Field Lattice-Gas Treatment of the System n-Alkane/Linear –Polyethylene. Colloid Polym. Sci. 258: 711-718.
[15] Panayiotou C, Vera J H (1982) Statistical Thermodynamics of r-Mer Fluids and Their Mixtures. Polym. J. 14: 681-694.
[16] Kiran E, Xiong Y, Zhuang W (1993) Modeling Polyethylene Solutions in Near and Supercritical Fluids Using the Sanchez Lacombe Model. J. Supercrit. Fluids 6: 193-203.
[17] Koak N, Heidemann R A (1996) Polymer-Solvent Phase Behavior Near the Solvent Vapor Pressure. Ind. Eng. Chem. Res. 35: 4301-4309.
[18] Gauter K, Heidemann R A (2000) A Proposal for Parametrizing the Sanchez-Lacombe Equation of State. Ind. Eng. Chem. Res. 39: 1115-1117.
[19] Krenz R A, Heidemann R A, de Loos Th W (2003) Correlation of the Cloud Point Behavior in Polyethylene & Hydrocarbon Systems, AIChE Spring National Meeting, New Orlenas, LA, U.S.A., Mar-30-Apr.3.
[20] Wertheim M S (1984a) Fluids with Highly Directional Attractive Forces. I Statistical Thermodynamics. J. Stat. Phys. 35: 19-34.
[21] Wertheim M.S (1984b) Fluids with Highly Directional Attractive Forces. II Thermodynamic Perturbation Theory And Integral Equations. J. Stat. Phys. 35: 35-74.

[22] Wertheim M.S (1986a) Fluids with Highly Directional Attractive Forces. III Multiple Attraction Sites. J. Stat. Phys. 42: 459-476.

[23] Wertheim M.S (1986b) Fluids with Highly Directional Attractive Forces. IV Equilibrium Polimerization. J. Stat. Phys. 42: 477-492.

[24] Chapman W G, Gubbins K E, Jackson G, Radosz M (1989) SAFT: Equation-of-State Solution Model for Associating Fluids. Fluid Phase Equilib. 52: 31-38.

[25] Chapman W G, Gubbins K E, Jackson G, Radosz M (1990) New Reference Equation of State for Associating Fluids. Mol. Phys. 29: 1709-1721.

[26] Huang S H, Radosz M (1990) Equation of State for Small, Large, Polydisperse, and Associating Molecules. Ind. Eng. Chem. Res. 29: 2284-2294.

[27] Huang S H, Radosz M (1990) Equation of State for Small, Large, Polydisperse, and Associating Molecules: extension to fluid mixtures. Ind. Eng. Chem. Res. 30: 1994–2005.

[28] Gross J, Sadowski G (2000) Application of Perturbed Theory to Hard-Chain Reference Fluid: An Equation of State for Square-Well Chains. Fluid Phase Equilib. 168: 183-199.

[29] Gross J, Sadowski G (2001) Perturbed-Chain SAFT: An Equation of State Based on a Perturbed Theory for Chain Molecules. Ind. Eng. Chem. Res. 40: 1244-1260.

[30] von Solms N, Michelsen M L, Kontogeorgis G M (2003) Computational and Physical Performance of a Modified PC-SAFT Equation of State for Highly Assimmetric and Associating Mixtures. Ind. Eng. Chem. Res. 42: 1098-1105.

[31] Soave G (1972) Equilibrium Constants from a Modifed Redlich-Kwong Equation of State. Chem. Eng. Sci. 27: 1197-1203.

[32] Peng D Y, Robinson D B (1976) A New Two-Constant Equation of State. Ind. Eng. Chem. Fundam. 15: 59-64.

[33] Tork T, Sadowski G, Arlt W, Haan A, Krooshof G (1999) Modeling of High-Pressure Phase Equilibria Using the Sako-Wu-Prausnitz Equation of State II. Vapor-Liquid Equilibria and Liquid-Liquid Equilibria in Polyolefin Systems. Fluid Phase Equilib. 163: 79-98.

[34] Prausnitz J M, Lichtenthaler R N, Azevedo E G (1999) Molecular Thermodynamics of Fluid-Phase Equilibria. Prentice Hall International Series, Third Edition.

[35] Adachi Y, Sugie H (1986) A New Mixing Rule-Modified Conventional Mixing Rule. Fluid Phase Equilib. 28: 103-118.

[36] Panagiotopoulos A Z, Reid R C (1986) New Mixing Rule for Cubic Equation of State for Highly Polar Asymmetric Systems, Equation of State - Theories and Applications, K. C. Chao and R. L. Robinson Jr., eds. ACS Symp. Ser. 300: 571-582.

[37] Schwartzentruber J, Renon H (1989) Development of a New Cubic Equation of State for Phase Equilibrium Calculations. Fluid Phase Equilib. 52: 127-134.

[38] Orbey N, Sandler S.I (1994) Vapour-Liquid Equilibrium of Polymer Solutions Using a Cubic Equation of State. AIChE J. 40: 1203-1209.

[39] Soave G (1984) Improvement of the Van der Waals Equation of State. Chem. Eng. Sci. 39: 357-369.

[40] Adachi Y, Sugie H (1985) Effects of Mixing Rules on Phase Equilibrium Calculations. Fluid Phase Equilib. 24: 353-362.

[41] Gupte P A, Daubert T E (1986) Extension of UNIFAC to High Pressure VLE Using Vidal Mixing Rules. Fluid Phase Equilib. 28: 155-170.

[42] Heidemann R A, Rizvi S S H (1986) Correlation of Ammonia-Water Equilibrium Data with Various Peng-Robinson Equation of State. Fluid Phase Equilib. 29: 439-446.

[43] Lermite C, Vidal J (1992) A Group Contribution Equation of State for Polar and Non-Polar Compounds. Fluid Phase Equilib. 72, 111-130.

[44] Soave G, Bertuco A, Vecchiato L (1994) Equation of State Group Contribution from Infinite-Dilution Activity Coefficients. Ind. Eng. Chem. Res. 33: 975-980.

[45] Mollerup J (1986) A Note on the Derivation of Mixing Rules from Excess Gibbs Energy Models. Fluid Phase Equilib. 25: 323-327.

[46] Heidemann R A, Kokal S L (1990) Combined Excess Free Energy Models and Equation of State. Fluid Phase Equilib. 56: 17-37.

[47] Michelsen M (1990) A Method for Incorporating Excess Gibbs Energy Models in Equations of State. Fluid Phase Equilib. 60: 47-58.

[48] Michelsen M (1990) Modified Huron-Vidal Mixing Rule for Cubic Equations of State. Fluid Phase Equilib. 60: 213-219.

[49] Dahl S, Michelsen M (1990) High-Pressure Vapor-Liquid Equilibrium with a UNIFAC-Based Equation of State. AIChE J. 36: 1829-1836.

[50] Zhong C, Masuoka H (1996b) Mixing Rules for Accurate Prediction of Vapor Liquid Equilibria of Gas/Large Alkane Systems Using SRK Equation of State Combined With UNIFAC. J. Chem. Eng. Jpn. 29: 315-322.

[51] Zhong C, Masuoka H (1996c) Prediction of Henry's Constants for Polymer-Containing Systems Using the SRK Equation of State Coupled with a New Modified UNIFAC Model. Fluid Phase Equilib. 126: 1-12.

[52] Stryjek R, Vera J H (1986) PRSV2: A Cubic Equation of State for Accurate Vapor-Liquid Equilibria Calculations. Can. J. Chem. Eng. 64: 323-333.

[53] Ahlers J, Gmehling J (2002) Development of a Universal Group Contribution Equation of State. 2. Prediction of Vapor-Liquid Equilibia for Asymmetric Systems. Ind. Eng. Chem. Res. 41: 3489-3498.

[54] Fredenslund Aa, Gmehling J, Rasmussen P (1977) Vapor-Liquid Equilibria Using UNIFAC a Group-Contribution Method, Elsevier Scientific Publishing Company , New York.

[55] Voutsas E, Magoulas K, Tassios D (2004) Universal Mixing Rule for Cubic Equations of State Applicable to Symmetric and Asymmetric Systems: Results with the Peng-Robinson Equation of State, Ind. Eng. Che. Res. 43: 6238-6246.

[56] Magoulas K, Tassios D (1990) Thermophysical Properties of n-Alkanes from C_1 to C_{20} and Their Prediction for Higher Ones. Fluid Phase Equilib. 56: 119-140.

[57] Orbey H, Bokis C P, Chen C (1998) Poymer-Solvent Vapor-Liquid Equilibrium: Equations of State versus Activity Coefficient Models. Ind. Eng. Chem. Res. 37: 1567-1573.

[58] Tochigi K (1998) Prediciton of Vapor-Liquid Equilibria in Non-Polymer and Polymer Solutions Using an ASOG-Based Equation of State (PRASOG). Fluid Phase Equilib. 144: 59-68.

[59] Tochigi K, Futakuchi H, Kojima K (1998) Prediction of Vapor-Liquid Equilibrium in Polymer Solutions Using a Peng-Robinson Group Contribution Model. Fluid Phase Equilib. 152: 209-217.
[60] Kang J W, Lee J H, Yoo K, Lee C S (2002) Evaluation of Equations of State Applicable to Polymers and Complex Systems. Fluid Phase Equilib. 194-197: 77-86.
[61] Louli V, Tassios D (2000) Vapor –Liquid Equilibrium in Polymer-Solvent Systems with a Cubic Equation of State. Fluid Phase Equilib. 168: 165-182.
[62] Haghtalab A, Espanani R (2004) A New Model and Extension of Wong-Sandler Mixing Rule for Prediciton of (Vapour+Liquid) Equilibrium of Polymer Solutins using EOS/GE. J. Chem. Thermodyn. 36: 901-910.
[63] Voutsas E, Louli V, Boukouvalas C, Magoulas K, Tassios D (2006) Thermodynamic Property Calculations with the Universal Mixing Rule for EOS/GE Models: Results with the Peng-Robinson EOS and a UNIFAC model. Fluid Phase Equilib. 241: 216-228.
[64] Hansen H K, Goto B, Kuhlmann B. (1992) UNIFAC with Linearly Temperature-Dependent Group-Interaction Parameters. Institut for Kemiteknik, DTH, Lingby, Denmark, SEP 9212, 1992.
[65] Costa G M N, Guerrieri Y, Kislansky S, Pessoa F L P, Vieira de Melo S A B, Embiruçu M (2009b) Simulation of Flash Separation in Polyethylene Industrial Processing: Comparison of SRK and SL Equations of State. Industrial Eng. Chem. Research 48: 8613-8628.
[66] Costa G M N, Guerrieri Y, Kislansky S.,Vieira de Melo S A B, Embiruçu M, Pessoa F L P (2009) Computational Aspects for Optimization of High Pressure Phase Equilibrium for Polymer Industrial Systems. Computer-Aided Chem. Eng. 27: 405-410.
[67] Bogdanic G, Vidal J A (2000) Segmental Interaction Model for Liquid–liquid Equilibrium Calculations for Polymer Solutions. Fluid Phase Equilib. 173: 241-252.
[68] Costa G M N, Kislansky S, Oliveira L C, Pessoa F L P, Vieira de Melo S A B, Embiruçu M (2011) Modeling Solid-Liquid Equilibrium for Polyethylene and Polypropylene Solutions with Equations of State. J. Appl. Polymer Science 121: 1832-1849.
[69] Wilson G M (1964) Vapor-Liquid Equilibrium. XI: A New Expression for the Excess Free Energy of Mixing. J. Am. Chem. Soc. 86: 127-130.
[70] Renon H, Prausnitz J M (1968) Local Composition in Thermodynamic Excess Function for Liquid Mixtures. AIChE J. 14: 135-144.
[71] Abrams D S, Prausnitz J M (1975) Statistical Thermodynamics of Liquid Mixtures: A New Expression for the Excess Gibbs Energy of Partly or Completely Miscible Systems. AIChE J. 21: 116-128.
[72] Kumar S K, Suter U W, Reid R C (1987) A Statistical Mechanics Based Lattice Model Equation of State. Ind. Eng. Chem. Res. 26: 2532-2542.
[73] Cheluget E L, Bokis C P B, Wardhaugh L, Chen C C, Fisher J (2002) Modeling Polyethylene Fractionation Using the Perturbed-Chain Statistical Associating Fluid Theory Equation of State. Ind. Eng. Chem. Res. 41: 968-988.
[74] Bae Y C, Shim J J, Soane D S, Prausnitz J M (1993) Representation of Vapor-Liquid and Liquid-Liquid Equilibria for Binary Systems Containing Polymers: Applicability of an Extended Flory-Huggins Equation. J. Appl. Polym. Sci. 47: 1193-1206.

[75] Sanchez I C, Lacombe R H (1976) An Elementary Molecular Theory of Classical Fluids: Pure Fluids. Journal of Physical Chemistry 80: 2352-2362.

[76] Gauter K, Heidemann R A (2001) Modelling Polyethilene-Solvent Mixtures with the Sanchez-Lacombe Equation. Fluid Phase Equilib. 183-184: 87-97.

[77] Kiszka M B, Meilchen M A, McHugh M A (1988) Modeling High-Pressure Gas-Polymer Mixtures Using the Sanchez-Lacombe Equation of State. J. Appl. Polymer Science 36: 583-597.

[78] Xiong Y, Kiran E (1994) Prediction of High-Pressure Phase Behaviour in Polyethylene/ n-Pentane/ Carbon Dioxide Ternary System with the Sanchez-Lacombe Model. Polymer 35: 4408-4415.

[79] Xiong Y, Kiran E (1994) High-Pressure Phase Behavior in Polyethylene/n-Butane Binary and Polyethylene/n-Butane/CO_2 Ternary Systems. J. Appl. Polymer Science 53: 1179-1190.

[80] Xiong Y, Kiran E (1995) Comparison of Sanchez-Lacombe and SAFT Model in Predicting Solubility of Polyethylene in High-Pressure Fluids. J. Appl. Polymer Science 55: 1805-1818.

[81] Song Y, Lambert S M, Prausnitz J M (1994a) A Perturbed Hard-Sphere Equation of State for Normal Fluids and Polymers. Ind. Eng. Chem. Res. 33: 1047-1057.

[82] Phoenix A V, Heidemann R A (1999) An Algorithm for Determining Cloud and Shadow Curves Using Continuous Thermodynamics. Fluid Phase Equilib. 158-160: 643-655.

[83] Wang W, Tree D A, High M S A (1996) Comparison of Lattice-Fluid Models for the Calculation of the Liquid-Liquid Equilibria of Polymer Solutions. Fluid Phase Equilib. 114: 47-62.

[84] Koak N, Visser R M, de Loos T W (1999) High-Pressure Phase Behavior of the Systems Polyethylene + Ethylene and Polybutene + 1-Butene. Fluid Phase Equilib. 158-160: 835-846.

[85] de Loos T W, Poot W, Diepen G A M (1983) Fluid Phase Equilibria in the System Polyethylene + Ethylene. 1. Sytems of Linear Polyethylene + Ethylene at High Pressure. Macromolecules 16: 111-117.

[86] Trumpi H, de Loos T W, Krenz R A, Heidemann R A (2003) High Pressure Phase Equilibria in the System Linear Low Density Polyethylene+Ethylene: Experimental Results and Modelling. J. Supercrit. Fluids 27: 205-214.

[87] Krenz R A (2005) Correlating the Fluid Phase Behaviour of Polydisperse Polyethylene Solutions Using the Modified Sanchez-Lacombe Equation of State. Calgary-AB: Department of Chemical and Petroleum Engineering-University of Calgary, (Ph.D. Thesis). Canada.

[88] Krenz R A, Heidemann R A, de Loos Th W (2003) The Impact of the Molar Mass Distribution on Predicting the Cloud Point Behavior of Polyethylene+Ethylene Systems, AIChE Annual Meeting San Francisco, CA, USA, Nov, 17-21.

[89] Schnell M, Stryuk S, Wolf, B A (2004) Liquid/Liquid Demixing in the System N-hexane/Narrowly Distributed Linear Polyethylene. Ind. Eng. Chem. Res. 43: 2852-2859.

[90] Voutsas E, Pappa G D, Boukouvalas C, Magoulas K, Tassios D (2004) Miscibility in Binary Polymer Blends: Correlation and Prediction. Ind. Eng. Chem. Res. 43: 1312-1321.

[91] Elbro S, Fredenslund A, Rasmussen P (1990) A New Simple Equation for the Prediction of Solvent Activities in Polymer Solutions. Macromolecules 23: 4707-4714.

[92] Chen X, Yasuda K, Sato Y, Takishima S, Masuoka H (2004) Measurement and Correlation of Phase Equilibria of Ethylene + n-Hexane+ Metallocene Polyethylene at Temperatures between 373 and 473 K and at Pressures up to 20 MPa. Fluid Phase Equilib. 215: 105-115.

[93] Nagy I, de Loos Th W, Krenz R A, Heidemann R A (2006) High Pressure Phase Equilibria in the Systems Linear Low Density Polyethylene +n-Hexane and Linear Low Density Polyethylene +n-Hexane +Ethylene: Experimental Results and Modelling with the Sanchez-Lacombe Equation of State. J. Supercrit. Fluids 37: 115-124.

[94] Krenz R, Laursen T, Heidemann R A (2009) The Modified Sanchez-Lacombe Equation of State Applied to Polydisperse Polyethylene Solutions. Ind. Eng. Chem. Res. 48: 10664-10681.

[95] Krenz R A, Heidemann R A (2007) Modelling the Fluid Phase Behavior of Polydisperse Polyethylene Blends in Hydrocarbons Using the Modified Sanchez-Lacombe Equation of State. Fluid Phase Equilib. 262: 217-226.

[96] Neau E (2002) A Consistent Method for Phase Equilibrium Calculation Using the Sanchez-Lacombe Lattice-Fluid Equation of State. Fluid Phase Equilib. 203:133-140.

[97] Kanellopoulos V, Mouratides D, Pladis P, Kiparissides C. (2006) Prediction of Solubility of α-Olefins in Polyolefins Using a Combined Equation of State-Molecular Dynamics Approach, Ind. Eng. Chem. Res. 45: 5870-5878.

[98] Nagy I, Ryan A K, Heidemann R A, de Loos T W (2007) High-Pressure Phase Equilibria in the System Linear Low Density Polyethylene + Isohexane: Experimental Results and Modelling. J. Supercrit. Fluids 40: 125-133.

[99] Beret S, Prausnitz J M (1975) Perturbed Hard-Chain Theory: Na Equation of State for Fluids Containing Small or Large Molecules. AIChE J., 26: 1123-1132.

[100] Carnahan N.F, Starling K.E (1969) Equation of State for Nonattracting Rigid Spheres. J. Chem. Phys. 51: 635-636.

[101] Vilmalchand P, Donohue M D (1985) Thermodynamics of Quadrupolar Molecules: The Perturbed Anisotropic -Chain Theory, Ind. Eng. Chem. Fundam. 24: 246-257.

[102] Vilmalchand P, Celmins I, Donohue M D (1986) VLE Calculations for Mixtures Containing Multipolar Compounds Using the Perturbed Anisotropic Chain Theory. AIChE J. 32: 1735-1738.

[103] Kim C H, Vilmalchand P, Donohue M D, Sandler S I (1986) Local Composition Model for Chainlike Molecules: A New Simplified Version of the Perturbed Hard Chain Theory. AIChE J. 32: 1726-1734.

[104] Ikonomou G D, Donohue M D (1988) Extension of the Associated Perturbed Anisotropic Chain Theory to Mixtures with More than One Associating Component. Fluid Phase Equilib. 39: 129-159.

[105] Economou I G (2002) Statistical Associating Fluid Theory: A Successful Model for the Calculation of Thermodynamic and Phase Equilibrium Properties of Complex Fluid Mixtures. Ind. Eng. Chem. Res. 41: 953-962.

[106] Cottermann R L, Schwarz B J, Prausnitz J M (1986) Molecular Thermodynamics for Fluids at Low and High Densities . part 1: Pure Fluids Containg Small or Large Molecules AIChE J. 32: 1787-1798.

[107] Chapman W G, Jackson G, Gubbins K E (1988) Phase Equilibria of Associating Fluids. Chain Molecules with Multiple Bonding Sites. Mol. Phys. 65: 1057-1079.

[108] Barker J A, Henderson D (1967) Perturbed Theory and Equation of State for Fluids: The Square-Well Potential. J. Chem. Phys. 47: 2856-2861.

[109] Muller E A, Gubbins K E (2001) Molecular- Based Equations of State for Associating Fluids: A Review of SAFT and Related Approaches. Ind. Eng. Chem, Res. 40: 2193-2211.

[110] Blas F J, Veja L F (1998) Prediction of Binary and Ternary Diagrams Using the Statistical Associating Fluid Theory (SAFT) Equation of State. Ind. Eng. Chem. Res. 37: 660-674.

[111] Gil-Villegas A, Galindo A, Whitehead P J, Mills S J, Jackson G, Burgess A N (1997) Statistical Associating Fluid Theory for Chain Molecules with Attractive Potentials of Variable Range. J. Chem. Phys. 106: 4168-4186.

[112] McCabe C, Galindo A, Garcia-Lisbona M N, Jackson G (2001) Examining the Adsorption (Vapor-Liquid Equilibria) of Short-Chain Hydrocarbons in Low-Density Polyethylene with the SAFT-VR Approach. Ind. Eng. Chem. Res. 40: 3835-3842.

[113] Paricaud P, Galindo A, Jackson G (2004) Modeling the Cloud Curves and the Solubility of Gases in Amorphous and Semicrystalline Polyethylene with the SAFT-VR Approach and Flory Theory of Crystallization. Ind. Eng. Chem. Res. 43: 6871-6889.

[114] Haslam A J, von Solms N, Adjiman C S, Galindo A, Jackson G, Paricaud P, Michelsen M L, Kontogeorgis G M (2006) Predicting Enhanced Absorption of Light Gases in Polyethylene Using Simplified PC-SAFT and SAFT-VR. Fluid Phase Equilib. 243: 74-91.

[115] von Solms N, Kouskoumvekaki I A, Michelsen M L, Kontogeorgis G M (2006) Capabilities, Limitations and Challenges of a Simplified PC-SAFT Equation of State. Fluid Phase Equilib. 241: 344-353.

[116] Dariva C (2000) Equilíbrio de Fases a Altas Pressões em Sistemas com Polipropilenos. Dados Experimentais e Modelagem SAFT (in portuguese). Rio de Janeiro-RJ: PEQ/COPPE/UFRJ (D.Sc. Thesis). Brazil.

[117] Chen S, Radosz M (1992) Density-Tuned Polyolefin Phase Equilibria. 1. Binary Solutions of Alternating Poly(ethy1ene-propylene) in Subcritical and Supercritical Propylene, 1-Butene, and 1-Hexene. Experiment and Flory-Patterson Model. Macromolecules 25: 3089-3096.

[118] Chen S J, Economou I.G, Radosz M (1992a) Density-Tuned Polyolefin Phase Equilibria. 2. Multicomponent Solutions of Alternating Poly(ethylene-propylene) in Subcritical and Supercritical Olefins. Experiment and SAFT Model. Macromolecules 25: 4987-4995.

[119] Han S J, Gregg C J, Radosz M (1997) How the Solute Polydispersity Affects the Cloud-Point and Coexistence Pressures in Propylene and Ethylene Solutions of Alternating Poly(ethylene-co-propylene). Ind. Eng. Chem. Res. 36: 5520-5525.

[120] Pan C, Radosz M (1998) Copolymer SAFT Modeling of Phase Behavior in Hydrocarbon-Chain Solutions: Alkane Oligomers, Polyethylene, Poly(ethylene-co-

olefin-1), Polystyrene, and Poly(ethylene-co-styrene). Ind. Eng. Chem. Res. 37: 3169-3179.

[121] Pan C, Radoz, M (1999) Modeling of Solid–Liquid Equilibria in Naphthalene, Normal-Alkane and Polyethylene Solutions Fluid Phase Equilib. 155: 57-73.

[122] Horst M.H, Behme S, Sadowski G, de Loos, Th.W (2002) The Influence of Supercritical Gases on the Phase Behavior of Polystyrene–Cyclohexane and Polyethylene–Cyclohexane Systems: Experimental Results and Modeling with the SAFT-Equation of State. J. Supercrit. Fluids 23: 181-194.

[123] Jog P K, Chapman W G, Gupta S K, Swindoll R D (2002) Modeling of Liquid-Liquid Phase Separation in Linear Low-Density Polyethylene-Solvent Systems Using the Statistical Associating Fluid Theory Equation if State, Ind. Eng. Chem. Res. 41: 887-891.

[124] de Loos T W, de Graaf L J, de Swaan Arons J (1996) Liquid-Liquid Phase Separation in Linear Low Density Polyethylene-Solvent Systems. Fluid Phase Equilib. 117: 40-47.

[125] Wiesmet V, Weidner E, Behme S, Sadowski G, Arlt W (2000) Measurement and Modelling of High-Pressure Phase Equilibria in the Systems Polyethyleneglycol (PEG)–Propane, PEG–Nitrogen and PEG–Carbon Dioxide. J. Supercrit. Fluids 17: 1-12.

[126] Gross J, Sadowski G (2002a) Modeling of Polymer Systems Using the Perturbed-Chain Statistical Associating Fluid Theory Equation of State. Ind. Eng. Chem. Res. 41: 1084-1093.

[127] Kinzl M, Luft G, Adidharma H, Radosz M (2000) SAFT Modeling of Inert-Gas Effects on the Cloud-Point Pressures in Ethylene Copolymerization Systems: Poly(ethylene-co-vinyl acetate) + Vinyl Acetate + Ethylene and Poly(ethylene-co-hexene-1) + Hexene-1 + Ethylene with Carbon Dioxide, Nitrogen, or n-Butane. Ind. Eng. Chem. Res. 39: 541-546.

[128] Bokis C P, Orbey H, Chen, C-C (1999) Properly Model Polymer Processes. Chem. Eng. Prog. April: 39-52.

[129] Pan C, Radosz M (1999) Phase Behavior of Poly(ethylene-co-hexene-1) Solutions in Isobutane and Propane. Ind. Eng. Chem. Res. 38: 2842-2848.

[130] Chen C K, Duran M A, Radosz M (1993) Phase Equilibria in Polymer Solutions. Block-Algebra, Simultaneous Flash Algorithm Coupled with SAFT Equation of State, Applied to Single-Stage Supercritical Antisolvent Fractionation of Polyethylene. Ind. Eng. Chem. Res. 32: 3123-3127.

[131] Tan S.P, Meng D, Plancher H, Adidharma H, Radosz M (2004) Cloud Points for Polystyrene in Propane and Poly (4-methyl styrene) in Propane. Fluid Phase Equilib. 226: 189-194.

[132] Wu C-S, Chen Y-P (1994) Calculation of Vapor-Liquid Equilibria of Polymer Solutions Using the SAFT Equation of State. Fluid Phase Equilib. 100: 103-119.

[133] Behme S, Sadowski G, Arlt W (1999) Modeling of the Separation of Polydisperse Polymer Systems by Compressed Gases. Fluid Phase Equilib. 158–160: 869-877.

[134] Albrecht K.L, Stein F.P, Han S.J, Gregg C.J, Radosz M (1996) Phase Equilibria of Saturated and Unsaturated Polyisoprene in Sub- and Supercritical Ethane, Ethylene, Propane, Propylene, and Dimethyl Ether. Fluid Phase Equilib. 117: 84-91.

[135] Gregg C.J, Stein F.P, Radosz M (1994a) Phase Behavior of Telechelic Polyisobutylene in Subcritical and Supercritical Fluids. 3. Three-Arm-Star PIB (4K) as a Model Trimer for

Monohydroxy and Dihydroxy PIB (1K) in Ethane, Propane, Dimethyl Ether, Carbon Dioxide, and Chlorodifluoromethane. Journal of Physical Chemistry 98: 10634-10639.

[136] Gregg C.J, Stein F.P, Radosz M (1994b) Phase Behavior of Telechelic Polyisobutylene (PIB) in Subcritical and Supercritical Fluids. 1. Inter- and Intra-Association Effects for Blank, Monohydroxy, and Dihydroxy PIB (1K) in Ethane, Propane, Dimethyl Ether, Carbon Dioxide, and Chlorodifluoromethane. Macromolecules 27: 4972-4980.

[137] Sadowski G, Mokrushina L.V, Arlt W (1997) Finite and Infinite Dilution Activity Coefficients in Polycarbonate Systems. Fluid Phase Equilib. 139: 391-403.

[138] Folie B, Gregg C, Luft G, Radosz M (1996) Phase Equilibria of Poly(ethylene-co-vinyl acetate) Copolymers in Subcritical and Supercritical Ethylene and Ethylene-Vinyl Acetate Mixtures. Fluid Phase Equilib. 120: 11-37.

[139] Lee S-H, Hasch B M, McHugh M A (1996) Calculating Copolymer Solution Behavior with Statistical Associating Fluid Theory. Fluid Phase Equilib., 117: 61-68.

[140] Chan K C, Adidharma H, Radosz M (2000) Fluid-Liquid and Fluid-Solid Transitions of Poly(ethylene-co-octene-1) in Sub- and Supercritical Propane Solutions. Ind. Eng. Chem. Res. 39: 3069-3075.

[141] Lora M, McHugh M A (1999) Phase Behavior and Modeling of the Poly(Methyl Methacrylate)–CO_2–Methyl Methacrylate System. Fluid Phase Equilib. 157: 285-297.

[142] Chen S-J, Economou I G, Radosz M (1992b) Phase Behavior of LCST and UCST Solutions of Branchy Copolymers: Experiment and SAFT Modeling. Fluid Phase Equilib. 83: 391-398.

[143] Luna-Barcenas G, Mawson S, Takishima S, DeSimone J M, Sanchez I C, Johnston K P (1998) Phase Behaviour of Poly(1,1-dihydroperfluorooctylacrylate) in Supercritical Carbon Dioxide. Fluid Phase Equilib. 146: 325-337.

[144] Tumakaka F, Gross J, Sadowski G (2002) Modeling of Polymer Phase Equilibria Using Perturbed-Chain SAFT. Fluid Phase Equilib. 194-197: 541-551.

[145] Gross J, Sadowski G (2002b) Application of the Perturbed-Chain SAFT Equation of State to Associating Systems. Ind. Eng. Chem. Res. 41: 5510-5515.

[146] Gross J, Spuhl O, Tumakaka F, Sadowski G (2003) Modeling Copolymer Systems Using Perturbed-Chain SAFT. Fluid Phase Equilib. 42: 1266-1274.

[147] Kouskoumvekaki I A, von Solms N, Michelsen M L, Kontogeorgis G M (2004b) Application of the Perturbed Chain SAFT Equation of State to Complex Polymer Systems Using Simplified Mixing Rules. Fluid Phase Equilib. 215: 71-78.

[148] Tumakaka F, Sadowski G (2004) Application of the Perturbed-Chain SAFT Equation of State to Polar Systems. Fluid Phase Equilib. 217: 233-239.

[149] Jog P K, Sauer S G, Blaesing J, Chapman W G (2001) Application of Dipolar Chain Theory to the Phase Behavior of Polar Fluids and Mixtures. Ind. Eng. Chem. Res. 40: 4641-4648.

[150] Kouskoumvekaki I.A, Krooshof G.P, Michelsen M.L, Kontogeorgis G.M (2004a) Application of the Simplified PC-SAFT Equation of State to the Vapor-Liquid Equilibria of Binary and Ternary Mixtures of Polyamide 6 with Several Solvents. Ind. Eng. Chem. Res. 43: 826-834.

[151] Arce P F, Aznar M (2005a) Phase Behavior of Polypropylene + n-Pentane and Polypropylene + n-Pentane + Carbon Dioxide: Modeling with Cubic and Noncubic Equations of State. J. Supercrit. Fluids 34: 177-182.

[152] Spyriouni T, Economou I G (2005) Evaluation of SAFT and PC-SAFT Models for the Description of Homo- and Co-polymer Solution Phase Equilibria. Polymer 46: 10772-10781.

[153] Pedrosa N, Vega L F, Coutinho J A P, Marrucho I M (2006) Phase Equilibria of Polyethylene Solutions from SAFT-Type Equation of State. Macromolecules 39: 4240-4246.

[154] Buchelli A, Call M L, Brown A L, Bokis C P, Ramanathan S, Franjione J (2004) Nonequilibrium Behavior in Ethylene/Polyethylene Flash Separators. Ind. Eng. Chem. Res. 43: 1768-1778.

[155] Guerrieri Y (2007) Modelagem e Simulação do Equilíbrio de Fases em Plantas de Polietileno Utilizando a Equação de Estado PC-SAFT (in portuguese). Campinas-SP: FEQ-UNICAMP (M.Sc. Dissertation). Brazil.

[156] Tihic A, Kontogeorgis G M, von Solms N, Michelsen M L, Constantinou L A (2008) Predictive Group-Contribution Simplified PC-SAFT Equation of State: Application to Polymer Systems. Ind. Eng. Chem. Res. 47: 5092-5101.

[157] Tihic A, von Solms N, Michelsen M L, Kontogeorgis G M, Constantinou L (2009) Application of sPC-SAFT and Group Contribution sPCSAFT to Polymer Systems. Capabilities and Limitations. Fluid Phase Equilib. 281: 70-77.

[158] Abbas S, Mukherjee r, De S, Ganguly S (2004) Real-Time Inferencing of Solid-Liquid Phase Equilibria in Ssolution Polymerization of Polyethylene Chem. Eng. Process. 43 1449-1458.

[159] Costa G M N, Kislansky S, Guerrieri Y, Pessoa F L P, Vieira de Melo S A B, Embiruçu M (2010) Calculation of Pressure-Temperature Diagrams and Distance for Phase Transition in Polyethylene Solutions. Ind. Eng. Chem. Res. 49: 12242–12253.

[160] Kleiner M, Tumakaka F, Sadowski G, Latz H, Buback M (2006) Phase Equilibria in Polydisperse and Associating Copolymer Solutions: Poly(ethene-co-(meth)acrylic acid)–Monomer Mixtures. Fluid Phase Equilib. 241:113-123.

[161] Costa G M N, Guerrieri Y, Kislansky S., Embiruçu M (2012) Phase-Dependent Binary Interaction Parameters in Industrial Low-Density Polyethylene Separators. J. Appl. Polymer Science, submitted.

[162] von Solms N, Kouskoumvekaki I A, Lindvig T, Michelsen M L, Kontogeorgis G M (2004) A Novel Approach to the Liquid-Liquid Equilibrium in Polymer Systems with Application to Simplified PC-SAFT. Fluid Phase Equilib. 222-223: 87-93.

[163] Dominik A, Chapman W G (2005) Thermodynamic Model for Branched Polyolefins Using the PC-SAFT Equation of State. Macromolecules 38: 10836-10843.

[164] Becker F, Buback M, Latz H, Sadowski G, Tumakaka F (2004) Cloud-Point Curves of Ethylene-(Meth)Acrylate Copolymers in Fluid Ethene up to High Pressures and Temperatures – Experimental Study and PC–SAFT Modeling. Fluid Phase Equilib. 215: 263-282.

[165] Arce P, Aznar M (2005b) Modeling the Phase Behavior of Commercial Biodegradable Polymers and Copolymer in Supercritical Fluids. Fluid Phase Equilib. 238: 242-253.

[166] van Schilt M A, van Meerendonk W J, Kemmere M F, Keurentjes J T F, Kleiner M, Sadowski G, de Loos Th W (2005) High-Pressure Phase Behavior of the System PCHC-CHO-CO_2 for the Development of a Solvent-Free Alternative toward Polycarbonate Production. Ind. Eng. Chem. Res. 44: 3363-3366.

[167] Krueger K-M, Pfohl O, Dohrn R, Sadowski G (2006) Phase Equilibria and Diffusion Coefficients in the Poly(dimethylsiloxane) + n-Pentane System. Fluid Phase Equilib. 241: 138-146.

Permissions

The contributors of this book come from diverse backgrounds, making this book a truly international effort. This book will bring forth new frontiers with its revolutionizing research information and detailed analysis of the nascent developments around the world.

We would like to thank Dr. Ailton de Souza Gomes, for lending his expertise to make the book truly unique. He has played a crucial role in the development of this book. Without his invaluable contribution this book wouldn't have been possible. He has made vital efforts to compile up to date information on the varied aspects of this subject to make this book a valuable addition to the collection of many professionals and students.

This book was conceptualized with the vision of imparting up-to-date information and advanced data in this field. To ensure the same, a matchless editorial board was set up. Every individual on the board went through rigorous rounds of assessment to prove their worth. After which they invested a large part of their time researching and compiling the most relevant data for our readers. Conferences and sessions were held from time to time between the editorial board and the contributing authors to present the data in the most comprehensible form. The editorial team has worked tirelessly to provide valuable and valid information to help people across the globe.

Every chapter published in this book has been scrutinized by our experts. Their significance has been extensively debated. The topics covered herein carry significant findings which will fuel the growth of the discipline. They may even be implemented as practical applications or may be referred to as a beginning point for another development. Chapters in this book were first published by InTech; hereby published with permission under the Creative Commons Attribution License or equivalent.

The editorial board has been involved in producing this book since its inception. They have spent rigorous hours researching and exploring the diverse topics which have resulted in the successful publishing of this book. They have passed on their knowledge of decades through this book. To expedite this challenging task, the publisher supported the team at every step. A small team of assistant editors was also appointed to further simplify the editing procedure and attain best results for the readers.

Our editorial team has been hand-picked from every corner of the world. Their multi-ethnicity adds dynamic inputs to the discussions which result in innovative outcomes. These outcomes are then further discussed with the researchers and contributors who give their valuable feedback and opinion regarding the same. The feedback is then collaborated with the researches and they are edited in a comprehensive manner to aid the understanding of the subject.

Apart from the editorial board, the designing team has also invested a significant amount of their time in understanding the subject and creating the most relevant covers. They scrutinized every image to scout for the most suitable representation of the subject and create an appropriate cover for the book.

The publishing team has been involved in this book since its early stages. They were actively engaged in every process, be it collecting the data, connecting with the contributors or procuring relevant information. The team has been an ardent support to the editorial, designing and production team. Their endless efforts to recruit the best for this project, has resulted in the accomplishment of this book. They are a veteran in the field of academics and their pool of knowledge is as vast as their experience in printing. Their expertise and guidance has proved useful at every step. Their uncompromising quality standards have made this book an exceptional effort. Their encouragement from time to time has been an inspiration for everyone.

The publisher and the editorial board hope that this book will prove to be a valuable piece of knowledge for researchers, students, practitioners and scholars across the globe.

List of Contributors

M. Petit
Université 20 Août 1955-Skikda, Algérie

Isabel Cristina Celerino de Moraes Porto
Federal University of Alagoas, CESMAC University Center, Brazil

Inger Martínez de Arenaza, Emiliano Meaurio and Jose-Ramon Sarasua
University of the Basque Country (UPV-EHU), Department of Mining-Metallurgy Engineering & Materials Science, School of Engineering, Bilbao, Spain

Marisol Ochoa-Villarreal, Emmanuel Aispuro-Hernández and Miguel A. Martínez-Téllez
Laboratorio de Fisiología Vegetal Molecular, Tecnología de Alimentos de Origen Vegetal, Centro de Investigación en Alimentación y Desarrollo, A.C., Hermosillo, Sonora, México

Irasema Vargas-Arispuro
Laboratorio de Ecología Química. Ciencia de los Alimentos, Centro de Investigación en Centro de Investigación en Alimentación y Desarrollo, A.C., Hermosillo, Sonora, México

Hao Ren, Zhen Niu, Juan Wang and Jing Ning
Shandong Provincial Key Laboratory of Fluorine Chemistry and Chemical Materials, School of Chemistry and Chemical Engineering, University of Jinan, Shandong, China

S. Damavandi
Department of Chemistry, Sarvestan Branch, Islamiz Azad University, Iran

S. Ahmadjo
Department of Catalyst, Iran Polymer and Petrochemical Institute, Tehran, Iran

R. Sandaroos
Department of Chemistry, Faculty of Science, Birjand University, Birjand, Iran

G.H. Zohuri
Department of Chemistry, Ferdowsi University of Mashhad, Mashhad, Iran

Mara Soares da Silva and Teresa Casimiro
REQUIMTE, Departamento de Química, Faculdade de Ciências e Tecnologia, FCT, Universidade Nova de Lisboa, Caparica, Portugal

Junxiao Yang, Yawen Huang and Ke Cao
School of Material Science and Engineering, Southwest University of Science and Technology, P. R. China

M. Khaddazh
Kurnakov Institute of General and Inorganic Chemistry Russian Academy of Sciences, Moscow, Russia

I. A. Gritskova
Lomonosov State University of Fine Chemical Technologies, Moscow, Russia

G. I. Litvinenko
Karpov Institute of Physical Chemistry, Moscow, Russia

Alexey Lyapkov, Elena Ionova, Vladimir Bondaletov and Alexey Pestryakov
National Research Tomsk Polytechnic University, Russia

Fabiana V. Motta, Carlos A. Paskocimas, Mauricio R. D. Bomio and Amélia S. F. Santos
Department of Materials Engineering, Federal University of Rio Grande do Norte, UFRN, Natal, RN, Brazil

Ana Paula A. Marques
Department of Exact and Earth Sciences, Federal University of São Paulo, UNIFESP, São Paulo, SP, Brazil

Edson R. Leite
Interdisciplinary Laboratory of Electrochemistry and Ceramics, Department of Chemistry, Federal University of São Carlos, UFSCar, São Carlos, SP, Brazil

José A. Varela and Elson Longo
Interdisciplinary Laboratory of Electrochemistry and Ceramics, Chemistry Institute, São Paulo State University, UNESP, Araraquara, SP, Brazil

Reza Sandaroos
Department of Chemistry, Faculty of Science, University of Birjand, Birjand, Iran

Tomás Cuenca
Departamento de Química Inorgánica, Universidad de Alcalá, Campus Universitario, Alcalá de Henares, Spain

Gholam Hossein Zohuri
Department of Chemistry, Faculty of Science, Ferdowsi University of Mashhad, Mashhad, Iran

Saman Damavandi
Department of Chemistry, Sarvestan Branch, Islamic Azad University, Iran

Saeid Ahmadjo
Department of Catalyst, Iran Polymer and Petrochemical Institute, Tehran, Iran

Hülya Arslan
Bülent Ecevit University, Department of Chemistry, Zonguldak, Turkey

Tomoharu Osada and Mitsuko Masutani
Mitsubishi Chemical Medience Corporation, Kashima Laboratory, Japan
Division of Genome Stability Research, National Cancer Center Research Institute, Japan

Hiroshi Murata
Department of Prosthetic Dentistry, Graduate School of Biomedical Sciences, Nagasaki University, Nagasaki, Japan

Daniel S. Correa
Embrapa Instrumentação, São Carlos, SP, Brazil

Leonardo De Boni, Adriano J. G. Otuka, Vinicius Tribuzi and Cleber R. Mendonça
Instituto de Física de São Carlos, Universidade de São Paulo, São Carlos, SP, Brazil

Yuri Guerrieri, Karen Valverde Pontes, Gloria Meyberg Nunes Costa and Marcelo Embiruçu
Programa de Engenharia Industrial (PEI) - Escola Politécnica - Universidade Federal da Bahia, Salvador-BA, Brazil